AF885494

Cover Page Designed by:

Ms. Gowri Vemuri

First Edition

May, 2019

Author:

MATH-KNOTS

Index

Preface

The word "Science"brings to mind many different pictures: a fat textbook, bunch of equations, space shuttle, medicines and microscopes etc., simply discovering and learning about the nature around us. There are many branches of science and physics the most basic area of sciences, which deals with energy, matter, space, time and their interactions. The word physics is derived from the Latin word *physica*, which means "natural thing."

Physics as a fundamental science helps to uncover the laws of nature. The language of expression is mathematics. During ancient times, humans studied Stars and celestial bodies that lead to the development of disciplines like astronomy and mathematics. Based on size and speed, Physics as a subject is categorized several categories, which you can find from the following picture.

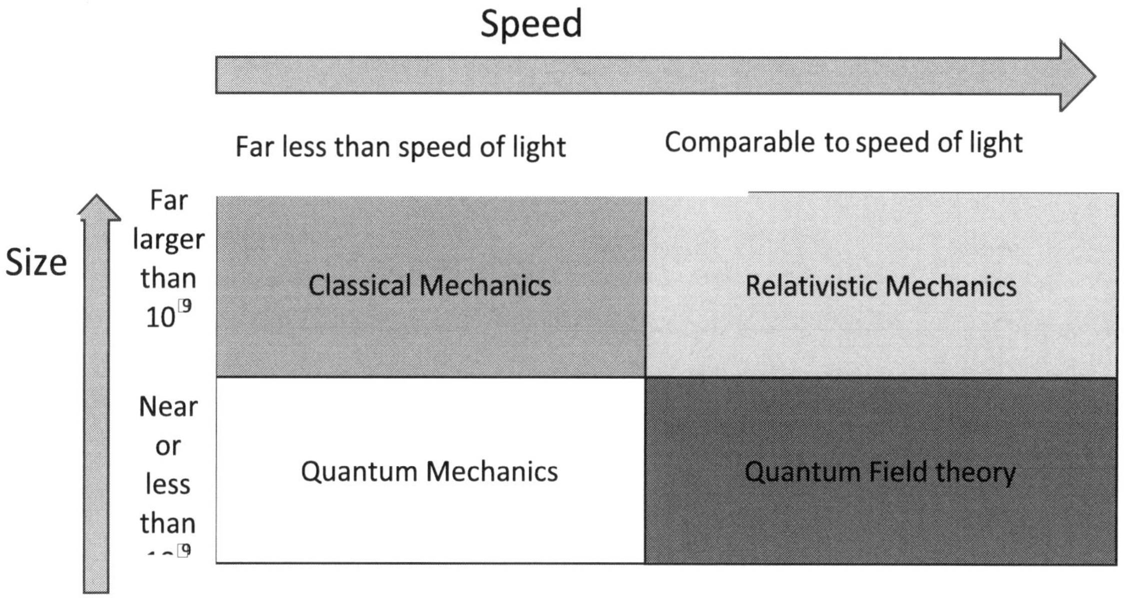

There are many ways to learn physics, but often students face difficulty in solving a problem by the concept they have learnt in the class. Our book is designed to help students in this area. Each chapter in the book, provides a summary of a concept, along with many practice problems.

Chapter - 1

ELECTRIC CHARGES

LEVEL - I (CW) - HINTS

1. $Q = \pm ne$ n is integer

2. $Q = \pm ne$ n is integer

3. $F^{|} = \frac{F}{K}$

4. $F = \frac{1}{4\pi\varepsilon_0}\frac{q_1 q_2}{r^2}$

5. $F = \frac{1}{4\pi\varepsilon_0}\frac{(q_1 + q_2)^2}{4d^2}$

6. $\frac{1}{4\pi\varepsilon_0}\frac{QQ}{l^2} + \frac{1}{4\pi\varepsilon_0}\frac{qQ}{\left(\frac{l}{2}\right)^2} = 0$

7. $F = \frac{1}{4\pi\varepsilon_0}\frac{q_1 q_2}{r^2}$

8. $x = \frac{d}{\sqrt{\frac{q_2}{q_1}} - 1}$

9. $E = \frac{1}{4\pi\varepsilon_0}\frac{q_1}{x_1^2} - \frac{1}{4\pi\varepsilon_0}\frac{q_2}{x_2^2}$

10. $\vec{E} = \frac{1}{4\pi\varepsilon_0}\frac{Q}{r^3}\vec{r}$

11. $qE = mg\tan\theta$

12. $s = \frac{1}{2}\frac{qE}{m}t^2$

13. $E = \frac{q}{4\pi\varepsilon_0}\left[\frac{1}{1^2} + \frac{1}{2^2} + \frac{1}{4^2} + \text{------}\right]$

14. Pressure $= \frac{\sigma^2}{2\varepsilon_o}$ and Force $= \frac{\sigma^2}{2\varepsilon_o} \times \pi R^2$

15. $F = Eq = \frac{Vq}{d}$

16. $V = \frac{Q}{4\pi\varepsilon_0}\left[\frac{1}{1} + \frac{1}{2} + \frac{1}{4} + \frac{1}{8} + \text{---}\right]$

17. Equipotential surface

18. $U = \frac{1}{4\pi\varepsilon_0}\left[\frac{q_1 q_2}{r_1} + \frac{q_2 q_3}{r_2} + \frac{q_1 q_3}{r_3}\right]$

19. $\frac{1}{2}mv^2 = eV$

20. $V = \frac{1}{4\pi\varepsilon_0}\frac{q}{d}$

21. Along the field direction potential decreases.

22. $V = -\vec{E}\cdot d\vec{r}$

23. $W_1 = pE(1 - \cos\theta)$ and $W_2 = pE(\cos\theta_1 - \cos\theta_2)$

24. $p_1 = lq = p_2$ and $P_R = \sqrt{3}ql$

LEVEL - I (HW)

COULOMB'S LAW

1. A charge Q is divided into two parts q_1 and q_2 such that they experience maximum force of repulsion when separated by certain distance. The ratio of Q, q_1 and q_2 is

1) 1 : 1 : 2 2) 1 : 2 : 2 3) 2 : 2 : 1 4) 2 : 1 : 1

2. Two charges each 1μC are at $P(2\hat{i} + 3\hat{j} + \hat{k})m$ and $Q(\hat{i} + \hat{j} - \hat{k})m$. Then the force between them is _____

1) 100N 2) 10N 3) 10^4 dyne 4) 100 dyne

3. **Two charges of $+200\mu C$ and $-200\mu C$ are placed at the corners B and C of an equilateral triangle ABC of side 0.1 m. The force on a charge of $5\mu C$ placed A is**

1) 1800 N 2) $1200\sqrt{3}N$ 3) $600\sqrt{3}N$ 4) $900N$

4. **Two equally charged pith balls 3 cm apart repel each other with a force of 4×10^{-5} newton. The charge on each ball is**

1) $2\times10^{9}C$ 2) $2\times10^{-9}C$ 3) $\frac{2}{3}\times10^{9}C$ 4) $\frac{2}{3}\times10^{-9}C$

ELECTRIC FIELD

5. **An electron $\left(mass = 9.1\times10^{-31}kg\right)$ is sent into an electric field of intensity 9.1×10^{6} newton/coulomb. The acceleration produced is**

1) $1.6\times10^{18}\ m/s^2$ 2) $1.6\times10^{6}\ m/s^2$ 3) $1.6\times10^{-18}\ m/s^2$ 4) $1.6\times10^{-6}\ m/s^2$

6. **The electric field at (30, 30) cm due to a charge of -8 nC at the origin in NC^{-1} is**

1) $-400\left(\bar{i}+\bar{j}\right)$ 2) $400\left(\bar{i}+\bar{j}\right)$ 3) $-200\sqrt{2}\left(\bar{i}+\bar{j}\right)$ 4) $200\sqrt{2}\left(\bar{i}+\bar{j}\right)$

7. **Two charges of 10 μ C and -90 μ C are separated by a distance of 24 cm. Electrostatic field strength from the smaller charge is zero at a distance of**

1) 12 cm 2) 24 cm 3) 36 cm 4) 48 cm

8. **Two electric charges of $+10^{-9}C$ and $-10^{-9}C$ are placed at the corners A and B of an equilateral triangle ABC side 5cm.The electric intensity at C is**

1) 1800N/C 2) 3600 N/C 3) 900N/C 4) 2700 N/C

ELECTRIC POTENTIAL AND POTENTIAL ENERGY

9. **If $4\times10^{20}eV$ is required to move a charge of 0.25 coulomb between two points, the potential difference between these two points is**

1) 256 volt 2) $\frac{1}{256}$ volt 3) $256\times10^{+19}$ volt 4) 250 volt

10. **Two electric charges of $9\mu C$ and $-3\mu C$ are placed 0.16m apart in air. There are two points A and B on the line joining the two charges at distances of (i) 0.04m from $-3\mu C$ and in between the charges and (ii) 0.08m from $-3\mu C$ and out side the two charges. The potentials at A and B are**

1) 0V, 5V 2) 0V, 0V 3) 5V, 0V 4) 5V, 10V

11. **Four charges $+3\mu C,\ -1\mu C,\ +5\mu C$ and $-7\mu C$ are arranged on the circumference of a circle of radius 0.5 m. The potential at the centre is**

1) Zero 2) $18\times10^{4}V$ 3) $-18\times10^{4}V$ 4) $288\times10^{3}V$

12. **A positive point charge 'q' is carried from a point 'B' to a point 'A' in the electric field of a point charge +Q. If the permittivity of free space is ϵ_0, the work done in the process is given by**

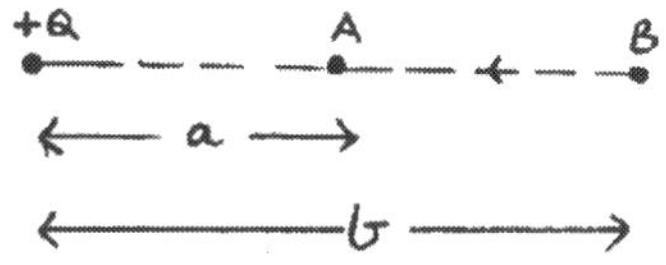

1) $\frac{qQ}{4\pi\epsilon_0}\left[\frac{1}{a}-\frac{1}{b}\right]$ 2) $\frac{qQ}{4\pi\epsilon_0}\left[\frac{1}{a}+\frac{1}{b}\right]$ 3) $\frac{qQ}{4\pi\epsilon_0}\left[\frac{1}{a^2}-\frac{1}{b^2}\right]$ 4) $\frac{qQ}{4\pi\epsilon_0}\left[\frac{1}{a^2}+\frac{1}{b^2}\right]$

13. An electric cell does 5 joules of work in carrying 10 Coulomb's of charge around a closed circuit. The emf of the cell is

1) 2V 2) 0.5V 3) 4V 4) 1V

14. Two positive charges $12\mu C$ and $10\mu C$ are initially separated by 10cm. The work done in bringing the two charges 4cm closer is

1) 7.2J 2) 3.6J 3) 8.4J 4) 12.4J

15. An insulated charged conducting sphere of radius 5 cms has a potential of 10V at the surface. What is the potential at centre?

1) 10V 2) zero

3) same as that at 5 cms from the surface 4) same as that at 25 cms from the surface

16. A positive charge 'Q' is fixed at a point. A negatively charged particle of mass 'm' and charge 'q' is revolving in a circular path of radius 'r_1' with 'Q' as the centre. The work to be done to change the radius of the circular path from r_1 to r_2 in Joules is

1) 0 2) $\frac{Qq}{4\pi\varepsilon_o}\left[\frac{1}{r_1}-\frac{1}{r_2}\right]$ 3) $\frac{1}{4\pi\varepsilon_o}\frac{Qq}{4\pi\varepsilon_o}\left[\frac{1}{r_1}-\frac{1}{r_2}\right]$ 4) $\frac{Qq}{4\pi\varepsilon_o}\left[\frac{1}{r_2}-\frac{1}{r_1}\right]$

17. Figure bellow shows a square array of charged particles, with distance d between adjacenet particle. What is the electric potential at point P at the centre of the square if the electric potential is zero at infinity ?

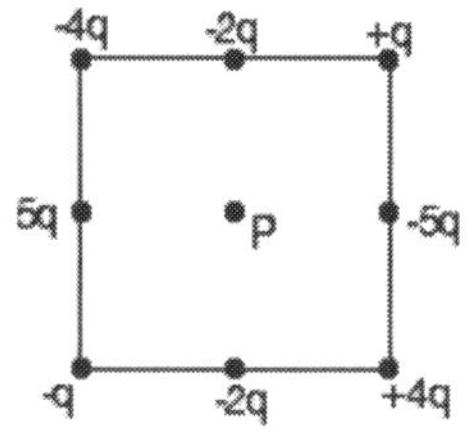

1) Zero 2) $\frac{-2q}{4\pi\epsilon_0 d}$ 3) $\frac{-4q}{4\pi\epsilon_0 d}$ 4) $\frac{q}{4\pi\epsilon_0 d}$.

18. The radii of two charged metal spheres are 5cm and 10cm both having the same charge 60mC. If they are connected by a wire

1) A charge of 20mC flows through the wire from larger to smaller sphere

2) A charge of 20mC flows through the wire from smaller to larger sphere

3) A charge of 40mC flows through the wire from smaller to larger sphere

4) No charge flows through the wire because both spheres have same charge.

19. The electric potential at a point (x, 0, 0) is given by $V=\left[\frac{1000}{x}+\frac{1500}{x^2}+\frac{500}{x^3}\right]$ then the electric field at x = 1 m is (in volt/m)

1) $-5500\hat{i}$ 2) $5500\hat{i}$ 3) $\sqrt{5500}\hat{i}$ 4) zero

DIPOLE

20. An electric dipole of moment p is placed in the position of stable equilibrium in uniform electric field of intensity E. It is rotated through an angle θ from the intial position. The potential energy of electric dipole in the position is

1) $pE\cos\theta$ 2) $pE\sin\theta$ 3) $pE(1-\cos)\theta$ 4) $-pE\cos\theta$

21. An electric dipole of moment $\vec{p}$ is placed normal to the lines of force of electric intensity $\vec{E}$, then the work done in deflecting it through an angle of 180^0 is

1) pE 2) $+2pE$ 3) $-2pE$ 4) zero

LEVEL - I (HW)- KEY

1) 4	2) 4	3) 4	4) 2	5) 1	6) 3	7) 1	8) 2	9) 1	10) 2
11) 1	12) 1	13) 2	14) 1	15) 1	16) 2	17) 3	18) 2	19)2	20) 4
21) 4									

LEVEL - I (HW) HINTS

1. $F\alpha q_1q_2$

2. $F=\frac{1}{4\pi\varepsilon_0}\frac{q_1q_2}{r^2}$

3. $F_1=F_2=\frac{1}{4\pi\varepsilon_0}\frac{q_1q_2}{r^2}$

$F_r=F_1=F_2$ because angle between then is 120^0

4. $F=\frac{1}{4\pi\varepsilon_0}\frac{q^2}{d^2}$

5. $a=\frac{eE}{m}$

6. $\vec{E}=\frac{1}{4\pi\varepsilon_0}\frac{Q}{r^3}\vec{r}$

7. $x=\frac{d}{\sqrt{\frac{q_2}{q_1}}-1}$

8. $E=\frac{1}{4\pi\varepsilon_0}\frac{Q}{a^2}$

9. $W=q\Delta V$

10. $V=\frac{1}{4\pi\varepsilon_0}\left[\frac{q_1}{r_1}+\frac{-q_2}{r_2}\right]$

11. $V=\frac{1}{4\pi\varepsilon_0}\sum\frac{Q}{r}$

12. $W=\frac{q_1q_2}{4\pi\varepsilon_0}\left[\frac{1}{r_1}-\frac{1}{r_2}\right]$

13. $emf=\frac{W}{q}$

14. $W=\frac{q_1q_2}{4\pi\varepsilon_0}\left[\frac{1}{r_1}-\frac{1}{r_2}\right]$

15. $V=\frac{1}{4\pi\varepsilon_0}\frac{Q}{R}$

16. $W=\frac{q_1q_2}{4\pi\varepsilon_0}\left[\frac{1}{r_1}-\frac{1}{r_2}\right]$

17. $V=\frac{1}{4\pi\varepsilon_0}\sum\frac{Q}{r}$

18. V=constant and $Q\alpha R$

19. $E=\frac{-dV}{dx}$

20. $U=-\bar{p}.\bar{E}$

21. $W_1=pE(1-\cos\theta)$

LEVEL - II (CW)

COULOMB'S LAW

1. **Two charges when kept at a distance of 1m apart in vacuum hava some force of repulsion. If the force of repulsion between these two charges be same, when placed in an oil of dielectric constant 4, the distance of separation is**

   1) 0.25m 2) 0.4m 3) 0.5m 4) 0.6m

2. **The excess (equal in number) number of electrons that must be placed on each of two small spheres spaced 3 cm apart with force of repulsion between the spheres to be $10^{-19}N$ is**

   1) 25 2) 225 3) 625 4) 1250

3. **Two small conducting spheres each of mass $9\times10^{-4}kg$ are suspended from the same point by non conducting strings of length 100 cm. They are given equal and similar charges until the strings are equally inclined at 45^0 each to the vertical. The charge on each sphere is ..... coulomb**

   1) 1.4×10^{-6} 2) 1.6×10^{-6} 3) 2×10^{-6} 4) 1.96×10^{-6}

4. **Two point charges of magnitude 4 μC and -9 μC are 0.5m apart. The electric intensity is zero at a distance 'x' m from ' A' and 'y' m from 'B'. 'x' and 'y' are respectively**

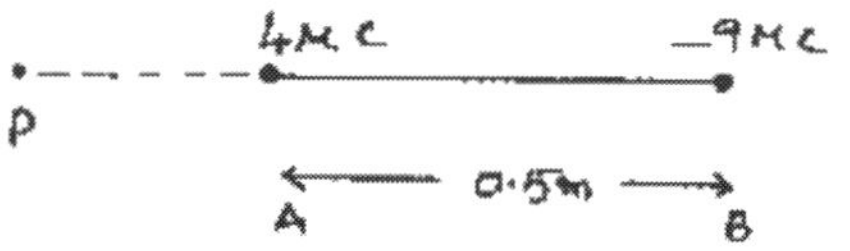

   1) 0.5m, 1.0m 2) 1.0m, 1.5m 3) 2.0m, 1.5m 4) 1.5m, 2.0m

5. **A charge +q is fixed to each of three corners of a square. On the empty corner a charge Q is placed such that there is no net electrostatic force acting on the diagonally opposite charge. Then**

   1) $Q=-2q$ 2) $Q=-2\sqrt{2}q$ 3) $Q=-\sqrt{2q}$ 4) $Q=-4q$

6. **Electrical force between two point charges is 200N. If we increase 10% charge on one of the charges and decrease 10% charge on the other, then electrical force between them for the same distance becomes**

   1) 198 N 2) 100 N 3) 200 N 4) 99 N

7. **N fundamental charges each of charge 'q' are to be distributed as two point charges seperated by a fixed distance, then the maximum to minimum force bears a ratio (N is even and greater than 2)**

   1) $\frac{(N-1)^2}{4N^2}$ 2) $\frac{4N^2}{(N-1)}$ 3) $\frac{N^2}{4(N-1)}$ 4) $\frac{2N^2}{(N-1)}$

8. **A particle A having a charge of $2\times10^{-6}C$ and a mass of 100g is placed at the bottom of a smooth inclined plane of inclination 30^0. The distance of another particle of same mass and charge, be placed on the incline so that it may remain in equilibrium is**

   1) 27 cm 2) 16 cm 3) 30 cm 4) 45 cm

9. **Two identical particles of charge q each are connected by a massless spring of force constant k. They are placed over a smooth horizontal surface.They are released when unstretched. If maximum extension of the spring is r, the value of k is : (neglect gravitational effect)**

   1) $k=\frac{q}{r}\sqrt{\frac{1}{\pi\varepsilon_0 r}}$ 2) $k=\frac{1}{4\pi\varepsilon_0}\frac{q^2}{l^2}\times\frac{1}{r}$ 3) $k=\frac{2q}{r}\sqrt{\frac{1}{\pi\varepsilon_0 r}}$ 4) $k=\frac{q}{r}\sqrt{\frac{2}{\pi\varepsilon_0 r}}$

ELECTRIC FIELD

10. In the figure shown, the electric field intensity at $r = 1m,\ r = 6m,\ r = 9m$ in Vm^{-1} is

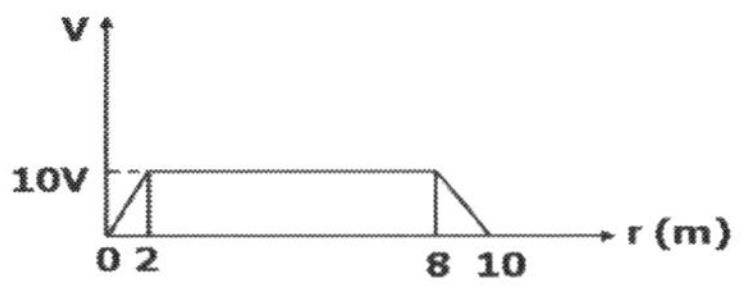

1) -5, -1.67, +5 2) -5, 0, +5 3) 0,1.67,0 4) +5, 1.67, -5

11. Point charges of $3\times10^{-9}C$ are situated at each of three corners of a square whose side is 15 cm. The magnitude and direction of electric field at the vacant corner of the square is

1) 2296 V/m along the diagonal 2) 9622 V/m along the diagonal
3) 22.0 V/m along the diagonal 4) zero

12. A large flat metal surface has uniform charge density $+\sigma$. An electron of mass m and charge e leaves the surface at an angle at point A with speed v, and return to it at point B. The maximum value of AB is ____

1) $\frac{vm\in_0}{\sigma e}$ 2) $\frac{v^2m\in_0}{e\sigma}$ 3) $\frac{v^2e}{\in_0\sigma m}$ 4) $\frac{v^2\sigma e}{\in_0 m}$

13. 'n' charges Q, 4Q, 9Q, 16Q ..... are placed at distances of 1, 2, 3 ..... metre from a point '0' on the same straight line. The electric intensity at '0' is

1) $\frac{Q}{4\pi\in_0 n^2}$ 2) $\frac{Q}{4\pi\in_0 n}$ 3) Infinity 4) $\frac{nQ}{4\pi\in_0}$

14. Two point charges $q_1 = 2\mu C$ and $q_2 = 1\mu C$ are placed at distances b=1 cm and a=2 cm from the origin on the y and x axes as shown in figure. The electric field vector at point (a, b) will subtend an angle θ with the x - axis given by

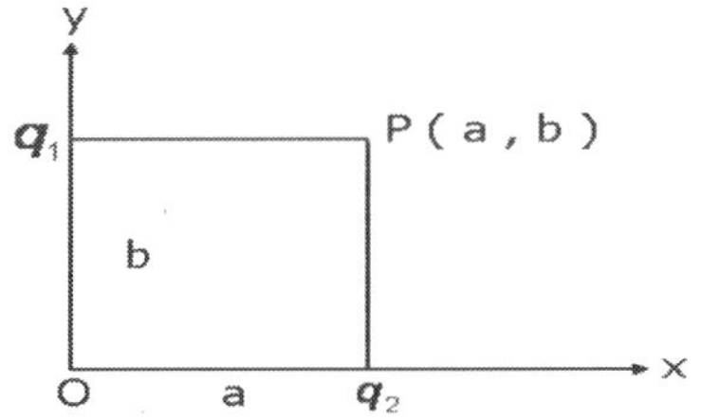

1) $\tan\theta = 1$ 2) $\tan\theta = 2$ 3) $\tan\theta = 3$ 4) $\tan\theta = 4$

15. A non-conducting ring of radius 0.5 m carries of total charge of 1.11x10^{-10}c distributed non-uniformly on its circumference producing an electric field E everywhere in space. The value of the integral $\int_{l=\infty}^{l=0} -\vec{E}.\vec{dl}$ (l=0 being centre of the ring) in volts is

1) +2 2) -1 3) -2 4) zero

ELECTRIC POTENTIAL AND POTENTIAL ENERGY

16. Three charges +q, -q and -q are kept at the vertices of an equilaterial triangle of 10cm side. The potential at the mid point in between -q, -q, if q = 5 μC is

1) -6.4×10^5V 2) -12.8×10^4V 3) -6.4×10^4V 4) -12.8×10^5V

17. Two charges each 'Q' are released when the distance between is 'd'. Then the velocity of each charge of mass 'm' each when the distance between them is '2d' is

1) $\frac{Q}{\sqrt{8\pi\varepsilon_0 dm}}$ 2) $\frac{Q}{\sqrt{4\pi\varepsilon_0 dm}}$ 3) $\frac{Q}{4\sqrt{\pi\varepsilon_0 dm}}$ 4) $\frac{Q}{\sqrt{2\pi\varepsilon_0 dm}}$

18. An oil drop carrying charge 'Q' is held in equilibrium by a potential difference of 600V between the horizontal plates. In order to hold another drop of twice the radius in equilibrium a potential drop of 1600V had to be maintained. The charge on the second drop is

1) $\frac{Q}{2}$ 2) 2Q 3) $\frac{3Q}{2}$ 4) 3Q

19. A body of mass one gram and carrying a charge $10^{-8}C$ passes through two points P and Q. The electrostatic potential at Q is OV. The velocity of the body at Q is $0.2ms^{-1}$ and at P is $\sqrt{0.028}ms^{-1}$. The potential at P is

1) 150V 2) 300V 3) 600V 4) 900V

20. Three charges each $20\mu C$ are placed at the corners of an equilateral triangle of side $0.4m$. The potential energy of the system is

1) $18\times10^{-6}J$ 2) $9J$ 3) $9\times10^{-6}J$ 4) $27J$

21. An electric field is expressed as $\vec{E} = 2\hat{i}+3\hat{j}$. The potential difference $(V_A - V_B)$ between two points A and B whose positions vectors are given by $r_A = \hat{i}+2\hat{j}$ and $r_B = 2\hat{i}+\hat{j}+3\hat{k}$ is

1) -1 V 2) 1 V 3) 2 V 4) 3 V

22. Figure shows three circular arcs, each of radius R and total charge as indicated. The net electric pontential at the centre of the curvature is

1) $\frac{Q}{2\pi\varepsilon_o R}$ 2) $\frac{Q}{4\pi\varepsilon_o R}$ 3) $\frac{2Q}{\pi\varepsilon_o R}$ 4) $\frac{Q}{\pi\varepsilon_o R}$

23. Two identical conducting very large plates P_1 and P_2 having charges $+4Q$ and $+6Q$ are placed very closed to each other at separation d. The plate area of either face of the plate is A. The potential difference between plates P_1 and P_2 is

1) $V_{P_1} - V_{P_2} = \frac{Qd}{A\varepsilon_o}$ 2) $V_{P_1} - V_{P_2} = \frac{-Qd}{A\varepsilon_o}$

3) $V_{P_1} - V_{P_2} = \frac{5Qd}{A\varepsilon_o}$ 4) $V_{P_1} - V_{P_2} = \frac{-5Qd}{A\varepsilon_o}$

DIPOLE

24. An electric dipole consists of two opposite charges of magnitude $1\mu C$ separated by a distance of 2cm. The dipole is placed in an electric filed $10^{-5}Vm^{-1}$. The maximum torque that the field exert on the dipole is

1) $10^{-3}Nm$ 2) $2\times10^{-13}Nm$ 3) $3\times10^{-3}Nm$ 4) $4\times10^{-3}Nm$

25. An electric dipole is formed two particles fixed at the ends of a light rigid rad of length l. The mass of each particle is m and charges are -q and +q The system is suspended by a torsionless thread in an electric field of intensity E such that the dipole axis is parallel to the field if it is slightly displaced, the period of angular motion is

1) $\frac{1}{2\pi}\sqrt{\frac{2qE}{ml}}$ 2) $2\pi\sqrt{\frac{ml}{qE}}$ 3) $2\pi\sqrt{\frac{ml}{2qE}}$ 4) $\frac{1}{2\pi}\sqrt{\frac{ml}{4qE}}$

26. Two point charges - q and +q are located at points (0,0,-a) and (0,0,a) respectively. The electric potential at point (0,0,z) is $(z > a)$

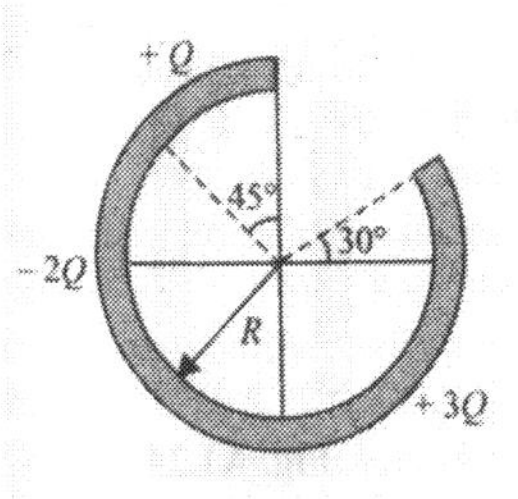

1) $\frac{qa}{4\pi\varepsilon_0 z^2}$ 2) $\frac{q}{4\pi\varepsilon_0 a}$ 3) $\frac{2qa}{4\pi\varepsilon_0\left(z^2-a^2\right)}$ 4) $\frac{2qa}{4\pi\varepsilon_0\left(z^2+a^2\right)}$

27. Two equal charges 'q' of opposite sign are separated by a small distance '2a'. The electric intensity 'E' at a point on the perpendicular bisector of the line joining the two charges at a very large distance 'r' from the line is

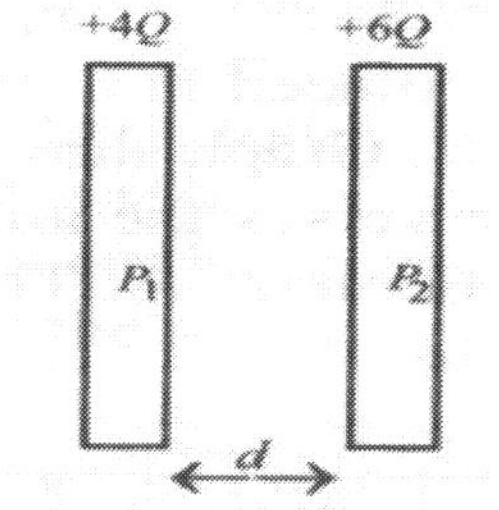

1) $\frac{1}{4\pi\varepsilon_0}\frac{qa}{r^2}$ 2) $\frac{1}{4\pi\varepsilon_0}\frac{2qa}{r^3}$ 3) $\frac{1}{4\pi\varepsilon_0}\frac{2qa}{r^2}$ 4) $\frac{1}{4\pi\varepsilon_0}\frac{qa}{r^3}$

KEY - LEVEL - II (CW)

1) 3	2) 3	3) 1	4) 2	5) 2	6) 1	7) 3	8) 1	9) 2	10) 2
11) 1	12) 2	13) 4	14) 1	15) 1	16) 4	17) 1	18) 4	19) 3	20) 4
21) 1	22) 1	23) 2	24) 2	25) 3	26) 3	27) 2			

LEVEL - II (CW) - HINTS

1. $t^1 - \frac{t}{\sqrt{k}}$

2. $F - \frac{1}{4\pi\varepsilon_0}\frac{q_1 q_2}{r^2}$ and q = ne

3. F = w $\tan\theta$ where $F = \frac{1}{4\pi\varepsilon_0}\frac{q_1 q_2}{r^2}$

4. Distance of null point $x = \dfrac{d}{\sqrt{\dfrac{Q_2}{Q_1}} \pm 1}$

+ve for like charges -ve for unlike charges

5. $\dfrac{1}{4\pi \in_0} \sqrt{2} \dfrac{q^2}{a^2} + \dfrac{1}{4\pi \in_0} \dfrac{Qq}{\left(\sqrt{2}a\right)^2} = 0$

6. $F = \dfrac{1}{4\pi\varepsilon_0} \dfrac{q_1 q_2}{r^2}$; $q_1^1 = \dfrac{110}{100} q_1$ and $q_2^1 = \dfrac{90}{100} q_2$

7. $\dfrac{F_{\max}}{F\min} = \dfrac{\left(N/2\right)^2}{(N-1)1}$

8. $mg\sin\theta = \frac{1}{4\pi\varepsilon_0} \dfrac{q^2}{r^2}$

9. $F_C = kx$

10. $E = -\dfrac{dV}{dr}$

11. $E = E\left(\sqrt{2} + 1/2\right) E = \dfrac{1}{4\pi\varepsilon_0} \dfrac{q}{r^2}$

r = length of the side

12. Field near metal surface $E = \dfrac{\sigma}{\in_0}$

Force on electron = eE $= \dfrac{e\sigma}{\in_0}$

Acceleration of electron $a = \dfrac{e\sigma}{m \in_0}$

It will act as projectile with max range $= \dfrac{u^2}{a} = \dfrac{u^2}{e\sigma} \times m \in_0$

13. $E = \dfrac{1}{4\pi \in_o} \cdot \left[\dfrac{Q_1}{x_1^2} + \dfrac{Q_2}{x_2^2} + ... + \dfrac{Q_n}{x_n^2} \right]$

14. $Tan\theta = \dfrac{E_2}{E_1}$

15. $V_o = K\dfrac{q}{R}, V_\infty = 0$; $\int_{t=\infty}^{t=0} -\vec{E}.\vec{dl} = V_o - V_\infty$

16. $V = V_1 + V_2 + V_3\ ;\ V_1 = V_2 = \frac{1}{4\pi\varepsilon_0}\frac{(-q)}{(a/2)}$

$$V_3 = \frac{1}{4\pi\varepsilon_0}\frac{(q)}{\left(\frac{\sqrt{3}a}{2}\right)}$$

17. gain in K.E = loss in P.E

18. $\frac{V_1}{V_2} = \left(\frac{R_1}{R_2}\right)^3 . \frac{Q_2}{Q_1}$

19. $\frac{1}{2}m\left[v_Q^2 - v_P^2\right] = q\left[V_P - V_Q\right]$

20. $U = \frac{1}{4\pi\varepsilon_0}\left(\frac{q_1q_2}{r_{12}} + \frac{q_2q_3}{r_{23}} + \frac{q_3q_1}{r_{13}}\right)$

21. $V_B - V_A = -\left[\int_1^2 2\,dx + \int_2^1 3\,dy\right]$

22. $V = V_1 + V_2 + V_3$

23. $V_{P_1} - V_{P_2} = \frac{-Q}{\varepsilon_0 A/d}$

24. $\tau_{max} = pE = 2aqE$

25. $\tau = PE\,\sin\theta$

$\tau = I\alpha$

$I\alpha = PE\,\sin\theta$

I = moment of inertia $= \frac{ml^2}{2}$

$\therefore$ Time period $= 2\pi\sqrt{\frac{I}{pE}}$

26. The distance of point P from charge +q is $r_1 = z - a$

and from charge -q is $r_2 = z + a$

Potential at P is $\frac{1}{4\pi\in_0}\left(\frac{q}{r_1} - \frac{q}{r_2}\right) = \frac{1}{4\pi\in_0}\frac{2qa}{(z^2 - a^2)}$

27. Similar to B on equitorial line of a short bar magnet

LEVEL - II (HW)

COULOMB'S LAW

1. Two equally charged identical metal spheres A and B repel each other with a force F. Another identical uncharged sphere C is touched to A and then placed midway between A and B. The net force on C is in the direction

   1) F towards A 2) F towards B 3) 2F towards A 4) 2F towards B

2. Two unlike charges seperated by a distance of 1m attract each other with a force of $0.108N$. If the charges are in the ratio $1:3$, the weak charge is

   1) $2\mu C$ 2) $4\mu C$ 3) $6\mu C$ 4) $5\mu C$

3. Three charges each equal to $10^{-9}C$ are placed at the corners of an equilateral triangle of side 1m. The force on one of the charges is

   1) $9\times10^{-9}N$ 2) $9\sqrt{3}\times10^{-9}N$ 3) $27\times10^{-9}N$ 4) $18\times10^{-9}N$

4. Two particles each of mass 'm' and carrying charge 'Q' are seperated by some distance. If they are in equilibrium under mutual gravitational and electro static forces, then Q/m (in c/Kg) is of the order of

   1) 10^{-5} 2) 10^{-10} 3) 10^{-15} 4) 10^{-20}

5. There point charges + q, – q and + q are placed at the vertices P, Q and R of an equilateral triangle as shown. If $F=\frac{1}{4\pi\varepsilon_0}\frac{q^2}{r^2}$, where 'r' is the side of the triangle, the force on charge at 'P' due to charges at Q and R is

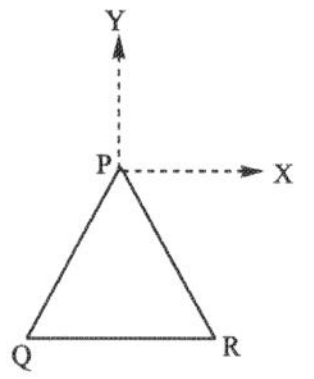

   1) F along positive x–direction 2) F along negative x–direction

   3) $\sqrt{2}$ F along positive x–direction 4) $\sqrt{2}$ F along negative x–direction

6. Three point charges +q, +q and –q are placed at the corners of an equilateral triangle of side 'a'. Another charge +Q is kept at the centroid. Force exerted on Q is:

   1) $\frac{1}{4\pi\varepsilon_o}\frac{2qQ}{a^2}$ 2) $\frac{1}{4\pi\varepsilon_o}\frac{6qQ}{a^2}$ 3) $\frac{1}{4\pi\varepsilon_o}\frac{8qQ}{a^2}$ 4) $\frac{1}{4\pi\varepsilon_o}\frac{14qQ}{a^2}$

7. Three charges $-q_1, +q_2$ and $-q_3$ are placed as shown in fig. The X-component of the force on $-q_1$ is proportional to

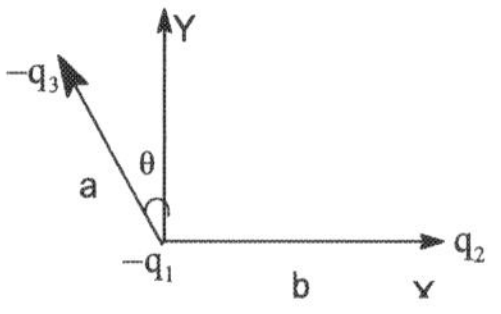

   1) $\frac{q_2}{b^2}-\frac{q_3}{a^2}\cos\theta$ 2) $\frac{q_2}{b^2}+\frac{q_3}{a^2}\sin\theta$ 3) $\frac{q_2}{b^2}+\frac{q_2}{a^2}\cos\theta$ 4) $\frac{q_2}{b^2}-\frac{q_2}{a^2}\sin\theta$

ELECTRIC FIELD

8. The breakdown electric intensity for air is 3×10^6 V/m. The maximum charge that can be held by a sphere of radius 1 mm is

1) 0.33 C 2) 0.33 nC 3) 3.3 C 4) 3.3 μC

9. There is a uniform electric field of strength $10^3 V/m$ along y-axis. A body of mass 1 g and charge $10^{-6} C$ is projected into the field from origin along the positive x-axis with a velocity 10 m/s. Its speed in m/s after 10s is (neglect gravitation)

1) 10 2) $5\sqrt{2}$ 3) $10\sqrt{2}$ 4) 20

10. The point charges $+1C, +1C$ and $-1C$ are placed at the vertices A, B and C of an equilateral triangle of side 1m. Then

(A) The force acting on the charge at A is $9\times10^9 N$

(B) The electric field strength at A is $9\times10^9 NC^{-1}$

1) A is correct but B is wrong 2) B is correct but A is wrong

3) Both A and B are wrong 4) Both A and B are correct

11. A pendulum bob of mass m carrying a charge q is at rest in a uniform horizontal electric field of intensity E. The tension in the thread is

1) $T=\sqrt{(Eq)^2+(mg)^2}$ 2) $T=\sqrt{\left(\frac{E}{q}\right)^2+(mg)^2}$

3) $T=\sqrt{\left(\frac{E}{q}\right)^2+\left(\frac{m}{g}\right)^2}$ 4) $T=mg+Eq$

ELECTRIC POTENTIAL AND POTENTIAL ENERGY

12. Four charges $10^{-8}; -2\times10^{-8}; +3\times10^{-8}$ and 2×10^{-8} coulomb are placed at the four corners of a square of side 1m the potential at the centre of the square is

1) zero 2) 360 volt 3) 180 volt 4) $360\sqrt{2}$ volt

13. Two metal spheres of radii R_1 and R_2 are charged to the same potential. The ratio of the charge on the two spheres is

1) 1 2) $\frac{1}{2}$ 3) R_1-R_2 4) $\frac{R_1}{R_2}$

14. Two concentric, thin metallic spherical shells of radii R_1 and R_2 $(R_1>R_2)$ bear charges Q_1 and Q_2 respectively. Then the potential at radius 'r' between R_1 and R_2 will be $\frac{1}{4\pi\in_0}$ times

1) $\frac{Q_1+Q_2}{r}$ 2) $\frac{Q_1}{R_1}+\frac{Q_2}{r}$ 3) $\frac{Q_1}{R_1}+\frac{Q_2}{R_2}$ 4) $\frac{Q_1}{R_2}+\frac{Q_2}{R_2}$

15. An electric charge $10^{-3}\mu C$ is placed at the origin (0, 0) of X-Y coordinate system. Two points A and B are situated at $\left(\sqrt{2},\sqrt{2}\right)$ and (2, 0) respecitvely. The potential difference between the points A and B will be:

1) 9 V 2) zero 3) 2 V 4) 4.5 V

16. A charge $-2\mu C$ at the origin, $-1\mu C$ at $+7cm$ and $1\mu C$ at $-7cm$ are placed on $X-$ axis. The mutual potential energy of the system is

1) $-0.051J$ 2) $-0.045J$ 3) $0.045J$ 4) $-0.064J$

17. Four equal charges Q are placed at the four corners of a square of side 'a' each. Work done in removing a charge $-Q$ from its centre to infinity is

1) zero 2) $\sqrt{2}Q^2/4\pi\in_0 a$ 3) $\sqrt{2}Q^2/\pi\in_0 a$ 4) $Q^2/2\pi\in_0 a$

18. The electrostatic potential V at any point (x,y,z) in space is given by $V=4x^2$

1) The y - and z - components of the electrostatic field at any point are not zero

2) The x - component of electric field intensity at any point is given by $\left(-8x\hat{i}\right)$

3) The x - component of electric field intensity at a point (2, 0,2) is $\left(-8\hat{i}\right)$

4) The y - and z - components of the field are constant in magnitude.

DIPOLE

19. The self potential energy of hydrogen chloride whose dipole moment is $3.44\times10^{-30}C$ - m and separation between hydrogen and chlorine atoms is $1.01\times10^{-10}m$ is

1) $1.036\times10^{-19}J$ 2) $3.2\times10^{5}J$ 3) $4.5\times10^{7}J$ 4) $1.65\times10^{6}J$

KEY LEVEL - II (HW)

1) 1	2) 1	3) 2	4) 2	5) 2	6) 2	7) 2	8) 2	9) 3	10) 4
11) 1	12) 4	13) 4	14) 2	15) 2	16) 4	17) 3	18) 2	19) 1	

LEVEL - II (HW) - HINTS

1. $F=\frac{1}{4\pi\varepsilon_0}\frac{q_1q_2}{r^2}$

2. $F=\frac{1}{4\pi\varepsilon_0}\frac{q_1q_2}{r^2}$

3. $F_1=F_2=\frac{1}{4\pi\varepsilon_0}\frac{q^2}{r^2}$; $F_R=\sqrt{3}.\frac{1}{4\pi\varepsilon_0}\frac{q^2}{r^2}$

4. $F_e=\frac{1}{4\pi\varepsilon_0}\frac{q^2}{r^2}$ and $F_g=\frac{Gm^2}{r^2}$

5. $F_1=F_2$ and angle between them is 120^0

6. $F=\left(\frac{1}{4\pi\varepsilon_0}\frac{q}{r^2}\right)2$ where $r=\frac{a}{\sqrt{3}}$

8. $E=\frac{1}{4\pi\in_o}.\frac{Q}{d^2}$

9. $v=u+at$ where $a=\frac{Eq}{m}$

10. $F=\frac{1}{4\pi\varepsilon_o}.\frac{q_1q_2}{r^2}$; $E=\frac{1}{4\pi\varepsilon_o}.\frac{q}{r^2}$

$F=\overline{F_1}+\overline{F_2}$; $E=\overline{E_1}+\overline{E_2}$

11. Conceptual

12. $V=\frac{1}{4\pi\varepsilon_0}\sum\frac{Q}{r}$

13. $V=\frac{1}{4\pi\varepsilon_0}\frac{Q}{R}$

14. Potential is constant within the sphere and is additive.

15. $V=\frac{q}{4\pi\varepsilon_o}\left(\frac{1}{r_1}-\frac{1}{r_2}\right)$

16. $PE=\frac{1}{4\pi\varepsilon_0}.\frac{q_1q_2}{r}$

17. Workdone = Electrostatic potential energy at the centre of the square

18. $E=\frac{-dV}{dx}$

19. $p=2qa\Rightarrow q=\frac{p}{2a}=3.41\times10^{-20}$

$\therefore PE=\frac{1}{4\pi\in_0}\frac{q^2}{2a}=1.036\times10^{-19}J$

LEVEL - III

ELECTROSTATIC FORCE

1. A ball of mass m = 0.5 kg is suspended by a thread and a charge q = 0.1 μC is supplied. When a ball with diameter 5cm and a like charge of same magnitude is brought close to the first ball, but below it, the tension decreases to 1/3 of its initial value. The distance between centres of the balls is

1) $0.12\times10^{-2}m$ 2) $0.51\times10^{-4}m$ 3) $0.2\times10^{-5}m$ 4) $0.52\times10^{-2}m$

2. Five point charges each +q, are placed on five vertices of a regular hexagon of side L, The magnitude of the force on a point charge of value – q placed at the centre of the hexagon (in newton) is

1) Zero 2) $\frac{\sqrt{3}q^2}{4\pi\in_0 L^2}$ 3) $\frac{q^2}{4\pi\in_0 L^2}$ 4) $\frac{q^2}{4\sqrt{3}\,\pi\in_0 L^2}$

3. Two small objects X and Y are permanently separated by a distance 1 cm. Object X has a charge of + 1.0 μC and object Y has a charge of - 1.0 μC. A certain number of electrons are removed from X and put onto Y to make the electrostatic force between the two objects an attractive force whose magnitude is 360 N. Number of electrons removed is

1) 8.4×10^{13} 2) 6.25×10^{12} 3) 4.2×10^{11} 4) 3.5×10^{10}

4. Two identical positive charges are fixed on the y-axis, at equal distance from the origin O, A partical with a negative charge starts on the negative x-axis at a large distance from O, moves along the x-axis passed through O and moves far away from O. Its acceleration a is taken as positive along its direction of motion. The particle's acceleration a is plotted against its x-co-ordinate. Which of the following best represents the plot?

1) 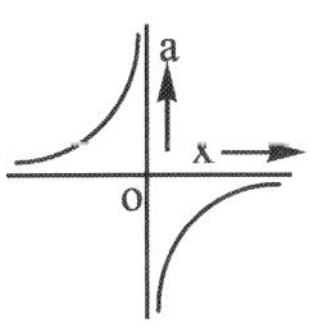

2) 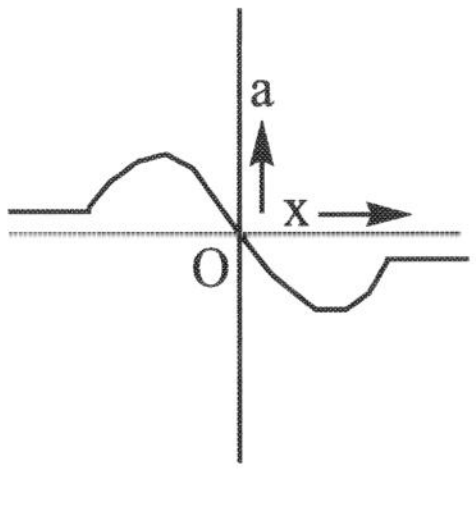

3) 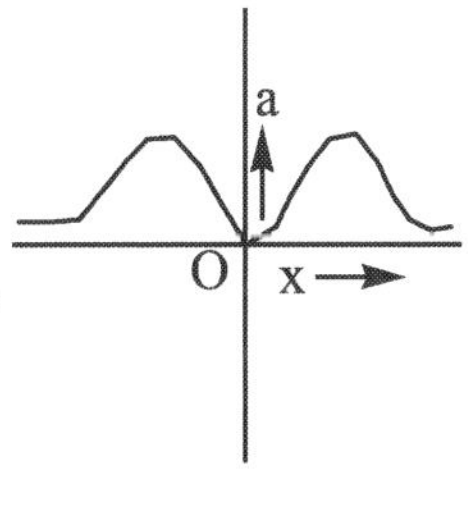

4) 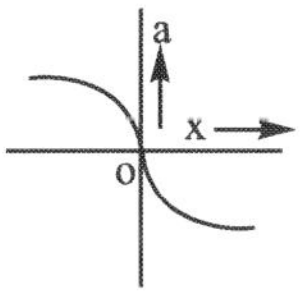

5. **Two equal negative charges –q each are fixed at points (0, –a) and (0,a) on y-axis. A positive charge Q is released from rest at the point (2a, 0) on the x-axis. The charge Q will**

1) execute simple harmonic motion about the origin

2) move to the origin and remain at rest

3) move to infinity

4) execute oscillatory but not simple harmonic motion

6. **In a liquid medium of dielectric constant K and of specific gravity 2, two identically charged spheres are suspended from a fixed point by threads of equal lengths. The angle between them is 90°. In another medium of unknown dielectric constant K^1, and specific gravity 4, the angle between them becomes 120°. If density of material of spheres is 8 gm/cc then K^1 is :**

1) $\frac{K}{2}$ 2) $\frac{\sqrt{3}}{K}$ 3) $\frac{\sqrt{3}}{2}K$ 4) $\frac{K}{\sqrt{3}}$

7. The force of attraction between two charges separated by certain distance in air is F_1. **If the space between the charges is completely filled with dielectric of constant 4 the force becomes F_2. If half of the distance between the charges is filled with same dielectric the force between the charges is F_3. Then $F_1 : F_2 : F_3$ is**

1) 16 : 9 : 4 2) 9 : 36 : 16 3) 4 : 1 : 2 4) 36 : 9 : 16

8. **Two small spheres of masses, M_1 and M_2 are suspended by weightless insulating threads of lengths L_1 and L_2. the sphere carry charges Q_1 and Q_2 respectively. The spheres are suspended such that they are in level with another and the threads are inclined to the vertical at angles of θ_1 and θ_2 as shown below, which one of the following conditions is essential , if $\theta_1 = \theta_2$.**

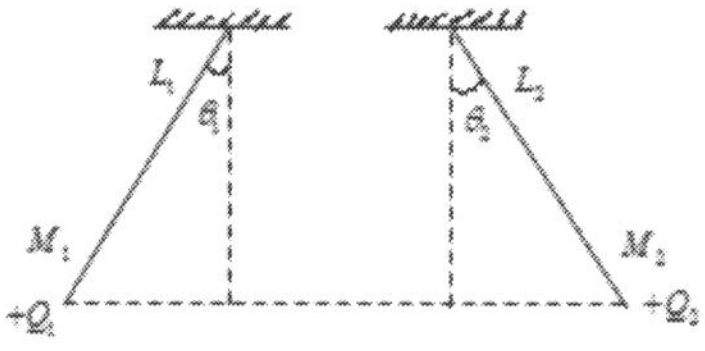

1) $M_1 \neq M_2$ but $Q_1 = Q_2$ 2) $M_1 = M_2$ 3) $Q_1 = Q_2$ 4) $L_1 = L_2$

ELECTRIC FIELD

9. **If the electric field between the plates of a cathode ray oscilloscope be $1.2\times10^4 N/C$, the deflection that an electron will experience if it enters at right angles to the field with kinetic energy 2000 eV is (The deflection assembly is 1.5cm long.)**

1) 0.34 cm 2) 3.4 cm 3) 0.034 mm 4) 0.34 mm

10. **A electric field of $1.5\times10^4 NC^{-1}$ exists between two parallel plates of length 2 cm. An electron enters the region between the plates at right angles to the field with a kinetic energy of $E_k = 2000eV$. The deflection that the electron experiences at the deflecting plates is**

1) 0.34 mm 2) 0.57 mm 3) 7.5 mm 4) 0.75 mm

11. **A bob of a simple pendulum of mass 40gm with a positive charge 4×10^{-6} C is oscillating with a time period T_1 .An electric field of intensity 3.6×10^4 N/C is applied vertically upwards.Now the time period is T_2 the value of $\frac{T_2}{T_1}$ is (g = 10m/s²)**

1)0.16 2) 0.64 3)1.25 4)0.8

12. **A particle of mass m and charge q is placed at rest in a uniform electric field E and then released. The kinetic energy attained by the particle after moving a distance y is**

1) qEy^2 2) qE^2y 3) qEy 4) q^2Ey

13. **Four equipotential curves in an electric field are shown in the figure. A,B,C are three points in the field. If electric intensity at A,B,C are E_A, E_B, E_C then**

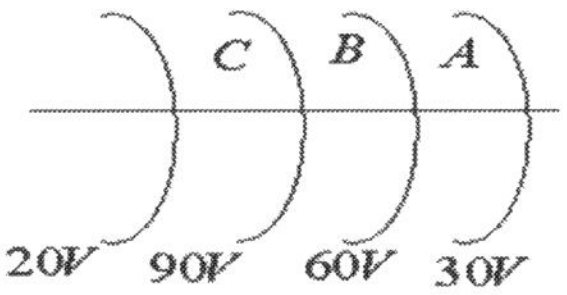

1) $E_A = E_B = E_C$ 2) $E_A > E_B > E_C$ 3) $E_A < E_B < E_C$ 4) $E_A > E_B < E_C$

14. **A particle of mass 1Kg and carrying 0.01C is at rest on an inclined plane of angle 30^0 with horizontal when an electric field of $\frac{490}{\sqrt{3}}NC^{-1}$ applied parllel to horizontal. The coefficient of friction is**

1) 0.5 2) $\frac{1}{\sqrt{3}}$ 3) $\frac{\sqrt{3}}{2}$ 4) $\frac{\sqrt{3}}{7}$

15. **Electric field on the axis of a small electric dipole at a distance r is $\vec{E}_1$ and $\vec{E}_2$ at a distance of 2r on a line of perpendicular bisector. Then**

1) $\vec{E_2} = -\vec{E_1}/8$ 2) $\vec{E_2} = -\vec{E_1}/16$ 3) $\vec{E_2} = -\vec{E_1}/4$ 4) $\vec{E_2} = \vec{E_1}/8$

16. **A particle having charge that on an electron and mass 1.6 x 10^{-30} kg is projected with an initial speed 'u' to the horizontal from the lower plate of a parallel plate capacitor as shown. The plates are sufficiently long and have separation 2cm. Then the maximum value of velocity of particle not to hit the upper plate. (E=10^3 V/m upwards).**

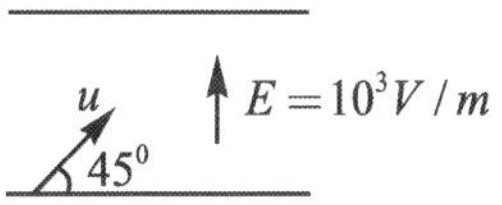

1) 2 x 10^6 m/s 2) 4 x 10^6 m/s 3) 6 x 10^6 m/s 4) 3x 10^6 m/s

17. **An electric field is acting vertically upwards. A small body of mass 1 gm and charge -1 μ C is projected with a velocity 10 m/s at an angle 45^0 with horizontal. Its horizontal range is 2m then the intensity of electric field is :(g = 10 m/s^2)**

1) 20,000 N/C 2) 10,000 N/C 3) 40,000 N/C 4) 90,000 N/C

18. **A thin copper ring of radius 'a' is charged with q units of electricity. An electron is placed at the centre of the copper ring. If the electron is displaced a little, it will have frequency.**

1) $\frac{1}{2\pi}\sqrt{\frac{eq}{4\pi\epsilon_0 ma^3}}$ 2) $\frac{1}{2\pi}\sqrt{\frac{q}{4\pi\epsilon_0 ema^3}}$ 3) $\sqrt{\frac{eq}{4\pi\epsilon_0 ma}}$ 4) $\sqrt{\frac{q}{4\pi\epsilon_0 ema^3}}$

19. **A thin fixed ring of radius 1 metre has a positive charge $1\times10^{-5}C$ uniformly distributed over it. A particle of mass 0.9gm and having a negative charge of $1\times10^{-6}C$ is placed on the axis at a distance of 1 cm from the centre of the ring. Assuming that the oscillations have small amplitude, the time period of oscillations is**

1) 0.23s 2) 0.39s 3) 0.49 s 4) 0.63s

20. **A sphere carrying charge 0.01 C is kept at rest without falling down, touching a wall by applying an electric field 100 N/C. If the coeffcient of friction between the sphere and the wall is 0.2, the weight of the sphere is**

1) 4N 2) 2 N 3) 20 N 4) 0.2 N

21. **A particle of mass 1kg and carrying positive charge 0.01 C is sliding down an inclined plane of angle 30^0 with the horizontal. An electric field E is applied to stop the particle. If the coefficient of friction between the particle and the surface of the plane is $\frac{1}{2\sqrt{3}}$, E must be**

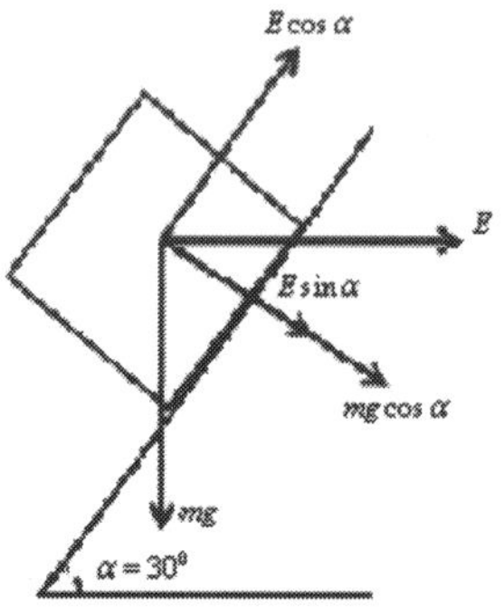

1) 1260 V/m 2) 245 V/m 3) $140\sqrt{3}$ V/m 4) $\frac{490}{\sqrt{3}}$ V/m

22. **Two identical point charges are placed at a separation of l. P is a point on the line joining the charges, at a distance x from any one charge. The field at P is E. E is plotted against x for values of x from close to zero to slightly less than l. Which of the following best represents the resulting curve?**

1)
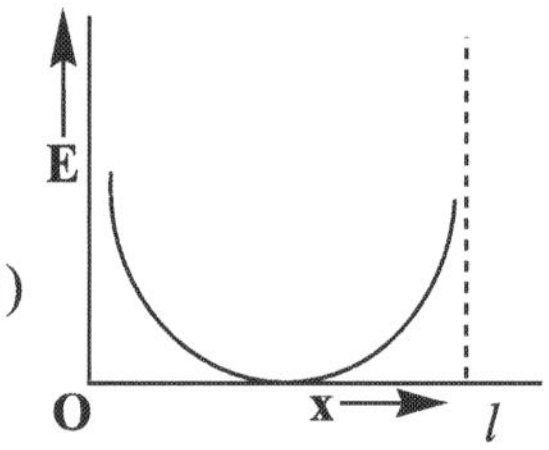

2)
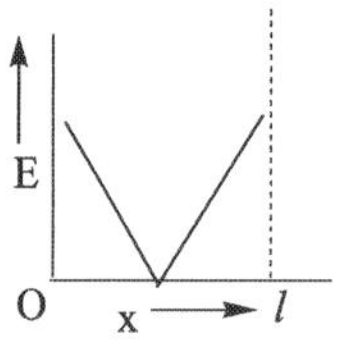

3)
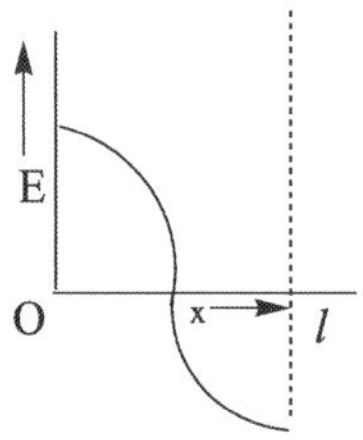

4)
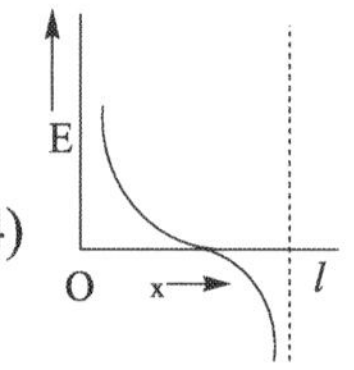

23. **A particle of charge $-q$ and mass m moves in a circular orbit of radius r about a fixed charge $+Q$. The relation between the radius of the orbit r and the time period T is**

1) $r=\frac{Qq}{16\pi^2 \epsilon_0 m}T^3$ 2) $r^3=\frac{Qq}{16\pi^3 \epsilon_0 m}T^2$ 3) $r^2=\frac{Qq}{16\pi^3 \epsilon_0 m}T^3$ 4) $r^2=\frac{Qq}{16\pi \epsilon_0 m}T^3$

24. **A thin semicircular ring of radius 'r' has a positive charge distributed uniformly over it. The net field E at the centre 'O' is**

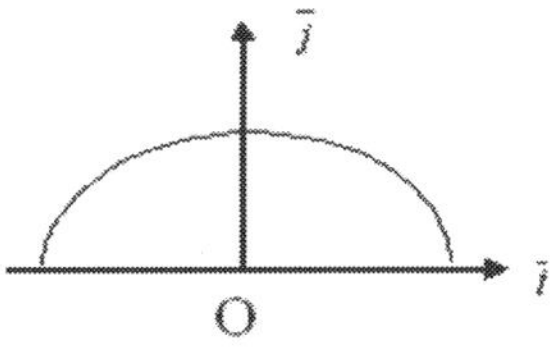

1) $\frac{q}{2\pi^2 \in_0 r^2}\bar{j}$ 2) $\frac{q}{4\pi^2 \in_0 r^2}\bar{j}$ 3) $-\frac{q}{4\pi^2 \in_0 r^2}\bar{j}$ 4) $-\frac{q}{2\pi^2 \in_0 r^2}\bar{j}$

ELECTRIC POTENTIAL

25. **Two thin rings each having a radius R are placed at distance d apart with their axes coinciding.The charges on the two rings are +q, -q. The potential difference between the rings**

1) $\frac{Q.R}{4\pi\varepsilon_0.d^2}$ 2) $\frac{Q}{2\pi\varepsilon_0}\left(\frac{1}{R}-\frac{1}{\sqrt{R^2+d^2}}\right)$ 3) $\frac{Q}{4\pi\varepsilon_0}\left(\frac{1}{R}-\frac{1}{\sqrt{R^2+d^2}}\right)$ 4) 0

26. **Two metal sphres A and B have their capacities in th ratio 3:4. They are put in contact with each other and an amount of charge $7\times10^{-6}C$ is given to the combination. Next, the two spheres are separated and kept wide the apart so that one has no electrical infuence on the other. The potential due to the smaller sphere at a distance of 50m from it is**

1) 540V 2) 270V 3) 1180V 4) zero

27. **A solid conducting sphere having a charge Q is surrounded by an uncharged concentric conducting spherical shell. The potential difference between the surface of solid sphere and the shell is V. The shell is now given a charge –3Q. The new potential difference between the same surfaces will be**

1) –2V 2) 4V 3) V 4) 2V

28. **A spherical charged conductor has surface charge density σ. The intensity of electric field and potential on its surface are E and V. Now radius of sphere is halved keeping the charge density as constant. The new electric field on the surface and potential at the centre of the sphere are**

1) 4E, V 2) E, V/2 3) E, V 4) 2E, 4V

29. **Two spherical conductors A and B of radii 1 mm and 2mm are seperated by a distance of 5 cm and are uniformly charged. If the spheres are connected by a conducting wire then in the equilibrium condition the ratio of electric fields at surfaces of A and B is** 1) 4: 1 2) 1 : 2 3) 2 : 1 4) 1 : 4

30. **A charge +q is fixed at each of the points $x=x_0, x=3x_0, x=5x_0 \ldots\ldots \infty$ on the x - axis and a charge-q is fixed at each of the points $x=2x_0, x=4x_0, x=6x_0 \ldots\ldots \infty$. Here x_0 is a positive constant. Take the electric potential at a point due to a charge Q at a distance r from it to be $\frac{Q}{4\pi \in_0 r}$. Then the potential at the origin due to the above system of charges is**

1) 0 2) $\frac{q}{8\pi \in_0 x_0 \log_e(2)}$ 3) ∞ 4) $\frac{q\log_e(2)}{4\pi \in_0 x_0}$

31. A non – conducting ring of radius 0.5 m carries a total charge of 1.11 x 10^{-10} C distributed non – uniformly in its circumference producing an electric field E everywhere in space. The value of the integral $\int_{l=\infty}^{l=0} -E\,dl$ ($l = 0$ being centre of the ring) in volt is :

1) +2 2) –1 3) –2 4) zero

32. Some equipotential surfaces are shown in figure. The electric field strength is

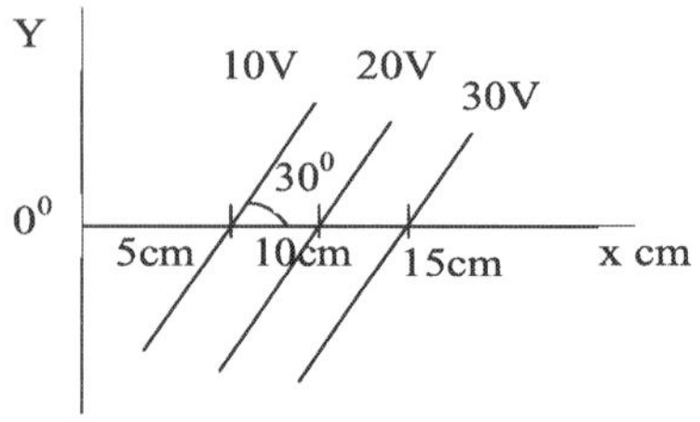

1) 100 V/m along x-axis

2) 100 V/m along y-axis

3) 400 V/m at an angle 120^0 with x-axis

4) $\frac{400}{\sqrt{3}}$ V/m an angle 120^0 with x-axis

33. A field of $100 Vm^{-1}$ is directed at 30^0 to positve x - axis. Find V_{BA} if OA = 2m and OB = 4m

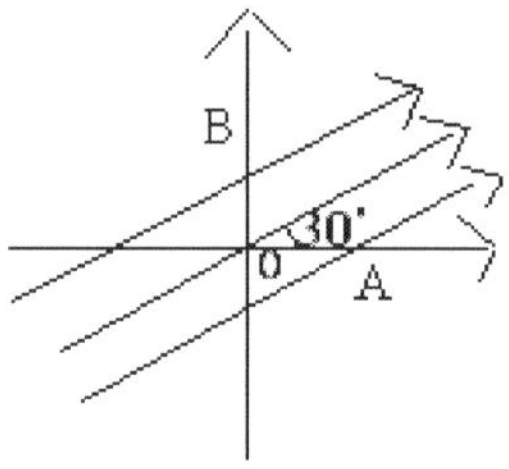

1) 100 $(\sqrt{3}-2)$ V

2) $100(2+\sqrt{3})$ V

3) 100 $(2-\sqrt{3})$ V

4) 200 $(2+\sqrt{3})$ V

34. Here is a special parallelogram with adjacent side lengths 2*a* and *a* and the one of the possible angles between them as 60°. Two charges are to be kept across a diagonal only. The ratio of the minimum potential energy of the system to the maximum potential energy is

1) $\sqrt{3} : \sqrt{7}$ 2) 3 : 7 3) 1 : 2 4) 1 : 4

35. Two concentric spherical conducting shells of radii R and 2R carry charges Q and 2Q respectively. Change in electric potential on the outer shell when both are connected by a conducting wire is : $\left(k = \frac{1}{4\pi\varepsilon_0}\right)$

1) zero 2) $\frac{3kQ}{2R}$ 3) $\frac{kQ}{R}$ 4) $\frac{2kQ}{R}$

36. The longer side of a rectangle is twice the length of its shorter side. A charge *q* is kept at one vertex. The maximum electric potential due to that charge at any other vertex is *V*, then the minimum electric potential at any other vertex will be

1) $2V$ 2) $\sqrt{3}\,V$ 3) $V/\sqrt{5}$ 4) $\sqrt{5}\,V$

37. **There are three uncharged identical metallic spheres 1,2 and 3 each of radius r and are placed at the vertices of an equilateral triangle of side d. A charged metallic sphere having charge q of same radius r is touched to sphere 1, after some time it is taken to the location of sphere 2 and is touched to it, then it is taken far away from spheres 1,2 and 3. After that the sphere 3 is grounded, the charge on sphere 3 is (neglect electrostatic induction by assuming d>> 2r)**

1) Zero 2) $\frac{-3qr}{4d}$ 3) $\frac{-qr}{2d}$ 4) $\frac{-qr}{4d}$

38. 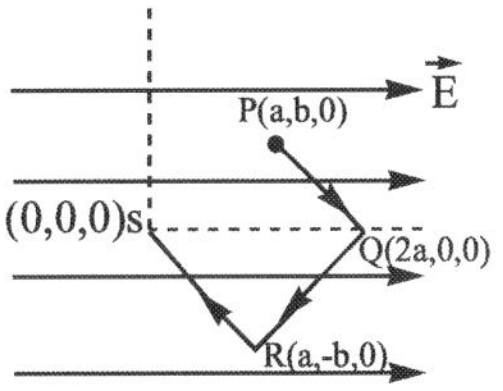

A point charge q moves from point P to point S along the path PQRS in a unifrom electric field $\vec{E}$ pointing parallel to the positive direction of the x-axis. The coordinates of the points P, Q, R and S are (a,b,0), (2a,0,0), (a, –b, 0) and (0,0,0) respectively. The work done by the field in the above process is given by the expression

1) qaE 2) –qaE 3) $q(\sqrt{a^2+b^2})E$ 4) $3qE\sqrt{a^2+b^2}$

39. **The potential at a point x (measured in μ m) due to some charges situated on the x-axis is given by $V(x)=\frac{20}{x^2-4}$ volt. The electric field E at x = 4 μ m is given by**

1) $\frac{5}{3}\frac{V}{\mu m}$ and in the positive x - direction
2) $\frac{10}{9}\frac{V}{\mu m}$ and in the negative x - direction
3) $\frac{10}{9}\frac{V}{\mu m}$ and in the positive x-direction
4) $\frac{5}{3}\frac{V}{\mu m}$ and in the negative x-direction

40. **Two points charges q_1 and q_2 (=q_1/2) are placed at points A(0, 1) and B (1, 0) as shown in the figure. The electric field vector at point P(1, 1) makes an angle q with the *x*–axis, then the angle q is**

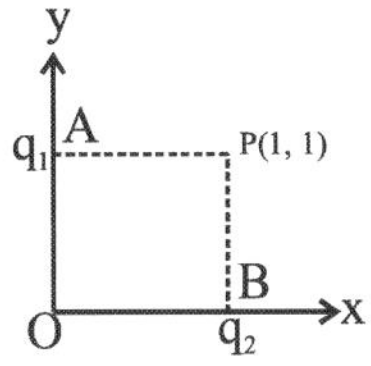

1) $\tan^{-1}\left(\frac{1}{2}\right)$ 2) $\tan^{-1}\left(\frac{1}{4}\right)$ 3) $\tan^{-1}(1)$ 4) $\tan^{-1}(0)$

41. **Figure shows three spherical and equipotential surfaces 1,2 and 3 round a point charge q. The potential difference $V_1–V_2 = V_2– V_3$. If t_1 and t_2 be the distance between them. Then**

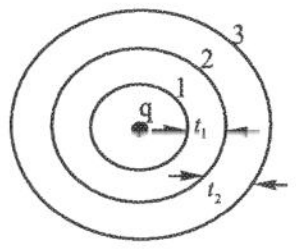

1) $t_1=t_2$ 2) $t_1>t_2$ 3) $t_1<t_2$ 4) $t_1 \le t_2$

42. **A half ring of radius ' r ' has a linear charge density λ. The potential at the centre of the half ring is**

1) $\frac{\lambda}{4\varepsilon_0}$ 2) $\frac{\lambda}{4\pi^2\varepsilon_0 r}$ 3) $\frac{\lambda}{4\pi\varepsilon_0 r}$ 4) $\frac{\lambda}{4\pi\varepsilon_0 r^2}$

43. **The distance between plates of a parallel plate capacitor is 5d. The positively charged plate is at x=0 and negativily charged plates is at x=5d. Two slabs one of conducotor and the other of a dielectric of same thickness d are inserted between the plates as shown in figre. Potential (V) versus distance x graph will be**

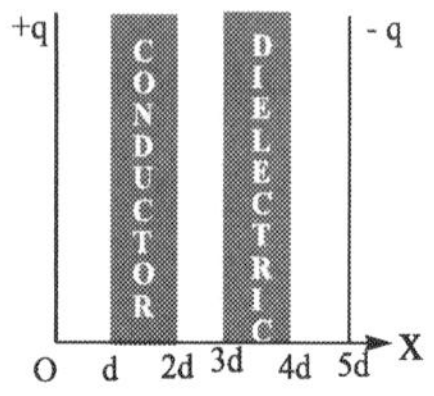

1) 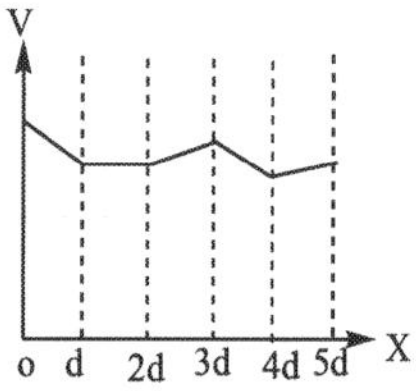

2) 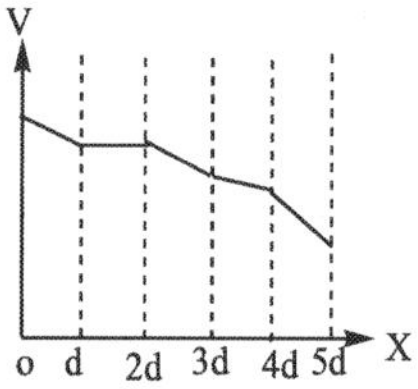

3) 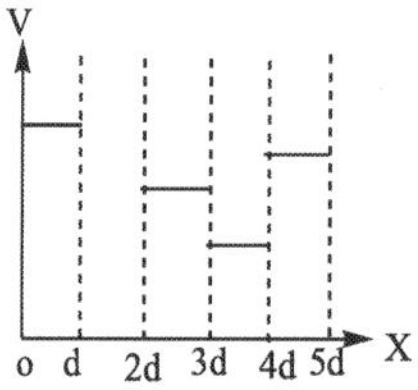

4) 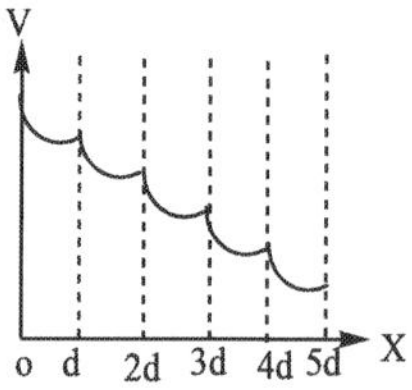

44. **A dielectric slab of thickness d is inserted in a parallel plate capacitor whose negative plate is at x=0 and positive plate is at x=3d. The slab is equidistant from the plates. The capacitor is given some charge. As one goes from 0 to 3d,**

1) the magnitude of the electric field remains the same

2) the direction of the electric field remains the same

3) the electric potential increases continuously'

4) the electric potential increases at first, then decreases and again increases.

45. **A solid sphere of radius R is charged uniformly. The electrostatic potential V is plotted as a function of distance r from the centre of the sphere. Which of the following best represents the resulting curve ?**

1) 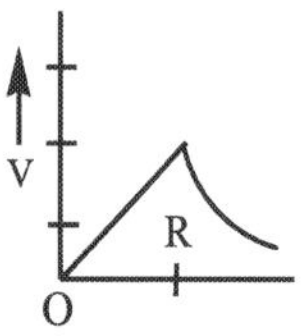

2) 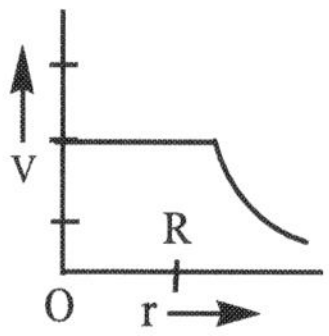

3) 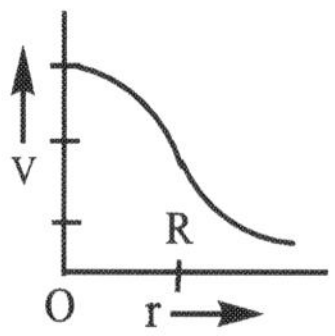

4) 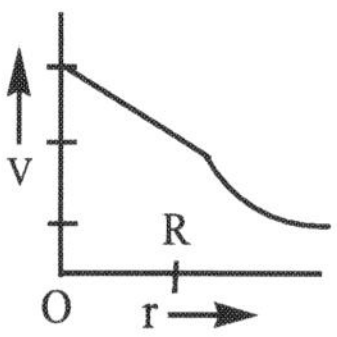

POTENTIAL ENERGY

46. **Along the X-axis, three charges $\frac{q}{2}$,-q and $\frac{q}{2}$ are placed at $x=0$, x=a and x=2a respectively.**

The resultant electric potential at x=a+r(if a ,<<r) is ($\in_0$ is the permittivity of free space

1) $\frac{qa}{4\pi\in_0 r^2}$ 2) $\frac{qa^2}{4\pi\in_0 r^3}$ 3) $\frac{q(a^2/4)}{4\pi\in_0 r^3}$ 4) $\frac{q}{4\pi\in_0 r}$

47. **An electron travelling from infinity with velocity 'v' into an electric field due to two stationary electrons separated by a distance of 2m. If it comes to rest when it reaches the mid point of the line joining the stationary electrons.The initial velocity ' v ' of the electron is**

1) 16m/s 2) 32m/s 3) $16\sqrt{2}m/s$ 4) $32\sqrt{2}m/s$

48. **Work performed when a point charge 2×10^{-8} C is transformed from infinity to a point at a distance of 1cm from the surface of the ball with a radius of 1cm and a surface charge density $\sigma=10^{-8}C/cm^2$**

1) $1.1\times10^{-4}J$ 2) $11\times10^{-4}J$ 3) $0.11\times10^{-4}J$ 4) $113\times10^{-4}J$

49. **A conducting sphere A of radius a, with a charge Q, is placed concentrically inside a conducting shell B of radius b. B is earthed. C is the common centre of A and B.**

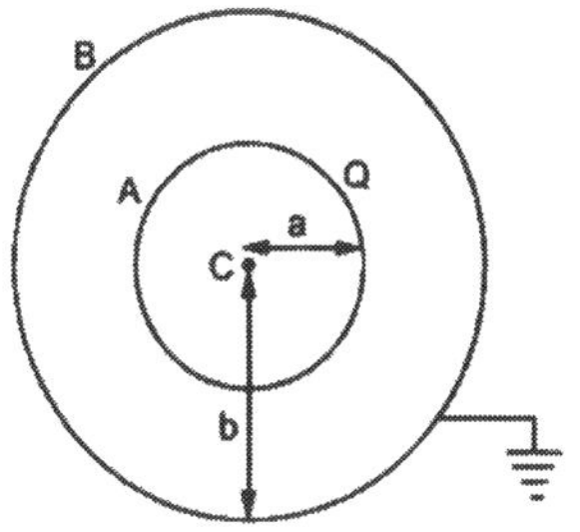

1) The field at a distance r from C, where $a \le r \le b$, is $k\frac{Q}{r^2}$

2) The potential at a distance r from C, where $a \le r \le b$, is $k\frac{Q}{r^2}$

3) The potential difference between A and B is

4) The potential at a distance r from C, where $a \le r \le b$, is

50. **Given figure shows an arrangement of six fixed charged particle. The net electrostatic force F acting on charge +q at the origin due to other charges is**

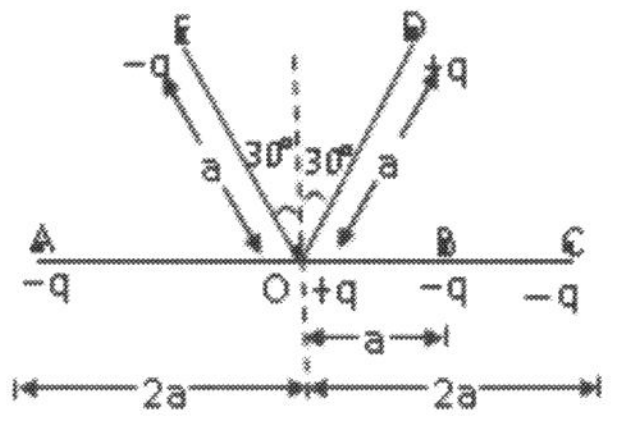

1) $\frac{6q^2}{4\pi\varepsilon_0 a^2}$ 2) zero 3) $\frac{7q^2}{2\pi\varepsilon_0 a^2}$ 4) $\frac{q^2}{4\pi\varepsilon_0 a^2}\left(\frac{3}{2}+\sqrt{3}\right)$

51. **2q and 3q are two charges separated by a distance 12 cm on x-axis. A third charge q is placed at 5 cm on y-axis as shown in figure. Find the change in potential energy of the system if 3q is moved from initial position to a point on X-axis in circular path:**

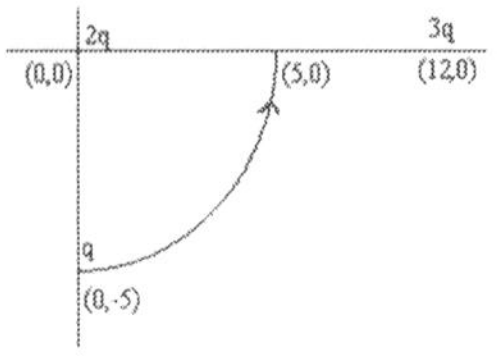

1) $\frac{q^2}{4\pi\varepsilon_o}$ 2) $\frac{6q^2}{4\pi\varepsilon_o(91)}$ 3) $\frac{18q^2}{4\pi\varepsilon_o(91)}$ 4) $\frac{3q^2}{4\pi\varepsilon_o}$

52. **An electron travelling from infinity with velocity 'V' into an electric field due to two stationary electrons seperated by a distance of 2m. If it comes to rest when it reaches the midpoint of the line joining the stationary electrons, the initial velocity 'V' of the electron is (in m/sec)**

1) 16 2) 32 3) $16\sqrt{2}$ 4) $32\sqrt{2}$

53. **A particle of mass m and charge q is projected vertically upwards. A uniform electric field $\vec{E}$ is acted vertically downwards. The most appropiate graph between potential energy U (gravitational plus electrostatic) and height h(<< radius of earth) is : (assume U to be zero on surface of earth)**

1)
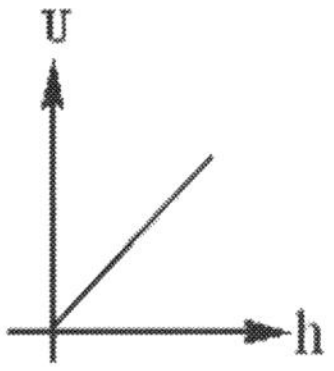

2)
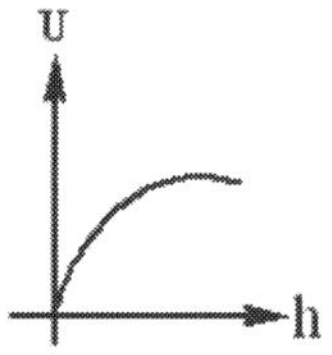

3)
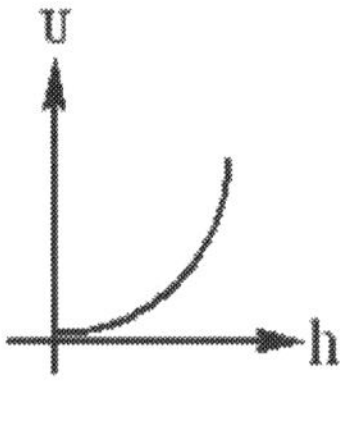

4)
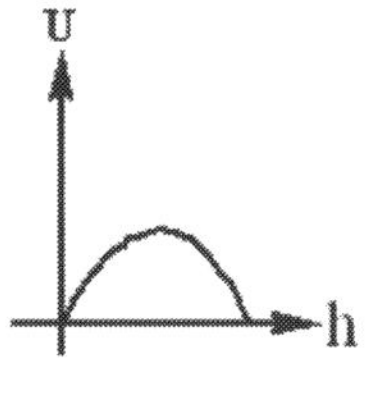

54. **Three charged particles are initially in position 1. They are free to move and they come in position 2 after some time. Let U_1 and U_2 be the electrostatic potential energies in position 1 and 2. Then :**

1) $U_1>U_2$ 2) $U_2>U_1$ 3) $U_1=U_2$ 4) $U_2 \geq U_1$

55. **Two identical thin rings, each of radius R, are coaxially placed a distance r apart. If Q_1 and Q_2 are respectively the charges uniformly spread on the two rings, the work done in moving a charge q from the centre of one ring to that of the orther is**

1) zero 2) $q(Q_1-Q_2)(\sqrt{2}-1)/(\sqrt{2}4\pi\varepsilon_0 R)$

3) $q\sqrt{2}(Q_1+Q_2)/(4\pi\varepsilon_0 R)$ 4) $q(Q_1/Q_2)(\sqrt{2}+1)(\sqrt{2}4\pi\varepsilon_0 R)$

56. **The electric potential at a point (x, 0, 0) is given by $V = \left[\frac{1000}{x} + \frac{1500}{x^2} + \frac{500}{x^3}\right]$ then the electric field at x = 1 m is (in volt/m)**

1) $-5500\hat{i}$ 2) $5500\hat{i}$ 3) $\sqrt{5500}\hat{i}$ 4) zero

57. **Six charges are placed at the vertices of a regular hexagon as shown in the figure. The electric field on the line passing through point O and perpendicular to the plane of the figure at a distance of x (>> a) from O is**

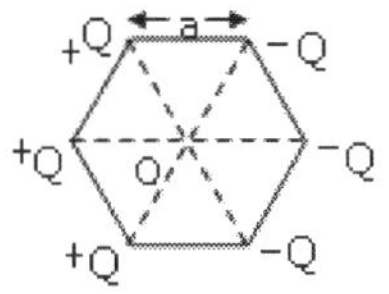

1) $\frac{Qa}{\pi\varepsilon_0 x^3}$ 2) $\frac{2Qa}{\pi\varepsilon_0 x^3}$ 3) $\frac{\sqrt{3}Qa}{\pi\varepsilon_0 x^3}$ 4) zero

DIPOLE

58. **A small electric dipole is placed at origin with its dipole moment directed along positive x - axis. The direction of electric field at point $(2, 2\sqrt{2}, 0)$ is**

1) along z - axis 2) along y - axis
3) along negative y -axis 4) along negative z-axis

59. **Two electric dipoles each of dipolemoment $p = 6.2\times10^{-30}C-m$ are placed with their axis along the same line and their centres at a distanced=$10^{-8}cm$. The force of attraction between dipoles is**

1) $2.1\times10^{-16}N$ 2) $2.1\times10^{-12}N$ 3) $2.1\times10^{-10}N$ 4) $2.1\times10^{-8}N$

60. **Two charges $+3.2\times10^{-19}C$ and $-3.2\times10^{-19}C$ placed $2.4\,A^0$ apart form an electric dipole. It is placed in a uniform electric field of intensity $4\times10^5\,V/m$ the work done to rotate the electric dipole from the equilibrium position by 180^0 is**

1) $3\times10^{-23}J$ 2) $6\times10^{-23}J$ 3) $12\times10^{-23}J$ 4) Zero

61. **Two opposite and equal charges 4 x10^{-8} coulomb when placed 2 x 10^{-2} cm away, from a dipole. If this dipole is placed in an external electric field 2 x 10^{-2} newton/coulomb, the value of maximum torque and the work done in rotating it through 180^0 will be**

1) $32x10^{-4}$Nm and $32x10^{-4}$J 2) $64x10^{-4}$Nm and $64x10^{-4}$J
3) $64x10^{-4}$Nm and $32x10^{-4}$J 4) $32x10^{-4}$J and $64x10^{-4}Nm$

62. **An electric dipole is made up of two particles having charges $+1\mu c$, mass 1 kg and other with charge $-1\mu c$ and mass 1 kg separated by distance 1m. It is in equilibrium in a uniform electric field of 20 x 10^3 V/m. If the dipole is deflected through angle 2^0, time taken by it to come again in equilibrium is**

1) 2.5 π s 2) 2.5 s 3) 5 π s 4) 4 π

63. **A point particle of mass M is attached to one end of a massless rigid non-conducting rod of length L. Another point particle of the same mass is attached to the other end of the rod. The two particle carry charges + q and – q respectively. This arrangement is held in a region of a** uniform electric field E such that the rod makes a small angle θ **(say of about 5^0) with the field direction (see figure). The expression for the minimum time needed for the rod to become parallel to the field after it is set free.**

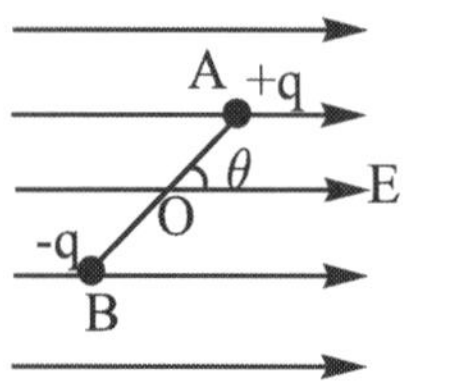

1) $t=\frac{\pi}{2}\sqrt{\frac{mL}{2qE}}$ 2) $t=\frac{\pi}{2}\sqrt{\frac{mL}{qE}}$ 3) $t=\frac{\pi}{2}\sqrt{\frac{2mL}{qE}}$ 4) $t=\frac{\pi}{2}\sqrt{\frac{3mL}{2qE}}$

KEY LEVEL - III

1) 4	2) 3	3) 2	4) 2	5) 4	6) 4	7) 4	8) 2	9) 4	10) 4
11) 3	12) 3	13) 3	14) 4	15) 2	16) 1	17) 3	18) 1	19) 4	20) 4
21) 3	22) 4	23) 2	24) 4	25) 2	26) 1	27) 3	28) 2	29) 3	30) 4
31) 1	32) 3	33) 1	34) 1	35) 1	36) 3	37) 2	38) 2	39) 3	40) 1
41) 3	42) 1	43) 2	44) 2	45) 3	46) 2	47) 2	48) 2	49) 1	50) 2
51) 3	52) 2	53) 1	54) 1	55) 2	56) 2	57) 1	58) 2	59) 4	60) 2
61) 4	62) 1	63) 1							

LEVEL - III HINTS

1. $T+\frac{1}{4\pi\varepsilon_o}\frac{q_1q_2}{r^2}=mg$

2. $F=\frac{1}{4\pi\in_0\in_r}\frac{q_1q_2}{d^2}$

3. $F=\frac{1}{4\pi\in_0\in_r}\frac{q_1q_2}{d^2}$

4. conceptual

5. $F=\frac{1}{4\pi\in_0\in_r}\frac{q_1q_2}{d^2}$

6. $F=mgTan\theta$

7. $F=\frac{1}{4\pi\in_0\in_r}\frac{q_1q_2}{d^2}$

8. There are three forces acting on each sphere are
(i) tension (ii) weight(w) (iii) electrostatic force of repulsion for sphere

In equailibirum, from figure

$\tan\theta_1 = F_1 / M_1 g$

From sphere 2, in equilibirum from figure

$\tan\theta_2 = F_2 / M_2 g$

for $F_1 = F_2$

or $\theta_1 = \theta_2$ **only for** $\frac{F_1}{M_1 g} = \frac{F_2}{M_2 g}$

But, $F_1 = F_2$ and then $M_1 = M_2$

9. Deflection $y = \frac{eEx^2}{4(K)}$ where K is kinetic energy.

10. $y = \frac{eE\alpha^2}{4k}(K = K.E)$

11. $T = 2\pi\sqrt{\frac{l}{g_{eff}}}$

12. K.E =FS

K.E – qEy

13. $E = -\frac{dV}{dx}$

14. N =mg $\sin\theta$ + qE $\sin\theta$

mg $\sin\theta = \mu N + qE\cos\theta$

15. $E_{axis} = \frac{2kp}{r^3}$ and $E_{bicector} = \frac{kp}{2r^3}$

16. Maximum height $= \frac{u^2 \sin^2\theta}{2\left(g \mp \frac{EQ}{m}\right)}$

17. Range $= \frac{u^2 \sin 2\theta}{g \mp \frac{EQ}{m}}$

18. $E = \frac{1}{4\pi \in_0} \frac{qx}{\left(a^2 + x^2\right)^{3/2}} = \frac{qx}{4\pi \in_0 a^3}$

$\therefore m\frac{d^2x}{dt^2} = -\frac{1}{4\pi \in_0} \frac{qex}{a^3}$

So motion is S.H.M.

$\omega^2 = \frac{1}{4\pi \in_0} \frac{qe}{ma^3}$

19. $F=\frac{Qq}{4\pi\epsilon_0}\frac{x}{R^3}=-kx$ and $T=2\pi\sqrt{\frac{m}{k}}$

20. $mg=\mu qE$

21. $mg(\sin\alpha-\mu\cos\alpha)-\mu qE\sin\alpha=qE\cos\alpha$

22. conceptual

23. $F=\frac{1}{4\pi\varepsilon_0}\frac{Qq}{r^2}=mr\omega^2$; $\omega=\frac{2\pi}{T}$

24. $E=\frac{1}{4\pi\epsilon_0}\frac{q\sin\pi/2}{r^2\,\pi/2}$ $\quad E=\frac{q\sin\pi/2}{2\pi^2\epsilon_0 r^2}(-\bar{j})$

25. $V_1=\frac{Q}{4\pi\varepsilon_0}\left(\frac{1}{R}-\frac{1}{\sqrt{R^2+d^2}}\right)$

$V_2=\frac{-Q}{4\pi\varepsilon_0}\left(\frac{1}{R}-\frac{1}{\sqrt{R^2+d^2}}\right)$

$\Delta V=V_1-V_2$

26. $q_1=\left(\frac{r_1}{r_1+r_2}\right)q$; $V_1=\frac{1}{4\pi\varepsilon_0}\frac{q_1}{r}$

27. Pd between the two spheres is independent of charge on outer shell.

28. $E=\frac{\sigma}{\varepsilon_0}$ and $V=\frac{\sigma R}{\varepsilon_0}$

29. $V=K.\frac{Q}{R}$ $\quad \frac{V}{2}=K.\frac{Q}{2R}$ $\quad =K.\frac{Q}{R}$

$\frac{1}{2}\left(K.\frac{Q}{R}\right)=K.\frac{Q}{d}$ $\quad d=2R$

When the two conducting spheres are connected by a conducting wire, charge will flow from one sphere (having higher potential) to other (having lower potential) till both acquire the same potential.

There fore, $E=\frac{V}{r}$

$\Rightarrow \frac{E_1}{E_2}=\frac{r_2}{r_1}=\frac{2}{1}=2:1$

30. $V=\frac{q}{4\pi\epsilon_0}\left[\frac{1}{x_0}-\frac{1}{2x_0}+\frac{1}{3x_0}-\frac{1}{4x_0}+---\right]$

$=\frac{q}{4\pi\epsilon_0 x_0}\left[1-\frac{1}{2}+\frac{1}{3}-\frac{1}{4}+---\right]$

$=\frac{q}{4\pi\epsilon_0 x_0}\log(2)$

31. $\Delta V = -\int \vec{E}.\vec{dr}$

32. $E = -\frac{dV}{dx}$

33. $V = -\vec{E}.\vec{r}$

34. long and short diagonal lengths are $\sqrt{p^2 + q^2 \pm 2pq\cos\theta}$

35. $V = \frac{1}{4\pi\varepsilon_0}\frac{3Q}{2R}$

36. If the charges is kept at 'A' then maximum and minimum potentials at D and C respectively
37. commen potential
38. w = Fs

W $= q\vec{E}.\vec{S}$

39. $E = -\frac{dv}{dx}$

40. concept of field

41. $V_1 - V_2 = kq\left(\frac{1}{r_1} - \frac{1}{r_2}\right)$

$r_2 - r_1 = \frac{(V_1 - V_2)r_1 r_2}{kq}$

but $(r_2 - r_1) = t$

$\therefore t \alpha r_1 r_2$

if P.D is constant then $(r_2 - r_1) = t$

42. potential due to small element 'p' at the centre

$dv = K.\lambda\frac{dl}{r}$

$v = \int dv = \int k\frac{\lambda}{r}dl = \frac{1}{4\pi\varepsilon_0}\frac{\lambda}{r}\int dl$

$= \frac{1}{4\pi\varepsilon_0}\frac{\lambda}{r}\pi r \Rightarrow \frac{\lambda}{4\varepsilon_0}$

43. $E = -\frac{dv}{dx}$

E inside the conductor is zero.

44. The direction of E is constant.

45. $V = \frac{1}{4\pi\varepsilon_0}\frac{Q}{r}$

46. Force of interaction $= \frac{2pq}{4\pi\epsilon_0}\left[\frac{1}{(d-l)^3} - \frac{1}{(d+l)^3}\right]$

47. $\frac{1}{2}mv^2 = \frac{1}{4\pi\varepsilon_0}\frac{q_1q_2}{r}$

48 . Potential at a distance 2cm from its centre $= \frac{Q}{4\pi \in_0 r} = \frac{4\pi r^2\sigma}{4\pi \in_0 r'} \Rightarrow \frac{r^2\sigma}{\in_0 r'} = \frac{\sigma}{2\in_0} \times \frac{1}{100}$

since r=1 cm and r' =2 cm

PD b/w the two points is equal to $\frac{\sigma}{200 \in_0}$

work done =VQ= $\frac{\sigma}{200\in_0}$ X $2\times10^{-8} = 11\times10^{-4} J$

49. field concept

50. concept of force

51. $\Delta U = U_f - U_i$ and $U = \frac{1}{4\pi\varepsilon_0}\frac{q_1q_2}{r}$

52. $u_i = 0$, $u_f = 2\times\frac{1}{4\pi\varepsilon_0}\cdot\frac{e^2}{(d/2)}$

$\Delta PE = KE = \frac{1}{2}mv^2$ calculate ' v '

53. conceptual.

54. Particles moves in a direction where potential energy of the system decreased.

55. $V_1 = \frac{Q_1}{4\pi\varepsilon_0 R} + \frac{Q_2}{4\pi\varepsilon_0\sqrt{2}R}$ and $V_2 = \frac{Q_1}{4\pi\varepsilon_0\sqrt{2}R} + \frac{Q_2}{4\pi\varepsilon_0 R}$

$W_{1\to2} = q(V_2 - V_1)$

56. $E = -\frac{dV}{dx}$

57. concept of field

58. Use vector representation

59. Force of interaction

$$= \frac{2pq}{4\pi \in_0}\left[\frac{1}{(d-l)^3} - \frac{1}{(d+l)^3}\right]$$

60. $W = PE_2 - PE_1 = 2PE = -2\times 2qdE$

61. $\tau = pESin\theta$

62. $T = 2\pi\sqrt{\frac{I}{pE}}$

63. $T = 2\pi\sqrt{\frac{I}{pE}}$

Chapter - 2

GAUSS LAW

GAUSS LAW

SYNOPSIS

Electric flux:

It is the measure of total number of electric lines of force crossing normally the given area.

➢ The total flux passes through the given surface is given by $\phi = \vec{E}.\vec{A}$

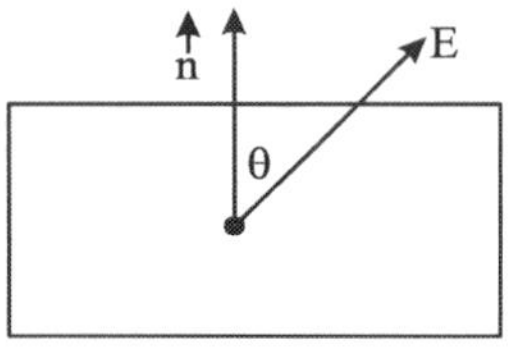

$\therefore \phi = EA\cos\theta$

where θ is the angle made by the normal with the electric field.

For a closed body outward flux is taken to be positive while inward flux is taken to be negative.

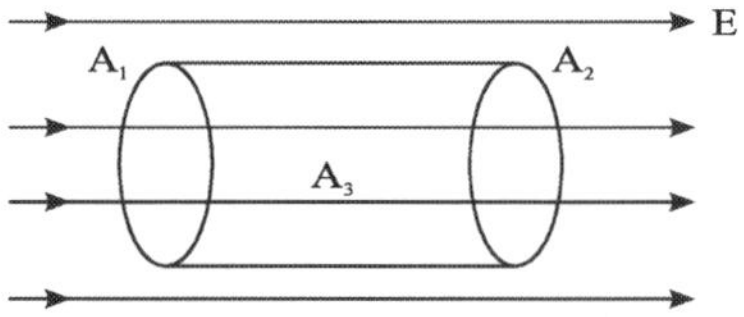

a) Flux through A_1 : negative b) Flux through A_2 : positive c) Flux through A_3 : 0

GAUSS LAW

i) According to this law, the total flux linked with a closed surface called Gaussian surface is $(1/\varepsilon_0)$ times the net charge enclosed by the closed surface.

ii) Alternatively, Gauss law can be stated as the surface integral of electric field $\overline{E}$ over a closed surface is equal to $1/\epsilon_0$ times the charge (q) enclosed by that closed surface.

i.e., $\phi = \oint \vec{E}.\overline{dA} = \dfrac{q}{\epsilon_0}$

q is the total charge enclosed by the Gaussian surface.

iii) Coulomb's law can be derived from Gauss law.

iv) The electric field $\vec{E}$ is the resultant field due to all charges, both those inside and those outside the Gaussian surface.

v) The electric field due to a charge outside the Gaussian surface contributes zero net flux through the surface, Because as many lines due to that charge enter the surface as leave it.

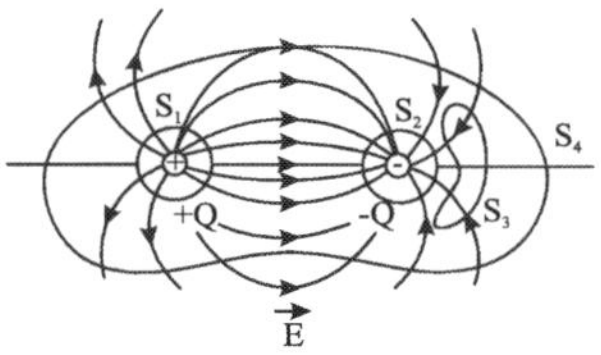

a) Flux from surface $S_1 = +\dfrac{Q}{\varepsilon_0}$ b) Flux from surface $S_2 = \dfrac{-Q}{\varepsilon_0}$

c) Flux from S_3 = flux from surface $S_4 = 0$

Applications of Gauss Law :

i) If a dipole is enclosed by a surface

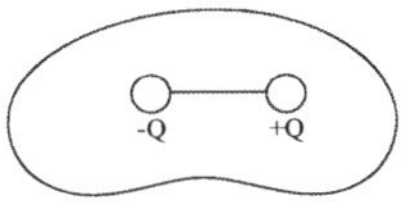

$\because Q_{enc} = 0 \quad \Rightarrow \phi = 0$

ii) The net charge Q_{enc} is the algebraic sum of the enclosed positive and negative charges. If Q_{enc} is positive then the net flux is outwards. If Q_{enc} is negative then the net flux is inwards.

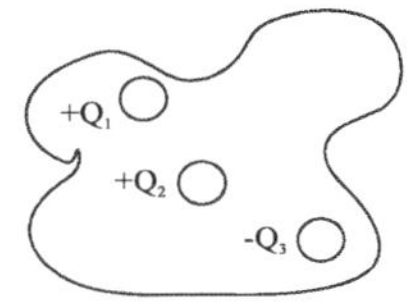

$$\phi = \frac{1}{\varepsilon_0}(Q_1 + Q_2 - Q_3)$$

iii) If a closed body (not enclosing any charge) is placed in an electric field (either uniform or non - uniform) total flux linked with it will be zero

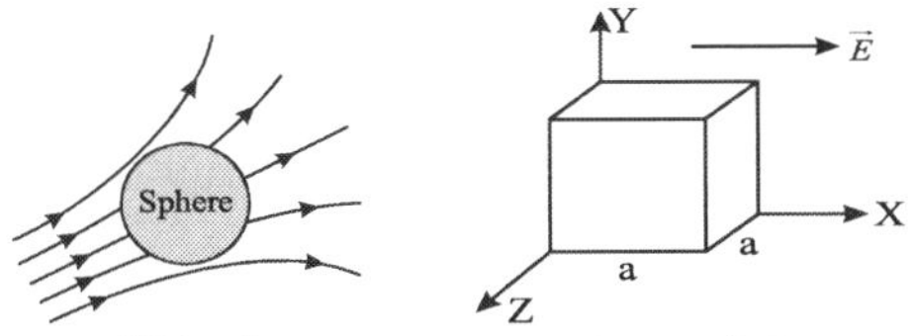

(A) $\phi_T = 0$ (B) $\phi_{in} = \phi_{out} = Ea^2 \Rightarrow \phi_T = 0$

iv) If charge is kept at the centre of cube

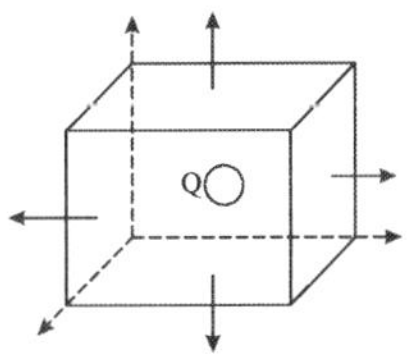

$$\phi_{total} = \frac{1}{\varepsilon_0}(Q) \; ; \; \phi_{face} = \frac{1}{6\varepsilon_0}(Q)$$

v) If charge is kept at the centre of a face, first we should enclose the charge by assuming a Gaussian surface (an identical imaginary cube)

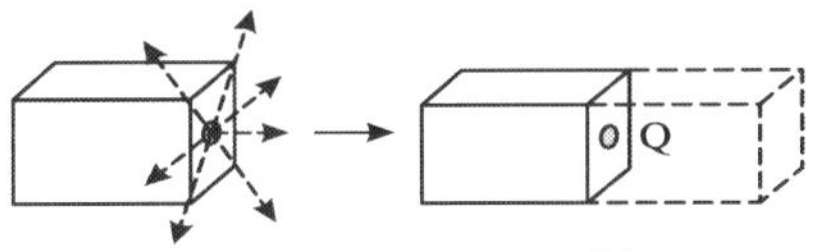

(a) (b)

Total flux emerges from the system (Two cubes) $\phi_{total} = \dfrac{Q}{\varepsilon_0}$

Flux from given cube (i.e., from left side 5 faces only) $\phi_{cube} = \dfrac{Q}{2\varepsilon_0}$

vi) If a charge is kept at the corner of a cube

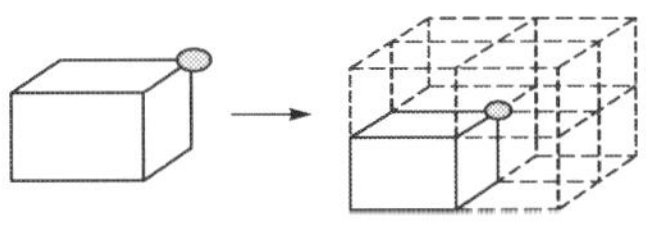

(a) (b)

For enclosing the charge seven more cubes are required so total flux from the 8 cube system is $\phi_T = \dfrac{Q}{\varepsilon_0}$.

Flux from given cube $\phi_{cube} = \dfrac{Q}{8\varepsilon_0}$.

Flux from one face opposite to the charge, of the given cube

$\phi_{face} = \dfrac{Q/8\varepsilon_0}{3} = \dfrac{Q}{24\varepsilon_0}$ (Because only three faces are seen).

Electric field at a point due to a line of charge: A thin straight wire over which 'q' amount of charge be uniformly distributed. λ be the linear charge density i.e, charge present per unit length of the wire.

$$E = \frac{q}{2\pi \in_0 rl} \quad E = \frac{\lambda}{2\pi \in_0 r}$$

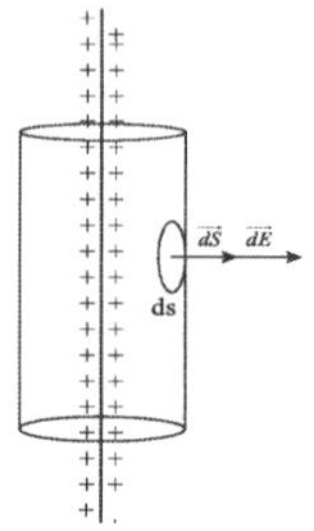

- This implies electric field at a point due to a line charge is inversely proportional to the distance of the point from the line charge.

Electric field intensity at a point due to a thin infinite charged sheet [Non conducting plate]

'q' amount of charge be uniformly distributed over the sheet. Charge present per unit surface area of the sheet be σ. i.e surface charge density σ

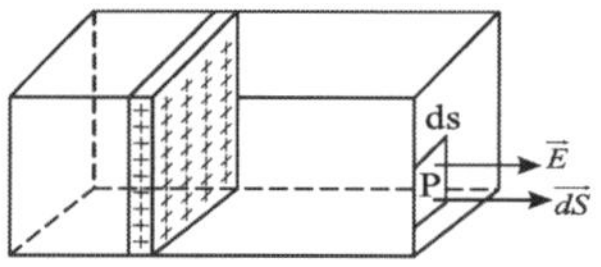

$$E = \frac{q}{2A \in_0}; \; E = \frac{\sigma}{2 \in_0} \; where \; \sigma = \frac{q}{A}$$

- E is independent of the distance of the point from the charged sheet.

Electric field intensity at a point due to a thick infinite charged sheet [Conducting plate] :

'q' amount of charge be uniformly distributed over the sheet. Charge present per unit surface area of the sheet be σ.

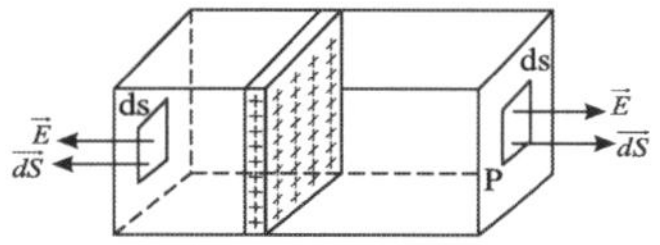

$$E = \frac{q}{A \in_0} = \frac{\sigma}{\in_0}$$

Electric field at a point due to a thick charged sheet is twice that produced by the thin charged sheet of same charge density.

APPLICATIONS OF GAUSS LAW

- **Electric field due to long uniformly charged cylinder:**

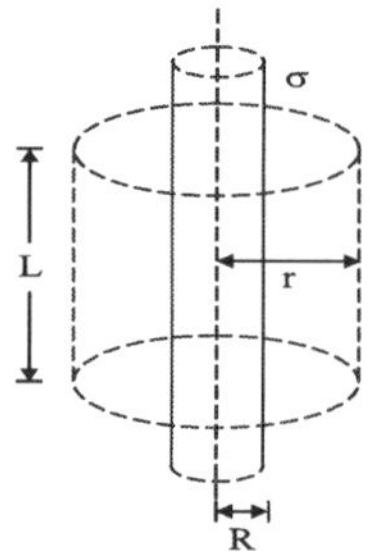

Consider a long cylinder of radius R which is uniformly charged on its surface with charge density σ. We know that at the interior points of a metal body electric field strength is zero. Let us find the electric field at a point and at a distance r from the axis of the cylinder. Consider a cylindrical Gaussian surface of radius r and length L as shown in the figure.

From Gauss's law, we can write

$$\oint \bar{E}.d\bar{s} = \frac{1}{\epsilon_0}(q_{en})$$

Here $q_{enclosed} = \sigma 2\pi RL$

Here electric flux through the circular faces is zero. So, from Gauss law

$$\oint \bar{E}.d\bar{s} = \frac{\sigma 2\pi RL}{\epsilon_0} \text{ or } E2\pi rL = \frac{\sigma 2\pi RL}{\epsilon_0}$$

$$\Rightarrow E = \frac{\sigma R}{\epsilon_0 r}$$

The variation of E with distance r from the axis is as shown in the graph.

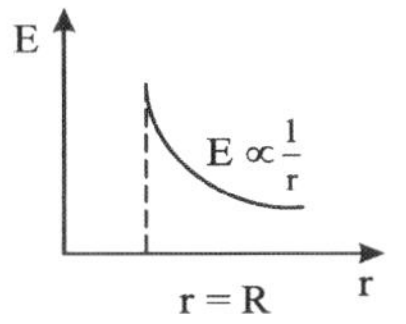

➢ **Electric field due to uniformly charged non- conducting cylinder:** Consider a long cylinder of radius R charged with volume charge density ρ uniformly. Let us find electric field at a distance r from the axis of the cylinder. Consider a cylindrical Gaussian surface of length L and radius r as shown,

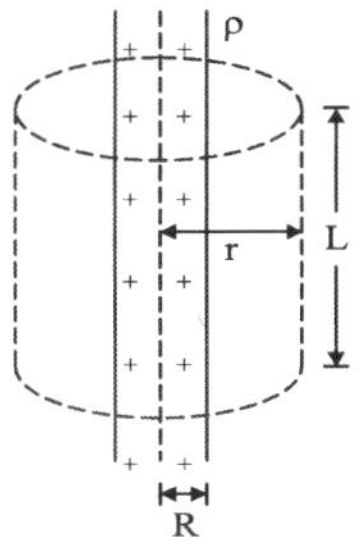

$$\oint \bar{E}.d\bar{s} = \frac{q_{encl}}{\epsilon_0}; \text{ where } q_{encl} = \rho\pi R^2 L$$

Here electric flux through the circular faces is zero.

Case (i): If r > R, then from Gauss's law

$$\oint \bar{E}.d\bar{s} = \frac{\rho\pi R^2 L}{\epsilon_0} \Rightarrow E2\pi rL = \frac{\rho\pi R^2 L}{\epsilon_0}$$

$$\text{or } E = \frac{\rho R^2}{2\epsilon_0 r} \Rightarrow E_{out} \propto \frac{1}{r}$$

Case (ii): If r = R, then $E = \frac{\rho R}{2\epsilon_0}$

Case (iii): If r < R, $q_{encl} = \rho\pi r^2 L$

from Gauss law $\oint \bar{E}.d\bar{s} = \frac{q_{encl}}{\epsilon_0}$

$$E2\pi rL = \frac{\rho\pi r^2 L}{\epsilon_0} \text{ (or) } E = \frac{\rho r}{2\epsilon_0} \Rightarrow E_{in} \propto r$$

In vector form $\vec{E} = \frac{\rho\vec{r}}{2\epsilon_0}$

The variation of E with distance r from the axis is as shown in the graph.

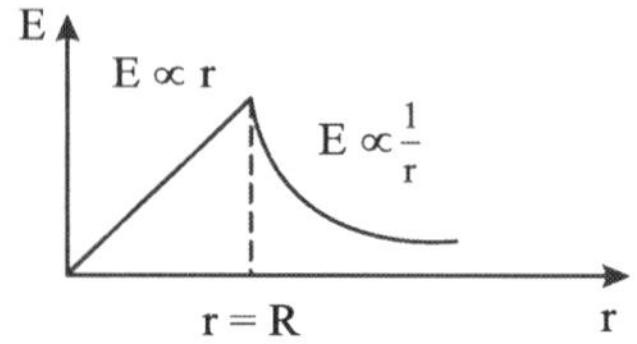

ELECTRIC INTENSITY AND ELECTRIC POTENTIAL DUE TO INFINITE PLANE SHEET OF CHARGE (NONCONDUCTING):

If E is the magnitude of electric field at point P, then electric flux crossing through the gaussian surface is given by

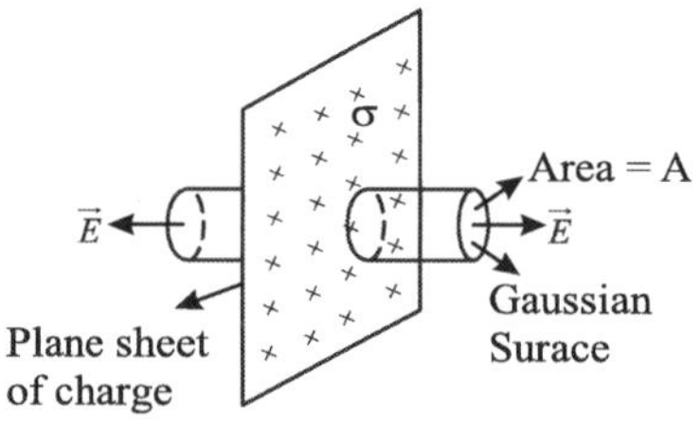

ϕ = E x area of the end face (circular caps) of the cylinder

or ϕ = E x 2A ....(i)

According to Gauss's theorem, we have

$$\phi = \frac{q}{\varepsilon_0}$$

Here, the charge enclosed by the gaussian surface, $q = \sigma A$

$$\therefore \phi = \frac{\sigma A}{\epsilon_0} \qquad ....(ii)$$

From equations (i) and (ii), we have

$$E \times 2A = \frac{\sigma A}{\varepsilon_0} \text{ or } E = \frac{\sigma}{2\varepsilon_0}\hat{n}$$

Where $\hat{n}$ is unit vector normal to the plane and away from it.

Thus, we find that the magnitude of the electric field at a point due to an infinite plane sheet of charge is independent of its distance from the sheet of charge.

Electric intensity due to two thin parallel charged sheets:

Two charged sheets A and B having uniform charge densities σ_A and σ_B respectively.

- **In region I** :

$$E = \frac{1}{2\epsilon_0}(\sigma_A + \sigma_B)$$

- **In region II:**

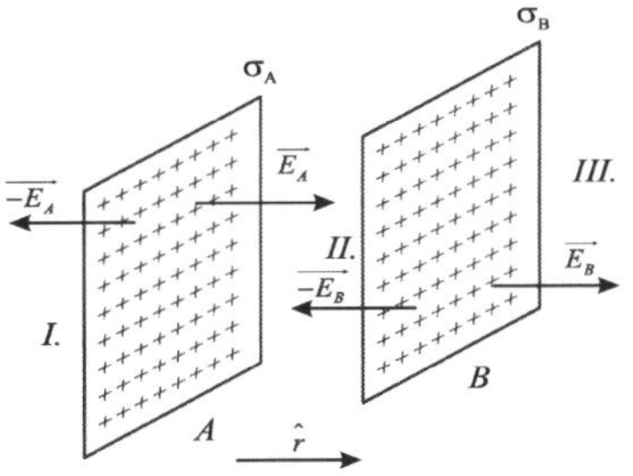

$$E_{II} = \frac{1}{2\epsilon_0}(\sigma_A - \sigma_B)$$

- **In region III** :

$$E_{III} = \frac{1}{2\epsilon_0}(\sigma_A + \sigma_B)$$

Electric field due to two oppositely charged parallel thin sheets :

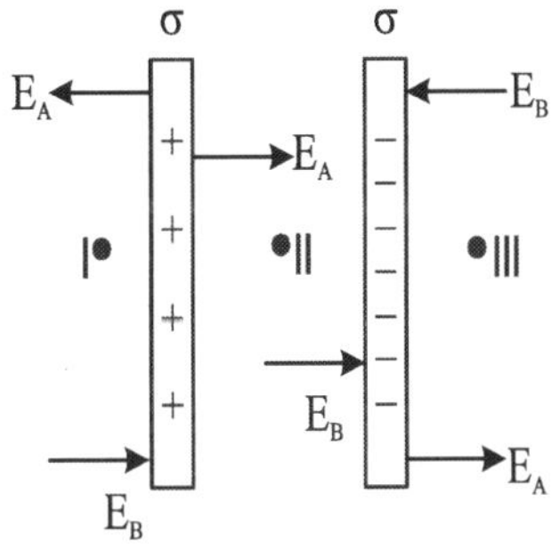

$$E_I = -\frac{1}{2\in_0}\left[\sigma + (-\sigma)\right] = 0$$

$$E_{II} = \frac{1}{2\in_0}[\sigma - (-\sigma)] = \frac{\sigma}{\in_0}$$

$$E_{III} = \frac{1}{2\in_0}(\sigma - \sigma) = 0$$

Electric field due to a charged Spherical shell

'q' amount of charge be uniformly distributed over a spherical shell of radius 'R'

$\sigma =$ Surface charge density, $\sigma = \dfrac{q}{4\pi R^2}$

When point 'P' lies outside the shell (r>R):

$$E = \frac{1}{4\pi\in_0} \times \frac{q}{r^2}$$

- This is the same expression as obtained for electric field at a point due to a point charge. Hence a charged spherical shell behaves as a point charge concentrated at the centre of it.

$$E = \frac{1}{4\pi\in_0}\frac{\sigma.4\pi R^2}{r^2}\left(\because \sigma = \frac{q}{4\pi R^2}\right)$$

$$E = \frac{\sigma . R^2}{\epsilon_0 r^2}$$

➢ **When point 'P' lies on the shell (r =R) :**

$$E = \frac{\sigma}{\epsilon_0}$$

➢ **When Point 'P' lies inside the shell (r <R):**

E = 0

➢ The electric intensity at any point due to a charged conducting solid sphere is same as that of a charged conducting sperical shell.

Electric Potential (V) due to a Uniformly Charged sphe-rical c onducting shell (Hollow sphere)

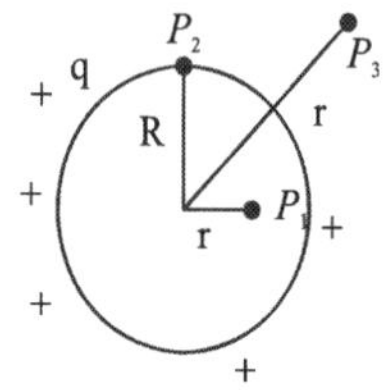

➢ When point (P_3) lies outside the sphere $(r > R)$, the electric potential, $V = \frac{1}{4\pi\varepsilon_0}\frac{q}{r}$

➢ When point (P_2) lies on the surface $(r = R)$, $V = \frac{1}{4\pi\varepsilon_0}\frac{q}{R}$

➢ When point (P_1) lies inside the surface $(r < R)$, $V = \frac{1}{4\pi\varepsilon_0}\frac{q}{R}$

➢ Note: The electric potential at any point inside the sphere is same and is equal to that on the surface.

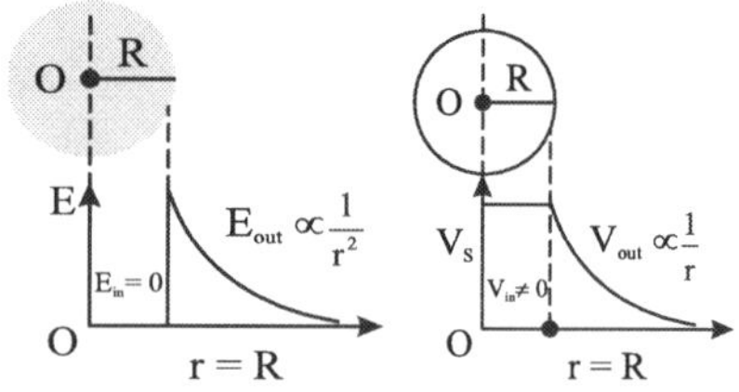

(A) (B)

Note: The electric potential at any point due to a charged conducting sphere is same as that of a charged conducting spherical shell

W.E-1: ***A particle that carries a charge '–q' is placed at rest in uniform electric field 10 N/C. It experiences a force and moves. In a certain time 't', it is observed to acquire a velocity $10\bar{i} - 10\bar{j}$ m/s. The given electric field intersects a surface of area 1m² in the x–z plane. Find the Electric flux through the surface.***

Sol: Force on charge $\bar{F} = q\bar{E}$

$\therefore$ particle moves opposite to $\bar{E}$ with $\bar{V}$

unit vector in the direction of $\bar{V}$ is $\frac{\bar{i}}{\sqrt{2}} - \frac{\bar{j}}{\sqrt{2}}$

unit vector in the direction of $\bar{E}$ is $\frac{\bar{i}}{\sqrt{2}} - \frac{\bar{j}}{\sqrt{2}}$

$\bar{E} = 10\left[\frac{-i}{\sqrt{2}} + \frac{j}{\sqrt{2}}\right]$ ie. $\bar{A} = 1 \times \bar{j}$

Electric flux $\phi = \bar{E} \cdot \bar{A} = 5\sqrt{2}\ \text{Nm}^2/\text{C}$

W.E-2: ***The electric field in a region is given by $\vec{E} = E_0 \frac{x}{L}\hat{i}$. Find the charge contained inside a cubical volume bounded by the surface x = 0, x = L, y =0 , y = L, z = 0 and z = L.***

Sol: At x = 0, E = 0 and at x = l, $\vec{E} = E_0\hat{i}$

The direction of the field is along the x-axis, so it will cross the yz-face of the cube. The flux of this field

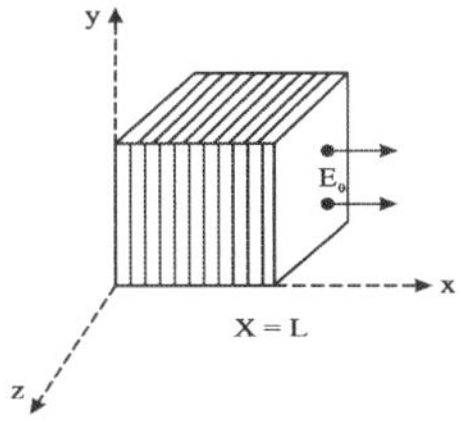

$\phi = \phi_{\text{left face}} + \phi_{\text{right face}}$; $= 0 + E_0L^2 = E_0L^2$

By Gauss's law, $\phi = \frac{q}{\epsilon_0}$ $\therefore q = \epsilon_0\ \phi = \epsilon_0\ E_0L^2$

W.E-3: ***A square surface of side lm in the plane of the paper. A uniform electric field E (V/m) also in the plane of the paper, is limited only to the lower half of the square surface, the electric flus (in SI units) associated with the surface is.***

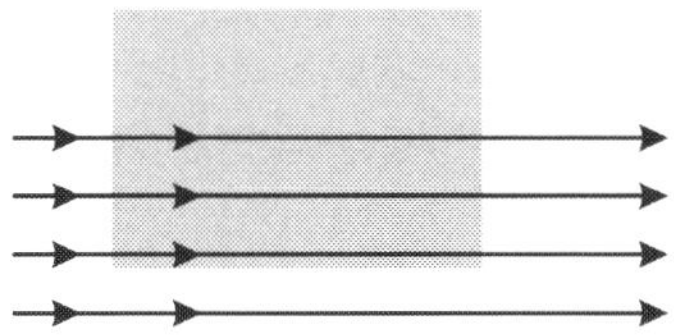

Sol: A, electric flux, $\phi_E = \int E.ds$

$= \int E\, ds \cos\theta = \int E\, ds \cos 90^0 = 0$

Thus, the lines are parallel to the surface.

W.E- 4 : ***A hollow cylinder has a chartge 'q' coulomb within it. If ϕ is the electric flux in unit of V-m, associated with the curved surface B, the electirc flux linked with the plane surface A in unit of V-m, will be***

Sol: We have, $\phi_{total} = \phi_A + \phi_B + \phi_c = \frac{q}{\varepsilon_0}$

$2\phi' + \phi = \frac{q}{\varepsilon_0}$ $\Rightarrow \phi' = \frac{1}{2}\left(\frac{q}{\varepsilon_0 - \phi}\right)$

W.E-5: ***The adjacent diagram shows a charge +Q held on an insulating suppot S and enclosed by a hollow spherical conductor, O represents the centre of the spherical conductor and P is a point such that OP=x and SP=r. The electric field at point, P will be***

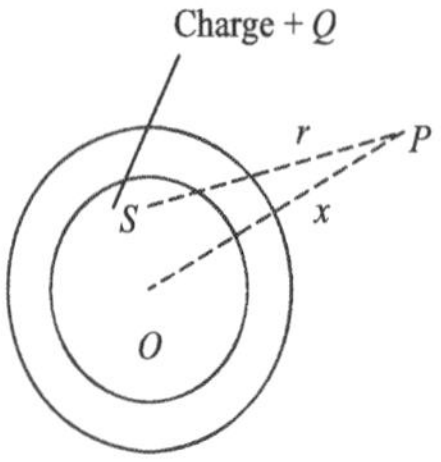

Sol: According to Gauss's' theorem,

$$\oint E.ds = \frac{Q_{in}}{\varepsilon_0}; \Rightarrow E.4\pi x^2 = \frac{Q}{\varepsilon_0} \text{ or } E = \frac{Q}{4\pi\varepsilon_0 x^2}$$

W.E-6: ***The electrostatic potential inside a charged spherical ball is given by $\phi = ar^2 + b$, where, 'r' is the distance from the centre, a and b are constants. Then the charge density the ball is***

Sol: Here, $\phi = ar^2 + b$; As $\phi = ar^2 + b \therefore \oint E.ds = \frac{q}{\varepsilon_0}$; $-2ar.4\pi r^2 = \frac{q}{\varepsilon_0} \Rightarrow q = -8\varepsilon_0 a\pi r^3$

$$\rho = \frac{q}{\frac{4}{3}\pi r^3}; \Rightarrow \rho = -6a\varepsilon_0$$

SOLID ANGLE

Solid angle is the three dimensional angle subtended by the lateral surface of a cone at its vertex

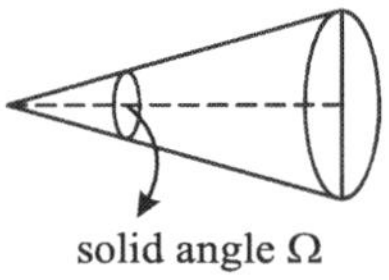

Let us calculate the solid angle subtended by a surface X at a point O. Join all the points of the periphery of the surface X to the point O by straight lines as shown. It gives a cone with vertex at O.

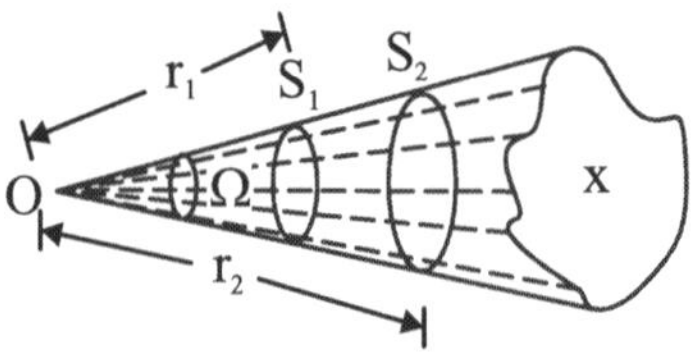

Fig. (b)

By taking centre at O, we draw several spherical sections on this cone of different radii as shown. Let the area of spherical section which is of radius r_1 be s_1 and the area of section of radius r_2 be s_2. **The ratios of area of any surface intersected by cone to the square of radius of that sphere is a constant** and it gives actually the solid angle Ω From the figure, solid angle subtended by surface X at the point O is given by $\Omega = \frac{s_1}{r_1^2} = \frac{s_2}{r_2^2}$.

Note: SI unit of solid angle is steradian and it is a dimensionless quantity.

one steradian is the solid angle subtended at the centre of the sphere by the surface of the sphere having area equal to square of the radius of the sphere.

The surface subtending solid angle need not be normal to the axis of the cone. For example consider a surface X of area $d\bar{s}$ as shown. The axis of cone formed by the surface at O is not normal to the surface. In this cone solid angle Ω subtended at point O can be given as

$$\Omega = \frac{ds\cos\theta}{r^2}$$

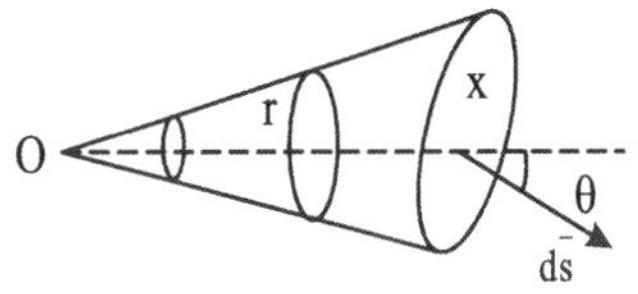

Here θ is the angle between $d\bar{s}$ and axis of the cone.

RELATION BETWEEN SEMI- VERTEX ANGLE OF A CONE AND SOLID ANGLE SUBSTENDED

Consider a spherical surface of radius R. Let X be a surface on that sphere which substends a semi vertex angle θ (in radian) at the centre of the sphere. Now consider an elemental strip of this section of radius $r = R\sin\alpha$ and angular width $d\alpha$ as shown. Then surface area of this strip is given by $ds = (2\pi R\sin\alpha)Rd\alpha$.

The total area of spherical section can be obtained by integrating this elemental area from 0 to θ.

Total area of spherical section is

$$S = \int ds = \int_0^\theta 2\pi R^2 \sin\alpha \, d\alpha$$

$$= 2\pi R^2(-\cos\alpha)_0^\theta \quad = 2\pi R^2(1-\cos\theta)$$

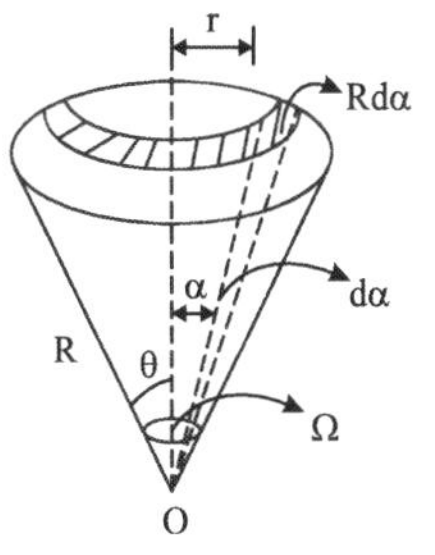

If Ω is solid angle subtended by this section at the centre O, then its area is given by $S = \Omega R^2$ (as discussed earlier) So, we can write $\Omega R^2 = 2\pi R^2(1-\cos\theta)$ and $\Omega = 2\pi(1-\cos\theta)$

Note : The solid angle substended by a hemispherical surface at its centre is given by

$$\Omega = 2\pi\left(1-\cos 90^0\right) = 2\pi \text{ steradians}$$

If $\theta = 180^0$ in the previous case, we get the solid angle substended by a closed surface

$$\Omega = 2\pi\left(1-\cos 180^0\right) = 4\pi \text{ steradians}$$

The total solid angle substanded by a closed surface is always 4π steradians, irrespective of the size and shape of the closed surface.

W.E-7: ***A point charge q is placed at a distance d from the centre of a circular disc of radius R. Find electric flux through the disc due to that charge***

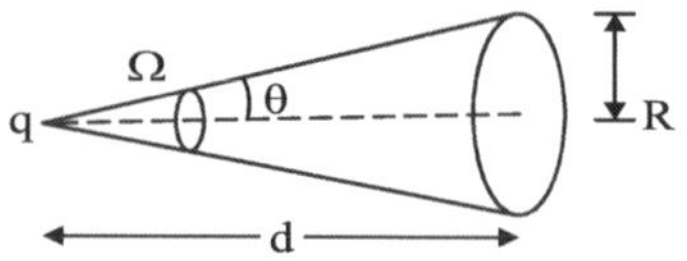

Sol : We know that total flux originated from a point charge q in all directions is $\frac{q}{\epsilon_0}$. This flux is originated in a solid angle 4π. In the given case solid angle subtended by the cone subtended by the disc at the point charge is $\Omega = 2\pi(1-\cos\theta)$

So, the flux of q which is passing through the surface of the disc is

$$\phi = \frac{q}{\epsilon_0}\frac{\Omega}{4\pi} = \frac{q}{2\epsilon_0}(1-\cos\theta)$$

From the figure, $\cos\theta = \frac{d}{\sqrt{d^2+R^2}}$ so

$$\phi = \frac{q}{2\epsilon_0}\left\{1-\frac{d}{\sqrt{d^2+R^2}}\right\}$$

W.E-8: ***Two point charges +Q_1 and -Q_2 are placed at A and B respectively. A line of force emanates from Q_1 at an angle θ with the line joining A and B. At what angle will it terminate at B?***

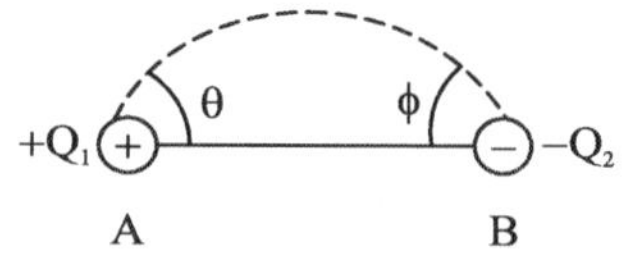

Sol: We know that number of lines of force emerge is proportional to magnitude of the charge. The field lines emanating from Q_1, spread out equally in all directions. The number of field lines or flux through cone of half angle θ is $\frac{Q_1}{4\pi}2\pi(1-\cos\theta)$. Similarly the number of lines of force terminating on -Q_2 at an angle ϕ is $\frac{Q_2}{4\pi}2\pi(1-\cos\phi)$. The total lines of force emanating from Q_1 is equal to the total lines of force terminating on Q_2

$$\Rightarrow \frac{Q_1}{4\pi}2\pi(1-\cos\theta) = \frac{Q_2}{4\pi}2\pi(1-\cos\phi)$$

or $\frac{Q_1}{2}(1-\cos\theta) = \frac{Q_2}{2}(1-\cos\phi)$; $Q_1\sin^2\theta/2 = Q_2\sin^2\phi/2$

$$\sin\phi/2 = \sqrt{\frac{Q_1}{Q_2}}\sin\theta/2 \quad \Rightarrow \phi = 2\sin^{-1}\left\{\sqrt{\frac{Q_1}{Q_2}}\sin\theta/2\right\}$$

CAVITY IN THE CONDUCTOR

We have discussed that there will be no electric field inside a charged conductor and all the charge resides on its outer surface only. Suppose that charged conductor has a cavity or cavities and there are no charges within the cavity or cavities, even then charge resides on the outer surface of the conductor. There will be no charge on the walls of the cavity or cavities. This can be verfied very easily using Gauss's law by enclosing the cavity with a Gaussian surface.

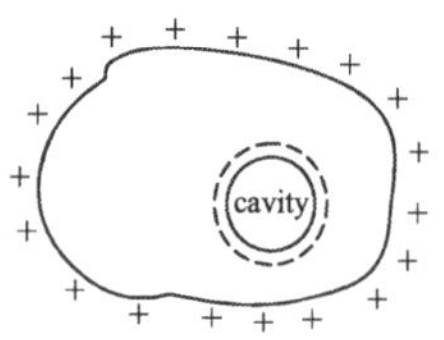

Fig.

$\oint \bar{E}.d\bar{s} = 0$ For the dotted surface.

$\Rightarrow q = 0$ inside cavity.

Consider a conductor with spherical cavity inside it. There is no charge on the conductor. Now a point charge +q is kept at the centre of the cavity. Due to this charge, a charge -q is induced on the inner surface of cavity . The total flux originated by +q will terminate on the cavity walls and no field lines enter into the conductor body

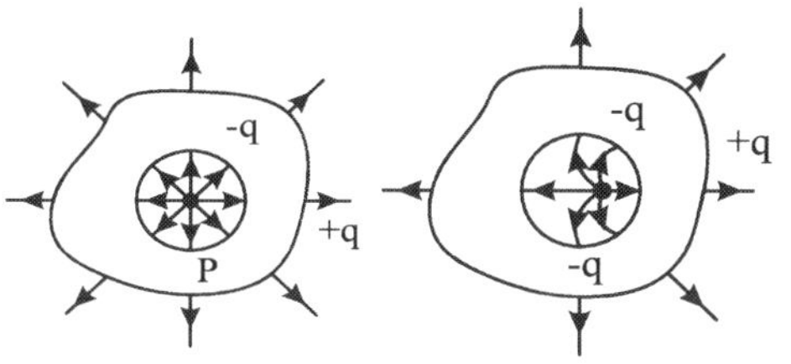

Fig. (a) Fig. (b)

We can consider a Gaussian surface around the cavity and prove that induced charge on the cavity walls is -q. The reason is electric field $(\bar{E})$ is zero inside the material of the conductor. The Total enclosed charge within the Gaussian surface is zero. Here the conductor is initially uncharged. From conservation of charge, we can say that on the outer surface of the conductor a charge +q will be induced. At any point inside the material of conductor, say at P, the electric field produced by +q in the cavity is cancelled by the field produced by charges induced on the walls of cavity and on the outer surface of the conductor. If the point charge is not at the centre of the spherical cavity, even then induced charges on the cavity walls and on the outer surface of the conductor are -q and +q only.

But the distribution of induced charges will change in such a way that at any point P in the material of the conductor resultant electric field is zero.

Suppose the conductor has charge q_0 on it initially. This charge resides on the outer surface of the conductor. If point charge q is kept inside the cavity, induced charges on the walls of cavity and on the outer surface of the conductor are the same as before. i.e., – q and + q. But the total charge on the outer surface of the conductor is $(q_0 + q)$ now.

If the charge inside the cavity is displaced, the induced charge distribution on inner surface of the body changes such that at any point inside the material of the conductor resultant field is zero. In this case the charge distribution on outer surface of the conductor does not change and only the charge distribution on the cavity walls will change.

Now the charge inside the cavity is fixed. If another charge is brought towards the conductor from outside., it will not affect the charge distribution inside the cavity and only the distribution of charge on the outer surface will be affected.

MECHANICAL FORCE ON THE CHARGED CONDUCTOR

We know that like charges repel each other. So, when a conductor is charged, the charge on any point of the conductor is repelled by the charge on its remaining part. It means surface of a charged conductor experiences mechanical force.

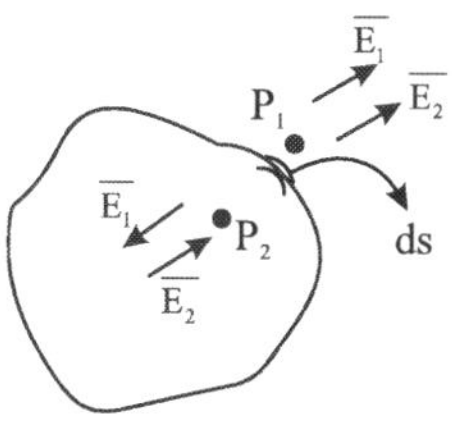

Consider a charged conductor as shown. Let ds be the surface area of a small element on the conductor. The electric field at point P_1 near the conductor surface can be considered as the superposition of fields $\bar{E}_1$ and $\bar{E}_2$. Here $\bar{E}_1$ is the field produced by that elemental surface and $\bar{E}_2$ is the field due to the remaing surface of the conductor.

$E = E_1 + E_2$

But we know that $E = \frac{\sigma}{\epsilon_0}$ at P_1 which is just outside the conductor and is zero at P_2 which is just inside the conductor

So at P_1 we have $E_1 + E_2 = \frac{\sigma}{\epsilon_0}$
and at P_2 we have $E_1 - E_2 = 0$

$\Rightarrow E_1 = E_2 = \frac{\sigma}{2\epsilon_0}$

Now the force experienced by small surface ds due to the charge on the rest of the surface is

$F = (dq)E_2 = (\sigma ds)E_2 = \frac{\sigma^2 ds}{2\epsilon_0}$

and $\frac{\text{Force}}{\text{Area}} = \frac{F}{ds} = \frac{\sigma^2}{2\epsilon_0} = \frac{1}{2}\epsilon_0 E^2$

ELECTRIC PRESSURE ON A CHARGED SURFACE

From the above derivation we observed that a small surface of a charged conductor will experience a force by the remaining surface. The force per unit area of the surface is $\frac{1}{2}\epsilon_0 E^2$ or $\frac{\sigma^2}{2\epsilon_0}$

This is known as electric pressure on the charged metal surface.

$\Rightarrow P_e = \frac{1}{2}\epsilon_0 E^2$

Suppose a charged body is in an external electric field. Let us find out the electric pressure on the surface of that charged body.
Consider a surface uniformly charged with charge density σ. On that surface 'ds' is the surface area of a small element. The charge on that element is $dq = \sigma ds$

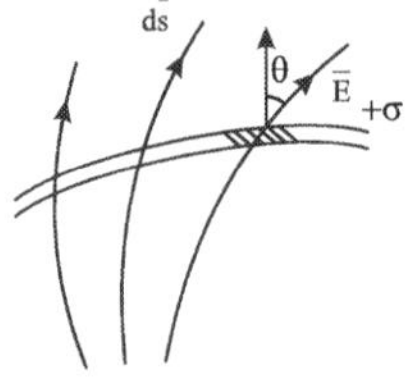

The given surface is in an external electric field represented by the field lines as shown.

Let E be the intensity of electric field on the elemental surface. Here angle between $\bar{E}$ and $d\bar{s}$ is θ. In this case $\bar{E}$ has two comperments.
Component parallel to the surface is

$E_{11} = E \sin\theta$

and component normal to the surface is

$E_{\perp} = E \cos\theta$

Here force due to E_{11} on the surface is tangential which tries to stretch the surface. Where as the force due to $E_{\perp}$ applies outward pressure on the surface. Now outward force on the elemental surface is

$dF = (dq)E_{\perp} = \sigma\, ds\, E_{\perp}$

So, the outwards electric pressure on the surface is

$P_e = \frac{dF}{ds} = \sigma E_{\perp} \Rightarrow P_e = \sigma E \cos\theta$

W.E-9: ***A thin spherical shell radius of r has a charge Q uniformly distributed on it. At the centre of the shell, a negative point charge -q is placed. If the shell is cut into two identical hemispheres, still equilibrium is maintained. Then find the relation between Q and q?***

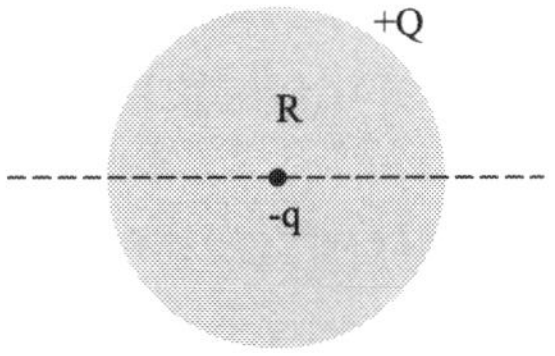

Sol : Here the outward electric pressure at every point on the shell due to its own charge is

$$P_1 = \frac{\sigma^2}{2\epsilon_0} = \frac{1}{2\epsilon_0}\left(\frac{Q}{4\pi r^2}\right)^2 ; \; P_1 = \frac{Q^2}{32\pi^2 \epsilon_0 r^4}$$

Due to -q, the electric field on the surface of the shell is

$$E = \frac{1}{4\pi\epsilon_0}\frac{q}{r^2}$$

This electric field pulls every point of the shell in inward direction. The inward pressure on the surface of the shell due to the negative charge is $P_2 = \sigma E$; $= \left(\frac{Q}{4\pi r^2}\right)\left(\frac{1}{4\pi\epsilon_0}\frac{q}{r^2}\right) = \frac{Qq}{16\pi^2 \epsilon_0 r^4}$

For equlibrium of the hemispherical shells $P_2 \geq P_1$ or $\frac{Qq}{16\pi^2 \epsilon_0 r^4} \geq \frac{Q^2}{32\pi^2 \epsilon_0 r^4}$; $\quad q \geq \frac{Q}{2}$

WE.10 : ***If r and T are radius and surface tension of a spherical soap bubble respectively then find the charge needed to double the radius of bubble***

Sol: For smaller bubble

$P_1 = \left(P_0 + \frac{4T}{r}\right)$ and $V_1 = \frac{4}{3}\pi r^3$ For larger bubble

$P_2 = P_0 + \frac{4T}{R} - \frac{\sigma^2}{2\epsilon_0}$ and $V_2 = \frac{4}{3}\pi R^3$ where $\sigma = \frac{q}{4\pi R^2}$

for air in the bubble, $P_1 V_1 = P_2 V_2$

$$\left(P_0 + \frac{4T}{r}\right) r^3 = \left[\left(P_0 + \frac{4T}{R}\right) - \frac{q^2}{16\pi^2 R^4 \times 2\epsilon_0}\right] R^3$$

$$P_0\left[R^3 - r^3\right] + 4T\left[R^2 - r^2\right] - \frac{q^2}{32\pi^2 \epsilon_0 R} = 0$$

But R = 2r

$$P_0\left[7r^3\right] + 4T\left[3r^2\right] - \frac{q^2}{32\pi^2 \epsilon_0 (2r)} = 0$$

$$\frac{q^2}{64\pi^2 \epsilon_0 r} = 7P_0 r^3 + 12Tr^2$$

$$q^2 = 64\pi^2 \epsilon_0 r^3\left[7P_0 r + 12T\right]$$

$$q = 8\pi r\left[\epsilon_0 r\left(7P_0 r + 12T\right)\right]^{1/2}$$

CONCEPTUAL QUESTIONS

ELECTRIC FLUX AND GAUSS LAW

1. **A cubical Gaussian surface encloses electric flux of 30 C per unit permittivity of a charge. The electric flux through each face of the cube per unit permittivity is**
   1) 30 C 2) 15 C 3) 10 C 4) 5 C
2. **As one penetrates uniformly charged conducting sphere, what happens to the electric field strength**
   1) decreases inversely as the square of the distance 2) decreases inversely as the distance
   3) becomes zero 4) increases inversely as the square of distance
3. **Mark the correct option**
   1) Gauss law is valid only for unsymmetrical charge distributions
   2) Gauss law is valid only for charge placed in vacuum
   3) The electric field calculated by Gauss law is the field due to the charges outside the Gaussian surface.
   4) The flux of the electric field through a closed surface due to all the charges is equal to the flux due to the charges enclosed by the surface
4. **If the flux of the electric field through a closed surface is zero**
   1) The electric field must be zero every where on the surface
   2) The electric field must not be zero everywhere on the surface
   3) The charge inside the surface must be zero
   4) The charge in the vicinity of the surface must be zero
5. **An infinite plane sheet of a metal is charged to charge density σ C/m^2 in a medium of dielectric constant K. Intensity of electric field near the metallic surface will be**
   1) $E=\dfrac{\sigma}{\varepsilon_o K}$ 2) $E=\dfrac{\sigma}{2\varepsilon_o}$ 3) $E=\dfrac{\sigma}{2\varepsilon_o K}$ 4) $E=\dfrac{K\sigma}{2\varepsilon_o}$
6. **The electric flux from a cube of edge l is ϕ. Its value if edge of cube is made $2l$ and charge enclosed is halved is**
   1) $\phi/2$ 2) 2ϕ 3) 4ϕ 4) ϕ
7. **If the electric flux entering and leaving an enclosed surface respectively is ϕ_1 and ϕ_2, the electric charge inside the surface will be**
   1) $(\phi_1+\phi_2)/\varepsilon_o$ 2) $(\phi_1-\phi_2)/\varepsilon_o$ 3) $(\phi_1+\phi_2)\varepsilon_o$ 4) $(\phi_2-\phi_1)\varepsilon_o$
8. **Electric flux at a point in an electric field is**
   1) positive 2) negative 3) zero 4) positive or negtive
9. **Electric flux over a surface in an electric field may be**
   1) positive 2) negative 3) zero 4) positive, negative, zero
10. **A charge Q is placed at the mouth of a conical flask. The flux of the electric field through the flask is**
    1)zero 2)Q/ε_0 3)$\dfrac{Q}{2\varepsilon_0}$ 4)$<\dfrac{Q}{2\varepsilon_0}$
11. **A charge Q is placed at the mouth of a conical flask. At the centre of the circular crossection flux of the electric field through it is**
    1) zero 2)Q/ε_0 3)$\dfrac{Q}{2\varepsilon_0}$ 4)$<\dfrac{Q}{2\varepsilon_0}$

12. **Electric field intensity at a point due to an infinite sheet of charge having surface charge density σ is E. If sheet were conducting electric intensity would be**

1) E/2 2) E 3) 2 E 4) 4 E

13. **Two thin infinite parallel sheets (non conducting) have uniform surface densities of charge $+\sigma$ and $-\sigma$. Electric field in the space between the two sheets is**

1) σ / ϵ_0 2) $\sigma / 2\epsilon_0$ 3) $2\sigma / \epsilon_0$ 4) zero

14. **In the above question, if the sheets were thick and conducting, value of E in the space between the two sheets would be**

1) $2\sigma / \epsilon_0$ 2) σ / ϵ_0 3) zero 4) $4\sigma / \epsilon_0$

15. **In the above problem the value of E in the space outside the sheets is.**

1) σ / ϵ_0 2) $\sigma / 2\epsilon_0$ 3) zero 4) $2\sigma / \epsilon_0$

16. **The Gaussian surface for calculating the electric field due to a charge distribution is**

1) any closed surface around the charge distribution

2) any surface near the charge distribution 3) a spherical surface

4) a closed surface at a every point of which electric field has a normal component which is zero or a fixed value

17. **The electric flux over a sphere of radius 1m is ϕ. If radius of the sphere were doubled without changing the charge enclosed, electric flux would become**

1) 2ϕ 2) $\phi/2$ 3) $\phi/4$ 4) ϕ

18. **A charge q is placed at the centre of a cube. What is the electric flux associated with one of the faces of cube**

1) q/ε_0 2) ε_0/q 3) $6q/\varepsilon_0$ 4) $q/6\varepsilon_0$

19. **A charge Q is placed at the corner of a cube. The electric flux through all the faces of the cube is**

1) Q/ε_0 2) $Q/6\varepsilon_0$ 3) $Q/8\varepsilon_0$ 4) $Q/3\varepsilon_0$

20. **A point charge +q is placed at mid point of a cube of side 'L'. The electric flux emerging from the cube is**

1) q/ε_0 2) $6qL^2/\varepsilon$ 3) $q/6L^2\varepsilon_0$ 4) zero

21. **A charge q is enclosed as shown below, the electric flux is**

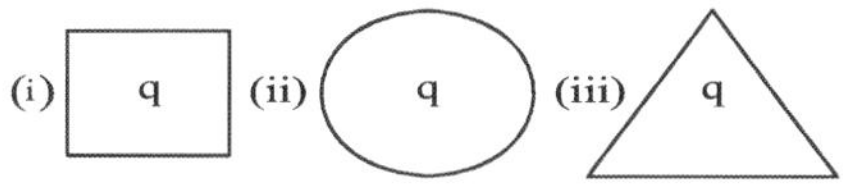

1) maximum in (i) 2) maximum in (ii)

3) maximum in (iii) 4) equal in all

22. **An ellipsoidal cavity is carved with in a perfect conductor. A positive charge q is placed at the centre of the cavity. The points A and B are on the cavity surface as shown in the figure then**

a) Electric field near A in the cavity = Electric field near B in the cavity

b) Charge density at A = Charge density at B

c) Potential at A – Potential at B

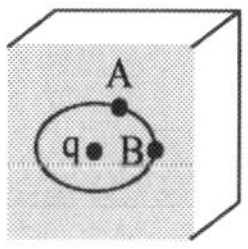

d) Total electric flux through the surface of the cavity is q/ε_0.

1) a,b,c,d are correct 2) a,b,c are correct

3) only a and b are correct 4) only c and d are correct

23. **A metallic shell has a point charge 'q' kept inside its cavity. Which one of the following diagrams correctly represents electric lines of forces**

1) 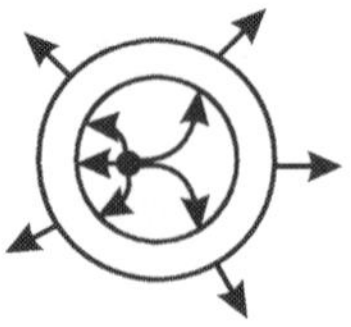

2) 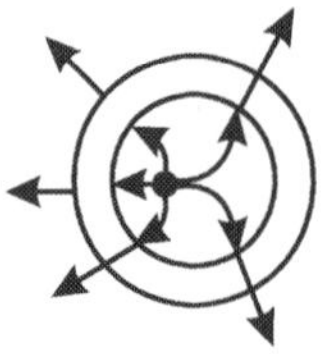

3) 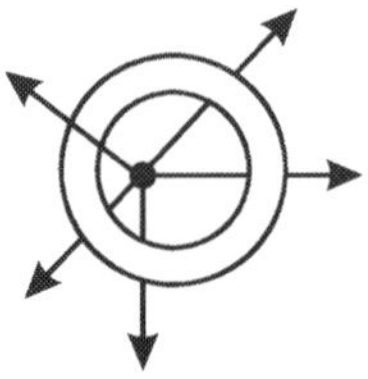

4) 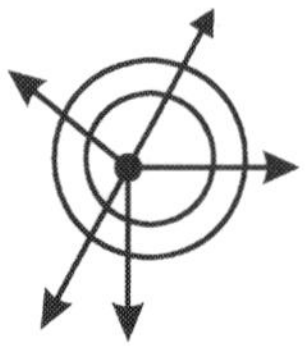

24. **Two infinitely long thin styraight wires having uniform linear charge densities λ and 2λ are arranged parallel to each other at a distance r apart . The intesity of the electric field at a point midway between them is**

1) $\dfrac{2\lambda}{\pi\varepsilon_0 r}$ 2) $\dfrac{\lambda}{\pi\varepsilon_0 r}$ 3) $\dfrac{\lambda}{2\pi\varepsilon_0 r}$ 4) $\dfrac{3\lambda}{2\pi\varepsilon_0 r}$

25. **Find the total flux due to charge q associated with the given hemispherical surface**

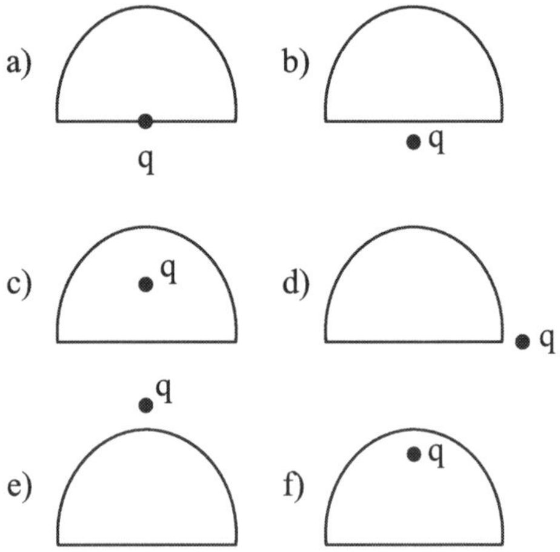

1) $(a)\dfrac{q}{2\epsilon_0},(b)0,(c)\dfrac{q}{\epsilon_0},(d)0,(e)0(f)\dfrac{q}{\varepsilon_0}$ 2) $(a)0,(b)\dfrac{q}{2\epsilon_0},(c)0,(d)\dfrac{q}{\epsilon_0},(e)0(f)\dfrac{q}{\varepsilon_0}$

3) $(a)\dfrac{q}{2\epsilon_0},(b)\dfrac{q}{\epsilon_0},(c)0,(d)\dfrac{q}{\epsilon_0},(e)0(f)\dfrac{q}{\varepsilon_0}$ 4) $(a)0,(b)\dfrac{q}{2\epsilon_0},(c)0,(d)\dfrac{q}{\epsilon_0},(e)\dfrac{q}{\epsilon_0}(f)\dfrac{q}{\varepsilon_0}$

26. **A : A metallic sheild in the form of a hollow shell may be built to block an electric field.**
R : In a hollow spherical shield, the electric field inside it is zero at every point.

1) Both 'A' and 'R' are true and 'R' is the correct explanation of 'A'.
2) Both 'A' and 'R' are true and 'R' is not the correct explanation of 'A'.
3) 'A' is true and 'R' is false 4) 'A' is false and 'R' is true

27. **A thin spherical shell of radius R has charge Q spread uniformly over its surface. Whicho f the following graphs most closely represents the electric field $E(r)$ produced by the shell in the range $0 \le r < \infty$, where r is the distance from the centre of the shell?**

1) 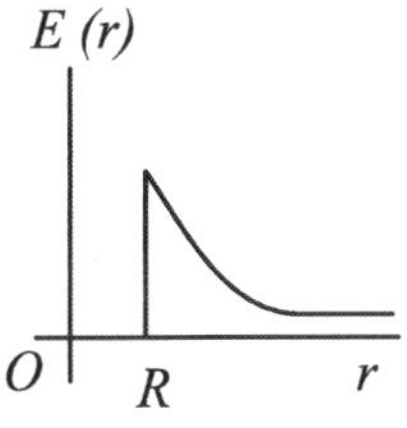

2) 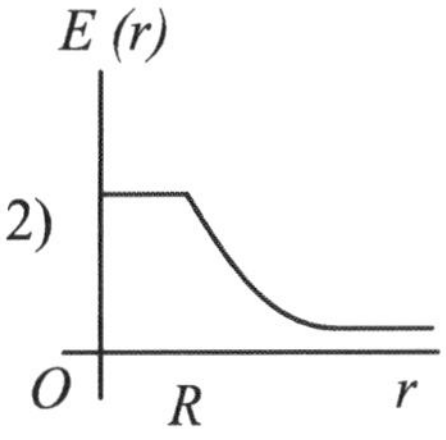

3) 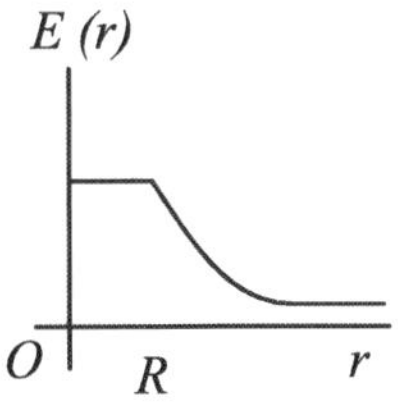

4) 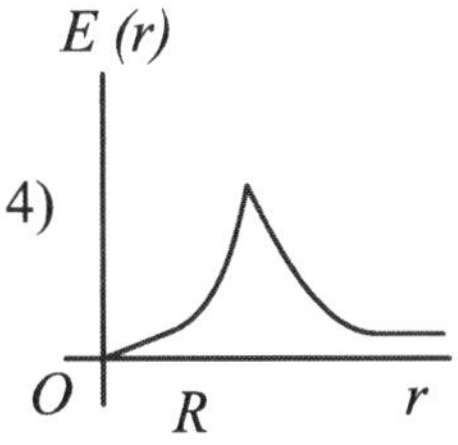

28. **Three positive charges of equal value q are placed at vertices of an equilateral triangle. The resulting lines of force should be sketched as in (3)**

1) 

2) 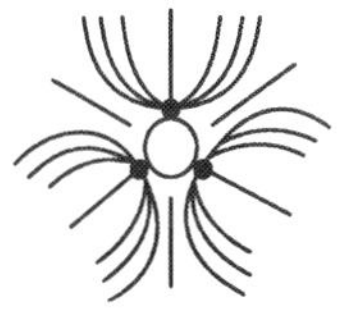

3) 

4) 

29. **An uncharged metal sphere is placed between two equal and oppositely Charged metal plates. The nature of line of force will be**

1) 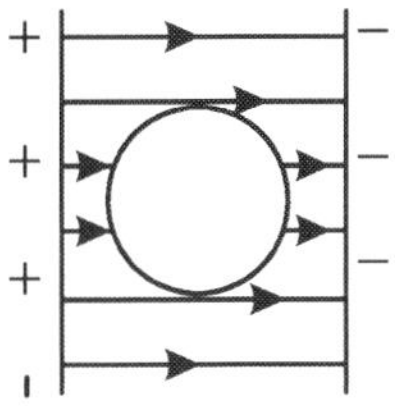

2) 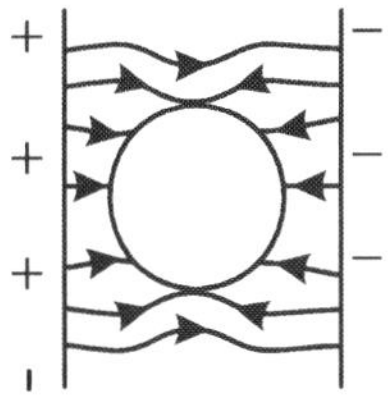

3) 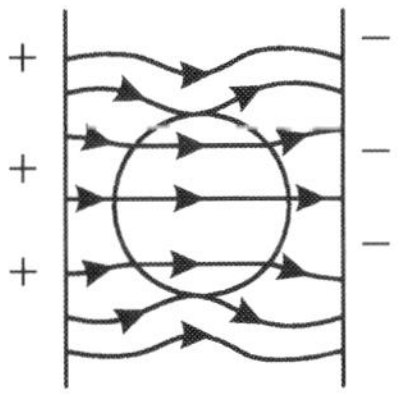

4) 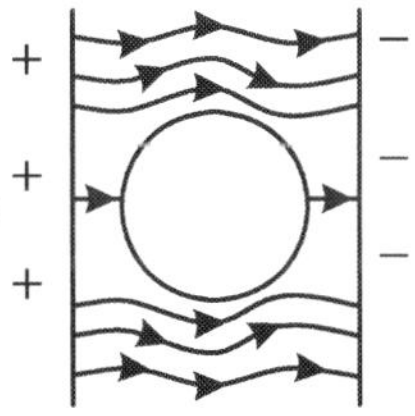

ASSERTION& REASON

1) Both (A) and (R) are true and (R) is the correct explanation of (A)

2) Both (A) and (R) are true and (R) is not the correct explanation of (A)

3) (A) is true but (R) is false 4) (A) is false but (R) is true

30. Assertion: A device used to measure $\vec{E}$ is located at some distance from a fixed point charge. In this situation, the device measures E_0 as the magnitude of electric field intensity. Now an uncharged conducting sphere with a very small hole is lowered by an insulating thread so that it surrounds the point charge. Now, the reading of the device becomes zero.

Reason: Electrostatic shielding is the phenomenon in which inside of hollow conductor is shielded for outside electric field

31. Assertion: E in outside vicinity of a conductor depends only on the local charge density σ

Reason: $\vec{E}$ in outside vicinity of a conductor is given by $\frac{\sigma}{\varepsilon_0}$

32. Assertion: Four point charges q_1, q_2, q_3 and q_4 are as shown in Fig. The flux over the shown Gaussian surface depends only on charges q_1 and q_2

Reason: Electric field at all points on Gaussian surface depends ony on charges q_1 and q_2

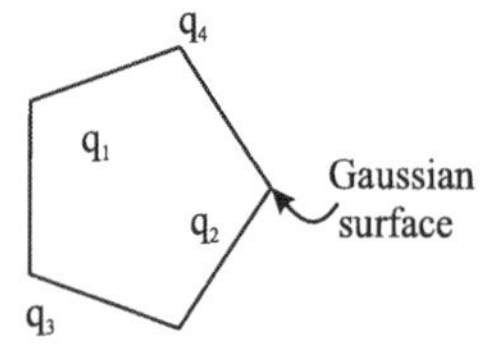

33. Assertion: A point charge q is placed near an arbitray shaped solid conductor as shown in figure. The potential difference between the points A and B within the conductor remain same irrespective of the magnitude of charge q.

Reason: The electric field inside a solid conductor is zero under electrostatic conditions.

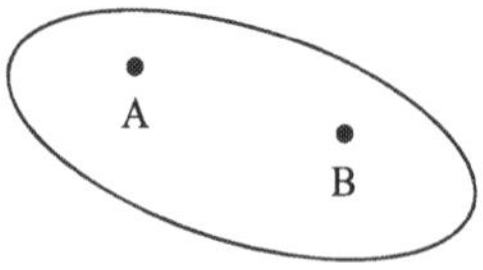

34. Assertion: Two point charges +Q and -Q are fixed at point A(+a,0,0) and point B(-a,0,0) respectively. Then the magnitude of electric flux due to electric field of either point charge through infinite y-z plane (that is x=0 plane) is less than magnitude of net electric flux due to electric field of both charges through that plane (x = 0 plane).

Reason: The magnitude of net electric flux through a surface due to a system of point charges is equal to sum of magnitude of electric flux through that surface due to each of the point charge of the system.

35. Assertion: In a region where uniform electric field exists, the net charge within volume of any size is zero.

Reason: The electric flux through any closed surface in region of uniform electric field is zero.

36. Assertion: A point charge q is placed at centre of spherical cavity inside a spherical conductor as shown. Another point charge Q is placed outside the conductor as shown in Fig. Now as the point charge Q is pushed away from conductor, the potential difference $(V_A - V_B)$ between two points A and B within the cavity of sphere remains constant

Reason: The electric field due to charges on outer surface of conductor and outside of the conductor is zero at all points inside the conductor.

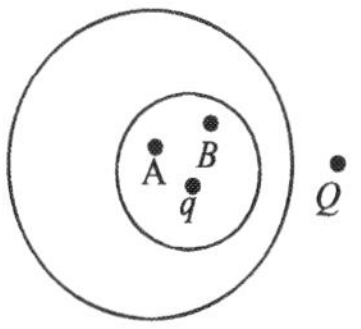

37. **Assertion: The electrostatic force on a charged particle located on a equipotential surface is zero**

    Reason: Componant of E along equipotential surface is zero.

38. **Assertion: We cannot produce electric field in a neutral conductor.**

    Reason: Neutral conductor cannot produce electric field.

39. **Two parallel metallic plates have surface charge densities σ_1 and σ_2 as shown in figure. Match the following:**

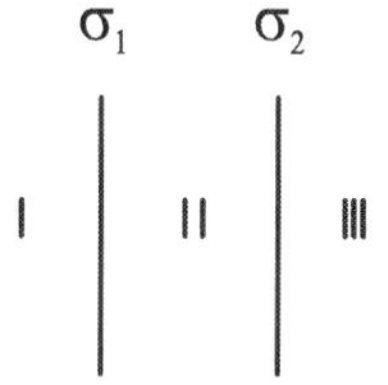

Column I	Column II
a) If $\sigma_1 + \sigma_2 = 0$ III is towards right	(p) Electric field in region
b) If $\sigma_1 + \sigma_2 > 0$ I is zero	(q) Electric field in region
c) If $\sigma_1 + \sigma_2 < 0$ I is towards right	(r) Electric field in region

1) a-p, b-q, c-r 2) a-q, b-p, c-r 3) a-r b-q, c-r 4) a-p, b-r, c-r

40. **Two spherical shells are as shown in figure. Suppose r is the distance of a point from their common centre. Then,**

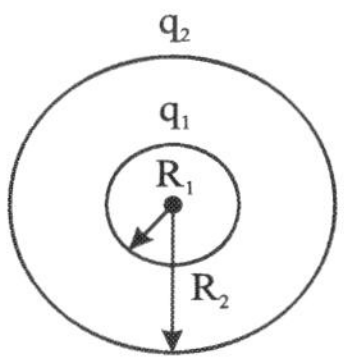

Column I	Column II
a) Electric field for $r < R_1$	(p) is constant for q_2 and vary for q_1
b) Electric potential for $r < R_1$	(q) is zero for q_2 and vary for q_1
c) Electron potential for $R_1 < r < R_2$	(r) is constant for both q_1 and q_2
d) Electric field for $R_1 < r < R_2$	(s) is zero

1) a-r, b-s, c-p, d-q 2) a-s, b-r, c-p, d-q 3) a-s, b-s, c-p ,d-q 4) a-r, b-q, c-p ,d-q

CONCEPTUAL KEY

1) 4	2) 3	3) 4	4) 3	5) 1	6) 1	7) 4	8) 3	9) 4	10) 3
11) 3	12) 3	13) 1	14) 1	15) 3	16) 4	17) 4	18) 4	19) 3	20) 1
21) 4	22) 4	23) 1	24) 2	25) 1	26) 1	27) 1	28) 3	29)2	30) 4
31) 2	32) 3	33) 1	34) 1	35) 2	36) 1	37) 4	38) 2	39) 2	40) 3

LEVEL - I (CW)

ELECTRIC FLUX AND GAUSS LAW

1. **A charged spherical conductor has a surface charge density of 0.7 C/m^2. When its charge is increased by 0.44C, the charge density changes by 0.14 C/m^2. The radius of the sphere is**
   1) 5 cm 2) 1 0 m 3) 0.5 m 4) 5 m

2. **The electric field in a region of space is given by $\vec{E} = 5\hat{i} + 2\hat{j}\ NC^{-1}$. The electric flux due to this field through an area $2\,m^2$ lying in the YZ plane in S.I. units is**
   1) 10 2) 20 3) $10\sqrt{2}$ 4) $2\sqrt{29}$

3. **Number of electric lines of force emerging from 1C of positive charge in vacuum is**
   1) 8.85×10^{-12} 2) 9×10^{9} 3) $1/4\pi\times9\times10^{9}$ 4) 1.13×10^{11}

4. **A charge of 5 C is placed at the centre of a spherical gaussian surface of radius 5 cm. The electric flux through the surface is $\frac{1}{\varepsilon_0}$ times of**
   1) 0.1 N-m²/C 2) 0.5 N-m²/C 3) 1 N-m²/C 4) 5 N-m²/C

5. **In a region where intensity of electric field is $5\ NC^{-1}$, 40 lines of electric force are crossing $10\ NC^{-1}$ will be**
   1) 20 2) 80 3) 100 4) 200

6. **A half ring of radius R has a charge of λ per unit length. The potential at the centre of the half ring is ($k = \frac{1}{4\pi\varepsilon_o}$)**
   1) $k\frac{\lambda}{R}$ 2) $k\frac{\lambda}{\pi R}$ 3) $k\frac{\pi\lambda}{R}$ 4) $k\pi\lambda$

7. **An electron is placed at the centre of a Conducting sphere of radius 0.2 metre having a charge 5×10^{-2} coulomb. The force on the electron is**
   1) zero 2) $11\times10^{9}N$ 3) $22.5\times10^{9}N$ 4) $2.5\times10^{9}N$

8. **Eightcharges, 1 μC, -7 μC, -4 μC, 10 μC, 2 μC, -5 μC, -3 μC and 6 μC are situated at the eight corners of a cube of side 20 cm. A spheric al surface of radius 80 cm encloses this cube. The center of sphere coincides with center of the cube. Then the outgoing flux from the spherical surface(in units Vm) is**
   1) $36\pi\times10^{3}$ 2) $684\pi\times10^{3}$ 3) zero 4) $72\pi\times10^{3}$

LEVEL- I (C. W) KEY

1) 3 2) 1 3) 4 4) 4 5) 2 6) 4 7) 1 8) 3

LEVEL - I (C. W) HINTS

1. $\Delta\sigma = \dfrac{\Delta Q}{A}$

2. $\phi_E = \int \vec{E}.\vec{ds} = E \times s = 5 \times 2$

4. $\phi_E = \dfrac{q}{\varepsilon_0}$

5. $\phi \propto q$

6. $q = \lambda \pi r \text{ and } V = \dfrac{kq}{r}$

7. $E_{inside} = 0, F = Eq$

LEVEL - I (HW)

1. Caluclate the net flux emerging from given enclosed surface - $Nm^2\ C^{-1}$

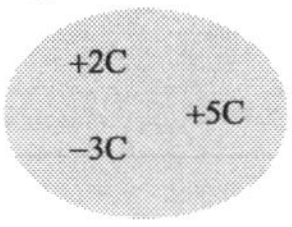

1) 4.5×10^{13} 2) 45×10^{12} 3) zero 4) 1.12×10^{12}

2. A charge Q is situated at the centre of a cube. The electric flux through one of the faces of the cube is

1) Q/ε_0 2) $Q/2\varepsilon_0$ 3) $Q/4\varepsilon_0$ 4) $Q/6\varepsilon_0$

3. The magnitude of the electric field on the surface of a sphere of radius r having a uniform surface charge density σ is

1) σ/ε_0 2) $\sigma/2\varepsilon_0$ 3) $\sigma/\varepsilon_0 r$ 4) $\sigma/2\varepsilon_0 r$

4. If the electric flux entering and leaving an enclosed surface respectively is ϕ_1 and ϕ_2, the electric charge inside the surface will be

1) $(\phi_2 - \phi_1)\,\varepsilon_0$ 2) $(\phi_1 + \phi_2)/\varepsilon_0$ 3) $(\phi_2 - \phi_1)/\varepsilon_0$ 4) $(\phi_1 + \phi_2)\varepsilon_0$

5. A charge q is placed at the centre of the open end of cylinderical vessel. Find the flux of the electric field through the surface of the vessel.

1) $\dfrac{q}{2\epsilon_0}$ 2) $\dfrac{q}{\epsilon_0}$ 3) $\dfrac{q}{3\epsilon_0}$ 4) zero

LEVEL- I (H. W) KEY

1) 1 2) 4 3) 1 4) 1 5) 1

LEVEL- I (H. W) HINTS

1. $\int \overline{E}.\overline{ds} = \dfrac{q}{\varepsilon_0} = \dfrac{2+5-3}{8.85 \times 10^{-12}} = 4.5 \times 10^{11}$

2. $\int \overline{E}.\overline{ds} = \dfrac{q}{\varepsilon_0} = \dfrac{q}{6\varepsilon_0}$ through on of the face

3. $E \times 4\pi r^2 = \dfrac{q}{\varepsilon_0}$; $E = \dfrac{q}{4\pi\varepsilon_0 r^2} = \dfrac{\sigma}{\varepsilon_0}$

4. $\int \overline{E}.\overline{ds} = \dfrac{(\phi_2 - \phi_1)}{\varepsilon_0}$

5. $\int \overline{E}.\overline{ds} = \dfrac{q}{\varepsilon_0}$

LEVEL - II (CW)

ELECTRIC FLUX AND GAUSS LAW

1. **The inward and outward electric flux for a closed surface in units of N-m²/C are respectively 8×10^3 and 4×10^3. Then the total charge inside the surface in S.I units is (where ε_0 =permitivity constant)**

   1) 4×10^3 2) -4×10^3 3) $-\frac{\pi R^2-\pi R}{E}$ 4) $-4\times10^3\varepsilon_0$

2. **A cylinder of radius R and length L is placed in the uniform electric field E parallel to the cylinder axis. The total flux from the two flat surfaces of the cylinder is given by**

   1) $2\pi R^2E$ 2) $\frac{\pi R^2}{E}$ 3) $\frac{\pi R^2-\pi R}{E}$ 4) zero

3. **A cube is arranged such that its length, breadth, height are along X,Y and Z directions. One of its corners is situated at the origin. Length of each side of the cube is 25cm. The components of electric field are $E_x = 400\sqrt{2}\,N/C$, $E_y = 0$ and $E_z = 0$ respectively. The flux coming out of the cube at one end will be**

   1) $25\sqrt{2}Nm^2/C$ 2) $5\sqrt{2}Nm^2/C$ 3) $250\sqrt{2}Nm^2/C$ 4) $25Nm^2/C$

4. **If a hemispherical body is placed in a uniform electric field E then the flux linked with the curved surface is**

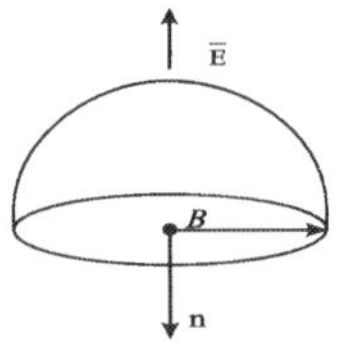

   1) $2\pi R^2E$ 2) πR^2E 3) $4\pi R^2E$ 4) $6\pi R^2E$

5. **A thin conducting ring of radius R is given a charge +Q. The electric field at the centre O of the ring due to the charge on the part AKB of the ring is E. The electric field at the centre due to the charge on the part ACDB of the ring is**

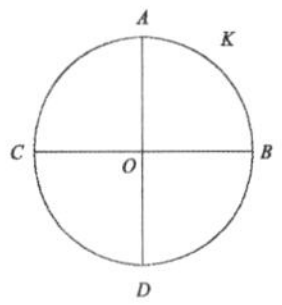

   1)3 E along OK 2)3E along KO

   3)E along OK 4) E along KO

6. **In a unifrorm electric field find the total flux associated with the given surfaces (R is radius)**

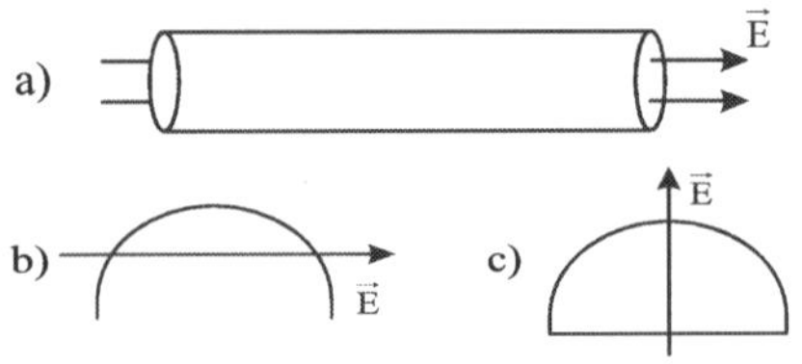

   1) $a=0,b=0,c=0$ 2) $a=0,b=\left(\pi R^2E\right),c=0$

   3) $a=2\pi RE,b=\left(\pi R^2E\right),c=0$ 4) $a=\pi R^2E,b=0,c=0$

7. **Surface charge density of soap bubble of radius 'r' and surface tension T is σ. If P is excess pressure, the value of σ is**

   1) $\in_0\left[\frac{4T}{r}-P\right]^{\frac{3}{2}}$ 2) $\left[2\in_0\left(\frac{4T}{r}-P\right)\right]^{\frac{1}{2}}$ 3) $\frac{4T}{r}$ 4) $\left[4\in_0\left(\frac{2T}{r}-P\right)\right]^{\frac{1}{2}}$

8. **An infinitely long thin straight wire has uniform linear charge density of 1/3 coul.m^{-1}. Then the magnitude of the electric intensity at a point 18 cm away is**

1) $0.33\times10^{11}\ NC^{-1}$ 2) $3\times10^{11}\ NC^{-1}$ 3) $0.66\times10^{11}\ NC^{-1}$ 4) $1.32\times10^{11}\ NC^{-1}$

9. **Consider two concentric spherical surface S_1 with radius a and S_2 with radius 2a, both centred on the origin. There is a charge +q at the origin, and no other charges. Compare the flux ϕ_1 through S_1 with the flux ϕ_2 through S_2**

1) $\phi_1 = 4\phi_2$ 2) $\phi_1 = 2\phi_2$ 3) $\phi_1 = \phi_2$ 4) $\phi_1 = \phi_2 / 2$

10. **The electric field on two sides of a large charged plate is shown in fig. The charge density on the plate in S.I. units is given by (ε_o is the permittivity of free space in S.I. units)**

1) $2\varepsilon_o$ 2) $4\varepsilon_o$ 3) $10\varepsilon_o$ 4) zero

LEVEL- II (C.W) KEY

1) 4 2) 4 3) 1 4) 2 5) 3 6) 1 7) 2 8) 1 9) 3 10) 2

LEVEL- II (C. W) HINTS

1. $\phi_E = \dfrac{q}{\varepsilon_0}$

3. $\overline{E}_1.\overline{A}_1 + \overline{E}_2.\overline{A}_2 + \overline{E}_3.\overline{A}_3 = \phi$

4. $\phi = \overline{E}.\overline{A}$

5. $\vec{E}_0 = 0$

9. Flux through both will be same as net charge enclosed by both is same

10. Electric field due to plate= $\dfrac{\sigma}{2\varepsilon_o}$

LEVEL - II (HW)

1. **The number of electric lines of force originating from a charge of 1C is**

1) 1.129×10^{11} 2) zero 3) 1.129×10^{-11} 4) 1.129×10^{10}

2. **A cube of side l is placed in a uniform field E, where $E = E\hat{i}$. The net electric flux through the cube is**

1) Zero 2) l^2E 3) $4l^2E$ 4) $6l^2E$

3. **A point charge +q is placed at the centre of a cube of side L. The electric flux emerging from the cube is**

1) $\dfrac{q}{\varepsilon_0}$ 2) Zero 3) $\dfrac{6qL^2}{\varepsilon_0}$ 4) $\dfrac{q}{6L^2\varepsilon_0}$

4 **A long thin flat sheet has a uniform surface charge density σ. The magnitude of the electric field at a distance ' r ' from it is given by**

1) σ/ε_0 2) $\sigma/2\varepsilon_0$ 3) $\sigma/\varepsilon_0 r$ 4) $\sigma/2\varepsilon_0 r$

5 **A charge of 8.85C is placed at the centre of a spherical Guassian surface of radius 5 cm. The electric flux through the surface is**

1) 10^{12} V/m 2) 10^{-12} V/m 3) 10^{8} V/m 4) 10^{10}V/m

6. The inward and outward electric flux for a closed surface in units of $N-m^2/C$ are respectively 8×10^3 and 4×10^3. Then the total charge inside the surface in S.I. units is (where ϵ_0 = permitivity in free space)

1) 4×10^3 2) -4×10^3 3) $\frac{(-4\times10^3)}{\epsilon_0}$ 4) $-4\times10^3\,\epsilon_0$

7. The total flux linked with unit negative charge put in air is

1) $\frac{1}{\varepsilon_0}$ out wards 2) $\frac{1}{\varepsilon_0}$ inwards 3) $\frac{1}{4\pi\varepsilon_0}$ outwards 4) $\frac{1}{4\pi\varepsilon_0}$ inwards

LEVEL- II (H. W) KEY

1) 4 2) 4 3) 1 4) 2 5) 4 6) 4 7) 1

LEVEL- II (H. W) HINTS

1. No of lines $= \frac{q}{\varepsilon_0} = \frac{1}{8.85\times10^{-12}} = 1.129\times10^{11}$
2. In ward equat to outward flux so net flux is zero
3. Total flux $= \frac{q}{\varepsilon_0}$
4. $E = \frac{\sigma}{2\varepsilon_0}$
5. $E = \frac{8.85}{8.85\times10^{-12}} = 10^{12}$
6. Net flux $= -4\times10^3$

   $-4\times10^3 = \frac{q}{\varepsilon_0}$

   $q = -4\times10^3\,\varepsilon_0$
7. Due to -ve charge flux is inward

LEVEL - III

ELECTRIC FLUX AND GAUSS LAW

1. A solid metallic sphere has a charge +3Q. Concentric with this sphere is a conducting spherical shell having charge +Q. The radius of the sphere is a and that of the spherical shell is b, (b>a). What is the electric field at a distance R (a<R<b) from the centre.

1) $\frac{Q}{2\pi\varepsilon_0 R}$ 2) $\frac{3Q}{2\pi\varepsilon_0 R}$ 3) $\frac{3Q}{4\pi\varepsilon_0 R^2}$ 4) $\frac{4Q}{2\pi\varepsilon_0 R^2}$

2. Two parallel plane sheets 1 and 2 carry uniform charge densities σ_1 *and* σ_2 as shown in fig. electric field in the region marked II is $(\sigma_1 > \sigma_2)$

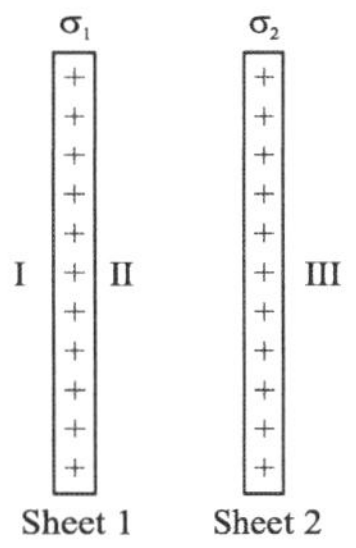

1) $-\frac{(\sigma_1+\sigma_2)}{2\varepsilon_0}$ 2) $\frac{-(\sigma_1\sigma_2)}{2\varepsilon_0}$

3) $\frac{(\sigma_1+\sigma_2)}{2\varepsilon_0}$ 4) $\frac{(\sigma_1-\sigma_2)}{2\varepsilon_0}$

3. Electric charge is uniformly distributed along a long straight wire of radius 1 mm. The charge per cm length of the wires is Q coulomb. Another cylindrical surface of radius 50 cm and length 1 m symmetrically encloses the wire as shown in the figure. The total electric flux passing through the cylindrical surface is

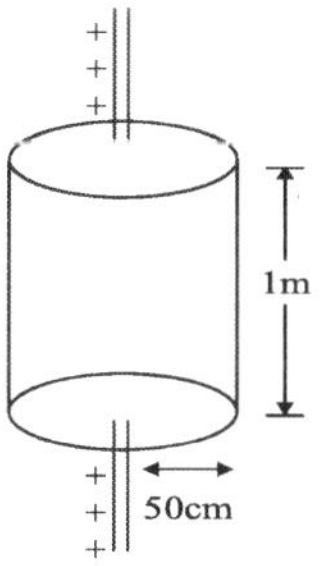

1) $\frac{Q}{\varepsilon_0}$ 2) $\frac{100Q}{\varepsilon_0}$

3) $\frac{10Q}{(\pi\varepsilon_0)}$ 4) $\frac{100Q}{(\pi\varepsilon_0)}$

4. A charge Q is distributed uniformly on a ring of radius r. A sphere of equal radius r is constructed with its centre at the periphery of the ring as shown in figure. Find the flux of the electric field through the surface of the sphere.

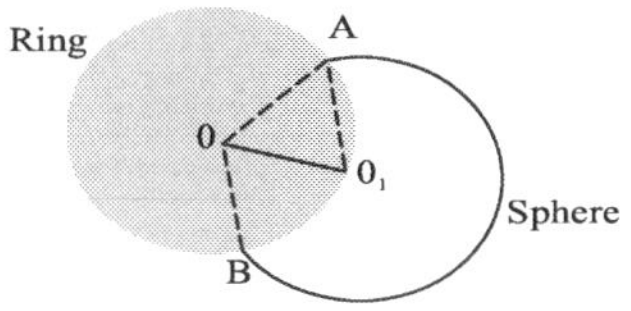

1) $\frac{Q}{3\varepsilon_0}$ 2) $\frac{q}{\varepsilon_0}$

3) $\frac{q}{2\varepsilon_0}$ 4) zero

5. q_1, q_2, q_3 *and* q_4 are point charges located at points as shown in the figure as S is a spherical Gaussian surface of radius R. Which of the following is true according to the Gauss's law

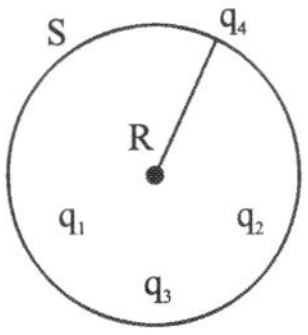

1) $\oint(\vec{E}_1+\vec{E}_2+\vec{E}_3).d\vec{A}=\frac{q_1+q_2+q_3}{2\varepsilon_0}$

2) $\oint(\vec{E}_1+\vec{E}_2+\vec{E}_3+\vec{E}_4).d\vec{A}=\frac{(q_1+q_2+q_3)}{\varepsilon_0}$

3) $\oint(\vec{E}_1+\vec{E}_2+\vec{E}_3+\vec{E}_4).d\vec{A}=\frac{(q_1+q_2+q_3+q_4)}{\varepsilon_0}$

4) None of the above

6. **Shown below is a distribution of charges. The flux of electric field due to these charges through the surface S is**

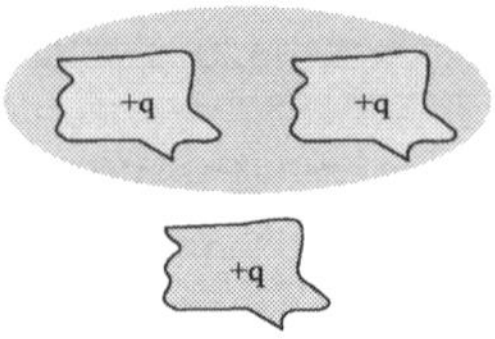

1) $3q/\varepsilon_0$ 2) $2q/\varepsilon_0$

3) q/ε_0 4) Zero

7. **A thin spherical conducting shell of radius R has a charge q. Another charge Q is placed at the centre of the shell. The electrostatic potential at a point P at a distance R/2 from the centre of the shell is**

1) $\frac{2Q}{4\pi\varepsilon_0 R}$ 2) $\frac{2Q}{4\pi\varepsilon_0 R}-\frac{2q}{4\pi\varepsilon_0 R}$ 3) $\frac{2Q}{4\pi\varepsilon_0 R}+\frac{q}{4\pi\varepsilon_0 R}$ 4) $\frac{(q+Q)}{4\pi\varepsilon_0}\frac{2}{R}$

8. **A charge $'q'$ is distributed over two concentric hollow conducting spheres of radii a and b (b > a) such that their surface charge densites are equal. The potential at their common centre is**

1) Zero   2) $\frac{q}{4\pi\epsilon_0}\frac{(a+b)}{\left(a^2+b^2\right)^2}$ 3) $\frac{q}{4\pi\epsilon_0}\left[\frac{1}{a}+\frac{1}{b}\right]$ 4) $\frac{q}{4\pi\epsilon_0}\left[\frac{a+b}{\left(a^2+b^2\right)}\right]$

9. **Two concentric sphere of radii a_1 and a_2 carry charges q_1 and q_2 respectively. If the surface charge density (σ) is same for both spheres, the electric potential at the common centre will be**

1) $\frac{\sigma}{\epsilon_0}\frac{a_1}{a_2}$ 2) $\frac{\sigma}{\epsilon_0}\frac{a_2}{a_1}$ 3) $\frac{\sigma}{\epsilon_0}(a_1-a_2)$ 4) $\frac{\sigma}{\epsilon_0}(a_1+a_2)$

10. **Assume three concentric conducting spheres where charge q_1 and q_2 have been placed on inner and outer sphere where as middle sphere has been earthed. Find the charge on the outer surface of middle spherical conductor**

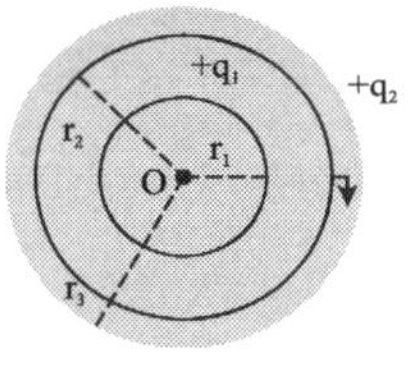

1) $-\frac{r_2}{r_3}q_2$ 2) $-q_1$

3) $-q_2$ 4) $\frac{r_2}{r_3}q_1$

11. **Three concentric metallic spheres A, B and C have radii a,b and c (a < b < c) and surface charge densities on them are $\sigma, -\sigma$ and σ respectively. The values of V_A and V_B will be**

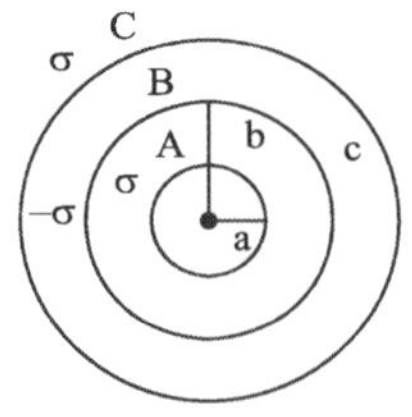

1) $\frac{\sigma}{\epsilon_0}(a-b+c), \frac{\sigma}{\epsilon_0}\left(\frac{a^2}{b}-b+c\right)$

2) $(a-b+c), \frac{a^2}{c}$

3) $\frac{\epsilon_0}{\sigma}(a-b+c), \frac{\epsilon_0}{\sigma}\left(\frac{a^2}{b}-b+c\right)$ 4) $\frac{\sigma}{\epsilon_0}\left(\frac{a^2}{c}-\frac{b^2}{c}+c\right), \frac{\sigma}{\epsilon_0}(a-b+c)$

12. A charged ball hangs from silk thread which makes an angle 'θ' with large charged conducting sheet 'P' as shown. The surface charge density (σ) of the sheet is proportional to

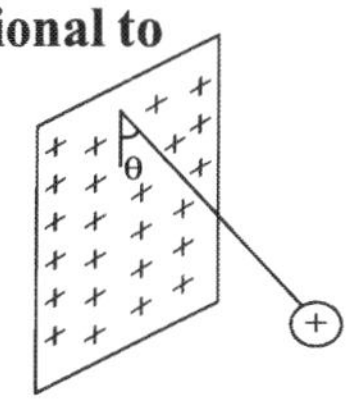

1) $\cos\theta$ 2) $\cot\theta$

3) $\sin\theta$ 4) $\tan\theta$

COMPREHENSION

The electric field in a region is given by $\overline{E}=(\alpha x)\bar{i}$. Here is α is a constant of proper dimensions.

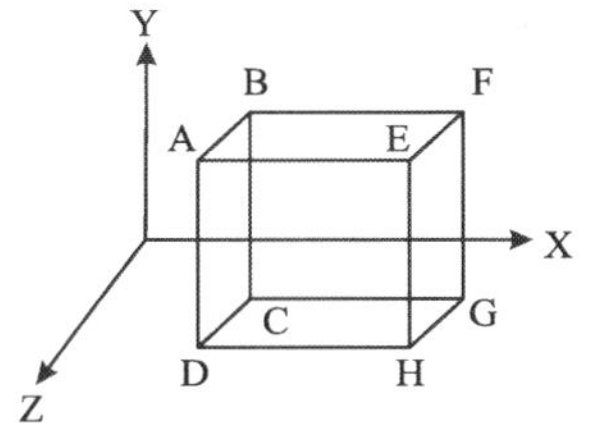

13. Find the total flux passing through a cube bounded by surfaces $x=l, x=2l, y=0, y=l, z=0, z=l.$

1) αl^3 2) $2\alpha l^3$ 3) $3\alpha l^3$ 4) $4\alpha l^3$

14. The charge contained inside the above cube is

1) $2\alpha \in_0 l^3$ 2) $\alpha \in_0 l^3$ 3) $4\alpha \in_0 l^3$ 4) $3\alpha \in_0 l^3$

15. Two point charges q *and* $-q$ are seperated by a distance 2a. Find the flux of the electric field vector acrossthe circle of radius r is shown.

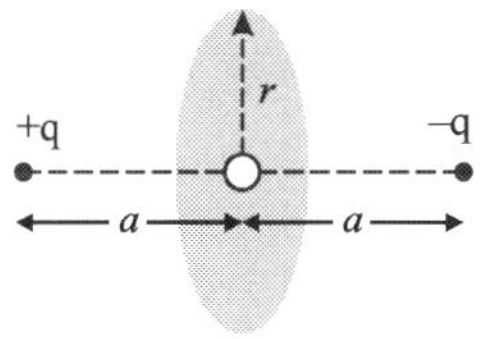

1) $\frac{q}{2\in_0}\left\{1-\frac{a}{\sqrt{a^2+r^2}}\right\}$ 2) $\frac{q}{\in_0}\left\{1-\frac{a}{\sqrt{a^2+r^2}}\right\}$

3) $\frac{2q}{\in_0}\left\{1-\frac{a}{\sqrt{a^2+r^2}}\right\}$ 4) Zero

16. A long string with a charge of λ per unit length passes through an imaginary cube of edge *a*. The maximum flux of the electric field through the cube will be

1) $\lambda a/\in_0$ 2) $\sqrt{2}\lambda a/\in_0$ 3) $6\lambda a^2/\in_0$ 4) $\sqrt{3}\lambda a/\in_0$

17. A rod with linear charge density λ is bent in the shape of circular ring. The electric potential at the centre of the circular ring is

1) $\frac{\lambda}{4\varepsilon_0}$ 2) $\frac{\lambda}{2\varepsilon_0}$ 3) $\frac{\lambda}{\varepsilon_0}$ 4) $\frac{2\lambda}{\varepsilon_0}$

18. **The electric field components in the figure are $E_x = \alpha x^{1/2}, E_y = 0, E_z = 0$ where α = 800 N / m². If a = 0.1 m is the side of cube then the charge within the cube is**

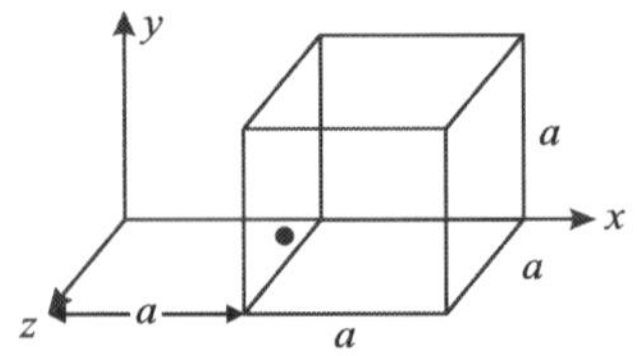

1) $9.27 \times 10^{-12} C$
2) $6 \times 10^{-12} C$
3) $2.5 \times 10^{-12} C$
4) Zero

19. **Three very large plates are given charges as shown in the figure. If the cross-sectional area of each plate is the same, the final charge distribution on plate C is**

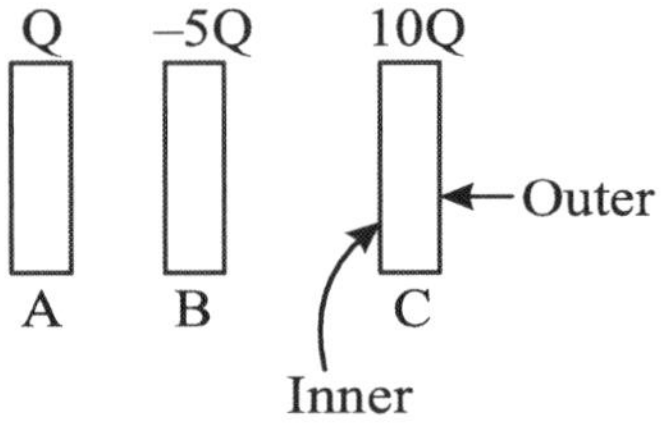

a) +5Q on the inner surface , +5Q on the outer surface
b) +6Q on the inner surface, +4Q on the outer surface
c) +7Q on the inner surface, +3Q on the outer surface
d) +8Q on the inner surface, +2Q on the outer surface

20. **An electric dipole of dipole moment P is kept at a distance r from an infinite long charged wire of linear charge density λ as shown. The force acting on the dipole is**

1) $\frac{P\lambda}{2\pi\varepsilon_0 r^2}$
2) $\frac{P\lambda}{\pi\varepsilon_0 r^2}$
3) $\frac{2P\lambda}{\pi\varepsilon_0 r^2}$
4) $\frac{P\lambda}{4\pi\varepsilon_0 r^2}$

21. **A point charge q is a distance r from the centre O of an uncharged spherical conducting layer, whose inner and outer radii equal to a and b respectively. The potential at the point**

O if r < a is $\frac{q}{4\pi \epsilon_0}$ times

1) $\left(\frac{1}{r} - \frac{1}{a} + \frac{1}{b}\right)$
2) $\left(\frac{1}{a} - \frac{1}{r} + \frac{1}{b}\right)$
3) $\left(\frac{1}{b} - \frac{1}{c} - \frac{1}{r}\right)$
4) $\left(\frac{1}{a} - \frac{1}{b} - \frac{1}{r}\right)$

22. **One-fourth of a sphere of radius R is removed as shown in fig. An electric field E exists parallel to x-y plane. Find the flux through the remaining curved part.**

1) $\pi R^2 E$
2) $\sqrt{2}\pi R^2 E$
3) $\pi R^2 E / \sqrt{2}$
4) $2\pi R^2 E$

LEVEL- III - KEY

1) 3	2) 4	3) 2	4) 1	5) 2	6) 2	7) 3	8) 4	9) 4	10) 1
11) 1	12) 4	13) 1	14) 2	15) 4	16) 4	17) 2	18) 1	19)3	20) 1
21) 3	22) 3								

LEVEL- III - HINTS

2. $E_{net} = E_1 - E_2 = \frac{\sigma_1}{2\in_0} - \frac{\sigma_2}{2\in_0}$

$E_{net} = \frac{\sigma_1 - \sigma_2}{2\in_0}$

3. The total flux passing through cylindrical surface

4. From the geometry of the figure. $OA = OO_1$ and $O_1A = O_1O$. Thus, OAO_1 is equilateral triangle.

Hence $\angle AOO_1 = 60^0$ or $\angle AOB = 120^0$.

The are AO_1B of the ring subtends an angle 120^0 at the centre O. Thus, third of the ring is inside the sphere.

The charge enclosed by the sphere $= \frac{Q}{3}$. From Gauss's law, the flux of the electric field through the surface of the sphere is $\frac{Q}{3\varepsilon_0}$.

5. $\oint(\bar{E}_1 + \bar{E}_2 + \bar{E}_3 + \bar{E}_4).\overline{ds} = \frac{q}{\in_0} = \frac{q_1 + q_2 + q_3}{3}$

6. The $\oint ds = \frac{q+q}{\in_1} = \frac{2q}{\in_0} = \frac{Q/cm}{\in_0} = \frac{100Q}{\in_0}$

8. $\sigma = \frac{q_1}{4\pi a^2} = \frac{q_2}{4\pi b^2}$

$\frac{q_1}{q_2} = \frac{a^2}{b^2}, q_1 + q_2 = q$

$q_1 + q_1\left[\frac{b^2}{a^2}\right] = q$

$q_1\left[1 + \frac{b^2}{a^2}\right] = q$; $q_1\left[\frac{a^2 + b^2}{a^2}\right] = q$

$q_1 = \frac{qa^2}{a^2 + b^2}$; $q_2 = \frac{qb^2}{a^2 + b^2}$

potential at commoncentre

$V = \frac{qa}{4\pi\in_0(a^2 + b^2)} \times \frac{qb}{4\pi\in_0(a^2 + b^2)}$

$V = \frac{q}{4\pi\in_0}\left[\frac{a}{a^2 + b^2} + \frac{b}{a^2 + b^2}\right]$

$V = \frac{q}{4\pi\in_0}\left[\frac{a+b}{a^2 + b^2}\right]$

10. $\frac{q_1}{r_2}+\frac{q-q_1}{r_2}+\frac{q_2}{r_3}=0$

$\therefore\ q=\frac{-r_2}{r_3}q_2$

11. $V_A=\frac{1}{4\pi\in_0}\left(\frac{4\pi a^2\sigma}{a}-\frac{4\pi b^2\sigma}{b}+\frac{4\pi c^2\sigma}{c}\right)$

$\Rightarrow V_A=\frac{\sigma}{\in_0}(a-b+c)$ and

$\Rightarrow V_B=\frac{1}{4\pi\in_0}\left(\frac{4\pi a^2\sigma}{b}-\frac{4\pi b^2\sigma}{b}+\frac{4\pi c^2\sigma}{c}\right)$

$\Rightarrow V_B=\frac{\sigma}{\in_0}\left(\frac{a^2}{b}-b+c\right)$

17. $\sigma=\frac{Q}{A}$

18. Magnitude of E at the left face $E_L=\alpha a^{1/2}$ at right face $E_R=\alpha\left(2a\right)^{1/2}$

$\phi=\left(E_R-E_L\right)a^2$ and $q=\phi\in_0$

22. $\phi=\vec{E}.\left(\vec{A_1}+\vec{A_2}\right)$

Chapter - 3

CAPACITORS

CAPACITORS

SYNOPSIS

Electric Capacity:

➢ The ratio of charge to potential of a conductor is called its capacity. $C = \frac{Q}{V}$

Unit : farad (F)

Parallel Plate Capacitor:

If two plates each of area A are seperated by a distance 'd' then its capacity $C = \frac{\varepsilon_0 A}{d}$ (air as medium),

$C = \frac{k\varepsilon_0 A}{d}$ (dielectric medium)

When a dielectric medium is introduced between the plates of a parallel plate capacitor, its capacity increases to 'k' times the original capacity.

➢ When a dielectric slab of thickness 't' is introduced between the plates of a parallel plate capacitor,

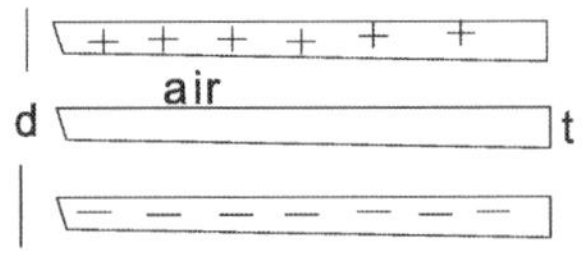

$$new\ capacity = \frac{\varepsilon_0 A}{d - t\left(1 - \frac{1}{k}\right)} = \frac{\epsilon_0\ A}{(d-t) + \frac{t}{k}}$$

GAUSS METHOD

Let us consider a case of parallel plate capacitor in which a medium of dielectric constant K is partially filled as shown in figure.

Then the field is uniform in air as well as in medium but they will have different values. let 't' be the thickness of the medium whose relative permittivity is K. The remaining space of (d - t) thickness be occupied by air.

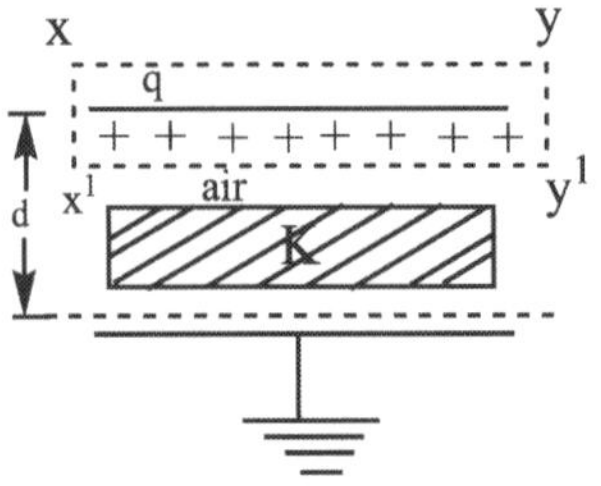

Fig. (b)

Imagine a Gaussian surface enclosing the plate as shown.

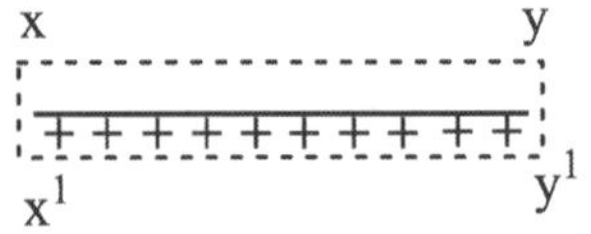

Fig. (c)

If E_0 is the field in air, then from Gauss law

$\int E_0 ds = \frac{q}{\varepsilon_0} \Rightarrow E_0 A = \frac{q}{\varepsilon_0}$ or

$E_0 = \frac{q}{A\varepsilon_0}$ .... (a)

Similarly by considering a Gaussian surface through the medium, then by Gauss law,

$\int E.ds = \frac{q}{\varepsilon_0 K} \Rightarrow EA = \frac{q}{\varepsilon_0 K}$

where E is a field in the medium

$\therefore E = \frac{q}{A\varepsilon_0 K}$ .... (b)

The P.D. between the two plates of the capacitor.

$V = E_0(d-t) + E.t$

$V = \frac{q}{A\varepsilon_0}(d-t) + \frac{q}{A\varepsilon_0 K}t$

$= \frac{q}{A\varepsilon_0}\left[(d-t) + \frac{t}{K}\right]$

or $C = \frac{q}{V} = \frac{q}{\frac{q}{A\varepsilon_0}[d-t+t/K]}$

$C = \frac{A\varepsilon_0}{\left[(d-t)+\left(\frac{t}{K}\right)\right]}$

- When a metal slab of thickness 't' is introduced between the plates of a parallel plate capacitor, new capacity $= \frac{\varepsilon_o A}{d-t}$.

  (for metal $k = \infty$)

- The method for the calculation of capacitance requires integration of the electric field between two conductors or the plates which are separated with a potential difference V_{ab}

  i.e. $V_{ab} = -\int_b^a \bar{E}.d\bar{r}$

  or $V_+ - V_- = -\int \bar{E}.d\bar{r}$ from this $C = \frac{q}{V_{ab}}$

- When a thin metal sheet $(t \approx 0)$ is introduced between the plates of a parallel plate capacitor, then capacity remains unchanged.
- A dielectric slab of thickness 't' is introduced between the plates, to restore the original capacity, if the distance between the plates is

increased by x, then $x = t\left(1 - \frac{1}{k}\right)$.

- Two dielectric slabs of equal thickness are introduced between the plates of a capacitor as shown in figure, then new capacity

$$= \frac{C}{2}(K_1 + K_2).$$

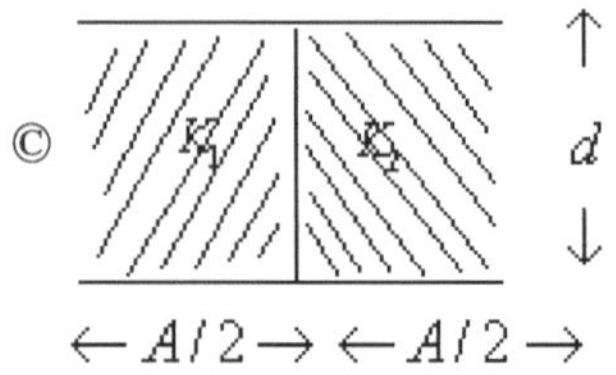

If the two dielectrics are of different face areas A_1 and A_2 but of same thickness, then capacity,

$$C = \frac{\epsilon_0}{d}(K_1 A_1 + K_2 A_2)$$

- If two dielectric slabs of constants k_1 and k_2 are introduced as shown in figure, new capacity $= \frac{2k_1 k_2}{(k_1 + k_2)}.C$

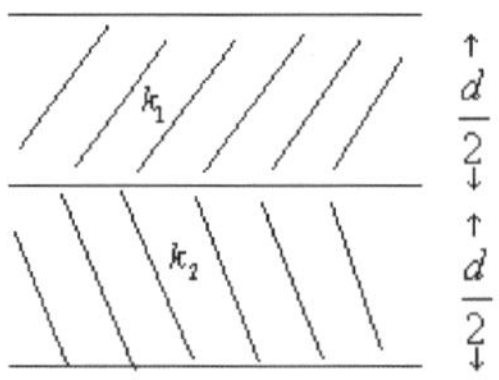

- If number of dielectric slabs of same cross sectional area 'A' and of thicknesses $t_1, t_2, t_3, ...t_n$ and constants $k_1, k_2.......k_n$ are introduced between the plates, effective capacity

$$C = \frac{\varepsilon_0 A}{d - (t_1 + t_2 + ...t_{3n}) + \left(\frac{t_1}{k_1} + ... + \frac{t_n}{k_n}\right)}$$

- In the above case if the dielectric media are completely filled between the plates, effective capacity

$$C = \frac{\varepsilon_0 A}{\left(\frac{t_1}{k_1} + ... \frac{t_n}{k_n}\right)}$$

- The capacity of a parallel plate capacitor is independent of the charge on it, potential difference between the plates and the nature of plate material.
- In a capacitor, the energy is stored in the electric field between the two plates.
- Capacity of a spherical conductor $= 4\pi \epsilon_0 r$, where r is the radius of the sphere.
- If we imagine earth to be a uniform solid sphere then capacity of earth is $4\pi \epsilon_0 R$
- Where R = Radius of the earth $= 6400 \times 10^3 m$

Note : For the earth, $R = 6.4 \times 10^6$ m

The capacity of earth is

$$C = 4\pi \epsilon_0 R = \frac{1}{9\times10^9} \times 6.4\times10^6 = 711\mu F$$

WE.1 A metal slab of thickness, equal to half the distance between the plates is introduced between the plates of a parallel plate capacitor as shown. Find its capacity.

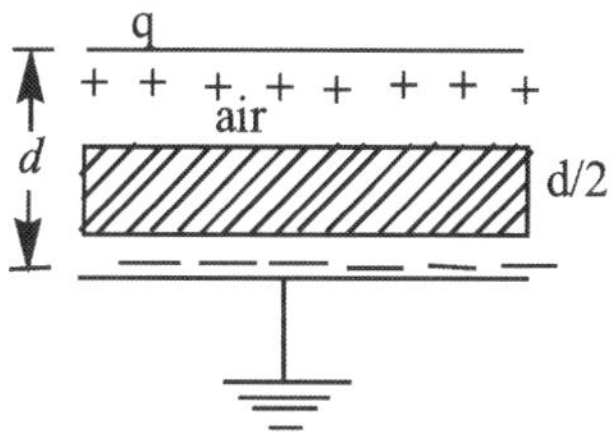

Sol. When capacitor is partially filled with dielectric capacity $C = \frac{\varepsilon_0 A}{\left[d - t(1-\frac{1}{k})\right]}$

For metal slab of thickness t = d/2,

$$C = \frac{\epsilon_0 A}{d - t} \quad (K = \infty \text{ for metal slab})$$

$$= \frac{\epsilon_0 A}{d - \frac{d}{2}} = 2\frac{\epsilon_0 A}{d}.$$

WE.2 Two conductors carrying equal and opposite charges produce a non uniform electric field along X - axis given by $E = \frac{Q}{\epsilon_0 A}(1 + Bx^2)$ where A and B are constants. Separation between the conductors along X-axis is X. Find the capacitance of the capacitor formed.

Sol. Potential difference between the conductors is given by $V = V_+ - V_- = \int_0^X E dx$

$$\Rightarrow V = \int_0^X \frac{Q}{\epsilon_0 A}(1 + Bx^2)dx$$

or $V = \frac{Q}{\epsilon_0 A}\left(x + \frac{Bx^3}{3}\right)_0^X = \frac{Q}{\epsilon_0 A}\left(X + \frac{BX^3}{3}\right)$

Capacity $C = \frac{Q}{V} = \frac{\epsilon_0 A}{X\left(1 + \frac{BX^2}{3}\right)}$

WE.3 Find the capacitance of a system of two identical metal balls of radius a if the distance between their centres is equal to b, with b>>a. The system is located in a uniform dielectric with permittivity K.

Sol. Let q and -q be the charges on two balls. Then

$V_1 = V_{ball} - V_\infty = V$

$V_2 = V_{ball} - V_\infty = -V$

The potential difference between the balls

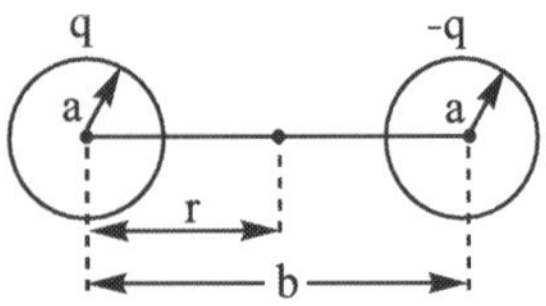

$V_1 - V_2 = 2V$

$$= 2\int_a^{b-a} E\,dr$$

$$= 2\int_a^{b-a}\left(\frac{1}{4\pi \epsilon_0 K}\frac{q}{r^2}\right)dr = \frac{2q}{4\pi \epsilon_0 K}\left[\frac{1}{a} - \frac{1}{b-a}\right]$$

$$C = \frac{q}{V_1 - V_2} = \frac{q}{\left[\frac{2q}{4\pi \epsilon_0 K}\frac{(b-2a)}{a(b-a)}\right]}$$

$$= \frac{2\pi \epsilon_0 K a(b-a)}{(b-2a)}$$

For b >> a, we can write $C = 2\pi \epsilon_0 K a$.

WE.4 Capacitor has square plates each of side 'l' making an angle 'α' with each other as shown. Then for small value of α, the capacitance 'C' is given by

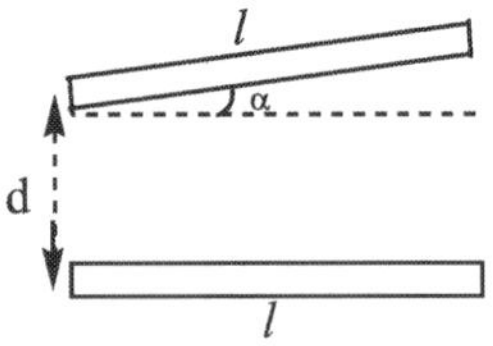

Sol. At one side, distance between plates d,

At another side,

distance $= d + l\sin\alpha \simeq d + l\alpha$

Mean distance between the plates

$$= \frac{d + (d + l\alpha)}{2} = d + \frac{l\alpha}{2}$$

Capacity $C = \frac{\epsilon_0 A}{d} = \frac{\epsilon_0 l^2}{d + \frac{l\alpha}{2}}$

$$= \frac{\epsilon_0 l^2}{d}\left[1 + \frac{l\alpha}{2d}\right]^{-1} = \frac{\epsilon_0 l^2}{d}\left[1 - \frac{l\alpha}{2d}\right]$$

Spherical condenser

$$V = V_p - V_q = \frac{1}{4\pi\varepsilon_0}\left(\frac{q}{a} + \frac{-q}{b}\right) - 0$$

$$= \frac{1}{4\pi\varepsilon_0} q\left(\frac{b-a}{ab}\right)$$

$$C = \frac{q}{V}$$

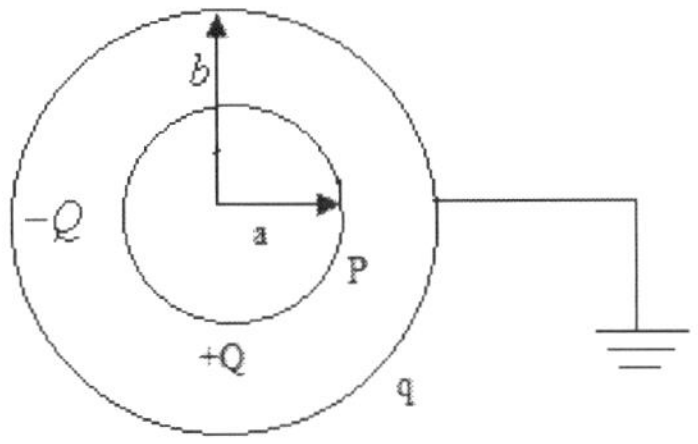

(a) $C = 4\pi \in_0 \dfrac{ab}{b-a}$, if inner sphere is charged and outer sphere is earthed.

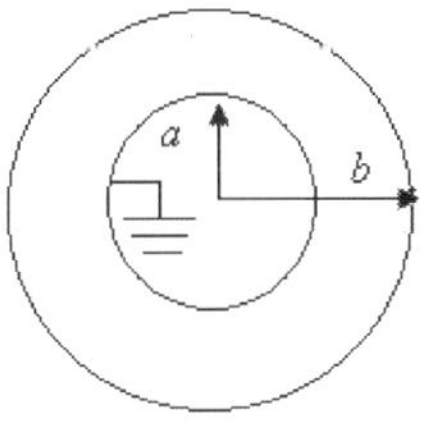

(b) $C = 4\pi \in_0 \dfrac{b^2}{b-a}$,

If inner sphere is earthed and outer sphere is charged.

Cylindrical Capacitor:

A cylinjdrical capacitor consists of two coaxial cylinders and its capacitance is given by

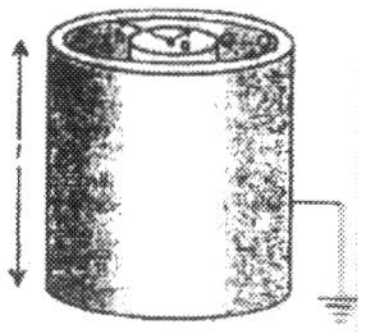

$$c = \frac{2\pi\varepsilon_0 l}{\log_e\left(\frac{b}{a}\right)}$$

Where l is the length of each of cylinder and a and b are the radii of the inner and outer cylinders.

FORCE BETWEEN THE PLATES OF A CAPACITOR

Consider a parallel plate capacitor with plate area A. Let Q and –Q be the charges on the plates of capacitor. Let F be the force of attraction between the plates. Let E be the field between the capacitor plates. The expression for the force can be derived by energy method. Let the distance between the plates be x.

So electric field energy between the plates is

$$U = \frac{1}{2} \in_0 E^2 (Ax)$$

$$\frac{dU}{dx} = \frac{1}{2} \in_0 E^2 A$$

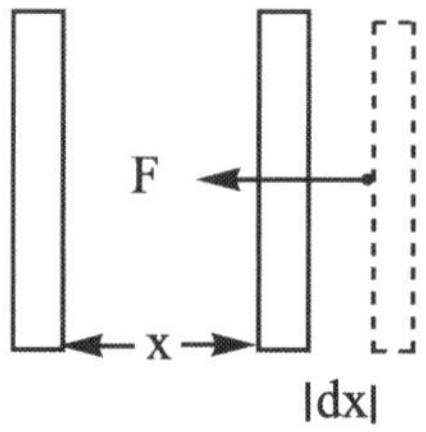

Fig.

By definition $F = -\frac{dU}{dx} = \frac{-1}{2} \in_0 E^2 A$

(Conservative force)

So the force of attraction between the plates is $F = \frac{1}{2} \in_0 E^2 A$

Note : For an isolated charged capacitor $F = \frac{Q^2}{2 \in_0 A}$. This force does not depend on the separation between the plates, and so the constant amount of force is needed to change the separation.

Note : For a capacitor having constant potential difference across the plates the force

$$F = \frac{C^2 V^2}{2 \in_0 A} = \frac{\in_0^2 A^2}{d^2} \frac{V^2}{2 \in_0 A}$$

$$F = \frac{1}{2} \in_0 \frac{V^2}{d^2} A$$

In this case force depends on the separation between the plates. Thus to change the separation variable force is needed.

CAPACITORS IN SERIES

In series combination, the capacitors are first arranged in a series order such that the second plate of first capacitor is connected to the first plate of second capacitor, the second plate of second capacitor is connected to first plate of third capacitor and so on. And finally the first plate of first capacitor and second plate of last capacitor are connected to opposite terminals of battery.

Let us consider three capacitors of capacities C_1, C_2 and C_3 connected in series across a source of potential difference 'V' as shown in figure.

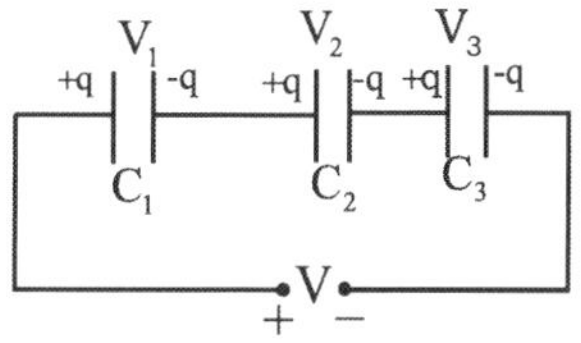

Fig.

At the moment, the system is connected to the source, left plate of first condenser acquires positive charge due to conduction. This inturn will produce negative charge of equal magnitude, on the left face of second plate of first condenser due to induction. The process continues for the remaining two condensers. Hence the charge acquired by all the three capacitors will be same.

As the capacitors are different, the potentials developed across them will be different.

$$q = C_1V_1 = C_2V_2 = C_3V_3$$

$$V_1 = \frac{q}{C_1}, V_2 = \frac{q}{C_2}, V_3 = \frac{q}{C_3}$$

But $V = V_1 + V_2 + V_3$

$$V = q\left[\frac{1}{C_1} + \frac{1}{C_2} + \frac{1}{C_3}\right] \quad .... (1)$$

If a single capacitor when connected across the same source draws the same charge, that capacitance is said to be the equivalent capacitance of the three capacitors. If C_S is the equivalent capacitance.

$$C_S = \frac{q}{V}$$

$$V = \frac{q}{C_S} \qquad \text{-----(2)}$$

Substituting (2) in (1)

$$\frac{q}{C_S} = \frac{q}{C_1} + \frac{q}{C_2} + \frac{q}{C_3}$$

$$\frac{1}{C_s} = \frac{1}{C_1} + \frac{1}{C_2} + \frac{1}{C_3}$$

In general $\frac{1}{C_S} = \sum \frac{1}{C_n}$

- The resultant capacity of series combination is smaller than the least capacity of the capacitors of the combination.
- In series, ratio of charges on three capacitors is 1 : 1 :1.
- The ratio of potential differences across three capacitors is

$$V_1 : V_2 : V_3 = \frac{Q}{C_1} : \frac{Q}{C_2} : \frac{Q}{C_3} = \frac{1}{C_1} : \frac{1}{C_2} : \frac{1}{C_3}$$

- P.D across first capacitor is $$V_1 = \frac{\frac{1}{C_1}}{\left(\frac{1}{C_1} + \frac{1}{C_2} + \frac{1}{C_3}\right)} V$$

similary we can find V_2 and V_3.

WE.5: The equivalent capacity between A and B in the given circuit is

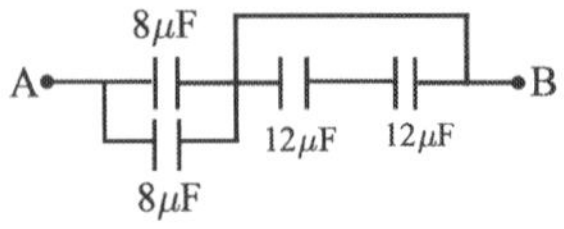

Sol. Here 12 μF and 12 μF are short circuited. Hence they are not charged.

∴ Take only 8 μF and 8 μF parallel combination.

$C = 8 + 8 = 16\,\mu F$

WE.6 : When the space between the plates of a parallel plate condenser is completely filled with two slabs of dielectric constants K_1 and K_2 and each slab having area A and thickness equal to $\frac{d}{2}$ as shown in the figure

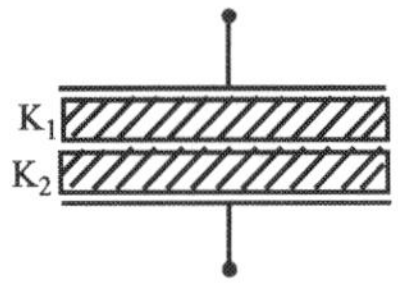

Fig.

The equivalent circuit is as shown

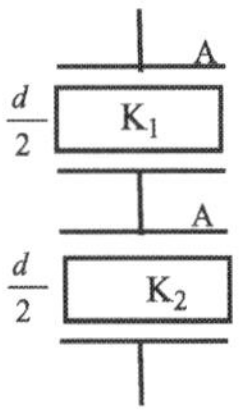

a) Capacity of the upper half $C_1 = \frac{2K_1 \epsilon_o A}{d}$

b) Capacity of the lower half $C_2 = \frac{2K_2 \epsilon_o A}{d}$

c) C_1 and C_2 may be supposed to be conencted in series.

d) Effective capacity

$$C = \frac{C_1C_2}{C_1 + C_2} = \frac{\varepsilon_0 A}{d}\left(\frac{2K_1K_2}{K_1 + K_2}\right) = C_0\left(\frac{2K_1K_2}{K_1 + K_2}\right)$$

Here C_0 is the capacity of the condenser with air medium.

e) Effective dielectric constant $K = \left(\frac{2K_1K_2}{K_1 + K_2}\right)$

CAPACITORS IN PARALLEL

Capacitors are said to be connected in parallel if the two plates of any capacitor are connected one to positive terminal and the other to negative terminal of the source, then the connection is said to be parallel connection.

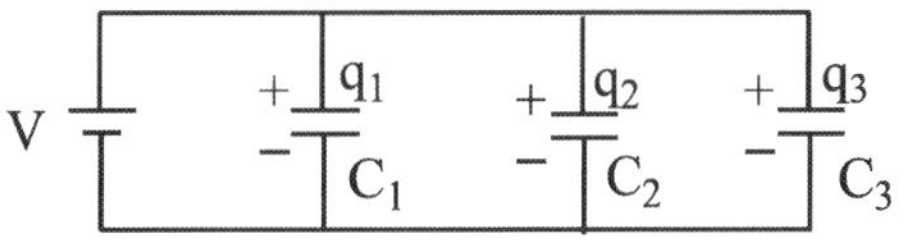

Fig.

Let us consider three capacitors of capacities C_1, C_2 and C_3 connected in parallel across a source 'V' as shown

The moment capacitors are connected, charge is drawn from the voltage source and this charge is drawn along three branches and thus gets shared. As all capacitors are connected in parallel, the potential across any of the capacitors is same . Here charge gets shared depending upon their capacitances for maintaining same potential.

$$V = \frac{q_1}{C_1} = \frac{q_2}{C_2} = \frac{q_3}{C_3}$$

$$\therefore q_1 + q_2 + q_3 = C_1V + C_2V + C_3V$$

$$q = V(C_1 + C_2 + C_3)$$

$$\frac{q}{V} = C_1 + C_2 + C_3 \qquad ... (1)$$

If a single capacitor when connected to the same source draws a charge q then that capacitor is said to be the effective or equivalent capacitor for the three parallel capacitors.

If the effective capactiance is C_p,

$$C_P = \frac{q}{V} \qquad ... (2)$$

from (1) and (2)

$$C_P = C_1 + C_2 + C_3$$

In general $C_P = \sum C_n$ s

- The resultant capacity of parallel combination is greater than the largest capacity of the capacitors of the combination.
- In parallel, ratio of P.D. on three capacitors is 1 : 1 :1.
- The ratio of charges on three capacitors is $Q_1 : Q_2 : Q_3 = C_1V : C_2V : C_3V = C_1 : C_2 : C_3$
- The charge on first capacitor is $Q_1 = \frac{C_1}{C_1 + C_2 + C_3} Q$

  similarly we can find Q_2 and Q_3.
- When n identical capacitors each of capacity **C** are first connected in series and next connected in parallel then the ratio of their effective capacities

$$C_s = \frac{C}{n}; C_p = nC$$

$$\frac{C_s}{C_p} = n^2 : 1$$

WE.7: In the net work three identical capacitors are connected as shown. Each of them can withstand to a maximum 100 V potential difference. What is the maximum voltage that can be applied across A and B so that no capacitor gets spoiled.

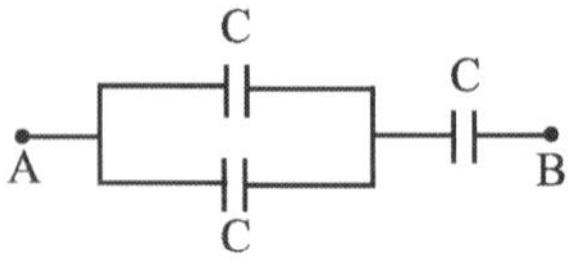

Sol. Let q_{max} be the max-charge supplied by the battery between A and B so that no capacitor gets spoiled.

For each capacitor

$$q_{max} = CV_0 = C(100) = 100C$$

For the combination $q_{max} = C_{equivalent}(V_{max})$

$$100C = \frac{2}{3}C(V_{max}) \Rightarrow V_{max} = 150\text{ V}$$

Among 150V, potential difference across parallel combination is 50V and the potential difference across the other capacitor is 100V.

WE.8 : Calculate the capacitance of a parallel plate capacitor, with plate area A and distance between the plates d, when filled with a dielectic whose permittivity varies as

$$\in(x) = \in_0 + kx\left(0 < x < \frac{d}{2}\right)$$

$$\in(x) = \in_0 + k(d-x)\left(\frac{d}{2} < x \le d\right)$$

Sol:

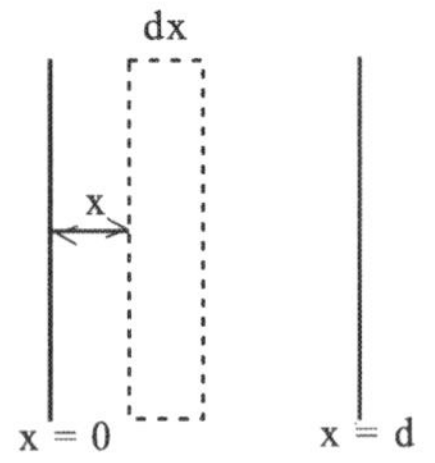

The given capacitor is equivalent to two capacitors in series. Let C_1 and C_2 be their capacities. Then

$$\frac{1}{C} = \int\left[\frac{1}{dC_1} + \frac{1}{dC_2}\right]$$

Consider an element of width dx at a distance x from the left plate. Then

$$dC_1 = \frac{(\varepsilon_0 + kx)A}{dx} \text{ for } 0 < x < \frac{d}{2}$$

$$\text{and } dC_2 = \frac{\{\in_0 + k(d-x)\}A}{dx} \text{ for } \frac{d}{2} < x \le d$$

on substituting these two values we get

$$\frac{1}{C} = \int\frac{1}{dC} = \frac{2}{KA}\ell n\left(\frac{2\in_0 + Kd}{2\in_0}\right)$$

$$\Rightarrow C = \frac{KA}{2}\ell n\left(\frac{2\in_0 + Kd}{2\in_0}\right)$$

WE.9: When the space between the plates of a parallel plate condenser is completely filled with two slabs of dielectric constants K_1 and K_2 and each slab having area $\frac{A}{2}$ and thickness equal to distance of seperation d as shown in the figure.

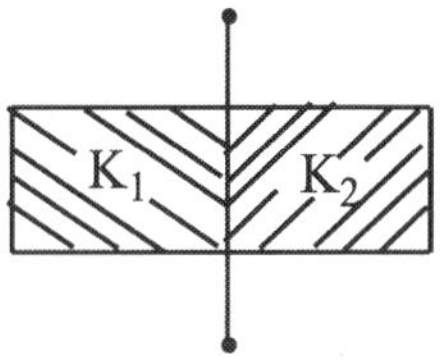

Fig.

The equivalent circuit is as shown

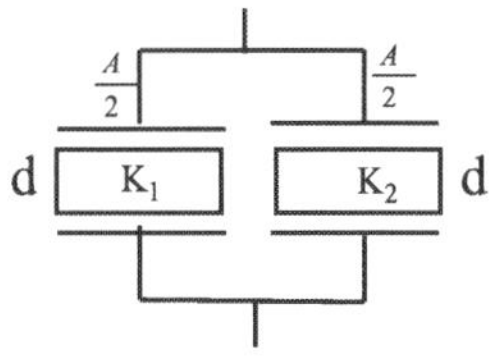

a) Capacity of the left half $C_1 = K_1 \frac{\varepsilon_0 A}{2d}$

b) Capacity of the right half $C_2 = K_2 \frac{\varepsilon_0 A}{2d}$

c) C_1 and C_2 may be supposed to be connected in parallel then effective capacity

$$C = C_1 + C_2 = \frac{\varepsilon_0 A}{d}\left(\frac{K_1 + K_2}{2}\right)$$

$$C = C_0\left(\frac{K_1 + K_2}{2}\right)$$ where C_0 is capacity of capacitor without dielectric.

d) Effective dielectric constant $K = \frac{K_1 + K_2}{2}$

WE.10: A parallel plate capacitor of area A, plate separation d and capacitance C is filled with three different dielectric materials having dielectric constants K_1, K_2 and K_3 as shown in fig. If a single dielectric material is to be used to have the same effective capacitance as the above combination then its dielectric constant K is given by :

Sol. Let $C = \frac{\epsilon_0 A}{d}$

$$C_1 = K_1 \frac{\epsilon_0 \frac{A}{2}}{\frac{d}{2}} = K_1 C$$

$$C_2 = K_2 \frac{\epsilon_0 \frac{A}{2}}{\frac{d}{2}} = K_2 C$$

$$C_3 = \frac{K_3 \epsilon_0 A}{\frac{d}{2}}, = 2K_3C$$

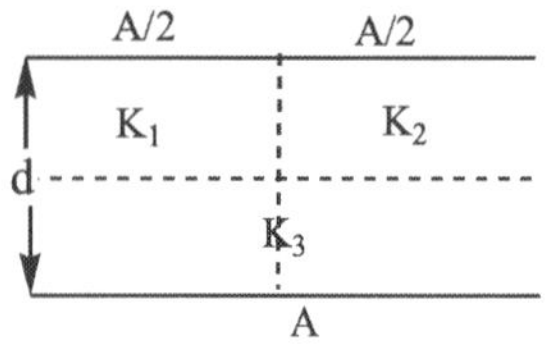

The equivalent circuit as shown

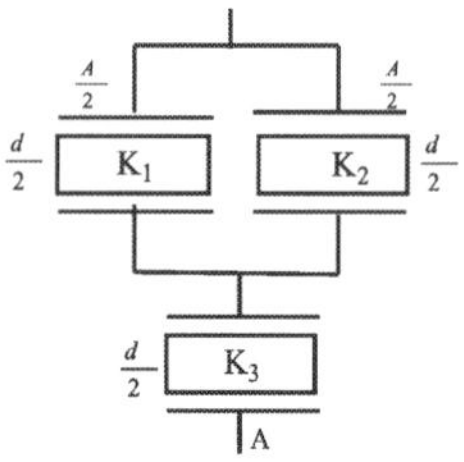

$$\frac{1}{C} = \frac{1}{C_3} + \frac{1}{C_1 + C_2},$$

$$\frac{1}{KC} = \frac{1}{2K_3C} + \frac{1}{(K_1 + K_2)C}$$

$$\frac{1}{K} = \frac{1}{K_1 + K_2} + \frac{1}{2K_3}$$

WE-11 Four identical metal plates are located in air at equal distance d from one another. The area of each plate is A. Find the equivalent capacitance of the system between X and Y.

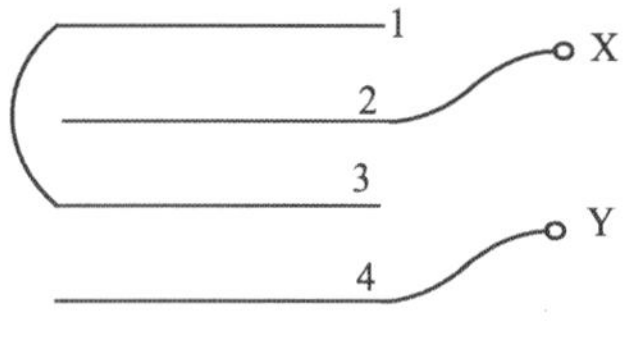

Sol. Let us give numbers to the four plates. Here X and Y are connected to the positive and negative terminals of the battery (say), then the charge distribution will be as shown

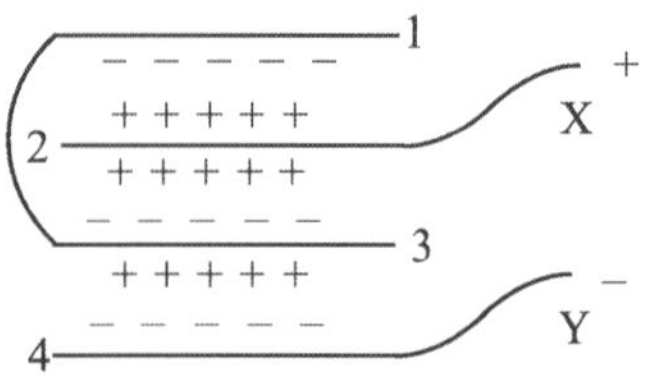

Here the arrangement can be represented as the grouping of three identical capacitors each of capacity $\frac{\epsilon_0 A}{d}$. The arrangement will be as shown

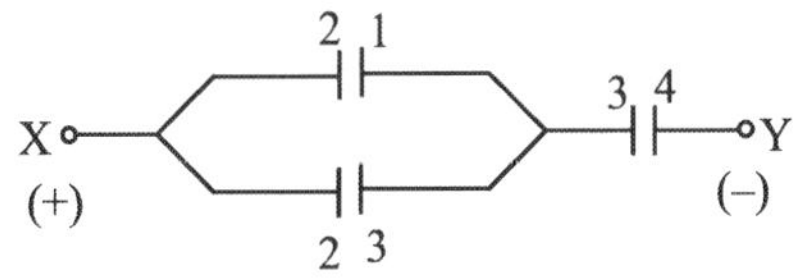

Now the equivalent capacitance between X and Y is $C_{XY} = \frac{(C+C)C}{C+C+C} = \frac{2C}{3} = \frac{2 \epsilon_0 A}{3d}$

WE.12 Find equivalent capacity between X and Y

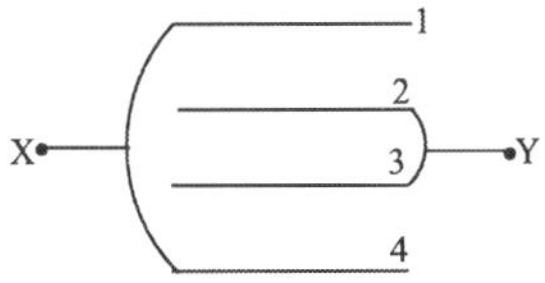

Sol. Let us give numbers to the four plates. Here X and Y are connected to the positive and negative terminals of the battery (say) ,

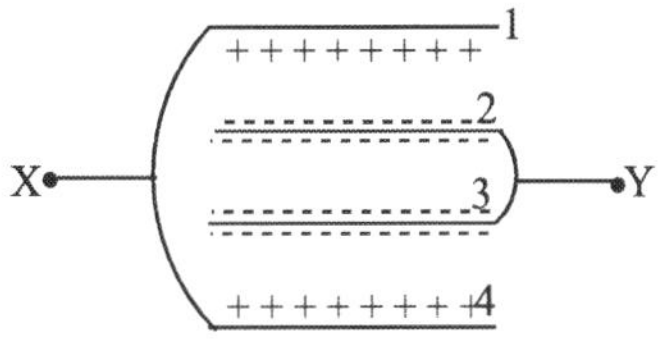

Here the arrangement can be represented as the grouping of two identical capacitors each of capacity $\frac{\epsilon_0 A}{d}$. The arrangement will be as shown

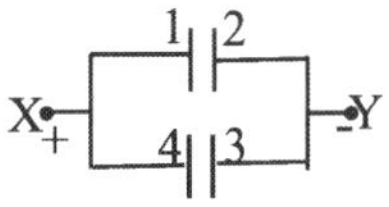

Now the equivalent capacitance between X and Y is

$C_{XY} = (C + C) = 2C = 2\frac{\varepsilon_0 A}{d}$

WE.13 Find equivalent capacity X and Y

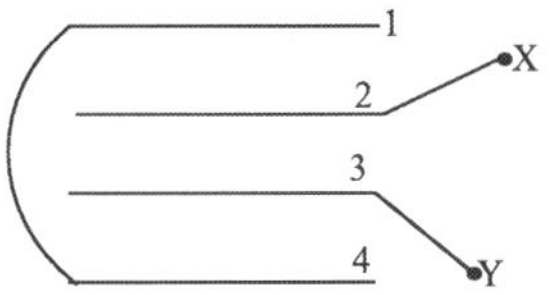

Sol. Let us give numbers to the four plates. Here X and Y are connected to the positive and negative terminals of the battery (say).

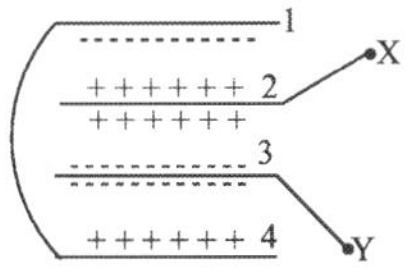

Here the arrangement can be represented as the grouping of three identical capacitors each of capacity $\frac{\epsilon_0 A}{d}$. The arrangement will be as shown

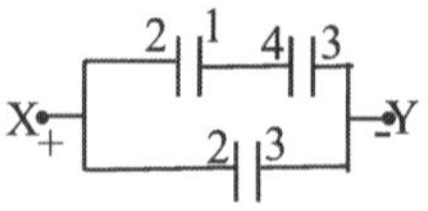

Now the equivalent capacitance between X and Y is $C_{XY} = \left(\frac{(C)(C)}{C+C}\right) + C = \frac{C}{2} + C = \frac{3}{2}C$

$$= C_{XY} = \frac{3}{2}C = \frac{3}{2}\frac{\varepsilon_0 A}{d}$$

TYPES OF DIELECTRICS :

- A dielectric is an insulating material in which electrons are tightly bound to the nuclei of the atoms. Ex. glass, mica, paper etc.
- There are two types of dielectrics
  1) Non-polar dielectrics
  2) Polar dielectrics
- In non polar dielectrics the centre of positive charge and centre of negative charge of each molecule coincide
- Under ordinary conditions Non-polar molecule will have zero dipole moment.
- When a Non-polar dielectric is subjected to electric field, the positive charge of each molecule is shifted in the direction of electric filed and negative charge in the opposite direction.
  Ex. : oxygen, nitrogen
- In polar dielectrics the centre of positive charges and centre of negative charges of each molecule do not coincide.
- Each molecule has a permanent dipole moment.
- When polar dielectric is subjected to external electric field, the electric field exerts torque on the dipoles, tending to align them in the direction of the field.
  Ex. : CO_2, NH_3,HCl, etc.
- If a dielectric is charged by induction then induced charge q^1 is less than inducing charge q. Induced charge, $$q^1 = -q\left[1 - \frac{1}{K}\right]$$
  where K is dielectric constant.
- Electric field due to induced charges on the dielectric is E_{ind} or $E_p = E_0 - \frac{E_0}{K} = E_0\left(1 - \frac{1}{k}\right)$.

Dielectric Strength of Air :

A conducting sphere cannot hold very large quantity of charge. It can hold a maximum charge Q such that the electric intensity on the surface is equal to dielectric strength of air $\left(3 \times 10^6 Vm^{-1}\right)$

i.e. $\frac{1}{4\pi \epsilon_0}\frac{Q}{R^2} = 3 \times 10^6 Vm^{-1}$

Energy stored in a condenser :

- Energy stored in a charged condenser $U = \frac{1}{2}CV^2 = \frac{1}{2}qV = \frac{q^2}{2C}$
- If a condenser is connected across a battery and U is the energy stored in the condenser then the work done by the battery in charging the condenser is $2U$ ($W = qV = 2U$)

For a parallel plate capacitor

$$U = \frac{1}{2}(Ad)\frac{\sigma^2}{\epsilon_0}\left(as\ E = \frac{\sigma}{\epsilon_0}\right)$$

Energy density

$$\frac{U}{V} = \frac{\sigma^2}{\epsilon_0} = \frac{1}{2}\epsilon_0 E^2$$ (here V is volume i.e. Ad)

- a) When three capacitors are in **series**, the ratio of energies is

$$U_1 : U_2 : U_3 = \frac{Q^2}{2C_1} : \frac{Q^2}{2C_2} : \frac{Q^2}{2C_3} = \frac{1}{C_1} : \frac{1}{C_2} : \frac{1}{C_3}$$

- b) When three capacitors are in **parallel**, the ratio of energies is

$$U_1 : U_2 : U_3 = \frac{1}{2}C_1V^2 : \frac{1}{2}C_2V^2 : \frac{1}{2}C_3V^2 = C_1 : C_2 : C_3$$

- c) Energy density (μ) = energy/ volume

$$\mu = \frac{1}{2}\varepsilon E^2 = \frac{1}{2}k\varepsilon_0 E^2$$

(Where K in the dielectric constant of medium between the plates)

Effect of Dielectric:

- A parallel plate capacitor is fully charged to a potential V. **Without disconnecting the battery** if the gap between the plates is completely filled by a dielectric medium,capacity increases to k times the original capacity.
- P.D. between the plates remains same.
- Charge on the plates increases to k times the original charge.
- Energy stored in the capacitor increases to k times the original energy.
- **After disconnecting the battery** if the gap between the plates of the capacitor is filled by a dielectric medium,capacity increases to k times the original capacity.
- P.D. between the plates decreases to $\frac{1}{k}$ times the original potential.
- Charge on the plates remains same.
- Energy stored in the capacitor decreases to $\frac{1}{k}$ times the original energy.
- A capacitor is fully charged to a potential 'v'. After disconnecting the battery, the distance between the plates of capacitors is increased by means of insulating handles. Potential difference between the plates increases. ($V = \frac{Q}{C}$, Q remains same, and C decreases)
- A capacitor with a dielectric is fully charged. Without disconnecting the battery if the dielectric slab is removed, then some charge flows back to the battery.

Mixed Grouping of Capacitors:

- Number of capacitors in a row $n = \frac{desired\ \ potential}{given\ \ potential}$

Ø Number of such rows $m = \frac{desired\ capacity}{original\ capacity} \times n$

Ø Total number of capacitors = $m \times n$.

Coalesence of a Charged Oil Drops:

There are 'n' charged drops of radius 'r' and charge 'q'. The drops are merge to form a bigger drop. If capicity of small drop is 'C' then

1) capacity of bigger drop is $C^1 = n^{\frac{1}{3}} \times C$

2) Potential of bigger drop is $V^1 = \frac{Q}{C^1} = \frac{nq}{n^{1/3}.C} = \frac{n^{2/3}q}{C} = n^{2/3}V.$

3) Energy of bigger drop is $U = \frac{Q^2}{2C^1} = \frac{n^2q^2}{2n^{1/3}.C} = \frac{n^{5/3}q^2}{2C} = n^{5/3}U.$

4) Surface charge density of bigger drop is $\sigma^1 = \frac{Q}{4\pi R^2} = \frac{nq}{4\pi n^{2/3}.r^2} = \frac{n^{1/3}q}{4\pi r^2} = n^{1/3}.\sigma$

S.No.	Quantity	For each charged small drop	For the big drop
1.	Radius r	$R = n^{1/3} r$	
2.	Charge q	$Q = n \times q$	
3.	Capacity	C	$C^1 = n^{1/3} \times C$
4.	Potential	V	$V^1 = n^{2/3} \times V$
5.	Energy	U	$U^1 = n^{5/3} U$
6.	Surface charge	σ	$\sigma = n^{1/3}.\sigma$

INTRODUCTION OF DIELECTRIC IN A CHARGED CAPACITOR

A dielectric slab (K) is introduced between the plates of the capacitor

Sno	Physical quantity	With battery permanently connected	with battery disconnected
1	Capacity	K times increases	K times increases
2	charge	K times increases	Remains constant
3	P.D	Remains constant	K times decreases
4	Electric Intensity	Remains constant	K times decrease
5	Energy stored in condenser	K times increases	K times decrease

The distance between the plates of condenser is increased by n times.

Sno	Physical quantitypermanently	With battery connected	with battery disconnected
1	Capacity n times	decreases	n times decreases
2	charge n times	decreases	Remains constant
3	P.D	Remains constant	n times increases
4	Electric Intensity	n times decrease	Remains constant
5	Energy stored in condenser	n times decreases	n times increases

REDISTRIBUTION OF CHARGE, COMMON POTENTIAL AND LOSS OF ENERGY

Two capacitors of capacities C_1 and C_2 are charged to potentials V_1 and V_2 separately and they are connect so that charge flows. Here charge flows from higher potential to lower potential till both capacitors get the same potential

a) Two capacitors are connected in parallel such that positive plate of one capacitor is connected to positive plate of other capacitor

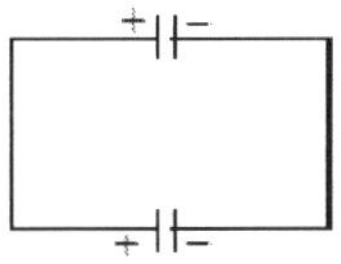

Fig.

Let V be the common potential

Then $Q = Q_1 + Q_2$ (charge conservation)

$(C_1 + C_2) V = C_1V_1 + C_2 V_2$

$$V = \frac{C_1V_1 + C_2V_2}{C_1 + C_2}$$

In this case there will be loss in energy of the system

$\Delta U = U_f - U_i$

where $U_f - \frac{1}{2}C_1V^2 + \frac{1}{2}C_2V^2$

$$U_i = \frac{1}{2}C_1V_1^2 + \frac{1}{2}C_2V_2^2$$

$$\Delta U = \frac{1}{2}\frac{C_1C_2}{C_1 + C_2}(V_1 - V_2)^2$$

b) If positive plate of one capacitor is connected to negative plate of other capacitor, common potential is given by

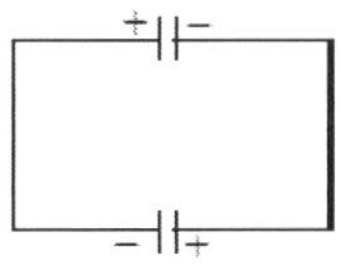

$$V = \frac{C_1V_1 \sim C_2V_2}{C_1 + C_2}$$

Here charge flow takes place if $V_1 \neq V_2$

In this case, the loss of energy $\Delta U = \frac{1}{2}\frac{C_1C_2}{C_1 + C_2}(V_1 + V_2)^2$

Charge transfered is $= q_1 - q_1^1 \ (or)(q_2 - q_2^1)$

$= C_1V_1 - C_1V \ (or) \ C_2V_2 - C_2V \quad = C_1(V_1 - V)\ (or)\ C_2(V_2 - V)$

WE.14: A capacitor of capacitance C_o is charged to a potential V_o and then isolated. A small capacitor C is then charged from C_o, discharged and charged again, the process being repeated n times. Due to this, potential of the large capacitor is decreased to V. Find the capacitance of the small capacitor:

Sol. When key is closed, common potential $V_1 = \frac{C_o V_o}{C_o + C}$ charge left on large capacitor after first sharing of charges $Q_o^1 = C_o V_1$

common potential after second sharing of charges in $V_2 = \frac{C_0}{C_0 + C} V_1$

$$V_2 = \frac{C_o^2 V_0}{(C_o + C)^2}$$

after n[th] sharing charges $V_n = \left(\frac{C_o}{C_o + C}\right)^n V_0$

But $V_n = V$

$$V = \left(\frac{C_0}{C_0 + C}\right)^n V_0$$

$$\therefore C = C_o \left[\left(\frac{V_o}{V}\right)^{1/n} - 1\right]$$

WE.15 : In the circuit shown in figure $C_1 = 1\mu F$ and $C_2 = 2\mu F$. The capacitor C_1 is charged to 100V and the capacitor C_2 is charged to 20V. After charging then are connected as shown. When the switches S_1, S_2 *and* S_3 are closed, the charge flowing through S_1 is

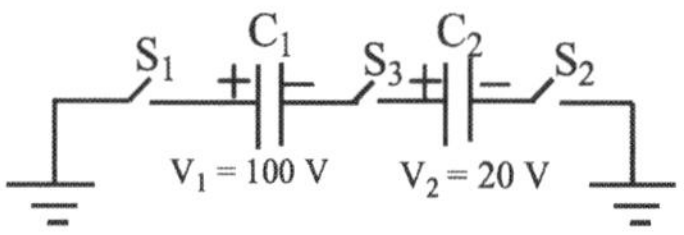

Sol. When S_1, S_2 and S_3 are closed, both the capacitors are in parallel with unlike charged plates together. So, they attain a common potential.

Before closing the switches,

Charge on C_1 is $q_1 = 100 \times 1 = 100\mu C$

Charge on C_2 is $q_2 = 20 \times 2 = 40\mu C$

After closing the switches

Common potential $V = \frac{q_1 - q_2}{C_1 + C_2} = \frac{100 - 40}{3} = 20\,V$

Now final charges $q_1^1 = C_1 V = 1 \times 20 = 20\mu C$

$q_2^1 = C_2 V = 2 \times 20 = 40\mu C$

The charge that flows through S_1 is $\Delta q = 100 - 20 = 80\mu C$

Application :

a) **Redistribution of charges when two conductors are connected by conduting wire :**

In charging a conductor, work is required to be done. This work done is stored up as the potential energy of the conductor.

Energy of a charged conductor,

$$U = \frac{1}{2}CV^2 = \frac{1}{2}QV = \frac{Q^2}{2C}$$

When two charged bodies are connected by a conducting wire then charge flows from a conductor at higher potential to that at lower potential until their potentials are equal.

Let the amounts of charge on two conductors A and B are Q_1 and Q_2 their capacities are C_1 and C_2 and their potentials are V_1 and V_2 respectively, then $Q_1 = C_1V_1$ and $Q_2 = C_2V_2$

Let the amount of charge after the conductors are connected, are $Q_1^{|}$ and $Q_2^{|}$ respectively, then $Q_1^1 = C_1V; Q_2^1 = C_2V$

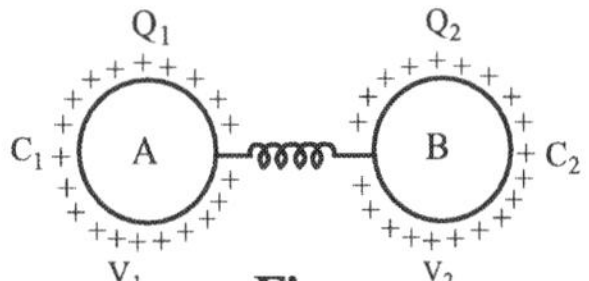

Fig.

Charges are redistributed in the ratio of their capacities.

$\therefore$ $Q^I{}_1 : Q^I{}_2 = C_1 : C_2$ (since V is same)

In case of spherical conductors, $C = 4\pi\varepsilon_0 r$

so, $Q^I{}_1 : Q^I{}_2 = r_1 : r_2$

Van De Graff Generator

Van De Graaff generator is used to develop very high voltages and resulting large electric fields and used to accelerate charged particles to high energies

Principle :- Whenever a charge is given to a metal body it will spread on the outer surface of it. if we put a charged metal body inside the hollow metal body and the two are connected by a wire, whole of the charge of inner body will flow to the outer surface of the hallow body. No matter how large the charge is on the inner body.

Consider a sperical conductor 1 of radius r_1 holding charge q_1 uniformly distributed on it. it is kept inside a hollow conductor 2 of radius r_2 which is uncharged.

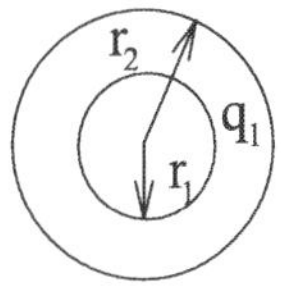

Electric potential of inner sphere is $V_1 = \frac{1}{4\pi\varepsilon_o}\frac{q_1}{r_1}$

Electric potential of outer sphere is $V_2 = \frac{1}{4\pi\varepsilon_o}\frac{q_1}{r_2}$

potential difference between the two conductors

$$V_1 - V_2 = \frac{q_1}{4\pi\varepsilon_o}\left(\frac{1}{r_1} - \frac{1}{r_2}\right)$$

If 'q_2' charge is on the outer shell

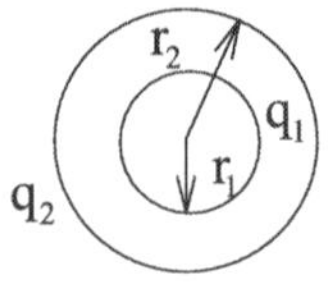

$$V_1 = \frac{1}{4\pi\varepsilon_o}\left(\frac{q_1}{r_1} + \frac{q_2}{r_2}\right)$$

$$V_2 = \frac{1}{4\pi\varepsilon_o}\left(\frac{q_1 + q_2}{r_2}\right)$$

$$V_1 - V_2 = \frac{q_1}{4\pi\varepsilon_o}\left(\frac{1}{r_1} - \frac{1}{r_2}\right)$$

potential difference $(V_1 - V_2)$ will remain the same for any value of q_2

CONCEPTUAL

1. **A condenser stores**
   1) potential 2) charge 3) current 4) energy in magnetic field
2. **Out of the following statements**
   A) The capacity of a conductor is affected due to the presence of an uncharged isolated conductor
   B) A conductor can hold more charge at the same potential if it is surrounded by dielectric medium.
   1) Both A and B are correct 2) Both A and B are wrong
   3) A is correct and B is wrong 4) A is wrong and B is correct
3. **If an earthed plate is brought near positively charged plate, the potential and capacity of charged plate**
   1) increases, decreases 2) decreases, increases
   3) decreases, decreases 4) increases, increases
4. **The plates of charged condenser are connected by a conducting wire. The quantity of heat produced in the wire is**
   1) Inversely proportional to the capacity of the condenser.
   2) Inversely proportional to the square of the potential of the condenser.
   3) proportional to the length of wire
   4) independent of the resistance of the wire
5. **A capacitor works in**
   1) A.C. circuits only 2) D.C. circuits only
   3) both A,C & D.C 4) neither A.C. nor in D.C. circuit.
6. **In order to increase the capacity of a parallel plate condenser one should introduce between the plates a sheet of (assume that the space is completely filled)**
   1) Mica 2) Tin 3) Copper 4) Stainless steel
7. **In a parallel plate capacitor, the capacitance**
   1) increases with increase in the distance between the plates
   2) decreases if a dielectric material is put between the plates
   3) increases with decrease in the distance between the plates
   4) increases with decrease in the area of the plates

8. When a dielectric material is introduced between the plates of a charged condenser, after disconnecting the battery the electric field between the plates

1) decreases 2) increases 3) does not change 4) may increase or decrease

9. A parallel plate capacitor is charged and the charging battery is then disconnected. If the plates of the capacitor are moved further apart by means of insulating handles

1) the charge in the capacitor becomes zero 2) the capacitance becomes infinite

3) the charge in the capacitor increases 4) the voltage across the plates increases

10. The ratio of charge to potential of a body is known as

1) conductance 2) capacitance 3) inductance 4) reactance

11. A parallel plate capacitor filled with a materail of dielectric constant K is charged to a certain voltage and is isolated. The dielectric material is removed. Then

a)The capacitance decreases by a factor K b) The electric field reduces by a factor K

c) The voltage across the capacitor increases by a factor K

d) The charge strored in the capacitor increases by a factor K

1) a and b are true 2) a and c are true 3) b and c are true 4) b and d are true

12. Force acting upon a charged particle kept between the plates of a charged condenser is F. If one of the plates of the condenser is removed, force acting on the same particle will become

1) zero 2) F/2 3) F 4) 2F

13. A condenser is charged and then battery is removed. A dielectric plate is put between the plates of condenser, then correct statement is

1) Q constant V and U decrease 2) Q constant V increases U decreases

3) Q increases V decreases U increases 4) Q, V and U increase

14. If an uncharged capacitor is charged by connecting it to a battery, then the amount of energy lost as heat is

1) 1/2 QV 2) QV 3) $1/2QV^2$ 4) QV^2

15. When air is replaced by a dielectric medium of constant K, the capacity of the condenser

1) increases K times 2) increases K^2 times 3) remains unchanged 4) decreases K times

16. If we increase the distance between two plates of the capacitor, the capacitance will

1) decrease 2) remain same 3) increase 4) first decrease then increase

17. In a charged capacitor the energy is stored in (r) is less than at B

1) both in positive and negative charges 2) positive charges

3) the edges of the capacitor plates 4) the electric field between the plates

18. A metal plate of thickness half the separation between the capacitor plates of capacitance C is inserted. The new capacitance is

1) C 2) C/2 3) zero 4) 2C

19. One plate of parallel plate capacitor is smaller than the other, the charge on the smaller plate will be

1) less than other 2) more than other

3) equal to other 4) will depend upon the medium between them

20. Two condensers of unequal capacities are connected in series across a constant voltage d.c. source. The ratio of the potential differences across the condensers will be

1) direct proportion to their capacities 2) inverse proportion to their capacities

3) direct proportion to the square of their capacities

4) inverse proportion to the square root of their capacities

21. **A parallel plate capacitor is first charged and then isolated , and a dielectric slab is introduced between the plates. The quantity that remains unchanged is**

1) Charge Q 2) Potential V 3) Capacity C 4) Energy U

ENERGY STORED IN A CONDENSER AND TYPES OF CAPACITORS

22. **The condenser used in the tuning circuit of radio receiver is**

1) paper condenser 2) electrolytic condenser 3) leyden jar 4) gang condenser

23. **Space between the plates of a parallel plate capacitor is filled with a dielectric slab. The capacitor is charged and then the supply is disconnected to it. If the slab is now taken out then**

1) work is not done to take out the slab 2) energy stored in the capacitor reduces

3) potential difference across the capacitor is decreased

4) potential difference across the capacitor is increased

24. **A parallel plate condenser is charged by connecting it to a battery. The battery is disconnected and a glas slab is introduced between the plates. Then**

1) potential increases 2) electric intensity increases

3) energy decreases 4) capacity decreases

25. **A parallel plate condenser is charged by connecting it to a battery. Without disconnecting the battery, the space between the plates is completely filled with a medium of dielectric constant k. Then**

1) potential becomes 1/k times 2) charge becomes k times

3) energy becomes 1/k times 4) electric intensity becomes k times.

26. **A parallel plate capacitor of capacity C_0 is charged to a potential V_0.**

A) The energy stored in the capacitor when the battery is disconnected and the plate separation is doubled is E_1

B) The energy stored in the capacitor when the charging battery is kept connected and the separation between the capacitor plates is doubled is E_2. Then $\frac{E_1}{E_2}$ value is

1) 4 2) 3/2 3) 2 4) 1/2

27. **Select correct Statements**

a) Charge cannot be isolated

b) Repulsion is the sure test to know the presence of charge

c) Waxed paper is dielectric in paper capacitor

d) Variable capacitor is used in tuning circuits in radio

1) a, b only 2) a, c only 3) a, b, c only 4) b,c,d only

28. **A variable parallel plate capacitor and an electroscope are connected in parallel to a battery. The reading of the electroscope would be decreased by**

1) increasing the area of overlap of the plates

2) placing a block of paraffin wax between the plates

3) decreasing the distance between the plates 4) decreasing the battery potential

29. **Three identical capacitors are connected together differently. For the same voltage to every combina-tion, the one that stores maximum energy is**

1) the three in series 2) the three in parallel

3) two in series and the third in parallel with it 4) two in parallel and the third in series with it

30. The magnitude of electric field E in the annular region of a charged cylindrical capacitor

1) is same throughout

2) is higher near the outer cylinder than near the inner cylinder

3) varies as $\frac{1}{r}$ where r is the distance from the axis

4) varies as r where r is the distance from the axis

31. Two identical capacitors are joined in parallel, charged to a potential V, separated and then connected in series i.e., the positive plate of one is connected to the negative plate of other.

1) the charges on the free plates are enhanced
2) the charges on the free plates are decreased

3) the energy stored in the system increases

4) the potential difference between the free plates is 2V

32. Two parallel plate air capacitors are constructed, one by a pair of iron plates and the second by a pair of copper plates of same area and same spacings. Then

1) the copper plate capacitor has a greater capacitance than the iron one

2) both capacitors will have equal non zero capacitances, in the uncharged state

3) both capacitors will have equal capacitances only if they are charged equally

4) the capacitances of the two capacitors are unequal even they are unequally charged

33. Select correct statement for a capacitor having capacitance C, is connected to a source of constant emf E

1) Almost whole of the energy supplied by the battery will be stored in the capacity, if resistance of connecting wire is negligibly small

2) Energy received by the capacitor will be half of energy supplied by the battery only when the capacitor was initially uncharged

3) Strain energy in the capacitor must increases even if the capacitor had an initial charge

4) Energy stored depends on type of the source of emf

34. Van de Graff genetor is used to :

1) supply electricity for industrial use
2) produce intense magnetic fields

3) generate high voltage
4) obtain highly penetrating X-rays

35. A number of spherical conductors of different radii have same potential. Then the surface charge density on them

1) is proportional to their radii
2) is inversely proportional to their radii

3) are equal
4) is proportional to square of their radii

36. Three charged particles are initially in position 1. They are free to move and they come in position 2 after some time. Let U_1 and U_2 be the electrostatic potential energies in position 1 and 2. Then

1) $U_1 > U_2$
2) $U_2 > U_1$
3) $U_1 = U_2$
4) $U_2 \geq U_1$

37. An insulator plate is passed between the plates of a capacitor. Then current

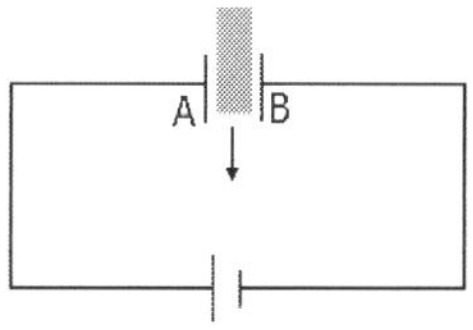

1) always flows from A to B
2) always flows from B to A

3) first flows from A to B and then from B to A
4) first flows from B to A and then from A to B

38. **Read the following statements**

a) Non polar molecules have uniform charge distribution

b) Polar molecules have non - uniform charge distribution

c) Polar molecules are already polarized

d) Molecules are not already polarized without electric field in Non - polar molecules

1) only a & b are correct 2) only c & d are correct 3) only c is wrong 4) all are correct

39. **The capacitance of a capacitor depends on**

1) the geometry of the plates 2) separation between plates

3) the dielectric between the plates 4) all the above

40. **The electric field $\left(\vec{E}\right)$ between two parallel plates of a capacitor will be uniform if**

1) the plate separation (d) is equal to area of the plate (A)

2) the plate separation (d) greater when compared to area of the plate (A)

3) the plate separation (d) is less when compared to area of the plate (A) 4) 2 (or) 3

41. **For metals the value of dielectric constant (K) is**

1) One 2) Infinity 3) Zero 4) Two

42. **A capacitor C is connnected to a battery circuit having two switches S_1 *and* S_2 and resistors R_1 *and* R_2. The capacitor will be fully charged when**

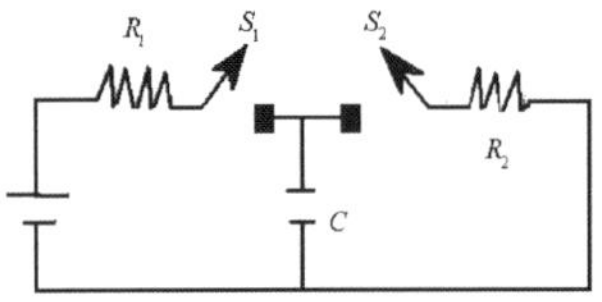

1) both S_1 *and* S_2 are closed 2) S_1 is closed and S_2 is open

3) S_1 is open and S_2 is closed 4) any one of the above

43. **Figure shows two capacitors connected in series and joined to a cell. The graph shows the variation in potential as one moves from left to right on the branch containing capacitors.**

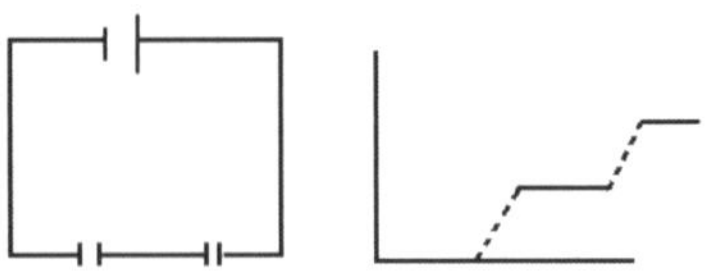

1) $C_1 > C_2$ 2) $C_1 = C_2$ 3) $C_1 < C_2$

4) data insufficient to conclude the answer

44. **Two condensers of unequal capacities are connected in parallel across a constant voltage d.c. source. The ratio of the charges stored in the condensers will be**

1) direct proportion to their capacities 2) inverse proportion to their capacities

3) direct proportion to the square root of their capacities

4) inverse proportion to the square of their capacities

45. **A parallel plate capacitor is charged and then isolated. Regarding the effect of increasing the plate separation, select the appropriate alternative.**

Charge	Potential	Energy
1) decreases constant	decreases	
2) increases	increases	increases
3) constant	decreases	decreases
4) constant	increases	increases

46. **A parallel plate capacitor is charged by connecting its plates to the terminals of a battery. The battery remains connected to the condenser plates and a glass plate is interposed between the plates of the capacitor, then**
1) the charge increases while the potential difference remains constant
2) the charge decreases while the potential difference remains constant
3) the charge decreases while the potential difference increases
4) the charge increases while the potential difference decreases

47. **A parallel plate capacitor is charged to a fixed potential and the charging battery is then disconnected. If now, the plates of the capacitor are moved further apart, then**
1) the charge on the capacitor increases
2) the voltage across the capacitor increases
3) the energy stored in the capacitor decreases
4) the capacitance increases

48. **A parallel plate air condenser is charged and then disconnected from the charging battery. Now the space between the plates is filled with a dielectric then, the electric field strength between the plates**
1) increases while its capacity increases
2) increases while its capacity decreases
3) decreases while its capacity increases
4) decreases while its capacity decreases

49. **When two identical condensers are connected in series choose the correct statement regarding the working voltage (the maximum p.d. that can be applied to a condenser) and the capacity**
1) working voltage increases, capacity increases
2) working voltage increases, capacity decreases
3) working voltage decreases, capacity increases
4) working voltage decreases, capacity decreases

50. **Two unequal capacitors, initially uncharged, are connected in series across a battery. Which of the following is true**
1) The potential across each is the same
2) The charge on each is the same
3) The energy stored in each is the same
4) The equivalent capacitance is the sum of the two capacitances

51. **Which of the following will not increase the capacitance of an air capacitor?**
1) adding a dielectric in the space between the plates
2) increasing the area of the plates
3) moving the plates closer together
4) increasing the voltage

52. **In a parallel-plate capacitor, the region between the plates is filled by a dielectric slab. The capacitor is connected to a cell and the slab is taken out. Then**
1) some charge is drawn from the cell
2) some charge is returned to the cell
3) the potential difference across the capacitor is reduced
4) no work is done by an external agent in taking the slab out

53. **Which of the following statements are correct?**

a) When capacitors are connected in parallel the effective capacitance is less than the individual capacitances

b) The capacitances of a parallel plate capacitor can be increased by decreasing the separation of plates

c) When capacitors are connected in series the effective capacitance is less than the least of the individual capacities

d) In a parallel plate capacitor the electrostatic energy is stored on the plates

1) (a) and (b) 2) (a) and (c) 3) (c) and (d) 4) (b) and (c)

54. **Three identical condensers are connected together in four different ways. First all of them are connected in series and the equivalent capacity is C_1. Next all of them are connected in parallel and the equivalent capacity is C_2. Next two of them are connected in series and the third one connected in parallel to the combination and the equivalent capacity is C_3. Next two of them are connected in parallel and the third one connected in series with the combination and the equivalent capacity is C_4. Which of the following is correct ascending order of the equivalent capacities?**

1) $C_1 < C_3 < C_4 < C_2$ 2) $C_1 < C_4 < C_3 < C_2$ 3) $C_2 < C_3 < C_4 < C_1$ 4) $C_2 < C_4 < C_3 < C_1$

55. **On a capacitor of capacitance C_0 following steps are performed in the order as given in column I.**

(A) Capacitor is charged by connecting it across a battery of emf E_0

(B) Dielectric of dielectric constant K and thickness d is inserted

(C) Capacitor is disconnected from battery

(D) Separation between plates is doubled

Column-I (Steps performed)	**Column-II** (Final value of Quantity (Symbols have usual meaning))
(a) (A)(D)(C)(B)	**(p)** $Q = \frac{C_0 E_0}{2}$
(b) (D)(A)(C)(B)	**(q)** $Q = \frac{KC_0 E_0}{K+1}$
(c) (B)(A)(C)(D)	**(r)** $C = \frac{KC_0}{K+1}$
(d) (A)(B)(D)(C)	**(s)** $V = \frac{E_0(K+1)}{2K}$

1) a-p,r,s, b-p,r,s, c-r, d-q,r 2) a-p, b-p,r c-r, d-q,

3) a-p,s, b-r,s, c-r, d-q, 4) a-r,s, b-s, c-r, d-q,r

56. **In the circuit, both capacitors are indentical. Column I indicates action done on capacitors 1 and Column II indicates effect on capacitor 2**

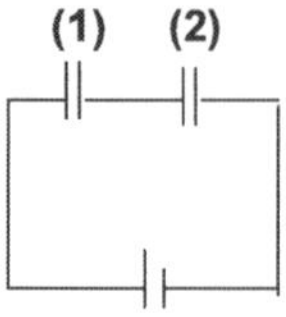

Column-I | **Column-II**

(a) Plates are moved further apart | **(p)** Amount of charge on left plate increases

(b) Area increased difference | **(q)** Potential increases

(c) Left plate is earthed | **(r)** Amount of charge on right plate decreases

(d) It's plates are short circuited | **(s)** None of the above effects

1.a-r, b-p,q, c-s, d-p,q 2.a-r, b-p, c-s, d-q 3.a-r, b-p, c-r, d-q 4.a-s, b-,q, c-s, d-q

57. The potential across a 3 μ F capacitor is 12 V when it is not connected to anything. It is then connected in parallel with an uncharged 6 μ F capacitor. At equilibrium, the charge and potential difference across the capacitor 3 μ F and 6 μ F are listed in column I. Match it with column III.

Column-I	**Column-II**
(a) charge on 3 μ F capacitor	**(p)** 12 μ C
(b) charge on 6 μ F capacitor	**(q)**24 μ F
(c) potential difference across 3 μ F capacitor	**(r)** 8 V
(d) potential difference across 6 μ F capacitor	**(s)** 4 V

1) a-r, b-p, c-s, d-q 2) a-p, b-q, c-s, d-s 3) a-r, b-p, c-q, d-q 4) a-r, b-q, c-s, d-q

58. Some events related to a capacitor are listed in column-I. Match these with their effect(s) in column - II

Column-I	**Column-II**
(a) Insertion of dielectric while battery	**(p)** Eelctric field between plates changes remain attached
(b) Removal of dielectric while battery	**(q)** Charge present on plates changes is not present
(c) Slow decrease in separation between	**(r)** Energy stored in capacitor increases plates while battery is attached
(d) Slow increase of separation between plates while battery is not present	**(s)** Work done by capacitor agent is positive

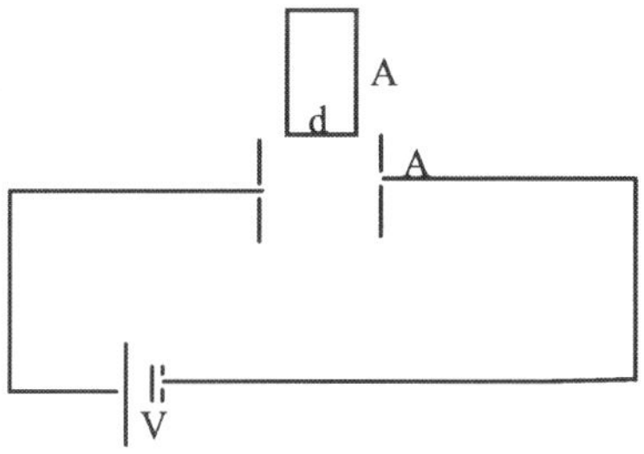

1) a-r, b-p,, c-p,q,s, d-q 2) a-p, b-p,, c-r,s, d-s

3) a-q,r, b-p,r,s c-p,q,r, d-r,s 4) a-r, p,b-q,, c-s, d-q

59. **The effective capacity of the following capacitors is ________**

a) 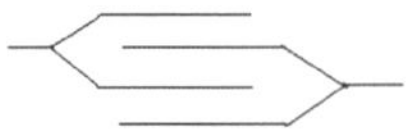

e) $\frac{2c}{3}$

b) 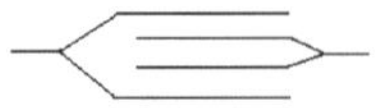

f) 2C

c) 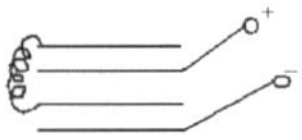

g) 3C

d) 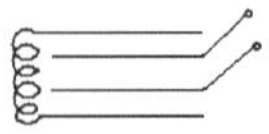

h) $\frac{5C}{2}$

i) $\frac{3C}{2}$

1) $a-g, b-f, c-e, d-i$ 2) $a-g, b-h, c-e, d-i$

3) $a-i, b-h, c-e, d-g$ 4) $a-g, b-e, c-h, d-i$

60. **The circuit involves two ideal cells connected to a 1 μ F capacitor via key K. Initially the key K is in position 1 and the capacitor is charged fully by 2V cell. The key is pushed to position 2. Column I gives physical quantities involving the circuit after the key is pushed from position 1. Column.II gives corresponding results. Match the column-I with Column-II**

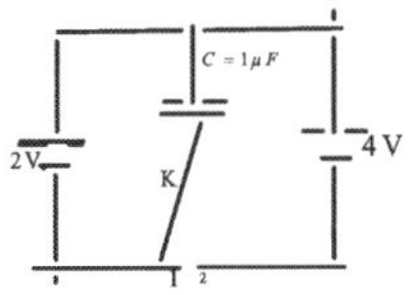

Column-I	Column-II
(a) The net charge crossing the 4 volt cell in μC is	**(p)** 2
(b) The magnitude of work done by 4 volt cell in μJ is	**(q)** 6
(c) The gain in potential energy of capacitor in μJ is	**(r)** 8
(d) The net heat produced in circuit in μJ is	**(s)** 16

1) a-r, b-p, c-s, d-q 2) a-p, b-r, c-q, d-p 3) a-r, b-p, c-q, d-q 4) a-r, b-q, c-s, d-q

61. **Two identical capacitors A and B are connected to a battery of emf E as shown in figure. Now a dielectric slab is inserted between the plates of capacitor B while battery remains connected. Due to this inserting some physical quantities may change which are mentioned in Column-I and the effect is mentioned in Column-II.Match the Column I with Column-II**

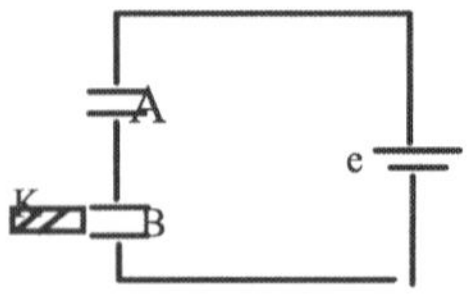

Column-I	Column-II
(a) Charge on A	**(p)** Increases
(b) Charge on B	**(q)** Decreases

(c) Potential difference across A
(d) Potential difference across B
(r) Remains constant
(s) Will change

1) a-r, b-p, c-s, d-q
2) a-p,s b-q,s, c-q,s d-q,s
3) a-r, b-p, c-q, d-q
4) a-r, b-q, c-s, d-q

62. **Match the following**

Set -I	**Set-II**
a) Electrolyte Capacitors	e) Radio circuits & cheap in cost
b) Paper Capacitor	f) Proper Polarity high capacitance of order $10^3 \mu F$
c) Multiple Capacitor	g)High frequency oscillating circuits
d) Variable Capacitor	h) Tuning circuits in radio & T.V receivers

1) $a-f, b-g, c-h, d-e$
2) $a-g, b-f, c-e, d-h$
3) $a-f, b-e, c-g, d-h$
4) $a-h, b-e, c-f, d-g$

63.

Column - I	**Column - II**
A) electrical potential	p) vector
B) energy stored in a condenser	q) $\frac{1}{2}CV^2$
C) force between two	r) scalar
D) electric capacity capacitor plates	s) $\frac{1}{2}\epsilon_0 E^2 A$

	A	B	C	D		A	B	C	D
1)	r	q,r	p,s	r	2)	r	q,r	p,q	s
3)	q,r	p,q	r,s	s	4)	p,q	r	q,r	s

64. **Assertion (A) : The strength of electric filed in the charged and isolated capacitor is decreased when the dielectric slab is inserted.**

Reason(R): When the dielctric slab is inserted between the plates of a charged capacitor, electricfield produced due to induced charge, opposite to the external field.

1) Both A and R false
2) Both A and R true and R is not correct reason for A
3) A is true and R is false
4) Both A and R are true and R is correct reason of A.

65. **Assertion: If temperature is increased, the dielectric constant of a polar dielectric decreases whereas that of a non-polar dielectric does not change significantly**

Reason: The magnitude of dipole moment of individual polar molecule decreases significantly with increase in temperature.

66. **Assertion: The heat produced by a resistor in any time t during the charging of a capacitor in a series circuit is half the energy stored in the capacitor by that time.**

Reason: Current in the circuit is equal to the rate of increase in charge on the capacitor.

67. **Assertion: A dielectric is inserted between the plates of an isolated fully-charged capacitor. The dielectric completely fills the space between the plates. The magnitude of electrostatic force on either metal plate decreases, as it was before the insertion of dielectric medium.**

Reason: Due to insertion of dielectric slab in an isolated parallel plate capacitor (the dielectric completely fills the space between the plates), the electrostatic potential energy of the capacitor decreases.

68. **Assertion: If the potential difference across a plane parallel plate capacitor is doubled then the potential energy of the capacitor is doubled then the potential energy of the capacitor becomes four times under all conditions**

Reason: The potential energy U stored in the capacitor is $U=\frac{1}{2}CV^2$, where C and V have usual meaning.

69. **Assertion: A parallel plate capacitor is charged to a potential difference of 100V, and disconnected from the voltage source. A slab of dielectric is then slowly inserted between the plates. Compared to the energy before the slab was inserted, the energy stored in the capacitor with the dielectric is decreased.**

Reason: When we insert a dielectric between the plates of a capacitor, the induced charges tend to draw in the dielectric into the field (just as neutral objects are attracted by charged objects due to induction). We resist this force while slowly inserting the dielectric, and thus do negative work on the system, removing electrostatic energy from the system.

70. **Statement 'A': The energy stored gets reduced by a factor 'K' when the battery is disconnected after charging the capacitor and then the dielectric is introduced**

Statement 'B': The energy stored in the capacitor increases by a factor 'k' when a dielectric is introduced between the plates with the battery present in the circuit

1) only 'A' is correct 2) only 'B' is correct 3) both A & B are wrong 4) both A & B are correct

71. **Assertion (A): A metallic sheild in form of a hollow shell may be built to block an electric field.**

Reason (R): In a hollow spherical sheild, the electric field inside it is zero at every point.

72. **Assertion (A): When two spheres carrying same charge but a different radii are connected by a conducting wire, the charge flows from smaller sphere to large sphere.**

Reason (R): Smaller sphere is at high potential when equal charges are imparted to both the spheres

73. **Assertion (A): Two capacitors are connected in par allel to a battery. If a dielectric medium is inserted between the plates of one of the capacitors then the energy stored in the system will increase**

Reason (R): On inserting dielectric medium between the plates of a capacitors, its capacity increases

74. **Assertion (A): When a charged capacitor is discharged through a resistor, heat is produced in the resistor**

Reason (R): In charging a capacitor, energy is stored in the capacitor.

75. **Assertion (A): A capacitor of capacitance C is connected across a battery of potential difference V. The energy stored in the capacitor is $\frac{1}{2}CV^2$**

Reason (R): The energy supplied by the battery is $\frac{1}{2}CV^2$

76. Assertion (A): Two metal plates each of area A form a parallel plate capacitor. Now one plate is displaced up, then the capacitance of capacitor decreases.

Reason (R): Due to displacing one plate, the overlapping area decreases, capacitance $C = \frac{\varepsilon_0 A}{d}$ decreases.

77. Assertion (A): Two plates of a parallel plate capacitor are drawn apart, keeping them connected to a battery. Next the same plates are drawn apart from the same initial condition, keeping the battery disconnected, then the work done in both cases are same.

Reason (R): Capacitor plates have same charge in both cases and displacements of plates in both cases are also same.

78. Assertion (A) : Two metallic plates placed side by side form three capacitors.

Reason (R) : The infinity and first face of first plate is one capacitor, the second face of first plate and first face of second plate forms second capacitor and the second face of second plate and infinity forms the third capacitor, but the capacitance of first and third capacitance are extremely small

79. Statement 'A': The energy stored gets reduced by a factor 'K' when the battery is disconnected after charging the capacitor and then the dielectric is introduced

Statement 'B': The energy stored in the capacitor increases by a factor 'k' when a dielectric is introduced between the plates with the battery present in the circuit

1) only 'A' is correct 2) only 'B' is correct

3) both A & B are wrong 4) both A & B are correct

CONCEPTUAL - KEY

1) 2	2) 1	3) 2	4) 4	5) 3	6) 1	7) 3	8) 1	9) 4	10) 2
11) 2	12) 2	13) 1	14) 1	15) 1	16) 1	17) 4	18) 4	19) 3	20) 2
21) 1	22) 4	23) 4	24) 3	25) 2	26) 1	27) 4	28) 4	29) 2	30) 3
31) 4	32) 2	33) 3	34) 3	35) 2	36) 1	37) 4	38) 4	39) 4	40) 3
41) 2	42) 2	43) 3	44) 1	45) 4	46) 1	47) 2	48) 3	49) 2	50) 2
51) 4	52) 2	53) 4	54) 2	55) 1	56) 1	57) 2	58) 3	59) 1	60) 2
61) 2	62) 3	63) 1	64) 4	65) 3	66) 4	67) 4	68) 4	69) 1	70) 4
71) 1	72) 1	73) 1	74) 2	75) 3	76) 1	77) 4	78) 1	79) 4	

LEVEL - I (CW)

CAPACITANCE

1. The capacity of a parallel plate condenser consisting of two plates each 10 cm square and are seperated by a distance of 2 mm is (Take air as the medium between the plates)

1) $8.85\times10^{-13}F$ 2) $4.42\times10^{-12}F$ 3) $44.25\times10^{-12}F$ 4) $88.5\times10^{-13}F$

2. Sixty four spherical drops each of radius 2 cm and carrying 5C charge combine to form a bigger drop. Its capacity is

1) $\frac{8}{9}\times10^{-11}F$ 2) $90\times10^{-11}F$ 3) $1.1\times10^{-11}F$ 4) $9\times10^{11}F$

3. A highly conducting sheet of aluminium foil of negligible thickness is placed between the plates of a parallel plate capacitor. The foil is parallel to the plates. If the capacitance before the insertion of foil was 10 μ F, its value after the insertion of foil will be

1) 20 μ F 2) 10 μ F 3) 5 μ F 4) Zero

4. **Two metal plates are separated by a distance d in a parallel plate condenser. A metal plate of thickness t and of the same area is inserted between the condenser plates. The value of capacitance increases by ...... times**

1) $\frac{d-t}{d}$ 2) $\left(1-\frac{t}{d}\right)$ 3) $\left(t-\frac{t}{d}\right)$ 4) $\frac{1}{\left(1-\frac{t}{d}\right)}$

5. **A radio capacitor of variable capacitance is made of n parallel plates each of area A and separated from each other by a distance d. The alternate plates are connected together. The capacitance of the combination is**

1) $\frac{n A \in_o}{d}$ 2) $\frac{(n-1) A \in_o}{d}$ 3) $\frac{(2n-1) A \in_o}{d}$ 4) $\frac{(n-2) A \in_o}{d}$

6. **The radius of the circular plates of a parallel plate condenser is 'r'. Air is there as the dielectric. The distance between the plates if its capacitance is equal to that of an isolated sphere of radius r' is**

1) $\frac{r^2}{4r'}$ 2) $\frac{r^2}{r'}$ 3) $\frac{r}{r'}$ 4) $\frac{r^2}{4}$

CAPACITORS IN SERIES AND IN PARALLEL

7. **When two capacitors are joined in series the resultance capacity is $2.4\mu F$ and when the same two are joined in parallel the resultant capacity is $10\mu F$. Their individual capacities are**

1) $7\mu F, 3\mu F$ 2) $1\mu F, 9\mu F$ 3) $6\mu F, 4\mu F$ 4) $8\mu F, 2\mu F$

8. **Three condensers $1\mu F$, $2\mu F$ and $3\mu F$ are connected in series to a p.d. of 330 volt. The p.d across the plates of $3\mu F$ is**

1) 180 V 2) 300 V 3) 60 V 4) 270 V

9. **The effective capacitance between the point P and Q in the given figure is**

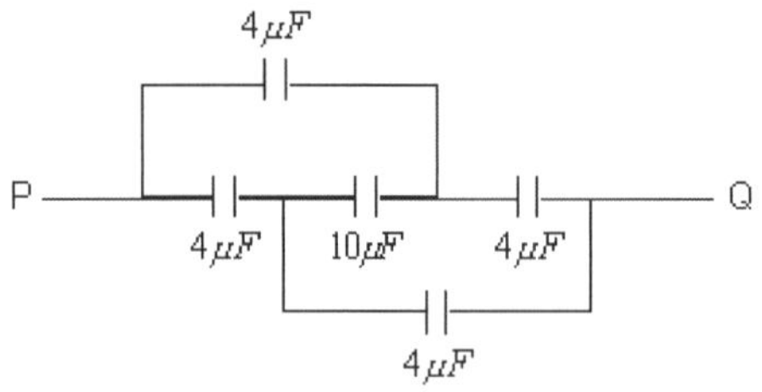

1) 4 μF 2) 16 μF 3) 26 μF 4) 10 μF

10. **The equivalent capacitance between P and Q is**

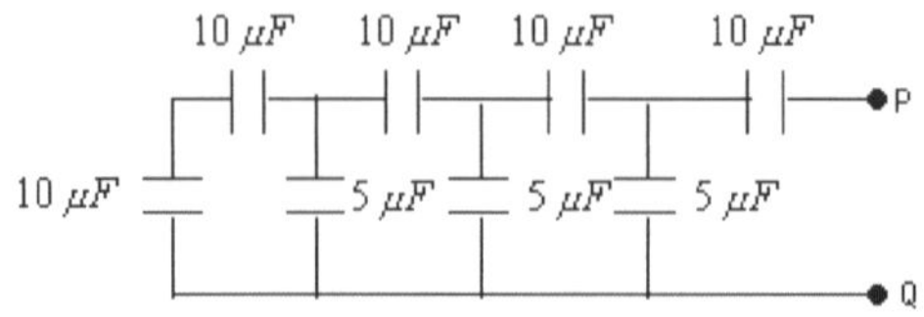

1) 10 μF 2) 20 μF 3) 5 μF 4) 15 μF

11. **The equivalent capacity between the points X and Y in the circuit with $C = 1\mu F$ (2007M)**

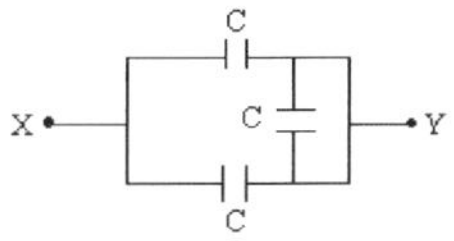

1) $2\mu F$ 2) $3\mu F$ 3) $1\mu F$ 4) $0.5\mu F$

12. **The equivalent capacitance of the network given below is 1 μ F. The value of 'C' is**

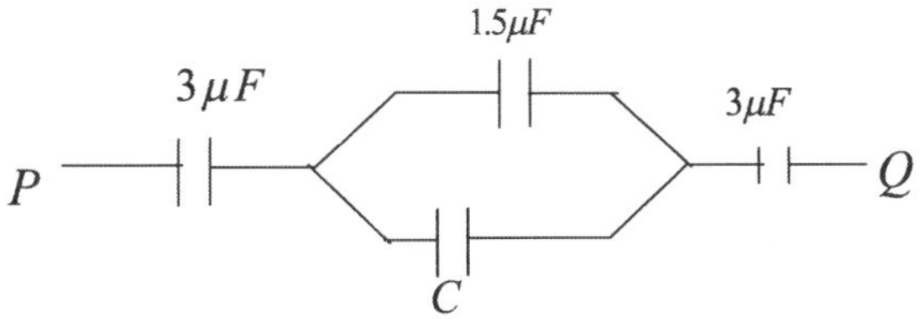

1) 3 μ F 2) 1.5 μ F 3) 2.5 μ F 4) 1 μ F

13. **Three capacitors of $3\mu F, 2\mu F$ and $6\mu F$ are connected in series. When a battery of 10V is connected to this combination then charge on $3\mu F$ capacitor will be**

1) $5\mu C$ 2) $10\mu C$ 3) $15\mu C$ 4) $20\mu C$

ENERGY STORED IN A CONDENSER, TYPES OF CAPACITORS

14. **Two spheres of radii 12 cm and 16 cm have equal charge. The ratio of their energies is**

1) 3 : 4 2) 4 : 3 3) 1 : 2 4) 2 : 1

15. **A condenser of capacity 10 μ F is charged to a potential of 500 V. Its terminals are then connected to those of an uncharged condenser of capacity 40 μ F. The loss of energy in connecting them together is**

1) 1J 2) 2.5J 3) 10J 4) 12 J

16. **A $2\mu F$ condenser is charged to 500V and then the plates are joined through a resistance. The heat produced in the resistance in joule is**

1) 50×10^{-2} Joule 2) 25×10^{-2} Joule 3) 0.25×10^{-2} Joule 4) 0.5×10^{-2} Joule

LEVEL - I (CW) - KEY

1) 2	2) 1	3) 2	4) 4	5) 2	6) 1	7) 3	8) 3	9) 1	10) 3
11) 1	12) 2	13) 2	14) 1	15) 1	16) 2				

LEVEL - I (CW) - HINTS

1. $C = \frac{\varepsilon_0 A}{d}$

2. $C^1 = n^{\frac{1}{3}} C$

3. Conceputal

4. $C = \frac{\varepsilon_0 A}{d - t + \frac{t}{k}}$ k = ∞

5. Due to n plates n-1 capacitors are formed

6. $\frac{\epsilon_o (\pi r^2)}{d} = 4\pi \epsilon_o r^1$ $\quad \therefore d = \frac{r^2}{4r^1}$

7. $C_S = \frac{C_1 C_2}{C_1 + C_2}$; $C_1 + C_2 = C_P$

8. $Q = C_{eff} V$ $\quad Q = C_1 V$

9. $C^1 = \frac{C_1 C_2}{C_1 + C_2}$ $\quad C^{11} = \frac{C_3 C_4}{C_3 + C_4}$

$C_{eff} = C^1 + C^{11}$

10. From Left $C = \frac{10 \times 10}{10 + 10} = 5\mu F$

$C^1 = 5 + 5 = 10\mu F$

$C^{11} = \frac{10 \times 10}{10 + 10} = 5\mu F$ and so on

finally $C_{eff} = \frac{10 \times 10}{10 + 10} = 5\mu F$

11. $C_{eff} = C_1 + C_2$

12. $1.5\mu c$, c are in parallel;
its effective capacitance $1.5 + c$
1.5+c, $3\mu F$, $3\mu F$ are in series

13. In series charge constant $Q = C_{eff} V$

14. $U = \frac{q^2}{2C}$, $U \alpha \frac{1}{r}$

15. $\Delta E = \frac{1}{2} \frac{C_1 C_2}{C_1 + C_2} (V_1 - V_2)^2$

16. Energy Stored $= \frac{1}{2} cv^2$

LEVEL - I (HW)

CAPACITANCE

1. **The charge stored in a capacitor is $20\mu C$ and the potential difference across the plates is 500 V. Its capacity is**

1) $0.04\mu F$ 2) $10^{-2}\mu F$ 3) $2\times10^{-6}\mu F$ 4) $250\mu F$

2. **An oil condenser has a capacity of $100\mu F$. The oil has dielectric constant 2. When the oil leaksout its new capacity is**

1) $200\mu F$ 2) $0.02\mu F$ 3) $50\mu F$ 4) $0.5\mu F$

3. **A dielectric of thickness 5cm and dielectric constant 10 is introduced between the plates of a parallel plate capacitor having plate area 500 sq. cm and separation between the plates 10cm. The capacitance of the capacitor with dielectric slab is** $\left(\varepsilon_0 = 8.8\times10^{-12} C^2/N-m^2\right)$

1) $4.4pF$ 2) $6.2pF$ 3) $8pF$ 4) $10pF$

4. **The capacitance of a sphere of radius 10cm situated in air is approximately**

1) $11\times10^{-6}F$ 2) 11×10^{-9}F 3) 11×10^{-12}F 4) Zero

CAPACITORS IN SERIES AND IN PARALLEL

5. **The ratio of the resultant capacities when three capacitors of** $2\mu F$, $4\mu F$ **and** $6\mu F$ **are connected first in series and then in parallel is**

1) 1 : 11 2) 11 : 1 3) 12 : 1 4) 1 : 12

6. **A condenser A of capacity** $4\mu F$ **has a charge** $20\mu C$ **and another condenser B of capacity** $10\mu F$ **has a charge** $40\mu C$. **If they are connected parallel, then**

1) charge flows from B to A till the charges on them are equal.
2) charge flows from B to A till common poten tial is reached
3) charge flows from A to B till common potential is reached
4) charge flows from A to B till charges on them are equal.

7. A capacitor of $30\mu F$ charged to 100 V is conncected in parallel to capacitor of $20\mu F$ charged to 50 volt. The common potential is

1) 75 V 2) 150 V 3) 50 V 4) 80 V

8. **The equivalent capacity between the points 'A' and 'B' in the following figure will be**

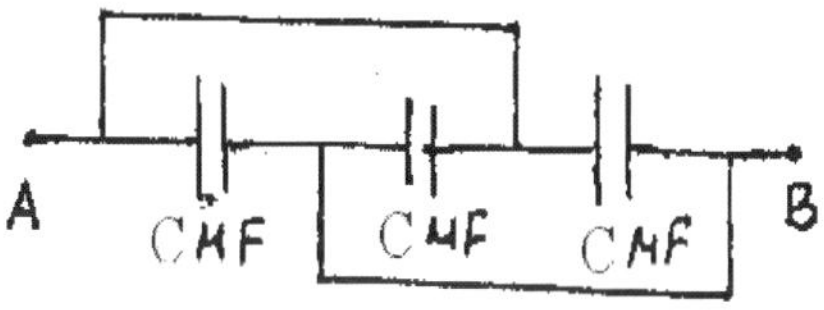

1) $3C$ 2) $\frac{C}{3}$ 3) $\frac{3}{C}$ 4) $\frac{1}{3C}$

9. **Two capacitors with capacitances** C_1 **and** C_2 **are charged to potentials** V_1 **and** V_2 **respectively. When they are connected in parallel the ratio of their respective charges is**

1) $\frac{C_1}{C_2}$ 2) $\frac{V_1}{V_2}$ 3) $\frac{V_1^2}{V_2^2}$ 4) $\frac{C_1^2}{C_2^2}$

10. **The equivalent capacitance between P and Q of the given figure is (the capacitance of each capacitor is** $1\mu F$**)**

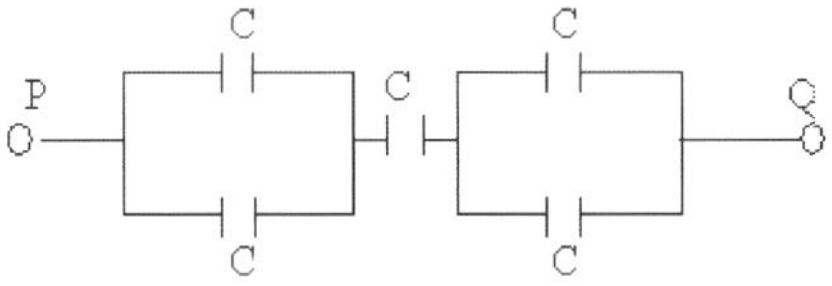

1) $2\mu F$ 2) $0.5\mu F$ 3) $5\mu F$ 4) $0.2\mu F$

11. **The resultant capacity between the points P and Q of the given figure is**

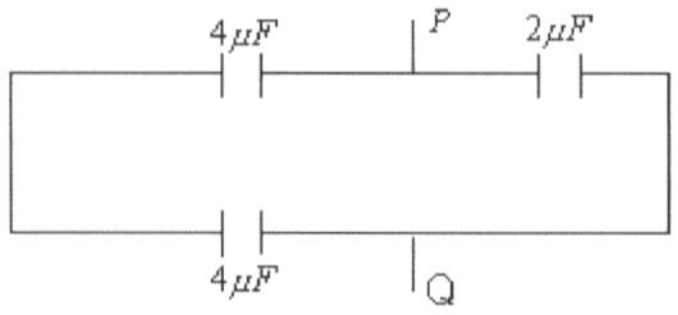

1) $4\mu F$ 2) $\frac{16}{3}\mu F$ 3) $1.6\mu F$ 4) $1\mu F$

12. **Charge 'Q' taken from the batteryof 12V in the circuit is**

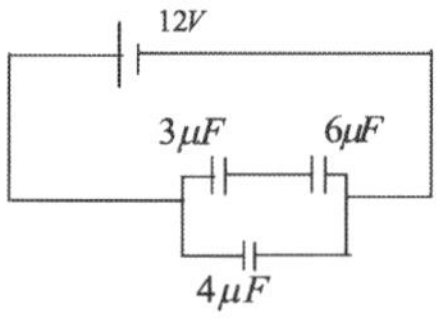

1) 72 μC 2) 36 μC 3) 156 μC 4)20 μC

13. **If 3 capacitors of values 1, 2 and 3 μF are available. The maximum and minimum values of capacitances one can obtain by different combinations of the three capacitors together are respectively .... and**

1) $6\mu F, \frac{6}{11}\mu F$ 2) $6\mu F, \frac{11}{6}\mu F$ 3) $3\mu F, 1\mu F$ 4) $4\mu F, 2\mu F$

ENERGY STORED IN A CONDENSER, TYPES OF CAPACITORS

14. **A capacitor of 8 micro farad is charged to a potential of 1000V. The energy stored in the capacitor is**

1) 8 J 2) 12 J 3) 2 J 4) 4 J

15. **A condenser is charged to a p.d. of 120 volt. Its energy is 1×10^{-5} *joule*. If the battery is there and the space between plates is filled up with a dielectric medium $(\varepsilon_r = 5)$, its new energy is**

1) $10^{-5}J$ 2) $2\times10^{-5}J$ 3) $3\times10^{-5}J$ 4) $5\times10^{-5}J$

16. The plates of a parallel plate capacitor have an area of $90\,cm^2$ each and are separated by 2 mm. The capacitor is charged by connecting if to a 400 V supply. Then the density of the energy stored in the capacitor $\left(\varepsilon_o = 8.8\times10^{-12}\,F/m\right)$

1) $0.113\,Jm^{-3}$ 2) $0.117\,Jm^{-3}$ 3) $0.152\,Jm^{-3}$ 4) $0.226\,Jm^{-3}$

LEVEL - I (HW) - KEY

1) 1 2) 3 3) 3 4) 3 5) 1 6) 3 7) 4 8) 1 9) 1 10) 2
11) 1 12) 1 13) 1 14) 4 15) 4 16) 2

LEVEL - I (HW) - HINTS

1. $C = \frac{q}{V}$

2. $C = \frac{C'}{K}$

3. $C = \dfrac{\varepsilon_0 A}{d - t + \dfrac{t}{k}}$

4. $C = 4\pi\varepsilon_0 R$

5. $\dfrac{1}{Cs} = \dfrac{1}{C_1} + \dfrac{1}{C_2} + \dfrac{1}{C_3}$ $\quad C_p = C_1 + C_2 + C_3$

6. In parallel potential constant $V = \dfrac{C_1V_1 + C_2V_2}{C_1 + C_2}$

then find charges $q_1^1 = C_1V$, $q_2^1 = C_2V$

7. $V = \dfrac{C_1V_1 + C_2V_2}{C_1 + C_2}$

8. Capacitors are in parallel

9. Parallel potential constant and $Q \alpha C$

10. $C_1 = C + C$

$C_2 = C$

$C_3 = C + C$

$\dfrac{1}{C_{eff}} = \dfrac{1}{C_1} + \dfrac{1}{C_2} + \dfrac{1}{C_3}$

11. $C_1 = \dfrac{4 \times 4}{4 + 4}$

$C_2 = 2\mu F$ $\quad C_{eff} = C_1 + C_2$

12. $C_{eff} = \dfrac{6 \times 3}{6 + 3} + 4$ $\quad Q = C_{eff}V$

13. For maximum capacitors are parallel
For minimum capacirots are in sseries

14. U = 1/2 CV^2

15. $E^1 = \varepsilon_r \times E$

16. $U = \dfrac{1}{2}\varepsilon_0 E^2, E = \dfrac{V}{d}$

LEVEL - II (CW)

CAPACITANCE

1. A parallel plate condenser has initially air medium between the plates. If a slab of dieletric constant 5 having thickness half the distance of seperation between the plates is introduced, the percentage increase in its capacity is

1) 33.3% 2) 66.7% 3) 50% 4) 75%

2. **When a dielectric slab of thickness 4 cm is introduced between the plates of parallel plate condenser, it is found the distance between the plates has to be increased by 3cm to restore to capacity to original value. The dielectric constant of the slab is**

1) $\frac{1}{4}$ 2) 4 3) 3 4) 1

3. **The area of the positive plate is A_1 and the area of the negative plate is $A_2\,(A_2 < A_1)$. They are parallel to each other and are separated by a distance d . The capacity of a condenser with air as dielectric is**

1) $\frac{\varepsilon_0 A_1}{d}$ 2) $\frac{\varepsilon_0 A_2}{d}$ 3) $\frac{\varepsilon_0 A_1 A_2}{d}$ 4) $\frac{\varepsilon_0 A_1}{A_2 d}$

4. **The cross section of a cable is shown in fig. The inner conductor has a radius of 10 mm and the dielectric has a thickness of 5 mm. The cable is 8 km long. Then the capacitance of the cable is $[\log_e 1.5 = 0.4]$**

1) 3.8 μF 2) 1.1 μF 3) 4.8×10^{-10} μF 4) 3.3 μF

CAPACITORS IN SERIES AND IN PARALLEL

5. **Two condensers of capacity C and 2C are connected in parallel and these are charged upto V volt. If the battery is removed and dielectric medium of constant K is put between the plates of first condenser, then the potential at each condenser is**

1) $\frac{V}{k+2}$ 2) $2+\frac{k}{3V}$ 3) $\frac{2V}{k+2}$ 4) $\frac{3V}{k+2}$

6. **Given a number of capacitors labelled as C,V. Find the minimum number of capacitors needed to get an arrangement equivalent to C_{net}, V_{net}**

1) $n=\frac{C_{net}}{C}\times\frac{V_{net}^2}{V^2}$ 2) $n=\frac{C}{C_{net}}\times\frac{V^2}{V_{net}^2}$ 3) $n=\frac{C}{C_{net}}\times\frac{V}{V_{net}}$ 4) $n=\frac{C_{net}}{C}\times\frac{V_{net}}{V}$

7. **Two capacitors of capacities $3\mu F$ and $6\mu F$ are connected in series and connected to 120V. The potential difference across $3\mu F$ is V_0 and the charge here is q_0. We have**

A) $q_0 = 40\mu C$ **B)** $V_0 = 60V$ **C)** $V_0 = 80V$ **D)** $q_0 = 240\mu C$

1) A, C are correct 2) A, B are correct 3) B, D are correct 4) C, D are correct

8. **n Capacitors of 2 μ F each are connected in parallel and a p.d of 200v is applied to the combination. The total charge on them was 1c then n is equal to**

1) 3333 2) 3000 3) 2500 4) 25

9. **An infinite number of identical capacitors each of capacitance 1 mF are connected as shown in the figure. Then the equivalent capacitance between A and B is**

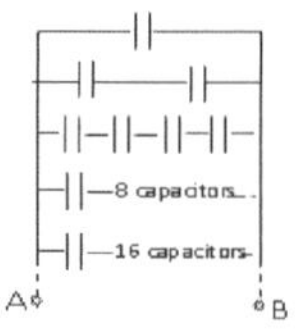

1) 1 mF 2) 2 mF 3) 1/2 mF 4) ¥

ENERGY STORED IN A CONDENSER TYPES OF CAPACITORS

10. Two capacitors of capacites 1 μF and C μF are connected in series and the combination is charged to a potential difference of 120 V. If the charge on the combition is $80\mu C$, the energy stored in the capacitor C in microjoules is :

1) 1800 2) 1600 3) 14400 4) 7200

11. A parallel capacitor of capacitance C is charged and disconnected from the battery. The energy stored in it is E. If a dielectric slab of dielectric constant 6 is inserted between the plates of the capacitor then energy and capacitance will become

1) 6E, 6C 2) E, C 3) $\frac{E}{6}, 6C$ 4) E, 6C

12. In the circuit diagram given below, the value of the potential difference across the plates of the capacitors are

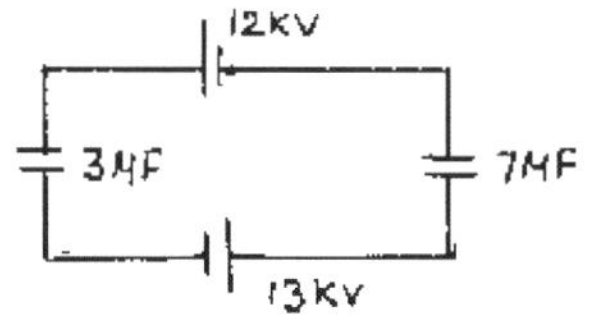

1) 17.5 KV, 7.5 KV 2) 10 KV, 15 KV 3) 5 KV, 20 KV 4) 16.5 KV, 8.5KV

13. The equivalent capacity of the infinite net work shown in the figure (across AB) is (Capacity of each capacitor is 1 μ F)

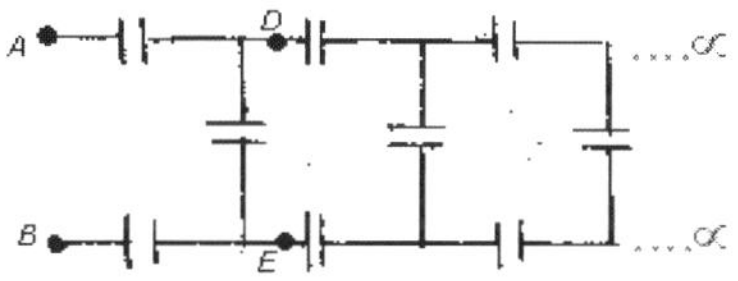

1) ∞ 2) $1\mu F$ 3) $\left(\frac{\sqrt{3}-1}{2}\right)\mu F$ 4) $\left(\frac{\sqrt{3}+1}{2}\right)\mu F$

14. The charge flowing through the cell on closing the key k is equal to

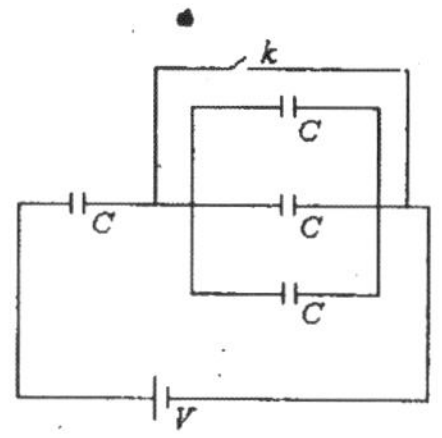

1) $\frac{CV}{4}$ 2) 4 CV 3) $\frac{4}{3}CV$ 4) $\frac{3}{4}CV$

LEVEL - II (CW) - KEY

1) 2	2) 2	3) 2	4) 1	5) 4	6) 4	7) 4	8) 3	9) 2	10) 2
11) 3	12) 1	13) 3	14) 1						

LEVEL - II (CW) - HINTS

1. $C_0 = \frac{\varepsilon_0 A}{d}$; $C = \frac{\epsilon_0 A}{d - t(1 - 1/k)}$

$\Delta C\% = \frac{C - C_0}{C_0} \times 100\%$

2. $C = \frac{\epsilon_0 A}{d - t(1 - 1/k)} = \frac{\epsilon_0 A}{d - d'}$

3. Effective area only $\therefore A_2$

4. $C = \frac{K\varepsilon_o 2\pi l}{\ln(b/a)}$

5. Q = constant , CV + 2CV = $KCV^{I} + 2CV^{I}$

6. $F = \frac{\sigma Q}{2\varepsilon_0} = \frac{Q^2}{2\varepsilon_0 A}$

7. $Q = \left(\frac{C_1 C_2}{C_1 + C_2}\right) V$

8. $Q = nCV$

9. $C_R = C + \frac{C}{2} + \frac{C}{4} + \ldots\ldots$

10. $C_1 V_1 = C_2 V_2$

11. $U = \frac{1}{2} CV^2$; $U_f = KU$; $C' = KC$

12. By Kirchoff loop theorem

$12 - \frac{2}{3} + 13 - \frac{q}{7} = 0$

$V_3 = \frac{q}{3}, V_4 = \frac{q}{7}$

13. Between D & F effective capacitance is x

$\frac{1}{x+1} + 1 + 1 = x$

14. $C_{net} = \frac{3}{4} C$

when key was open $q = \frac{3}{4} CV$

when key was closed 3C becomes short circuited. Net charge on C is now $q' = CV$

$$\Delta q = q' - q = \frac{CV}{4}$$

LEVEL - II (HW)

CAPACITANCE

1. **The capacity of a condenser A is $10\mu F$ and it is charged to a battery of 100 volt. The battery is disconnected and the condenser A is connected to a condenser B the common potential is 40V. The capacity of B is**

1) $8\mu F$ 2) $15\mu F$ 3) $2\mu F$ 4) $1\mu F$

2. **A parallel plate capacitor has the space between its plates filled by two slabs of thickness $\frac{d}{2}$ each and dielectric constant K_1 and K_2. d is the plate separation of the capacitor. The capacitance of the capacitor is**

1) $\frac{2\varepsilon_0 A}{d}\left(\frac{K_1+K_2}{K_1K_2}\right)$ 2) $\frac{2\varepsilon_0 A}{d}(K_1+K_2)$ 3) $\frac{2\varepsilon_0 A}{d}\left(\frac{K_1K_2}{K_1+K_2}\right)$ 4) $\frac{2\varepsilon_0 d}{A}\left(\frac{K_1+K_2}{K_1K_2}\right)$

3. **An isolated capacitor of capacitance 'C' is charged to a potential 'V'. Then a dielectric slab of dielectric constant K is inserted as shown in fig. The net charge on four surfaces 1,2,3 and 4 would be respectively.**

1) 0, CV,-CV, 0
2) 0, $\frac{CV}{K}, \frac{-CV}{K}$, 0
3) CV, 0, 0, -CV
4) CV, $\frac{-CV}{K}, \frac{CV}{K}$, -CV

CAPACITORS IN SERIES AND IN PARALLEL

4. **A parallel plate capacitor with plate area 'A' and separation 'd' is filled with two dielectrics of dielectric constants K_1 and K_2. If the permittivity of free space is ϵ_0, the capacitance of the capacitor is given by**

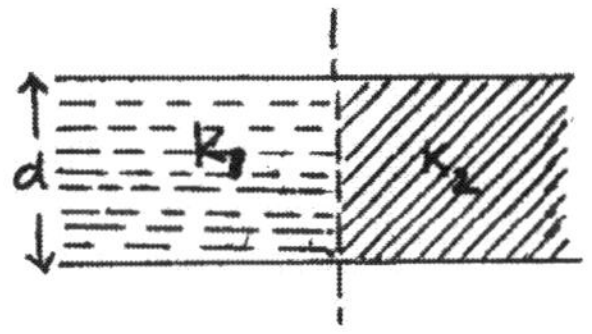

1) $\frac{A\epsilon_0}{d}(K_1+K_2)$
2) $\frac{2A\epsilon_0}{d}(K_1+K_2)$
3) $\frac{A\epsilon_0}{2d}(K_1+K_2)$
4) $\frac{2A\epsilon_0}{d}\cdot\frac{K_1K_2}{K_1+K_2}$

5. **'A' and 'B' are two condensers of capacities 2 μ F and 4 μ F. They are charged to potential differences of 12V and 6V respectively. If they are now connected (+ve to +ve), the charge that flows through the connecting wire is**

1) 24 μ C from A to B
2) 8 μ C from A to B
3) 8 μ C from B to A
4) 24 μ C from B to A

6. **Force of attraction between the plates of a parallel plate capacitor is**

1) $\frac{q^2}{2\varepsilon_0 A}$ 2) $\frac{q^2}{\varepsilon_0 A}$ 3) $\frac{q}{2\varepsilon_0 A}$ 4) $\frac{q^2}{2\varepsilon_0 A^2}$

7. Seven capacitors each of capacitance 2 μ F are to be connected in a configuration to obtain an effective capacitance of $\frac{10}{11}\mu F$. Which of the combination shown in figure will achieve the desired result

1) 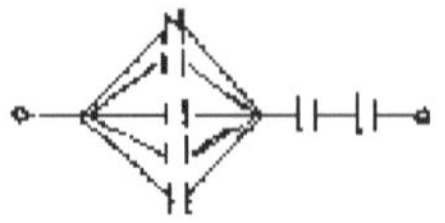

2) 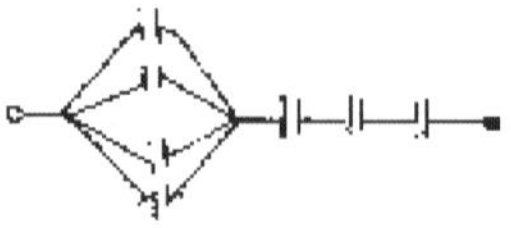

3) 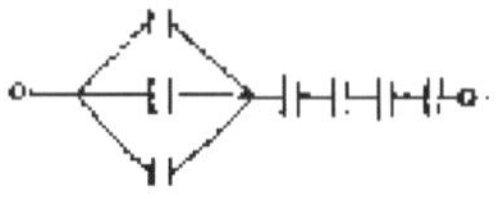

4) 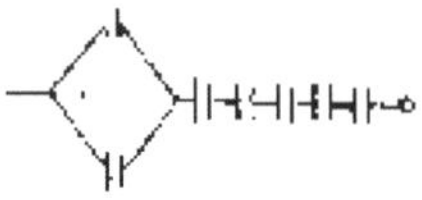

8. The equivalent capacitance between 'A' and 'B' in the adjoining figure is

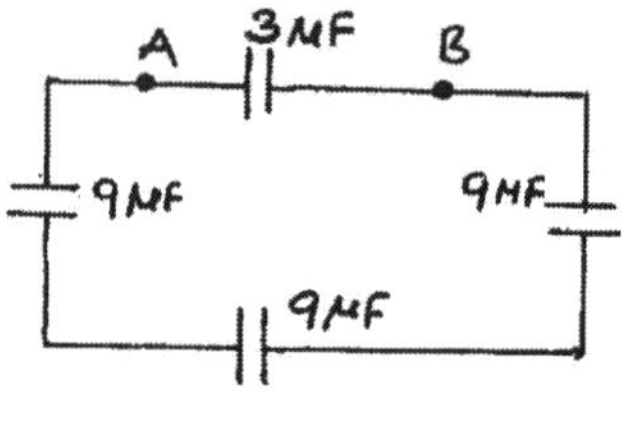

1) $\frac{51}{30}\mu F$ 2) $6\mu F$ 3) $30\mu F$ 4) $12\mu F$

NERGY STORED IN A CONDENSER TYPES OF CAPACITORS

9. A capacitor $4\mu F$ charged to $50V$ is connected to another capacitor $2\mu F$ charged to $100V$. The total energy of combination is

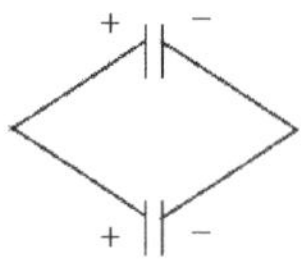

1) $13.3\times10^{-3}J$ 2) $20\times10^{-3}J$ 3) $5\times10^{-3}J$ 4) $10\times10^{-3}J$

10. A $4\mu F$ capacitor is charged by a 200V battery. It is then disconnected from the supply and is connected to another uncharged 2 μF capacitor. During this process, Loss of energy (in J) is:

1) Zero 2) 5.33 x 10^{-2} 3) 4 x 10^{-2} 4) 2.67 x 10^{-2}

11. A capacitor of capacitance C has charge Q and stored energy W. If the charge is increased $2Q$ the stored energy would be

1) $W/4$ 2) $W/2$ 3) $2W$ 4) $4W$

12. The equivalent capacitance between points M and N is

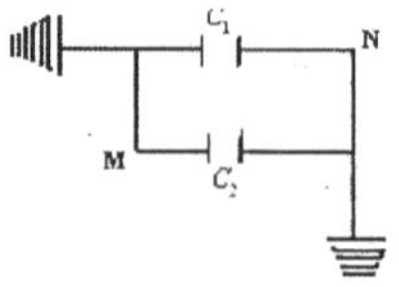

1) Infinity 2) $C_1+\frac{C_2}{C_1}$ 3) $\frac{C_1C_2}{C_1+C_2}$ 4) $\frac{C_1C_2}{C_1-C_2}$

LEVEL - II (HW) - KEY

1) 2 2) 3 3) 2 4) 3 5) 2 6) 1 7) 1 8) 2 9) 1 10) 4
11) 4 12) 1

LEVEL - II (HW) - HINTS

1. $V = \frac{c_1 v_1 + c_2 v_2}{c_1 + c_2}$

2. $\frac{1}{C} = \frac{1}{C_1} + \frac{1}{C_2}$

$C_1 = \frac{A\varepsilon_o K k_1}{d/2} = \frac{2A\varepsilon_o}{d} k_1$

$C_2 = \frac{2A\varepsilon_o}{d} k_2$

3. Due to polarization, charge on dielectric slab would be $CV\left[1 - \frac{1}{K}\right]$

4. $C_1 = \frac{k_1 \in_o \left(\frac{A}{2}\right)}{d}$; $C_2 = \frac{k_2 \in_o \left(\frac{A}{2}\right)}{d}$

C_1 and C_2 connected in parallel..

5 Charge flow $\Delta Q = C_1[V_1 - V_o]$

where $V_o \rightarrow$ common potential

$t \rightarrow$ thickness of metal sheet

6. $q = CV = \frac{C_1 C_2}{C_1 + C_2}.V. = V_o = \frac{q}{C_1}$

7. Verify the cases individually

$C_{eff} = nc$; $Q = C_{eff} V$

8. Resultant capacitance of $9\,\mu F, 9\,\mu F$ and $9\,\mu F$ is in parallel to $3\,\mu F$.

9. $V = \frac{C_1 V_1 + C_2 V_2}{C_1 + C_2}$

$U = \frac{1}{2}(C_1 + C_2)V^2$

10. $V = \frac{C_1 V_1 + C_2 V_2}{C_1 + C_2}$

$U = \frac{1}{2}(C_1 + C_2)V^2$

11. $E \propto Q^2$

12. $C_{eq} = \frac{q}{V_M - V_N} = \frac{q}{0} = \infty$

LEVEL - III

1. The time in seconds required to produce a P.D at 20V across a capacitor at $1000\ \mu F$ when it is charged at the steady rate of $200\ \mu C/\sec$ is

1) 50 2) 100 3) 150 4) 200

2. A parallel plate capacitor of capacity $5\mu F$ and plate separation 6cm is connected to a 1V battery and is charged. A dielectric of dielectric constant 4 and thickness 4 cm is introduced into the capacitor. The additional charge that flows into the capacitor from the battery is

1) $2\mu C$ 2) $3\mu C$ 3) $5\mu C$ 4) $10\mu C$

3. The force between the plates of a parallel plate capacitor of capacitance C and distance of separation of the plates d with a potential difference V between the plates, is

1) $\frac{CV^2}{2d}$ 2) $\frac{C^2V^2}{2d^2}$ 3) $\frac{C^2V^2}{d^2}$ 4) $\frac{V^2d}{C}$

4. Two identical capacitors are connected as show in the figure. A dielectric slab is introduced between the plates of one of the capacitors so as to fill the gap, the battery remaining connected. The charge on each capacitor will be (charge on each condenser is q_0 ; k = dielectric constant)

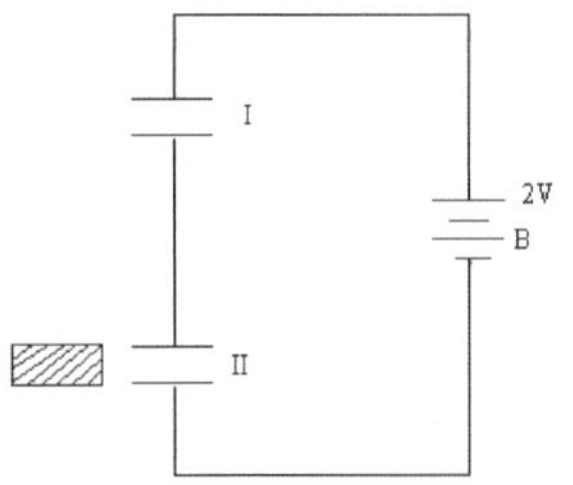

1) $\frac{2q_0}{1+\frac{1}{k}}$ 2) $\frac{q_0}{1+\frac{1}{k}}$ 3) $\frac{2q_0}{1+k}$ 4) $\frac{q_0}{1+k}$

5. Two identical capacitors 1 and 2 are connected in series to a battery as shown in figure. Capacitor 2 contains a dielectric slab of dielectric constant K as shown. Q_1 and Q_2 are the charges stored in the capacitors. Now the dielectric slab is removed and the corresponding charges are Q'_1 and Q'_2. Then

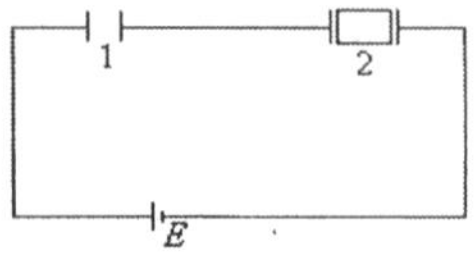

1) $\frac{Q'_1}{Q_1}=\frac{K+1}{K}$ 2) $\frac{Q'_2}{Q_2}=\frac{K+1}{2}$ 3) $\frac{Q'_2}{Q_2}=\frac{K+1}{2K}$ 4) $\frac{Q'_2}{Q_2}=\frac{K}{2}$

6. A capacitor of capacitance 1 μ F withstands a maximum voltage of 6 kV, while another capacitor of capacitance 2 μ F withstands a maximum voltage of 4 kV. If they are connected in series, the combination can withstand a maximum voltage of

1) 3 kV 2) 6 kV 3) 10 kV 4) 9 kV

7. Energy 'E is stored in a parallel plate capacitor 'C_1'. An identical uncharged capacitor 'C_2' is connected to it, kept in contact with it for a while and then disconnected, the energy stored in C_2 is

1) E/2 2) E/3 3) E/4 4) Zero

8. A parallel plate capacitor has area of each plate A, the separation between the plates is d . It is charged to a potential V and then disconnected from the battery. The amount of work done in the filling the capacitor Completely with a dielectric constant k is

1) $\frac{1}{2}\frac{\varepsilon_0 AV^2}{d}\left[1-\frac{1}{k^2}\right]$ 2) $\frac{1}{2}\frac{V^2\varepsilon_0 A}{kd}$ 3) $\frac{1}{2}\frac{V^2\varepsilon_0 A}{k^2 d}$ 4) $\frac{1}{2}\frac{\varepsilon_0 AV^2}{d}\left[1-\frac{1}{K}\right]$

9. A capacitor of capacitance 10 μ F is charged to a potential 50 V with a battery. The battery is now disconnected and an additional charge 200 μ C is given to the positive plate of the capacitor. The potential difference across the capacitor will be

1) 50 V 2) 80 V 3) 100V 4) 60 V

10. A capacitor is filled with an insulator and a certain potential difference is applied to its plates. The energy stored in the capacitor is U. Now the capacitor is disconnected from the source and the insulator is pulled out of the capacitor. The work performed against the forces of electric field in pulling out the insulator is 4U. Then dielectric constant of the insulator is

1) 4 2) 8 3) 5 4) 3

11. A capacitor of capacitance C is charged to a potential difference V from a cell and then disconnected from it. A charge +Q is now given to its positive plate. The potential difference across the capacitor is now

1) V 2) $V+\frac{Q}{C}$ 3) $V+\frac{Q}{2C}$ 4) $V-\frac{Q}{C}$, $ifV < CV$

12. A parallel plate capacitor with plates separated by air acquires 1 μ C of charge when connected to a battery of 500V. The plates still connected to the battery are then immersed in benzene [$k=2.28$]. Then a charge that flows from the battery is

1) 1.28 μ C 2) 2.28 μ C 3) 1 / 4 μ C 4) 4.56 μ C

13. An air capacitor with plates of area 1 m^2 and 0.01 metre apart is charged with 10^{-6} C of electricity. When the capacitor is submerged in oil of relative permittivity 2, then the energy decreases by

1) 20 % 2) 50 % 3) 60 % 4) 75 %

14. Three uncharged capacitors of capacities C_1,C_2 and C_3 are connected as shown in the figure to one another and the point. A, B and C are at potentials V_1,V_2 and V_3 respectively. Then the potential at O will be

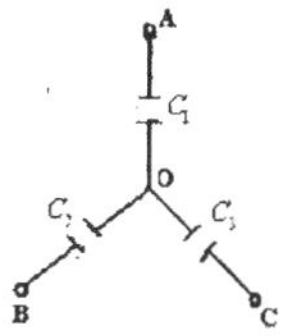

1) $\frac{V_1C_1+V_2C_2+V_3C_3}{C_1+C_2+C_3}$ 2) $\frac{V_1+V_2+V_3}{C_1+C_2+C_3}$ 3) $\frac{V_1(V_2+V_3)}{C_1(C_2+C_3)}$ 4) $\frac{V_1V_2V_3}{C_1C_2C_3}$

15. **In the given figure the capacitor of plate area A is charged upto charge q. The ratio of elongations (neglect force of gravity) in springs C and D at equilibrium position is**

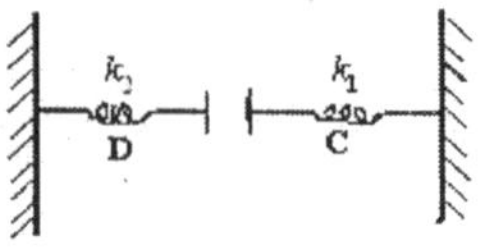

1) $\frac{k_1}{k_2}$ 2) $\frac{k_2}{k_1}$ 3) k_1k_2 4) $\sqrt{\frac{k_1}{k_2}}$

16. **If metal section of shape H is inserted in between two parallel plates as shown in figure and A is the area of each plate then the equivalent capacitance is**

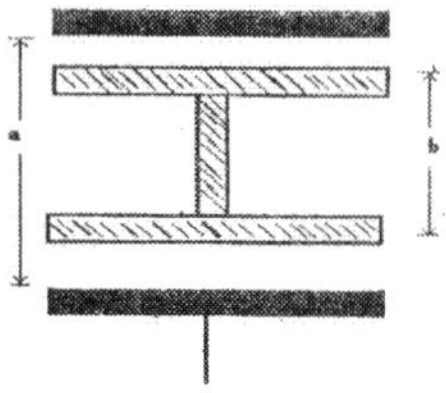

1) $\frac{A\in_0}{a}-\frac{A\in_0}{b}$ 2) $\frac{A\in_0}{a+b}$ 3) $\frac{A\in_0}{a}+\frac{A\in_0}{b}$ 4) $\frac{A\in_0}{a-b}$

17. **The equivalent capacitance C_{AB} of the circuit shown in the figure is**

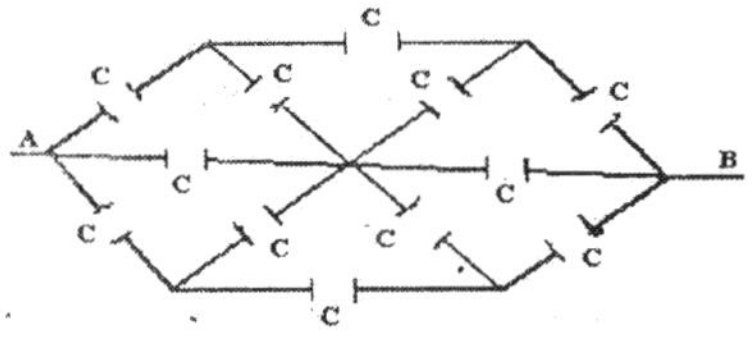

1) $\frac{5}{4}C$ 2) $\frac{4}{5}C$ 3) $2\,C$ 4) C

18. **A solid conducting sphere of radius 10cm is enclosed by a thin metallic shell of radius 20cm. A charge q=20 μC is given to the inner sphere. The heat generated in the process is**

1) 12 J 2) 9 J 3) 24 J 4) zero

19. **A condenser of capacity $500\mu F$ is charged at the rate of $400\mu C$ per second. The time required to raise its potential by $40V$ is**

1) 50 s 2) 100 s 3) 20 s 4) 10 s

20. **In the figure shown the effective capacity across P and Q is (the area of each plate is 'a')**

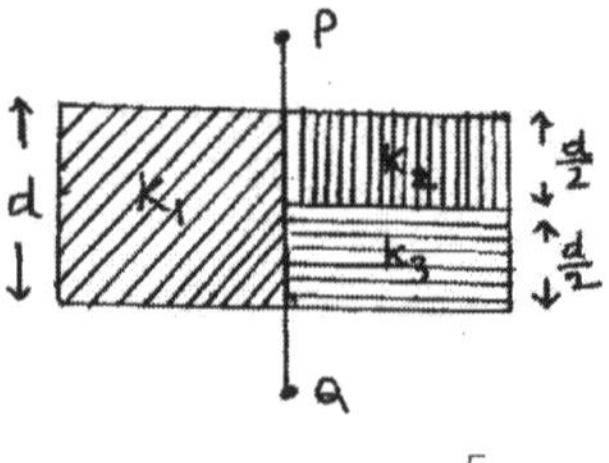

1) $\frac{a\in_0}{d}\left[\frac{K_1}{2}+\frac{K_2K_3}{K_2+K_3}\right]$ 2) $\frac{a\in_0}{2d}\left[\frac{K_2}{2}+\frac{K_1K_3}{K_1+K_3}\right]$

3) $\frac{a\in_0}{3d}\left[\frac{K_3}{2}+\frac{K_1K_2}{K_1+K_2}\right]$ 4) $\frac{a\in_0}{d}\left[\frac{K_1}{2}+\frac{K_1+K_2}{K_2K_3}\right]$

21. **Two capacitors $C_1 = 2\mu F$ and $C_2 = 6\mu F$ in series, are connected in parallel to a third capacitor $C_3 = 4\mu F$. This arrangement is then connected to a battery of e.m.f.=2 V, as shown in figure. The energy lost by the battery in charging the capacitors is**

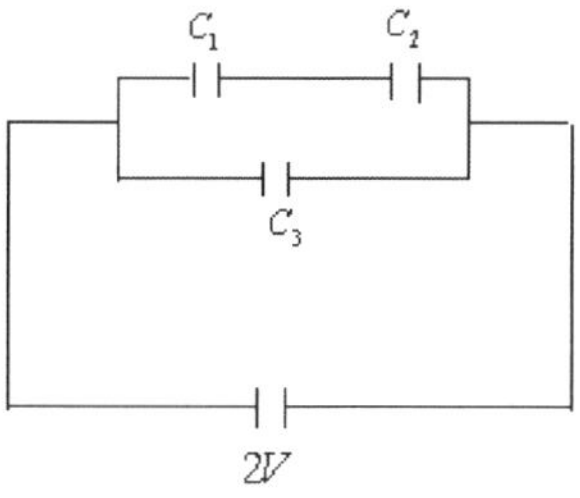

1) $22\times10^{-6} J$ 2) $11\times10^{-6} J$ 3) $\left(\frac{32}{3}\right)\times10^{-6} J$ 4) $\left(\frac{16}{3}\right)\times10^{-6} J$

22. **A capacitor is connected with a battery and stores energy U. After removing the battery, it is connected with another similar capacitor in parallel. The new stored energy in each capacitor will be**

1) $\frac{U}{2}$ 2) U 3) $\frac{U}{4}$ 4) $\frac{3U}{2}$

23. **A parallel plate condenser with a dielectric of dielectric constant K between the plates has a capacity C and is charged to a potential V volts. The dielectric slab is slowly removed from between the plates and then reinserted. The net work done by the system in this process is**

1) $\frac{1}{2}(K-1)CV^2$ 2) $CV^2(K-1)/K$ 3) $(K-1)CV^2$ 4) zero

24. **A fully charged capacitor has a capacitance C. It is discharged through a small coil of resistance wire embedded in a thermally insulated block of specific heat capacity s and mass m. If the temperature of the block is raised by ΔT, the potential difference V across the capacitor is**

1) $\sqrt{\frac{2mC\Delta T}{s}}$ 2) $\frac{mC\Delta T}{s}$ 3) $\frac{ms\Delta T}{C}$ 4) $\sqrt{\frac{2ms\Delta T}{C}}$

25. **A parallel plate capacitor of capacity $100\mu F$ is charged by a battery at 50 volts. The battery remains connected and if the plates of the capacitor are separated so that the distance between them is halved the original distance, the additional energy gives by the battery to the capacitor in Joules is ......**

1) 125×10^{-3} 2) 12.5×10^{-3} 3) 1.25×10^{-3} 4) 0.125×10^{-3}

26. **The equivalent capacity between the points A and B in the adjoining circuit will be**

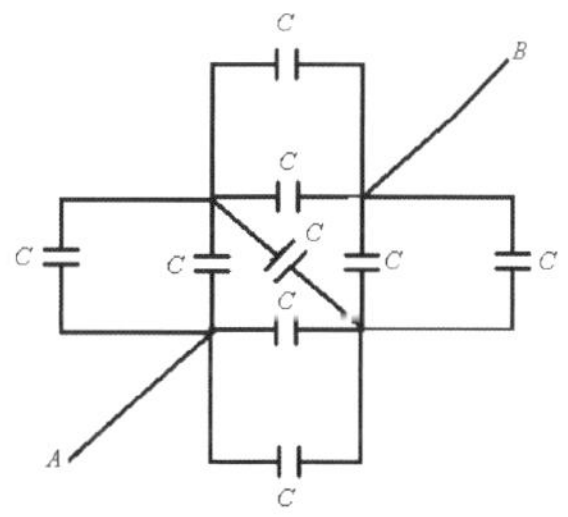

1) C 2) 2C 3) 3C 4) 4

27. **A parallel plate capacitor with air as medium between the plates has a capacitance of 10 μ F. The area of the capacitor is divided into two equal halves and filled with two media having dielectric constant $K_1 = 2$ and $K_2 = 4$. The capacitance will now be**

1) 10 μ F 2) 20 μ F 3) 30 μ F 4) 40 μ F

28. **The capacity of a parallel plate condenser with air medium is $60\mu F$ having distance of seperation d. If the space between the plates is filled with two slabs each of thinckness $d/2$ and dielectric constants 4 and 8, the effective capacity becomes**

1) $160\mu F$ 2) $320\mu F$ 3) $640\mu F$ 4) $360\mu F$

29. **In the adjoining diagram, the condenser C will be fully charged to potential V if**

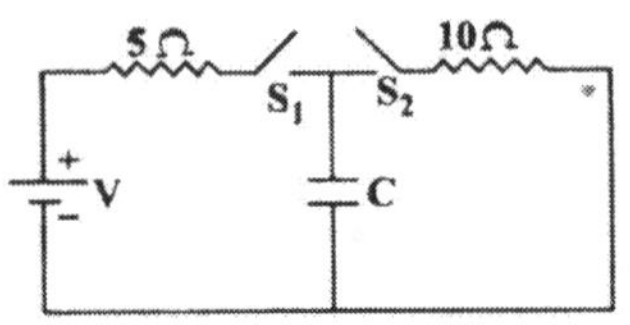

1) S_1 and S_2 both are open 2) S_1 and S_2 both are closed
3) S_1 is closed and S_2 is open 4) S_1 is open and S_2 is closed.

30. **The capacity between the point A and B in the adjoining circuit wil be**

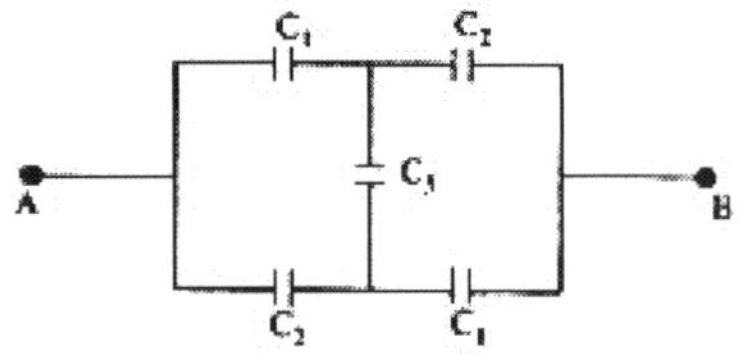

1) $\frac{2C_1C_2 + C_3(C_1 + C_2)}{C_1 + C_2 + 2C_3}$ 2) $\frac{C_1C_2 + C_2C_3 + C_3C_1}{C_1 + C_2 + C_3}$

3) $\frac{C_1(C_2 + C_3) + C_2(C_1 + C_3)}{C_1 + C_2 + 3C_3}$ 4) $\frac{C_1C_2C_3}{C_1C_2 + C_2C_3 + C_3C_1}$

31. **The capacitance C_{AB} in the given network**

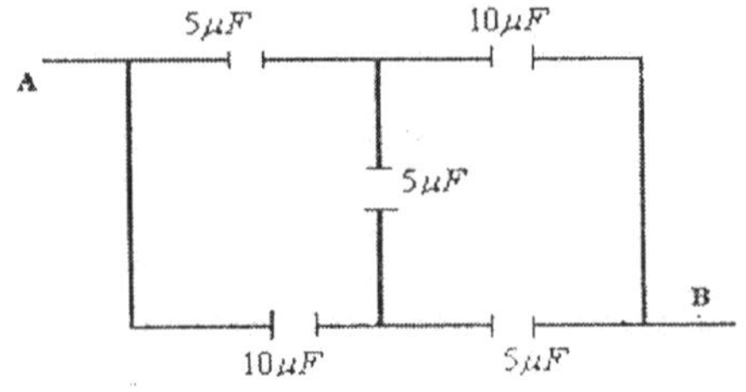

1) 7 μF 2) $\frac{50}{7}\mu F$ 3) $7.5\mu F$ 4) $\frac{7}{50}\mu F$

32. **In the following circuit; find the potentials at points A and B is**

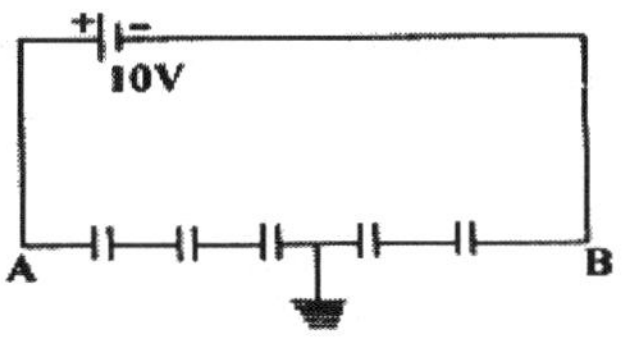

1) 10V, 0V 2) 6 V, -4V 3) 4V, -6V 4) 5V, -5V

33. **The potential difference between the points A and B in the following circuit in steady state will be**

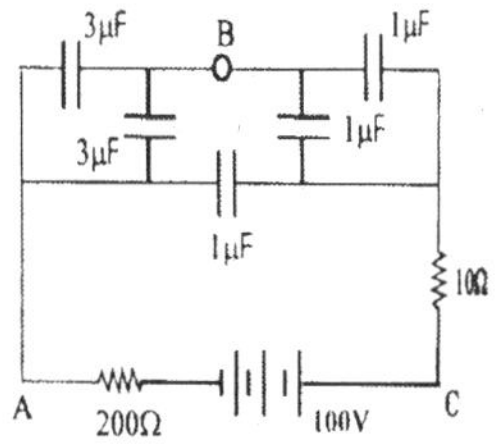

1) $V_{AB} = 100$ volt 2) $V_{AB} = 75$ volt 3) $V_{AB} = 25$ volt 4) $V_{AB} = 50$ volt

34. **In the following circuit two identical capacitors, a battery and a switch(s) are connected as shown. the switch(s) is opened and dielectric of constant $(K = 3)$ are inserted in the condensers. The ratio of electrostatic energies of the system before and after filling the dielectric will be**

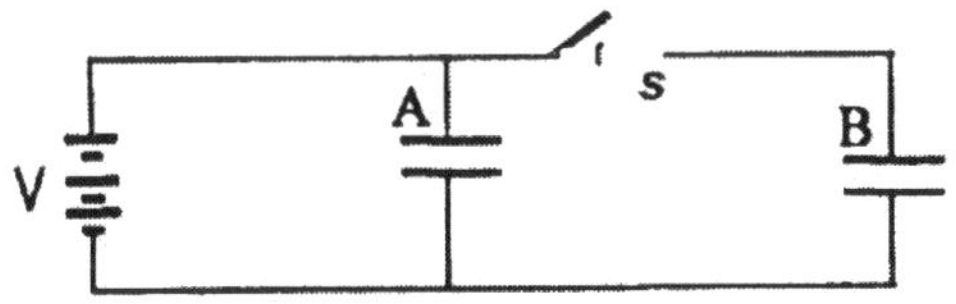

1) 3: 1 2) 5 : 1 3) 3:5 4) 5 : 3

35. **In the given figure a capacitor of plate area A is charged upto charge q. The mass of each plate is m_2. The lower plate is rigidly fixed. The value of m_1 if the system remains in equilibrium is**

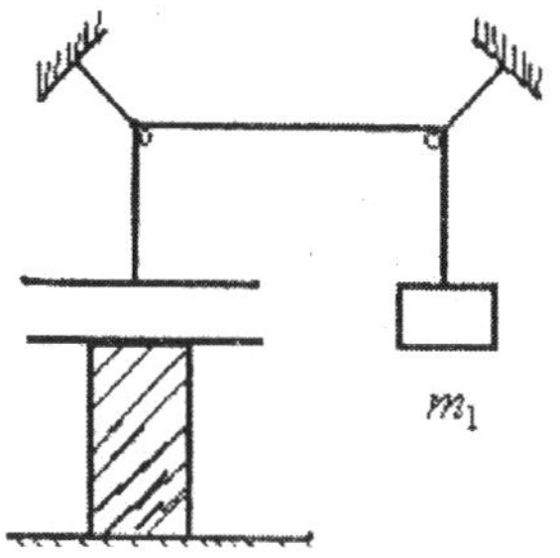

1) $m_2 + \dfrac{q^2}{\epsilon_0 Ag}$ 2) m_2 3) $\dfrac{q^2}{2\epsilon_0 Ag} + m_2$ 4) $2m_2$

36. One plate of a capacitor is connected to a spring as shown in figure. Area of both the plates is A. In steady state; separation between the plates is 0.8d (spring was unstretched and the distance between the plates was d, when the capacitor was uncharged). The force constant of the spring is approximately

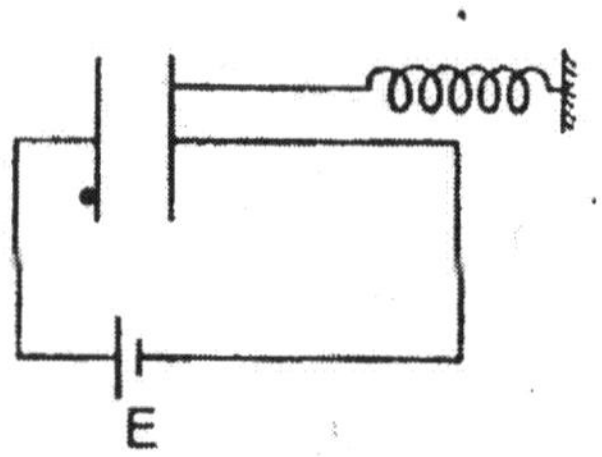

1) $\frac{4\epsilon_0 AE^2}{d^3}$ 2) $\frac{2\epsilon_0 AE}{d^2}$ 3) $\frac{6\epsilon_0 E^2}{Ad^3}$ 4) $\frac{\epsilon_0 AE^3}{2d^3}$

37. A capacitor is made of a flat plate of area A and second plate having a stair-like structure as shown in figure. The width of each plate is 'a' and the height is 'b'. The capacitance of the capacitor is

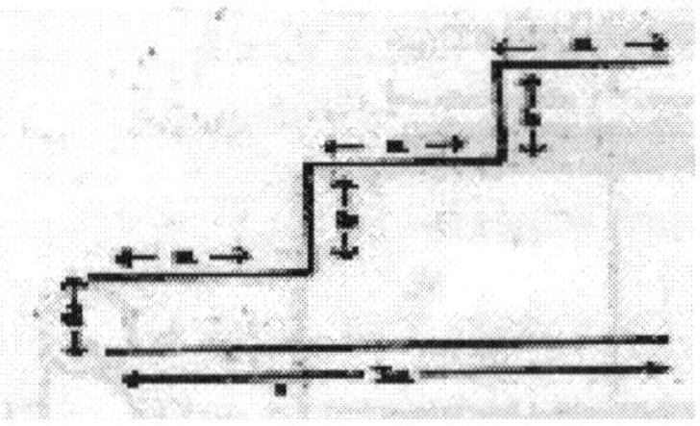

1) $\frac{2A\epsilon_0}{3(d+b)}$ 2) $\frac{A\epsilon_0(3d^2+6bd+2b^2)}{3d(b+d)(d+2b)}$ 3) $\frac{A\epsilon_0(d^2+2bd+b^2)}{3d(d+b)(d+2b)}$ 4) $\frac{2A\epsilon_0(d^2+2bd+b^2)}{3d(d+b)(d+2b)}$

38. A parallel plate capacitor of capacitance C is connected to a battery and is charged to a potential difference V. Another capacitor of capacitance 2C is similarly charged to a potential difference 2V. The charging battery is now disconnected and the capacitors are connected in parallel to each other in such a way that the positive terminal of one is connected to the negative terminal of the other. The final energy of the configuration is

1) zero 2) $\frac{3}{2}CV^2$ 3) $\frac{35}{6}CV^2$ 4) $\frac{9}{2}CV^2$

39. Two identical capacitors, have the same capacitance C. One of them is charged to potential V_1 and the other to V_2.The negative ends are also connected, the decrease in energy of the combined system is

1) $\frac{1}{4}C(V_1^2-V_2^2)$ 2) $\frac{1}{4}C(V_1^2+V_2^2)$ 3) $\frac{1}{4}C(V_1-V_2)^2$ 4) $\frac{1}{4}C(V_1+V_2)^2$

40. Consider the situation shown in the figure. The capacitor A has a charge q on it whereas B is uncharged. The charge appearing on the capacitor B a long 7 time after the switch is closed is:

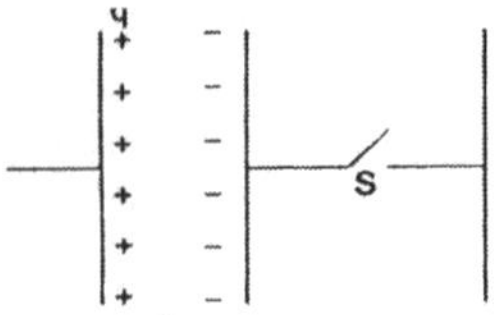

1) Zero 2) q /2 3) q 4) 2q

41. Find the capacitance of a system of two identical metal balls of radius a if the distance between their centres is equal to b, with b>>a. The system is located in a uniform dielectric with permittivity K.

1) $\pi \in_0 Ka$ 2) $4\pi \in_0 Ka$ 3) $2\pi \in_0 Ka$ 4) $\frac{2}{3}\pi \in_0 Ka$

LEVEL - III - KEY

1) 2	2) 3	3) 1	4) 1	5) 3	6) 4	7) 3	8) 4	9) 4	10) 3
11) 3	12) 1	13) 2	14) 1	15) 2	16) 4	17) 1	18) 2	19) 1	20) 1
21) 2	22) 3	23) 4	24) 4	25) 1	26) 2	27) 3	28) 2	29) 3	30) 1
31) 1	32) 2	33) 3	34) 3	35) 3	36) 1	37) 2	38) 2	39) 3	40) 1
41) 3									

LEVEL - III - HINTS

1. $\frac{dq}{dt} = \frac{c.dv}{dt}$

2. $q_1 = cv$; $q_2 = C^{/}V, \frac{c}{c^{/}} = \frac{d - t\frac{t}{K}}{d}$

   $\Delta q = q_2 - q_1$

3. $F = \frac{QE}{2} = \frac{CV}{2}\left[\frac{V}{d}\right]$

4. $C_{eff} = \frac{C_0 k}{1+k}$

   $q = C_{eff}V$

5. $Q_1^{'} = Q_2^{'} = \frac{CE}{2}$

   Before the slab is removed

   $C_1 = C \ and \ C_2 = kC$

   $C_{net} = \left(\frac{k}{k+1}\right)C$

   $\frac{Q_2^{'}}{Q_2} = \frac{k+1}{2k}$

6. $V_1 \le V_{max1}, V_2 \le V_{max2}$

7. $U_2^{/} = \frac{1}{2}C_2 V^2_{common}$

8. Work done = decrease in energy

   ie $w = E_1 - E_2 = \frac{1}{2}\frac{\varepsilon_0 A}{d}v^2 - \frac{\varepsilon_0 A v^2}{2d}\left[1 - \frac{1}{k}\right]$

9. $q_0 = CV = 500\mu C$

$$\frac{700-q}{2A\in_0}+\frac{q}{2A\in_0}+\frac{500-q}{2A\in_0}=\frac{q}{2A\in_0}$$

$q = 600\mu C$

$\Delta V = \frac{q}{C}$ =60V

10. $U = \frac{1}{2}\frac{q^2}{C}$

$U + 4U = \frac{1}{2}\frac{q^2}{C_0}$

$5U = \frac{1}{2}\frac{q^2}{C_0}$

$k = \frac{C}{C_0} = 5$

11. $E = \frac{\sigma}{\in_0} = \frac{\frac{Q}{2}+CV}{A\in_0}$

$$V' = Ed = \frac{\frac{Q}{2}+CV}{\frac{A\in_0}{d}}$$

$$= \frac{\frac{Q}{2}+CV}{C} = V + \frac{Q}{2C}$$

12. $Q_0 = C_0V_0$

$Q = CV_0 = KC_0V_0$

$\Delta Q = Q - Q_0 = (K-1)C_0V_0$

13. $\text{U}' = \frac{\text{U}}{\text{K}}$

14. $q_1 = q_2 + q_3$

$\Rightarrow (V_1 - V_0)C_1 = (V_0 - V_2)C_2 + (V_0 - V_3)C_3$

$$\therefore V_0 = \frac{C_1V_1 + C_2V_2 + C_3V_3}{C_1 + C_2 + C_3}$$

15. $F_e = k_1x_1 = k_2x_2 \quad \therefore \frac{x_1}{x_2} = \frac{k_2}{k_1}$

16. Net space between metal plates is a-b

17. 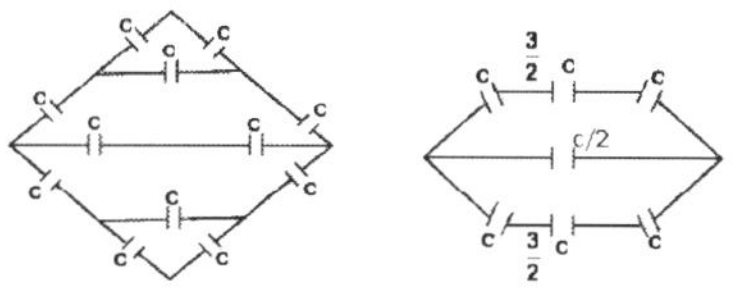

18. $H = U_i - U_f = \dfrac{q^2}{2C_1} - \dfrac{q^2}{2C_2}$

$C_1 = 4\pi \in_0 R_1, C_2 = 4\pi \in_0 R_2$

19. $\dfrac{dq}{dt} = \dfrac{c.dv}{dt}$

20. $C_1 = \dfrac{K_1 \in_o \left(\dfrac{A}{2}\right)}{d}$; $C_2 = \dfrac{K_2 \in_o A}{d}$; $C_3 = \dfrac{K_3 \in_o A}{d}$

$\therefore C = \dfrac{C_2C_3}{C_2 + C_3} + C_1$

21. $E_{lost} = \dfrac{1}{2} C_{eff} V^2$

22. $V = \dfrac{C_1V_1 + C_2V_2}{C_1 + C_2}$; $U = \dfrac{1}{2}(C_1 + C_2)V^2$

23. On introduction and removal and again on introduction, the capacity and potential remain same. So, net work done by the system in this process.

$W = U_f - U_i$

$\dfrac{1}{2}CV^2 - \dfrac{1}{2}CV^2 = 0$

24. $E = \left(\dfrac{1}{2}\right)CV^2$

The energy stored in capacitor is lost in form of heat energy.

$H = ms\Delta T$

$\therefore ms\Delta T = \left(\dfrac{1}{2}\right)CV^2$

$$C = \frac{q_1 + q_2}{V_A - V_B} = \frac{q_1 + q_2}{\dfrac{q_2}{C_2} + \dfrac{q_1}{C_1}}$$

$$\therefore C = \frac{2C_1C_2 + C_3(C_1 + C_2)}{C_1 + C_2 + 2C_3}$$

$$\therefore V = \sqrt{\frac{2ms\Delta T}{C}}$$

25. $E_1 = {}_{2}C_1V^2$

$E_2 = \frac{1}{2}C_2V^2$

$C_2 = 2C_1$

26. Wheat stone's Bridge

27. $c = \frac{c_0}{2}(k_1 + k_2)$

28. $C = \frac{2K_1K_2}{K_1 + K_2}C_0$

29. If S_1 and S_2 both are closed then charge and discharge processes with simoultaneously take place. Hence to charge the condenser fully the key S_1must be closed and S_2 must remain open

30. According to the symmetry of the circuit charges on two condensers fo capacity C_1 will be same and charges on condensers of capacity C_2 will be same.

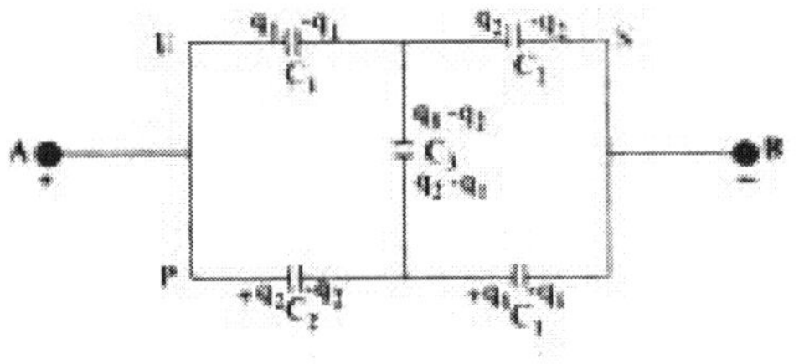

$$\frac{q_2}{C_2} + \frac{q_2 - q_1}{C_3} - \frac{q_1}{C_1} = 0$$

$$\therefore \frac{q_2}{q_1} = \frac{C_1(C_2 + C_3)}{C_2(C_1 + C_3)}$$

Capacity of whole circuit

31. $\therefore C = \frac{2C_1C_2 + C_3(C_1 + C_2)}{C_1 + C_2 + 2C_3}$

32. P.D across each condenser = 2V

Potential at earth = 0V

$\therefore V_A = +6V \quad V_B = -4V$

33 $V_{AB} = \frac{VC_2}{C_1 + C_2}$

$\therefore V_{AB} = \frac{1}{4} \times 100 = 25V$

34. $\therefore U_1 = \frac{1}{2}CV^2 + \frac{1}{2}CV^2 = CV^2$

$$\therefore U_2 = \frac{1}{2}\times 3\times CV^2 + \frac{1}{2}\times 3C\frac{V^2}{9} = \frac{5}{3}CV^2$$

$$\therefore \frac{U_1}{U_2} = \frac{3}{5}$$

35. $T = m_1 g$ and $T = m_2 g + F_e$

$\therefore m_1 g = m_2 g + F_e$

Here, $F_e = \dfrac{q^2}{2 \in_0 A}$

36. $\dfrac{q^2}{2A \in_0} = kx$

$\dfrac{(CE)^2}{2A \in_0} = k(d - 0.8d)$, $C = \dfrac{\in_0 A}{0.8d}$

$K = \dfrac{4 \in_0 AE^2}{d^3}$

37. In Equivalent circuit C_1, C_2, C_3 are parallel

$C_1 = \dfrac{A \in_0}{3d}, C_2 = \dfrac{A \in_0}{3(d+b)}, C_3 = \dfrac{A \in_0}{3(d+2b)}$

38. Net charge $Q = Q_2 - Q_1$ potential is V_t

$\therefore V_1 = \left(\dfrac{C_0}{C + C_0}\right)V_0$

Similarly after nth operation

$E = 1/2C^1V^2$

39. $Q = CV$

$U_i = 1/2CV^2$

$U_f = 1/2CV^2$ work done $= U_i - U_f$

40 Due to attraction with positive charge, the negative charge on capacitor A will not flow thrugh the switch S.

41. $V_1 - V_2 = 2V = 2\int_a^{b-a} E\, dr$

$C - \dfrac{q}{V_1 - V_2}$

Chapter - 4

(Part - 1)
CURRENT ELECTRICITY

CURRENT ELECTRICITY

SYNOPSIS

STRENGTH OF ELECRRIC CURRENT:

The strength of electric current is defined as rate of flow of charge through any cross section of a conductor.

1. The instantaneous current is defined by the equation, $I = \underset{\Delta t \to 0}{Lt} \frac{\Delta Q}{\Delta t} = \frac{dQ}{dt}$

Average current $i = \frac{q}{t}$

Ampere : If one coulomb of charge passes through a cross-section of the conductor per second then the current is one ampere.

$1 \text{ ampere} = \frac{1 \text{coulomb}}{1 \text{second}}$

current is a scalar quantity.

APPLICATIONS ON ELECTRIC CURRENT

1. If the current is varying with time t, then the charge flowing in a time interval from t_1 to t_2 is $q = \int_{t_1}^{t_2} I dt$

2. If n particles, each having a charge q, pass through a given cross sectional area in time t, then average current is $i = \frac{nq}{t}$

3. If a point charge q is revolving in a circle of radius r with speed v then its time period is

$T = (2\pi r / v)$

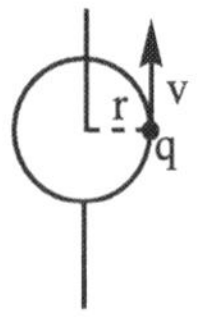

4. The average current associated with this revolving charge is

$$I = \frac{q}{T} = fq = \frac{\omega}{2\pi} q = \frac{vq}{2\pi r}$$

Where f is the frequency of revolution in Hz.
ω is the angular frequency in rad/sec
v is linear velocity of the charge q
r is radius of the circular path

5. If in a discharge tube n_1 protons are moving from left to right in t seconds and n_2 electrons are moving simultaneously from right to left in t seconds, then the net current in any crossection of the discharge tube is $I = \frac{(n_1 + n_2)e}{t}$ (from left to right)

here e is the magnitude of charge of electron (or) proton.

DRIFT VELOCITY :

Drift velocity is the average velocity acquired by free electrons inside a metal by the application of an electric field which results in current.

Drift velocity $v_d = \frac{J}{ne} = \frac{I}{Ane}$

where, $J = I/A$ is current density

n is number of free electrons per unit volume

e is charge of electron

The drift velocity is related to relaxation time is

$$v_d = \frac{eE}{m}\tau$$

NOTE :

1. The drift velocity of electrons is of the order of $10^{-4}\ ms^{-1}$.
2. Greater the electric field, greater will be the drift velocity $v_d \propto E$
3. The direction of drift velocity for electrons in a metal is opposite to that of electric field applied $\vec{E}$

W.E - 1 : In a hydrogen atom, electron moves in an orbit of radius 5 × 10^{-11} m with a speed of 2.2 × 10^{6} m/s. Calculate the equivalent current.

Solution : Current $i = f.e = \frac{v}{2\pi r}.e$

$$= \frac{2.2\times10^6}{2\pi\times5\times10^{-11}} \times 1.6 \times 10^{-19}$$

$= 1.12 \times 10^{-3}$ amp $= 1.12$ mA.

W.E - 2 : The current through a wire depends on time as $i = i_0 + \alpha t$, where $i_0 = 10A$ and $\alpha = 4A/s$. Find the charge that crossed through a section of the wire in 10 seconds.

Solution : $i = i_0 + \alpha t$; but $i = \frac{dq}{dt}$

$$\Rightarrow dq = (i_0 + \alpha t)dt$$

$$q = \int_{t=0}^{t=10} dq \Rightarrow q = \left[i_0 t + \frac{\alpha t^2}{2} \right]_0^{10}$$

$= (10i_0 + 50\alpha) = 300$ coloumb

CURRENT DENSITY $(\vec{J})$

Current density at a point is defined as a vector having magnitude equal to current per unit area.

$$\vec{J} = \underset{\Delta s\to0}{Lt} \frac{\Delta I}{\Delta s} = \frac{dI}{ds}\hat{n}$$

If the normal to the area makes an angle θ with the direction of the current, then the current density is

$$J = \frac{\Delta I}{\Delta s\cos\theta},\ dI = Jds\cos\theta \quad \text{(or)}$$

$dI = \vec{J}.\vec{ds}$ i.e., $I = \int \vec{J}.\vec{ds}$

SI unit of $\vec{J}$ is Am^{-2}

Dimensional formula of J is $[AL^{-2}]$

Current is the flux of current density.

RELAXATION TIME (τ) :

1. It is the time interval between two successive collisions of electrons with +ve ions in the metallic lattice.

The resistance of a conductor is given by

$$R = \frac{2ml}{ne^2\tau A}$$

where n = number density of electrons

e = electron charge

m = mass of electron

τ = relaxation time.

W.E - 3 : Consider a wire of length 4m and cross-sectional area I mm² carrying a current of 2A. If each cubic metre of the material contains 10^{29} free electrons, find the average time taken by an elctron to cross the length of the wire.

Solution : $\Rightarrow v_d = \frac{i}{nAe} = \frac{2}{10^{29} \times 10^{-6} \times 1.6 \times 10^{-19}} ms^{-1}$

$= 12.5 \times 10^{-4}\ ms^{-1}$

Average time taken by an electron to cross the length of wire

$$t = \frac{l}{v_d} = \frac{4}{1.25 \times 10^{-4}}\ s = 3.2 \times 10^4\ s$$

MOBILITY (μ) :

Mobility (μ) of a charge carrier (like electron) is defined as the average drift velocity resulting from the application of unit electric field strength.

$$\mu = \frac{drift\ velocity}{electric\ field}$$

$$\therefore \quad \mu = \frac{|v_d|}{E}$$

Mobility depends on pressure and temperature.

OHM'S LAW :

For a given conductor, at a given temperature the strength of electric current through it is directly proportional to the potential difference applied across at its ends".

i.e. $I \alpha V \Rightarrow I = \frac{V}{R}$

$V = IR$

Where R is electrical resistance of the conductor

NOTE :

- ohm's law is neither a basic law nor a deriavable one
- ohm's law is just an empherical relation.

- Microscopically Ohm's law is expressed as

  $J = nev_d \Rightarrow J = \sigma E$ where σ is the electrical conductivity of the material.
- The conductors which obey Ohm's law are called Ohmic conductors.
  Ex : all metals
- For Ohmic conductors V – i graph is a straight line passing through origin (metals).

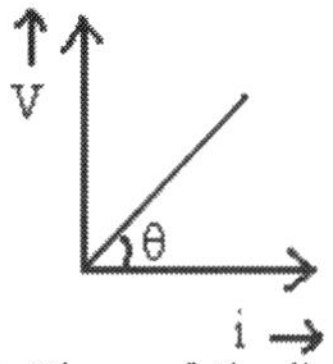

(A) Slope of the line $\tan\theta = \frac{V}{i} = R$

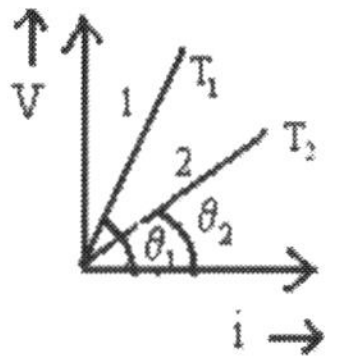

(B) Here $\tan\theta_1 > \tan\theta_2$
So $R_1 > R_2$
i.e $T_1 > T_2$

- The substances which do not obey Ohm's law are called non-Ohmic conductors.
  Ex: Thermistor, Electronic Valve, Semi-conductor devices, gases, crystal rectifier etc.,
- The V – i graph for a non – Ohmic conductor is non-linear.

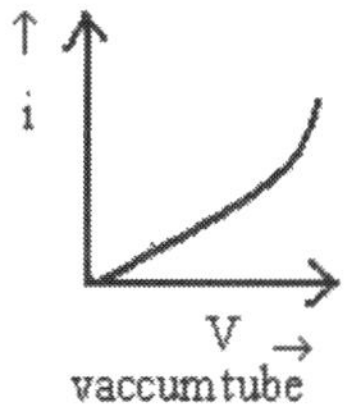

vaccum tube

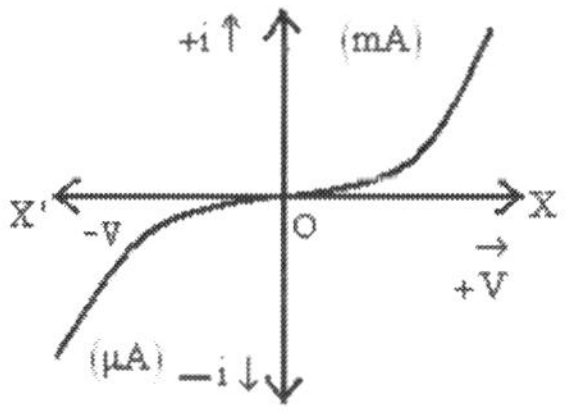

semi conductors

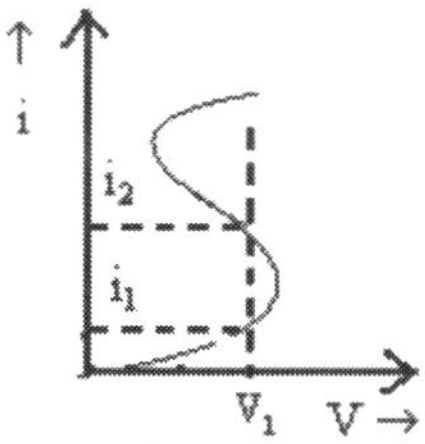

Thermistor

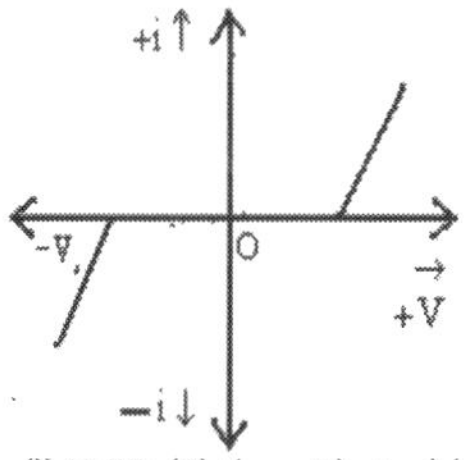

dil H_2SO_4 (Platinum Electrode)

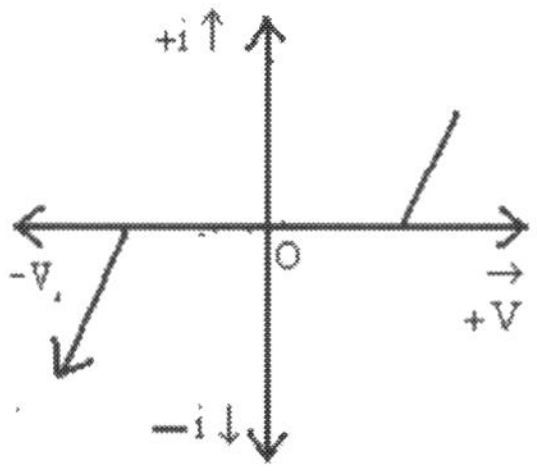

Neon Gas(With tungsten Electrode)

NON-OHMIC CIRCUITS:

- The circuits in which Ohm's law is not obeyed are called non-ohmic circuits. The V-I graph is a curve, e.g. torch bulb, electrolyte,
semiconductors, thermonic valves etc. as shown by curves (a), (b), (c).

a) 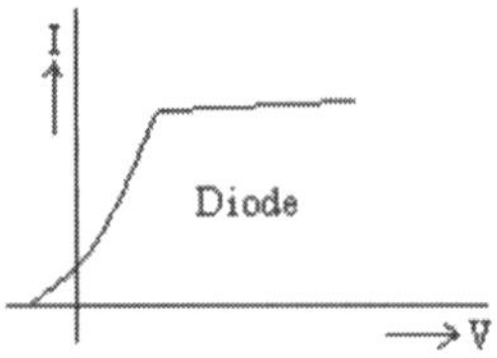

b) 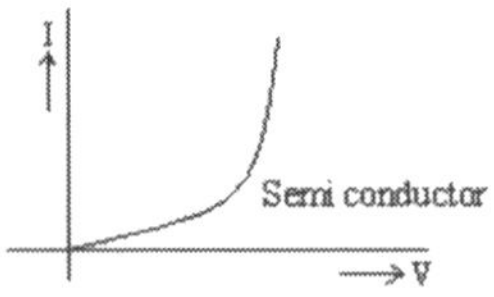

c) 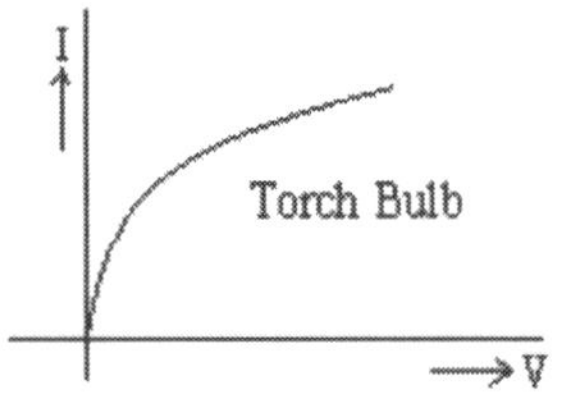

RESISTANCE :

Definiton : The resistance of a conductor is defined as the ratio of the potential difference 'V' across the condutor to the current 'i' flowing through the conductor.

Resistance $R = \frac{V}{i}$

- The resistance of a conductor depends upon
1) shape (dimensions) 2) nature of material 3) impurities 4) Temperature
- The resistance of a conductor increases with impurities.
- The resistance of a semi conductor decreases with impurities.

FACTORS EFFECTING THE RESISTANCE OF A CONDUCTOR :

1. The resistance of the conductor is directly proportional to the length (l) of the conductor i.e.

$R \alpha l$ (or) $\frac{R_1}{R_2} = \frac{l_1}{l_2}$

For small changes in the length, $\frac{\Delta R}{R} = \frac{\Delta l}{l}$

2. The resistance of a conductor is inversely proportional to the area of cross-section (A)

i.e., $R \alpha \frac{1}{A}$ (or) $R \propto \frac{1}{r^2}$

$$\frac{R_1}{R_2} = \left(\frac{A_2}{A_1}\right) = \left(\frac{r_2^2}{r_1^2}\right)$$

For small changes in area (or) radius we have $\frac{\Delta R}{R} = \frac{\Delta A}{A} = -\frac{2\Delta r}{r}$

3. As the temperature increases resistance of metallic conductors increases and that of semiconductors decreases.

CONDUCTANCE :

The reciprocal of resistance (R) is called conductance.

conductance, $G = \frac{1}{R}$.

The S.I unit of conductance is mho or siemen or ohm^{-1}.

RESISTIVITY :

As we know, that the resistance of the conductor is directly propertional to its length and inversely proportional to its area of cross section, we can write

$$R \propto \frac{l}{A} \Rightarrow R = \frac{\rho l}{A}$$

where ρ is specific resistance or resisitivity of the material of the conductor.

NOTE :

1. Resistitivity is the specific property of a material but Resistance is the bulk property of a conductor.
2. Resistivity is independent of dimensions of the conductor such as length, area of the cross section.
3. Resistivity depends on the nature of the material of the conductor, temperature and impurities.
4. Resistivity of any alloy is more than resistivity of its constituent elements.

   i) $R_{alloys} > R_{conductors}$ ii) $\alpha_{metals} > \alpha_{alloys}$

SPECIAL CASES :

1. The alternate forms of resistance is

$$R = \rho\frac{l^2}{V} = \rho\frac{l^2 d}{m} = \frac{\rho V}{A^2} = \frac{\rho m}{d\,A^2}$$

Where d is density of material of conductor

V is volume of the conductor

m is mass of the conductor

2. If a conductor is streched or elongated or drawn or twisted, then the volume of the conductor is constant. Hence

a. $R = \frac{\rho l^2}{V} \Rightarrow R\alpha l^2$

b. $R = \frac{\rho V}{A^2} \Rightarrow R\alpha\frac{1}{A^2}\alpha\frac{1}{r^4}$

c. Interms of mass of the wire $R\alpha\frac{l^2}{m}$

and $R\alpha\frac{m}{A^2}\alpha\frac{m}{r^4}$

3. For small changes in the length or radius during the stretching

$$\frac{\Delta R}{R} = 2\frac{\Delta l}{l}$$

$$\frac{\Delta R}{R} = -2\frac{\Delta A}{A} = -4\frac{\Delta r}{r}$$

4. In case of a cuboid of dimensions $l \times b \times h$ is

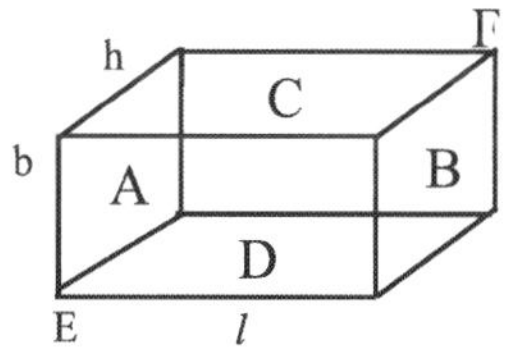

Resistance across AB, $R_{AB} = \frac{\rho l}{b \times h}$

Resistance across CD, $R_{CD} = \frac{\rho b}{l \times h}$

Resistance across EF, $R_{EF} = \frac{\rho h}{l \times b}$

If $l > b > h$, then

$R_{max} = \frac{\rho l}{b \times h}$ $R_{min} = \frac{\rho h}{l \times b}$

W.E - 4:

A rectangular block has dimensions 5 cm × 5 cm × 10 cm. Calculate the resistance measured between (a) two square ends and (b) the opposite rectangular ends. Specific resistance of the material is $3.5 \times 10^{-5}\,\Omega m$.

Solution :

a) Resistance between two square ends $R_1 = \frac{\rho \ell}{A}$

$$R_1 = \frac{3.5 \times 10^{-5} \times 10 \times 10^{-2}}{5 \times 5 \times 10^{-4}} = 1.4 \times 10^{-3}\,\Omega$$

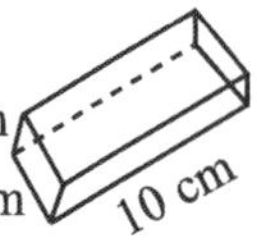

b) Resistance between the opposite rectangular ends $R_2 = \frac{\rho \ell}{A}$

$$R_2 = \frac{3.5 \times 10^{-5} \times 5 \times 10^{-2}}{5 \times 10 \times 10^{-4}} = 1.4 \times 10^{-4}\,\Omega$$

5. If a wire of resistance R is stretched to 'n' times its original length, its resistance becomes n^2 R.
6. If a wire of resistance R is stretched until its radius becomes $\frac{1}{n}$ th of its original radius then its resistance becomes **n^4R.**
7. When a wire is stretched to increase its length by x% (where x is very small) its resistance increases by 2x %.
8. When a wire is stretched to increase its length by x% (where x is large) its resistance increases by $\left(2x + \frac{x^2}{100}\right)$.
9. When a wire is stretched to reduce its radius byx% (where x is very small), its resistance increases by 4x%.

CONDUCTIVITY :

Conductivity is the measure of the ability of a material to conduct electric current through it. It is reciprocal of resistivity.

$$\sigma = \frac{1}{\rho} = \frac{l}{RA}$$

S.I unit : sieman / m : (Sm^{-1})

For perfect insulators $\sigma = 0$

For perfect conductors, σ is infinity.

Temperature dependence of resistance:

For conductors i.e metals resistance increases with rise in temperature

$R_t = R_o(1 + \alpha t + \beta t^2)$ for t > 300°C

$R_t = R_o(1 + \alpha t)$ for t < 300°C or $\alpha = \frac{R_t - R_o}{R_o t}$ $/^0 C$

If R_0 = resistance of conductor at 0°C

If R_t = resistance of conductor at t°C

And α, β = temperature co-efficients of resistance

If R_1 and R_2 are the resistances at t_1°C to t_2°C respectively then $\frac{R_1}{R_2} = \frac{1 + \alpha t_1}{1 + \alpha t_2}$

$\therefore \alpha = \frac{R_2 - R_1}{R_1 t_2 - R_2 t_1}$

The value of α is different at different temperatures.

At a given temperature $\alpha = \frac{1}{R_t}\left(\frac{dR}{dt}\right)$ at t°C

➤ 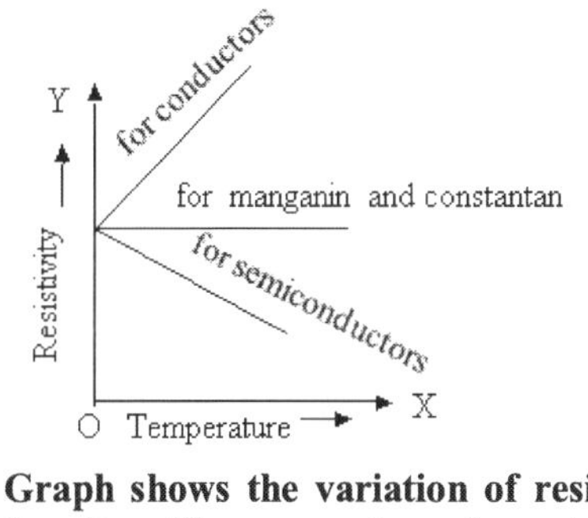

Graph shows the variation of resistivity with temperature for conductors, semiconductors and for alloys like manganin and constantan. Since the resistivity of manganin and constantan remains constant with respect to change in temperature, these materials are used for the bridge wires and resistance coils.

➤ The resistivity of manganin and constantan is almost independent of temperature.

➤ **Two resistors having resistances R_1 and R_2 at $0^o C$ are connected in series. The condition for the effective resistance in series in same at all temperatures**

$R_1 + R_2 = R_1' + R_2'$

$R_1 + R_2 = R_1(1 + \alpha_1 t) + R_2(1 + \alpha_2 t)$

$R_1\alpha_1 = -R_2\alpha_2$

Variation of resistance of some materials

Material	Temp. coefficient of resistance (α)	Variation of resistance with temperature rise
Metals	Positive	Increases
Solid non-metal	Zero	Independent
Semi-conductor	Negative	Decreases
Electrolyte	Negative	Decreases
Ionised gases	Negative	Decreases
Alloys	Small positive value	Almost constant

Variation of Resistivity with Temperature:

➢ If ρ_1 is the resistivity of a material at temperature t_1 and ρ_2 is the resistivity of the same material at temperature t_2, then

$$\rho_2 = \rho_1\left[1+\alpha\left(t_2 - t_1\right)\right]$$

W.E - 5 : The temperature coefficient of resistance of platinum is $\alpha = 3.92\times10^{-3}K^{-1}$ at 0° C. Find the temperature at which the increase in the resistance of platinum wire is 10% of its value at 0^0C.

Solution:

$$R_2 = \frac{110R_1}{100} = 1.1R_1; \alpha = 3.92\times10^{-3}K^{-1}$$

$$\Delta t = \frac{R_2 - R_1}{R_1\alpha} \Rightarrow = \frac{1.1R_1 - R_1}{R_1\alpha}$$

$$= \frac{R_1(1.1-1)}{R_1\alpha} = \frac{0.1R_1}{R_1\alpha} = \frac{0.1}{3.92\times10^{-3}}$$

$$\Delta t = 25.51^0C; \; t_2 = 25.51 + 20 = 45.51^0C$$

W.E - 6 : The resistance of iron wire is 10Ω and $\alpha = 5\times10^{-3}/°C$. If a current of 30A is flowing in it at $20°C$, keeping the potential difference across its length constant, if the temperature is increased to $120°C$, what is the current flowing through that wire ?

Solution :

$$\alpha = \frac{R_{120} - R_{20}}{R_{20}\left(120-20\right)}; \quad 5\times10^{-3} = \frac{R_{120} - 10}{10\times100}$$

$$\therefore R_{120} = 15\Omega$$

But V = IR

Here V is constant. Hence,

$$\frac{I_{120}}{I_{20}} = \frac{R_{20}}{R_{120}}; \quad \frac{I_{120}}{30} = \frac{10}{15}$$

$$\therefore I_{120} = 20A$$

W.E - 7: Resistance of a wire at temperature t^0C is $R = R_0(1 + at + bt^2)$

Here, R_0 is the temperature at 0^0C. Find the temperature coefficient of resistance at temperature t is

Sol : $\alpha = \frac{1}{R}.\frac{dR}{dt} = \frac{1}{R_0(1+at+bt^2)}[R_0(a+2bt)]$

$$\therefore \alpha = \left(\frac{a+2bt}{1+at+bt^2}\right)$$

W.E - 8 : A silver wire has a resistance of 2.1 Ω at $27.5°C$ & 2.7Ω at 100°C. Determine the temperature coefficient of resistivity of silver.

Sol: $R_t = R_0(1+\alpha\theta)$

$2.1 = R_0\ (1+\alpha \times 27.5)$ ......(1)

$2.7 = R_0\ (1+\alpha \times 100)$ ......(2)

Solve equation (1) and (2) $\alpha = 0.0039°C^{-1}$

W.E - 9 : V- I graph of a conductor at temperature T_1 and T_2 are shown in the figure $(T_2 - T_1)$ is proportional to

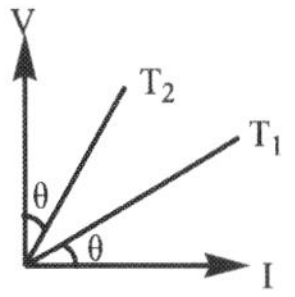

Solution : Slope of line gives resistance

So, $R_1 = \tan\theta = R_0(1+\alpha T_1)$

$$R_2 = \tan(90-\theta) = \cot\theta = R_0(1+\alpha T_2)$$

$$\cot\theta - \tan\theta = R_0\alpha(T_2 - T_1)$$

or $\frac{\cos\theta}{\sin\theta} - \frac{\sin\theta}{\cos\theta} = R_0\alpha(T_2 - T_1)$

$$R_0\alpha(T_2 - T_1) = \frac{\cos 2\theta}{\frac{(\sin 2\theta)}{2}}$$

or $T_2 - T_1 \ \alpha\ \cot 2\theta$

**** W.E - 10 : Figure shows a conductor of length l having a circular cross -section. The radius of cross-section varies linearly from a to b. The resistivity of the material is ρ. Assuming that b - a <<l, find the resistance of the conductor.**

Sol :

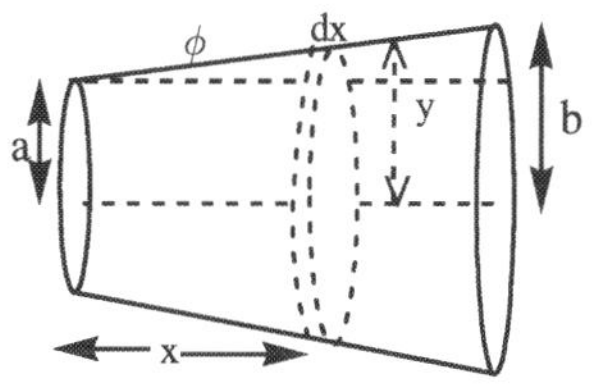

$$\tan\phi = \frac{b-a}{l} = \frac{y-a}{x}$$

$yl - al = bx - ax$

$$l\left(\frac{dy}{dx}\right) = (b-a) \Rightarrow dx = \left(\frac{l}{b-a}\right)dy \rightarrow (1)$$

Resistance across the elemental disc under consideration $dR = \rho\frac{dx}{A} \rightarrow (2)$

from (1) and (2) $dR = \rho\left(\frac{l}{b-a}\right)\frac{dy}{\pi y^2}$

$\Rightarrow$ Resistance across the given conductor, $R = \int_{y=a}^{b} dR \Rightarrow R = \rho\frac{l}{\pi(b-a)}.\int_{y=a}^{y=b}\frac{dy}{y^2}$

$\therefore R = \rho\frac{l}{\pi ab}$

**** W.E - 11 : A hollow cylinder of specific resistance ρ, inner radius R, outer radius 2R and length l is as shown in figure. What is the net resistance between the inner and outer surfaces ?**

Sol : Consider a ring of width 'dr' and radius 'r'.

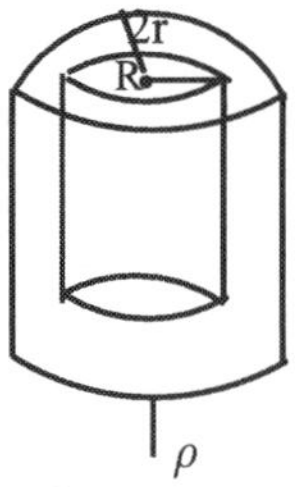

Resistance accross the ring is $dR = \frac{\rho\, dr}{dA} = \frac{\rho\, dr}{2\pi rl}$

Net resistance $= \int_{R}^{2R}\frac{\rho(dr)}{(2\pi rl)} = \left(\frac{\rho}{2\pi l}\right)ln(2)$

W.E - 12 : There are two concentric spheres of radius a and b respectively. If the space between them is filled with medium of resistivity ρ, then the resistance of the intergap between the two spheres will be (Assume b > a)

Solution : Consider a cocentric spherical shell of radius x and thickness dx, its resistance is

dR, $dR = \frac{\rho dx}{4\pi x^2}$

Total resistance

$$\mathbf{R} = \int_{a}^{b} dR = R = \left(\frac{\rho}{4\pi}\right)\int_{a}^{b}\frac{dx}{x^2} = \frac{\rho}{4\pi}\left[\frac{1}{a} - \frac{1}{b}\right]$$

W.E - 13 : A hollow copper cylinder is of inner radius 4cm and outer radius 5cm. Now hollow portion is completely filled with suitable copper wires. Find percentage change in its electric resistance.

Solution: A hollow cylidner of inner radius 'r' and outer radius 'R' has specific resistance 'ρ'. If its length is 'l' then its resistance

$$= \frac{\rho l}{\pi\left(R^2 - r^2\right)}$$

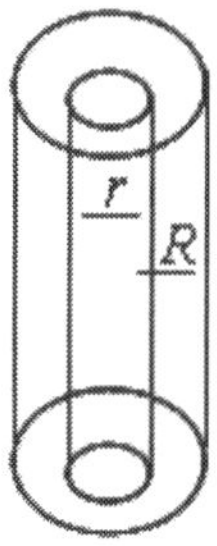

$$R_1 = \frac{\rho l}{\pi\left(5^2 - 4^2\right)} = \frac{\rho l}{9\pi} = \frac{k}{9}$$

Final Resistance $R_2 = \frac{\rho l}{\pi\left(5^2\right)} = \frac{\rho l}{25\pi} = \frac{k}{25}$

Percentage of change = $\frac{R_2 - R_1}{R_1} \times 100 \quad = \frac{\frac{k}{25} - \frac{k}{9}}{\frac{k}{9}} \times 100 = -64\%$

** **W.E - 14 : If resistivity of the material of a conductor of uniform area of cross-section varies along its length as $\rho = \rho_0\left[1 + \alpha x\right]$. Find then the resistance of the conductor if its lengths is 'L' and area of cross-section is 'A'**

Ans : $\frac{\rho_0}{A}\left[L + \frac{1}{2}\alpha L^2\right]$

[Hint : $dR = \rho\frac{dx}{A} = \rho_0\left(1 + \alpha x\right)\frac{dx}{A}$

$\therefore R = \int_0^L dR$]

W.E - 15 : How many number of turns of nichrome wire of specific resistance $10^{-6}\Omega m$ and diameter 2mm that should be wound on a cylinder of diameter 5cm to obtain a resistance of 40Ω?

Solution :

If R is the radius of the cylinder

r is the radius of the wire

N is the number of turns

then $R' = \frac{\rho\ell}{A} \quad \therefore \quad R' = \frac{\rho(2\pi R)N}{\pi r^2}$

$$40 = \frac{10^{-6}\left(2 \times 2.5 \times 10^{-2} \times N\right)}{1 \times 10^{-6}} = \therefore \; N = 800$$

THERMISTOR:

A thermistor is a heat sensitive and non-ohmic device.

- This is made of semiconductor compounds as oxides of Ni, Fe, Co etc.
- This will have high +ve (or) -ve temperature coefficient of resistance.

- Thermistor with -ve 'α' are used as resistance
  themometers which can measure low temperature of order of 10K and small changes of in the order of 10^{-3}K.
- Having -ve α, these are widely used in measuring the rate of energy flow in micro wave beam.
- Thermistor can also be used to serve as thermostat.

RESISTOR COLOUR CODES:

Colour	Number	Multiplier	Tolerance(%)
Black	0	$\times 10^0$	
Brown	1	$\times 10^1$	
Red	2	$\times 10^2$	
Orange	3	$\times 10^3$	
Yellow	4	$\times 10^4$	
Green	5	$\times 10^5$	
Blue	6	$\times 10^6$	
Violet	7	$\times 10^7$	
Gray	8	$\times 10^8$	
White	9	$\times 10^9$	–
Gold	–	$\times 10^{-1}$	$\pm 5\%$
Silver	–	$\times 10^{-2}$	$\pm 10\%$
No clour	–	$\pm 20\%$	

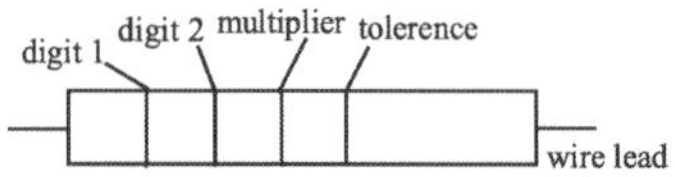

Colour bands on a resistor

B.B.ROY of **G**reat **B**ritain having **V**ery **G**ood **W**ife with **G**old and **S**ilver

- Resistors in the higher range are made mostly from carbon. Carbon resistors are compact, inexpensive and thus find extensive use in electronic circuits.

** **W.E - 16 :**

Suppose the colours on the resistor as shown in Figure are brown, yellow, green and gold as read from left to right. Using the table, find the resistance of the resistor

Solution :

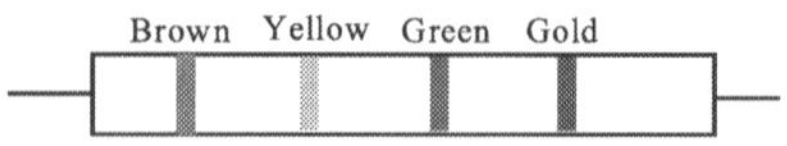

Brown	Yellow	Green	Gold
1	4 ⟶	$\times 10^5$ ⟶	+5% ⟶

$$= 14 \times 10^5 \left(1 \pm \frac{5}{100}\right)\Omega = (1.4 \pm 0.07)10^6\,\Omega = (1.4 \pm 0.07)\text{M}\Omega$$

Some times tolerance is missing from the code and there are only three bands. Then the tolerance is 20%.

SUPER CONDUCTOR:

- There are certain metals for which the resistance suddenly falls to zero below certain temp. called critical temperature.
- Critical temperature depends on the nature of material. The materials in this state are called super conductors.

- Without any applied emf steady current can be maintained in super conductors.
  Ex: Hg below 4.2 K or Pb below 8.2K

RESISTANCES IN SERIES :

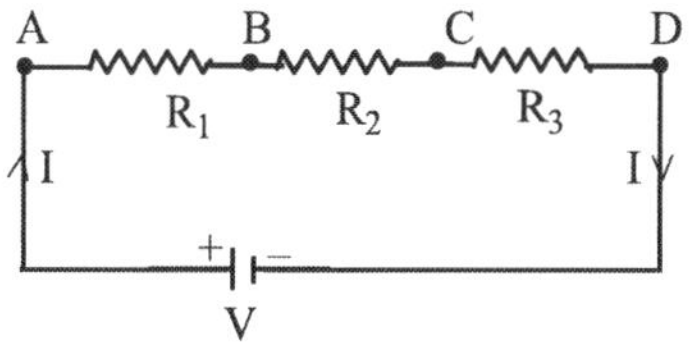

1. If resistors of resistances R_1, R_2, R_3, ... are connected in series, the resultant resistance $R = R_1 + R_2 + R_3 + \ldots\ldots$
2. When resistances are connected in series, same current passes through each resistor. But the potential differences are in the ratio $V_1 : V_2 : V_3 \ldots.. = R_1 : R_2 : R_3 \ldots..$
3. When resistors are joined in series, the effective resistance is greater than the greatest resistance in the circuit.
4. When two resistances are connected in series then $V_1 = \dfrac{VR_1}{R_1 + R_2}$ and $V_2 = \dfrac{VR_2}{R_1 + R_2}$

RESISTANCES IN PARALLEL

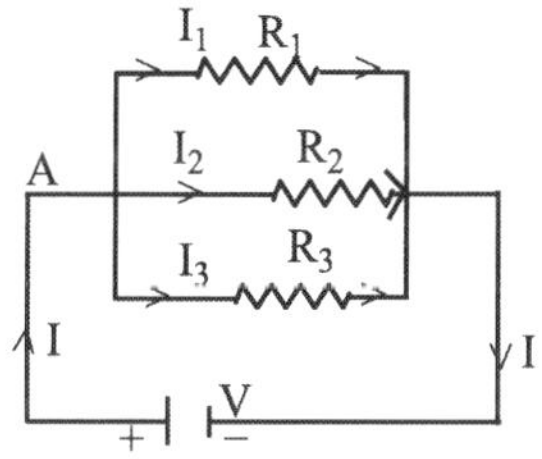

1. If resistors of resistance R_1, R_2, R_3 ....are connected in parallel, the resultant resistance R is given by

$$\frac{1}{R} = \frac{1}{R_1} + \frac{1}{R_2} + \frac{1}{R_3} + \ldots\ldots\ldots$$

2. If resistances R_1 and R_2 are connected in parallel, the resultant resistance. $R = \dfrac{R_1 R_2}{R_1 + R_2}$
3. When resistors are joined in parallel the potential difference across each resistor is same. But the currents are in the ratio $i_1 : i_2 : i_3 : \ldots\ldots\ldots = \dfrac{1}{R_1} : \dfrac{1}{R_2} : \dfrac{1}{R_3} : \ldots\ldots\ldots$
4. When two resistances are parallel then

$$I_1 = \frac{IR_2}{R_1 + R_2} \text{ and } I_2 = \frac{IR_1}{R_1 + R_2}$$

NOTE :

1. When resistors are joined in parallel, the effective resistance is less than the least resistance in the circuit.
2. A wire of resistance 'R' is cut into 'n' equal parts and all of them are connected in parallel, equivalent resistance becomes $\dfrac{R}{n^2}$.
3. In 'n' wires of equal resistances are given, the number of combinations that can be made to give different resistances is 2^{n-1}.
4. If 'n' wire of unequal resistances are given, the number of combinations that can be made to give different resistances is 2^n (If n >2).

5. If R_s and R_p be the resultant resistances of R_1 and R_2 when connected in series and parallel then

$$R_1 = \frac{1}{2}\left(R_s + \sqrt{R_s^2 - 4R_sR_p}\right)$$

$$R_2 = \frac{1}{2}\left(R_s - \sqrt{R_s^2 - 4R_sR_p}\right)$$

6. **If a uniform wire of resistance R is, stretched to 'm' times its initial length and bent into a regular polygon of 'n' sides**

a) Resistance of the wire after stretching is $R_1 = m^2 R (R' \alpha l^2)$

b) Resistance of each side $R_2 = \dfrac{m^2 R}{n}$

c) Resistance across diagonally opposite points $R_0 = \left(\dfrac{\frac{n}{2} R_2}{2}\right) \Rightarrow R_0 = \dfrac{m^2 R}{4}$

d) Resistance across one side $R_3 = \dfrac{(n-1)}{n} R_2 = \dfrac{(n-1)m^2 R}{n^2}$

7. 12 wires each of resistance 'r' are connected to form a cube. Effective resistance across

a) Diagonally opposite corners $= \dfrac{5r}{6}$. b) face diagonal $= \dfrac{3r}{4}$.

c) two adjacent corners $= \dfrac{7r}{12}$.

8. If two wires of resistivities ρ_1 and ρ_2, lengths l_1 and l_2 are connected in series, the equivalent resistivity

$$\rho = \frac{\rho_1 l_1 + \rho_2 l_2}{l_1 + l_2}.$$

If $l_1 = l_2$ *then* $\rho = \dfrac{\rho_1 + \rho_2}{2}$.

If $l_1 = l_2$ then conductivity $\sigma = \dfrac{2\sigma_1\sigma_2}{\sigma_1 + \sigma_2}$.

9. If two wires of resistivities ρ_1 and ρ_2, Areas of cross section A_1 and A_2 are connected in parallel, the equivalent resistivity $\rho = \dfrac{\rho_1\rho_2(A_1 + A_2)}{\rho_1 A_2 + \rho_2 A_1}$.

If $A_1 = A_2$ then $\rho = \dfrac{2\rho_1\rho_2}{\rho_1 + \rho_2}$.

and conductivity $\sigma = \dfrac{\sigma_1 + \sigma_2}{2}$.

10. If 'n' wires each of resistance 'R' are connected to form a closed polygon, equivalent resistance across two adjacent corners is $R_{eff} = \left(\dfrac{n-1}{n}\right) R$

11. If a uniform wire of resistance R is, stretched to m time its initial length and bent into a regular polygon of n sides

a) Resistance of the wire after stretching is $R_1 = m^2R \; (\because R' \propto l^2)$

b) Resistance of each side $R_2 = \frac{m^2R}{n}$

c) Resistance accross diagonally opposite points $R_0 = \left(\frac{\frac{n}{2}R_2}{2}\right) \Rightarrow R_0 = \frac{m^2R}{4}$

d) Resistance accross one side $R_3 = \frac{(n-1)}{n}R_2 \quad = \frac{(n-1)m^2R}{n^2}$

W.E - 17 : For a circuit shown in Fig find the value of resistance R_2 and current I_2 flowing through R_2

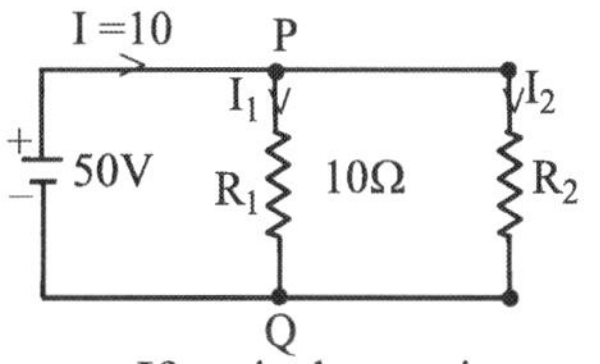

Solution : If equivalent resistance of parallel combination of R_1 and R_2 is R, then

$$R = \frac{R_1R_2}{R_1 + R_2} = \frac{10R_2}{10 + R_2}$$

According to Ohm's law, $R = \frac{V}{I}$

$$R = \frac{50}{10} = 5\Omega \Rightarrow \frac{10R_2}{10 + R_2} = 5 \Rightarrow R_2 = 10\Omega .$$

The current is equally divided into R_1 and R_2.

Hence I_2 = 5A.

W.E - 18 : Two wires of equal diameters of resistivities ρ_1 and ρ_2 and length x_1 and x_2 respectively are joined in series. Find the equivalent resistivity of the combination.

Solution : Resistance, $R_1 = \frac{\rho_1\ell_1}{A_1}; R_2 = \frac{\rho_2\ell_2}{A_2}$

$\ell_1 = x_1, \ell_2 = x_2$

As the wires are of equal diameters $A_1 = A_2 = A$.

$R_1 = \frac{\rho x_1}{A}, R_2 = \frac{\rho x_2}{A_2}; \; R = \frac{\rho x}{A}$

where $x = x_1 + x_2;$ $R = R_1 + R_2$

$\frac{\rho x}{A} = \frac{\rho_1 x_1}{A} + \frac{\rho_2 x_2}{A}; \; \rho x = \rho_1 x_1 + \rho_2 x_2$

$\rho(x_1 + x_2) = \rho_1 x_1 + \rho_2 x_2 \; [\because x = x_1 + x_2]$

$$\therefore \rho = \frac{\rho_1 x_1 + \rho_2 x_2}{x_1 + x_2} \text{ also } \frac{1}{\sigma} = \frac{\frac{x_1}{\sigma_1} + \frac{x_2}{\sigma_2}}{x_1 + x_2}$$

W.E - 19 : Find equivalent resistance of the network in Fig. between points (i) A and B and (ii) A and C.

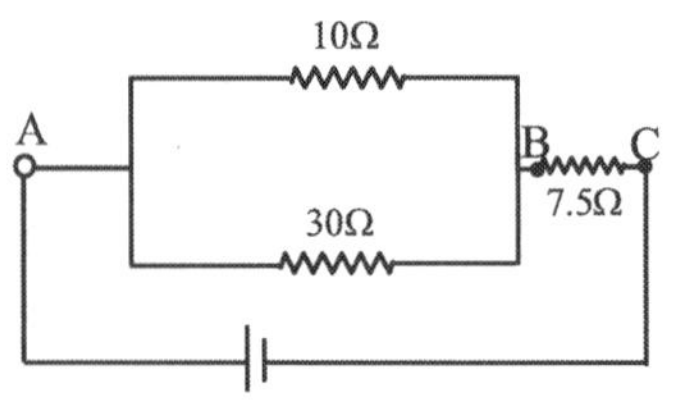

Solution :

(i) The 10Ω and 30Ω resistors are connected in parallel between points A and B. The equivalent resistance between A and B is $R_1 = \frac{10 \times 30}{10+30}\text{ohm} = 7.5\Omega$

(ii) The resistance R_1 is connected in series with resistor of 7.5Ω, hence the equivalent resistance between points A and C is, R_2=(R_1+7.5) ohm = (7.5+7.5) ohm = 15 Ω.

W.E - 20 : Find potential difference between points A and B of the network shown in Fig. and distribution of given main current through different resistors.

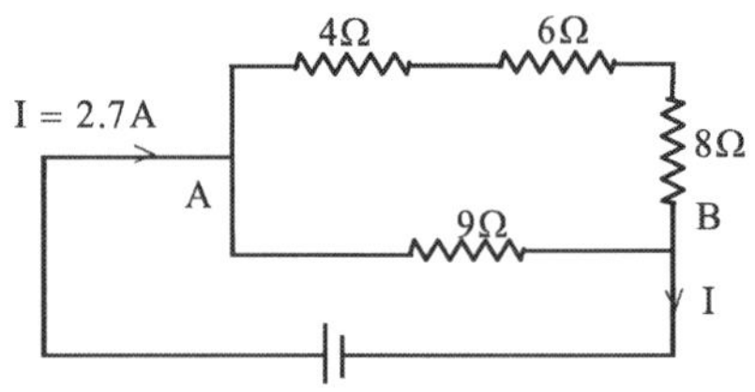

Solution :

Between points A and B resistors of 4Ω, 6Ω and 8Ω resistances are in series and these are in parallel to 9Ω resistor.

Equivalent resistance of series combinaiton is

$R_1 = (4 + 6 + 8)$ ohm = 18

If equivalent resistance between A and B is

$R = 9 \times 18 / (9 + 18)$ ohm = $6\,\Omega$

Potential difference between A and B is $V = IR = 2.7 \times 6V = 16.2V$

Current through 9 Ω resistor = 16.2/9=1.8A

Current through 4 Ω ,6 Ω and 8 Ω resistors = 2.7 – 1.8 = 0.9A.

W.E - 21 : P and Q are two points on a uniform ring of resistance R. The equivalent resistance between P and Q is

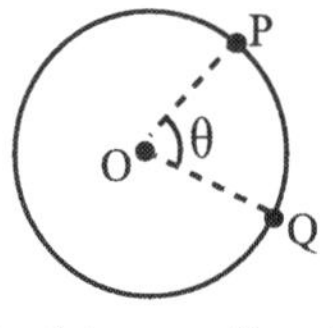

Solution : Resistance of section PSQ

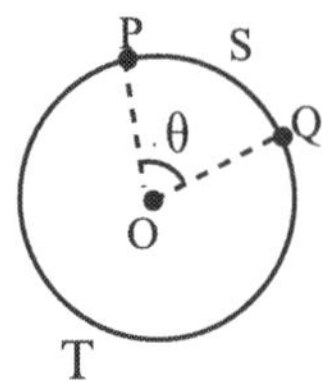

$$R_1 = \frac{R}{2\pi r}.r\theta = \frac{R\theta}{2\pi}$$

Resistance of section PTQ

$$R_2 = \frac{Rr(2\pi-\theta)}{2\pi r} \qquad R_2 = \frac{R(2\pi-\theta)}{2\pi}$$

As R_1 and R_2 are in parallel

$$\text{So, } R_{eq} = \frac{R_1R_2}{R_1+R_2} = \frac{R\theta}{4\pi^2}(2\pi-\theta)$$

W.E - 22 : Determine the current drawn from a 12V supply with internal resistance 0.5 Ω. by the infinite network shown in Fig. Each resistor has 1 Ω. resistance.

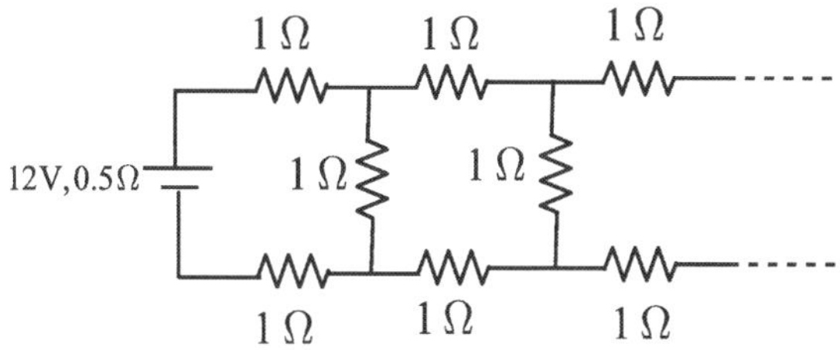

Solution: First calculate net resistance of ∞ network

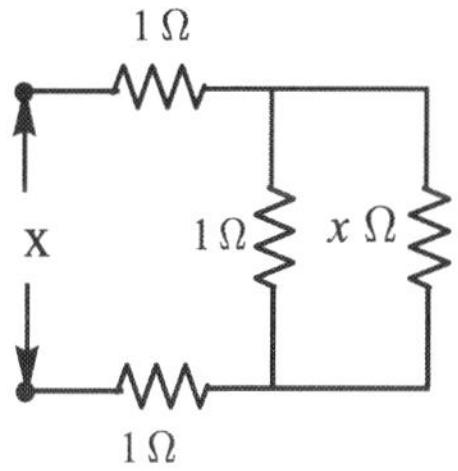

$$x = 2 + \frac{x}{x+1}; \quad x^2 - 2x - 2 = 0$$

on solving, $x = 1 + \sqrt{3} = 2.73\Omega$

Total resistance = 2.73 + 0.5 = 3.23 Ω

$$I = \frac{12}{3.23} = 3.73A$$

JOULE'S LAW

According to **Joule's law**, the current passing through a conductor produces heat.

W = vit   Now, work done, W = (iR) i t

$$W = i^2 R t = \frac{v^2}{R} t = v i t$$

This work is converted into energy in the conductor.

$\therefore$ Thermal energy produced, $Q = i^2 Rt$ in Joules

OR

$$Q = \frac{i^2 Rt}{4.2} \text{ in cal.}$$

As $H \propto i^2$, heating effect of current is common to both A.C and D.C.

Joule's effect is irreversible.

ELECTRICAL ENERGY:

- The electric energy consumed in a circuit is defined as the total workdone in maintaing the current in an electric circuit for a given time.

  Electrical Energy = Vit = Pt = $i^2Rt = \frac{V^2t}{R}$

  S.I. unit of electric energy is **joule**

  1 K.W.H. = 36×10^5 J

ELECTRICAL POWER:

- The rate at which work is done in maintaining the current in electric circuit. Electrical power

  $P = \frac{W}{t} = Vi = i^2R = \frac{V^2}{R}$ watt (or) joule/sec

- Heat energy produced due to the electric current $H = \frac{W}{J} = \frac{Pt}{J} = \frac{Eit}{J} = \frac{i^2Rt}{J} = \frac{E^2t}{RJ}$

  $H = ms\Delta t$

  Where $s = 4200 J/Kg^0C$

  where J is mechanical equivalent of heat.

** **FUSE WIRE :**

A fuse wire generally prepared from tin - lead alloy (63% tin + 37% lead). **It should have high resistivity, low melting point.**

Let R be the resistance of fuse wire.

We know that $R = \frac{\rho L}{\pi r^2}$

(L and r denote length and radius)

The heat produced in the fuse wire is

$$H = i^2R = \frac{i^2\rho L}{\pi r^2}$$

If H_0 is heat loss per unit surface area of the fuse wire, then heat radiated per second is = $H_0 2\pi rL$

At thermal equilibrium,

$$\frac{i^2\rho^2L}{\pi r^2} = H_0 2\pi rL \qquad \text{(or) } H_0 = \frac{i^2\rho}{2\pi^2 r^3}$$

According to Newton's law of cooling.

$$H_0 = K\theta$$

Where θ is the increase in temperature of fuse wire and K is a constant.

$$\theta = \frac{i^2\rho}{2\pi^2 r^3 K}$$

Here θ is independent of length L of the fuse wire provided i remains constant.

For a given material of fuse wire $i^2 \alpha r^3$.

W.E - 24 :

A fuse wire with radius of 0.2mm blows off with a current of 5 Amp. The fuse wire of same material, but of radius 0.3mm will blow off with a current of

1) $5\times\frac{3}{2}$ Amp 2) $\frac{5\sqrt{3}}{2}$ Amp 3) $5\sqrt{\frac{27}{8}}$ Amp 4) 5 Amp

Solution : $i^2 \propto r^3$; $\frac{i_1}{i_2}=\left(\frac{r_1}{r_2}\right)^{3/2}=\left(\frac{0.2}{0.3}\right)^{3/2}$

$i_2 = 5\sqrt{\frac{27}{8}}$ Amp

➢ **If radiation losses are neglected, due to heating effect of current the temperature of fuse wire will increase continuously, and it melt in time 't' such that**

$$H = ms\Delta\theta; \frac{I^2Rt}{J} = ms(\theta_{mp}-\theta_r)$$

$$t = \frac{\pi^2 r^4 s\left(\theta_{mp}-\theta_r\right)J}{I^2\rho}; t\,\alpha\, r^4$$

i.e., in absence of radiation lossess, the time in which fuse will melt is also independent on length and varies with radius as r^4.

NOTE :

a) If resistances are connected in series , i.e.., I is same

$P\alpha R$ with $V\alpha R$ [$as\ V = IR$]

i.e.., in series potential difference and power consumed will be more in larger resistance.

However, if resistances are connected in parallel, i.e.., V is same

$P\alpha\frac{1}{R}$ with $I\alpha\frac{1}{R}$ [$as\ V = IR$]

i.e.., in parallel current and power consumed will be more in smaller resistance. This in turn implies that more power is consumed in larger resistance if reistances are in series and in smaller reistance if reistances are in parallel.

b) A reistance R under a potential difference V dissipates power.

$P = \left(V^2/R\right)$

So If the resistance is changed from R to (R/n) keeping V same, the power consumed will be

$$P^1 = \frac{V^2}{(R/n)} = n\frac{V^2}{R} = nP$$

i.e.., if for a given voltage, resistance is changed from R to (R/n) , power consumed changes from P to nP.

c) If n equal resistances are connected in series with a voltage source, the power dissipated will be

$$P_s = \frac{V^2}{nR} \quad [\ as\ R_s = nR\ ]$$

And if the same resistances are connected in parallel with the same voltage source

$$P_p = \frac{V^2}{(R/n)} = \frac{nV^2}{R} \quad [as\ R_p = (R/n)]$$

So, $\frac{P_p}{P_s} = n^2$ i.e.., $P_P = n^2 P_S$.

i.e.., power consumed by n equal resistors in parallel is n^2 times that of power consumed in series if V remains same.

d) As resistance of a given electric appliance (e.g.., bulb , heater, geyser or press) is constant and is given by, $R = \frac{V_S}{I} = \frac{V_S}{(W/V_S)} = \frac{V_s^2}{W}$ $[as\ I = \frac{W}{V}]$

Where V_s and W are the voltage and wattage specified on the appliance. So if the applied voltage is different from specified, the ' actual power consumption' will be $P = \frac{V_A^2}{R} = \left(\frac{V_A}{V_S}\right)^2 \times W$ [as $R = \frac{V_S^2}{W}$].

W.E - 23 : A 1 kW heater is meant to operate at 200 V. (a) What is its resistance ? (b) How much power will it consume if the line voltage drops to 100 V ? (c) How many units of electrical energy will it consume in a month (of 30 days) if it operates 10 hr daily at the specified voltage ?

Solution: a) The resistance of an electric appliance is given by , $R = \frac{V_S^2}{W}$ so, $R = \frac{(200)^2}{1000} = 40\Omega$

b) The ' actual power ' consumed by an electric appliance is given by ,

$$P = \left(\frac{V_A}{V_S}\right)^2 \times W$$

$$\text{so, } P = \left(\frac{100}{200}\right)^2 \times 1000 = 250W$$

c) The total electrical energy consumed by an electric appliance in a specified time is given by,

$$E = \frac{\Sigma W_1 h_1}{1000} kWh$$

$$\text{so, } E = \frac{1000 \times (10 \times 30)}{1000} = 300kWh$$

BULBS CONNECTED IN SERIES:

➢ If Bulbs (or electrical appliances) are connected in series, the current through each resistance is same. Then power of the electrical appliance $P \propto R$ & $V \propto R\left[\therefore P = i^2Rt\right]$

i.e. In series combination; the potential difference and power consumed will be more in larger resistance.

➢ When the appliances of power P_1, P_2, P_3.... are in series, the effective power consumed (P) is

$\frac{1}{P} = \frac{1}{P_1} + \frac{1}{P_2} + \frac{1}{P_3} + \dots\dots\dots$ i.e. effective power is less than the power of individual appliance.

➢ If 'n' appliances, each of equal resistance 'R' are connected in series with a voltage source 'V', the power dissipated ' P_s ' will be $P_s = \frac{V^2}{nR}$.

BULBS CONNECTED IN PARALLEL:

- If Bulbs (or electrical appliances) are connected in parallel, the potential difference across each resistance is same. Then $P \propto \frac{1}{R}$ and $I \propto \frac{1}{R}$.

  i.e. The current and power consumed will be more in smaller resistance.
- When the appliances of power $P_1, P_2, P_3....$ are in parallel, the effective power consumed(P) is

  $P = P_1 + P_2 + P_3 + ...$

  i.e. the effective power of various electrical appliance is more than the power of individual appliance.
- If 'n' appliances, each of resistance 'R' are connected in parallel with a voltage source 'V', the power dissipated 'Pp' will be

  $$P_P = \frac{V^2}{(R/n)} = \frac{nV^2}{R}$$

  $$\frac{P_P}{P_S} = n^2 \ (or) \ P_P = n^2 P_S$$

  This shows that power consumed by 'n' equal resistances in parallel is n^2 times that of power consumed in series if voltage remains same.
- In parallel grouping of bulbs across a given source of voltage, the bulb of greater wattage will give more brightness and will allow more current through it, but will have lesser resistance and same potential difference across it.
- For a given voltage V, if resistance is changed from 'R' to $\left(\frac{R}{n}\right)$, power consumed changes from 'P' to 'nP' $P' = \frac{V^2}{R'}$ where $R' = \frac{R}{n}$, then

  $$P' = \frac{V^2}{(R/n)} = \frac{nV^2}{R} = nP.$$
- If t_1, t_2 are the time taken by two different coils for producing same heat with same supply, then

  If they are connected in series to produce same heat, time taken $t = t_1 + t_2$

  If they are connected in parallel to produce same heat, time taken is $t = \frac{t_1 t_2}{t_1 + t_2}$.

W.E - 25 : A lamp of 100W works at 220 volts. What is its resistance and current capacity ?

Solution: Power of the lamp, P = 100W

Operating voltage, V = 220V

Current capacity of the lamp,

$$i = \frac{P}{V} = \frac{100}{220} = 0.455A$$

Resistance of the lamp, $R = \frac{V^2}{P} = \frac{(220)^2}{100} = 484\Omega$

W.E - 26 : A 100W - 220V bulb is connected to 110V source. Calculate the power consumed by the bulb.

Solution: Power of the bulb, P = 100W

Operating voltage, V = 200V

Resistance of the bulb, $R = \frac{V^2}{P} = \frac{(220)^2}{100} = 484\Omega$

Actual operating voltage, $V^1 = 110$ V

Therefore, power consumed by the bulb,

$$P^1 = \frac{(V^1)^2}{R} = \frac{(110)^2}{484} = 25W.$$

W.E - 27 : A 100W and a 500W bulbs are joined in series and connected to the mains. Which bulb will glow brighter ?

Solution : Let R_1 and R_2 be the resistances of the two bulbs. If each bulb is connected separately to the mains of voltage V, then $P_1 = \frac{V^2}{R_1}$ and $P_2 = \frac{V^2}{R_2}$

$$\therefore \frac{P_1}{P_2} = \frac{R_2}{R_1} \text{ (or) } \frac{R_1}{R_2} = \frac{P_2}{P_1} = \frac{500}{100} = 5$$

If the two bulbs are in series with the mains, the same current 'i' flows through each of them.

Let P_1 and P_2 be the powers dissipated by two bulbs, turn

$P_1^1 = i^2R_1$ and $P_2^1 = i^2R_2$

$$\therefore \frac{P_1^1}{P_2^1} = \frac{R_1}{R_2} = 5 \text{ or } P_1^1 = 5P_2^1$$

Since 100 watt bulb dissipates more power, it glows brighter

W.E - 28 : A cell develops the same power across two resistances R_1 and R_2 separately. The internal resistance of the cell is

Solution : Let r be the internal resistance of the cell and E its EMF. When connected across the resistance R_1 in the circuit, current passing through the resistance is $i = \frac{E}{R_1 + r}$

$$\therefore P_1 = i^2R_1 = \left(\frac{E}{R_1 + r}\right)^2 R_1$$

Similarly $P_2 = \left(\frac{E}{R_2 + r}\right)^2 R_2$

Given that $P_1 = P_2$

Substituting the values, we get $r = \sqrt{R_1R_2}$

W.E - 29 : A 100 W bulb B_1 and two 60 W bulbs B_2 and B_3, are connected to a 250V source, as shown in the figure. Now W_1 , W_2 and W_3 are the output powers of the bulbs B_1, B_2 and B_3 respectively. Then

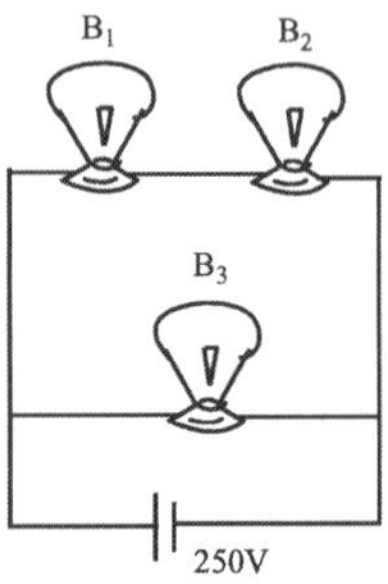

Solution: A bulb is essentially a resistance $R = \frac{V^2}{P}$ where P denotes the power of the bulb.

$\therefore$ Resistance of $B_1(R_1) = V^2 / 100$

Resistance of $B_2\ (R_2) = V^2 / 60$

Resistance of $B_3(R_3) = V^2 / 60$

$\therefore I_1$ = Current in $B_1 = \frac{250}{(R_1 + R_2)} = \frac{250 \times 300}{8V^2}$

I_2 = Current in $B_2 = \frac{250}{(R_1 + R_2)} = \frac{250 \times 300}{8V^2}$

I_3 = Current in $B_3 = I_1$ as B_1, B_2 are in series

$\therefore W_1$ output power of $B_1 = I_1^2 R_1$

$$\therefore W_1 = \left(\frac{250 \times 300}{8V^2}\right)^2 \times \frac{V^2}{100}$$

$$W_2 = I_2^2 R_2 \text{ or } W_2 = \left(\frac{250 \times 300}{8V^2}\right)^2 \times \frac{V^2}{60}$$

$$W_3 = I_2^2 R_3 \text{ or } W_3 = \left(\frac{250 \times 300}{8V^2}\right)^2 \times \frac{V^2}{60}$$

$\therefore W_1 : W_2 : W_3 = 15 : 25 : 64$ or $W_1 < W_2 < W_3$

MAXIMUM POWER TRANSFER THEOREM

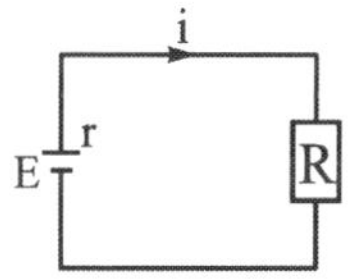

Fig.

Consider a device of resistance R connected to a source of e.m.f E and internal resistance r as shown.

Current in the circuit is $i = \left(\frac{E}{R + r}\right)$.

Power dissipated in the device is $P = i^2R \Rightarrow P = \frac{E^2R}{(R + r)^2}$

For maximum power dissipated in the device

$$\frac{dP}{dR} = 0 \Rightarrow \frac{d}{dR}\left[\frac{E^2R}{(R + r)^2}\right] = 0$$

On simplification, we can get $R = r$

So, the power dissipated in an external resistance is maximum if that resistance is equal to internal resistance of the source supplying the current to that device.

W.E - 30 : A battery if internal resistance 4Ω is connected to the network of ressitances as shown. What must be the value of R so that maximum power is delivered to the network ? Find the maximum power ?

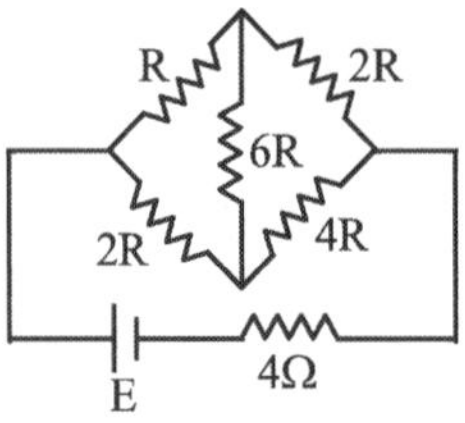

Solution :

i) According to maximum power transfer theorem

$$R_{ext} = R_{int}$$

$$\frac{3R \times 6R}{9R} = 4 \Rightarrow R = \frac{4}{2} = 2\Omega$$

ii) $P_{max} = i^2 R_{ext} \quad = \left(\frac{E}{4+4}\right)^2 \times 4 = \frac{E^2}{16}$

CONSUMPTION OF ELECTRICAL ENERGY:

➢ Units of electrical energy consumed by an electrical appliance =

$$\frac{\text{Number of watts} \times \text{Number of hours}}{1000}$$

It is in KWH.

CELLS

PRIMARY CELLS:

➢ Voltaic, Leclanche, Daniel and Dry cells are primary cells. They convert chemical energy into electrical energy. They can't be recharged. They supply small currents.

SECONDARY CELLS (OR) STORAGE CELLS:

➢ Electrical energy is first converted into chemical energy and then the stored chemical energy is converted into electrical energy due to these cells.

➢ These cells can be recharged.

➢ The internal resistance of a secondary cell is low where as the internal resistance of a primary cell is large.

EMF OF A CELL:

➢ The energy supplied by the battery to drive unit charge around the circuit is defined as electro motive force of the cell.

➢ EMF is also defined as the absolute potential difference between the terminals of a source when no energy is drawn from it. i.e., in the open circuit of the cell. It depends on the nature of electrolyte used in the cell.

Unit :J/C (or) Volt

➢ **emf of a cell depends on**

a) metal of electrodes b) nature of electrolyte c) temperature

➢ **emf of the cell is independent of**

a) area of plate b) quantitiy of electrolyte c) distance between plate

d) size of the cell

Internal Resistance of a Cell

- It is the resistance offered by the electrolyte of the cell.

It depends on

- area of the electrodes used ($r \alpha \frac{1}{A}$)
- nature of electrolyte , concentration($r \alpha C$)
- area of cross section of the electrolyte through which the current flows and
- age of the cell.
- Internal resistance of an ideal cell is zero.

TERMINAL VOLTAGE :

When no current flows through the cell, the circuit is said to be an open circuit. This is shown in figure.

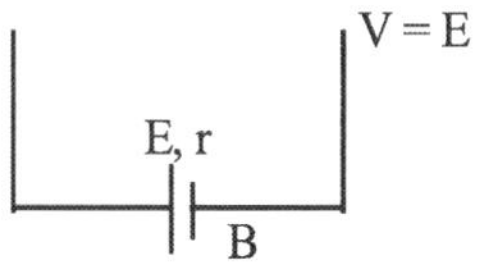

In such a case, the potential difference (p.d) across the terminals of the cell, called the terminal voltage (V) will be equal to the emf (E) of the cell.

If an external resistance R is connected across the two terminals of the cell, as in figure then current flows in the closed circuit.,

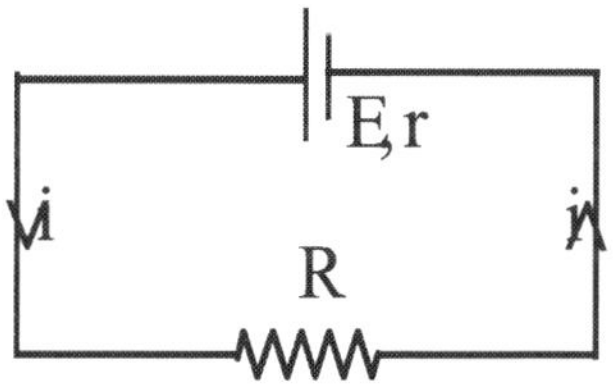

$$i = \frac{V}{R} \qquad \text{..........} (1)$$

and also $$i = \frac{E}{(R+r)} \qquad \text{..........} (2)$$

$iR + ir = E,\ V + ir = E,\ V = E - ir$

Lost volts :

It is the difference between emf and P.D. of a cell It is used in driving the current between terminals of the cell.

Lost volts **E - V = *i* r**

NOTE : Formulae related with cells

$$i = \frac{E-V}{r} \qquad \text{..................}(A)$$

$$r = \frac{E-V}{i} \qquad \text{....................}(B)$$

$$r = \left(\frac{E-V}{V/R}\right) = \left(\frac{E-V}{V}\right)R = \left(\frac{E}{V}-1\right)R \quad \text{....} (C)$$

- $V = iR = \dfrac{ER}{(R+r)}$

- Fractional energy useful $= \dfrac{V}{E} = \dfrac{R}{R+r}$

- % of fractional useful energy

  $= \left(\dfrac{V}{E}\right)100 = \left(\dfrac{R}{R+r}\right)100$

- Fractional energy lost, $\dfrac{V'}{E} = \dfrac{r}{R+r}$

- % of lost energy, $\left(\dfrac{V'}{E}\right)100 = \left(\dfrac{r}{R+r}\right)100$

- internal resistance, $r = \left[\dfrac{E-V}{V}\right]R$

DIFFERENT CONCEPTS WITH A CELL

- When the cell is charging, the EMF is less than the terminal voltage (E < V) and the direction of current inside the cell is from + ve terminal to the –ve terminal.

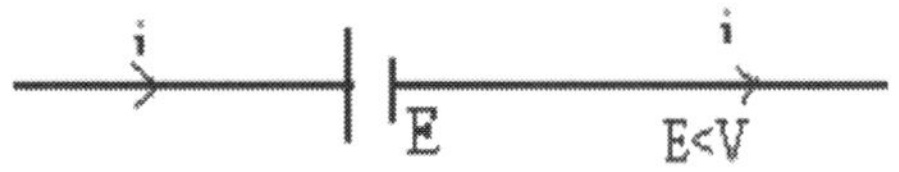

$V = E + ir$

- When the cell is discharging, the EMF is greater than the terminal voltage (E >V) and the direction of current inside the cell is from – ve terminal to the +ve terminal.

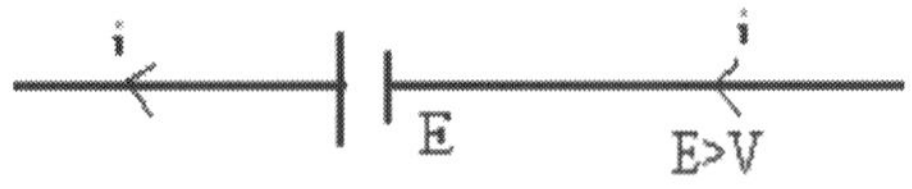

$V = E - ir$; Hence $E > V$

- Power delivered will be maximum when R = r. So $P_{max} = \dfrac{E^2}{4r}$

- This statement in generalized form is called 'maximum power transfer theorem'

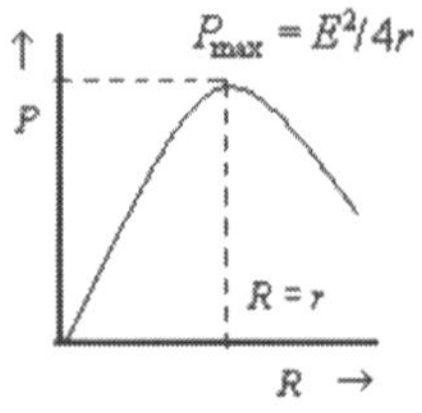

Here the % of energy lost and energy useful are each equal to 50%

BACK EMF:

When current flows through the electrolyte solution, electrolysis takes place with a layer of hydrogen and this hinders the flow of current. In the neighbourhood of both electrodes, the concentrations of ions get altered. This opposing EMF is called back EMF and the phenomenon is called Electrolytic polarisation.

To reduce back emf manganese dioxide (or) potassium dichromate is added to electrolyte of cell.

W.E - 31 : When a current drawn from a battery is 0.5A, its terminal potential difference is 20V. And when current drawn from it is 2.0A, the terminal voltage reduces to 16 V. Find out. e.m.f and internal resistance of the battery.

Solution : We know

$V = E --- Ir$; $I = 0.5$ A, $V = 20$ Volt, we have

$20 = E - 0.5r$ .... (i)

$I = 2$ A, $V = 16$ Volt, we have

$16 = E - 0.2r$ .... (ii)

From eqs (i) and (ii)

$2E - r = 40$ and $E - 2r = 16$

Solving we get $E = 21.3V$, $r = 2.675\,\Omega$.

W.E - 32 : An ideal battery passes a current of 5A through a resistor. When it is connected to another resistance of 10Ω in parallel, the current is 6A. Find the resistance of the first resistor.

Solution :

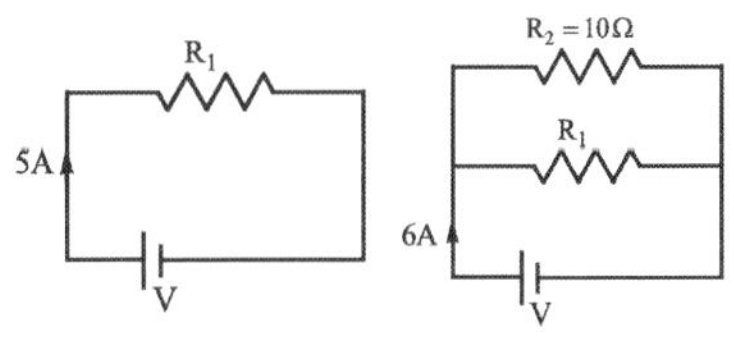

Current through R_1 in the first case $i_1 = 5A$

Current in the second case $i_2 = 6A$

Effective resistance in the second case $R = \frac{R_1R_2}{R_1+R_2}$; $V = I_1R_1$ and $V = I_2\frac{R_1R_2}{R_1+R_2}$

$$I_1R_1 = I_2\frac{R_1R_2}{R_1+R_2} \Rightarrow I_1 = I_2\frac{R_2}{R_1+R_2}$$

$$5 = 6\times\frac{10}{R_1+10} \Rightarrow 5(R_1+10) = 60$$

$5R_1 + 50 = 60$, $5R_1 = 10$

$$R_1 = \frac{10}{5} = 2\Omega \Rightarrow R_1 = 2\Omega\cdot$$

W.E - 33 : When a battery is connected to the resistance of 10Ω the current in the circuit is 0.12A. The same battery gives 0.07A current with 20Ω. Calculate e.m.f. and internal resistance of the battery.

Solution : We know that $E = Ir + IR$

$$I_1r + I_1R_1 = I_2r + I_2R_2;\ r = \frac{I_2R_2 - I_1R_1}{I_1 - I_2}$$

$$r = \frac{0.07\times20 - 0.12\times10}{0.12-0.07} = \frac{1.4-1.2}{0.05} = \frac{0.2}{0.05} = 4\Omega$$

Internal resistance $r = 4\Omega$

e. m. f $E = Ir + IR$

$0.12 \times 4 + 0.12 \times 10 = 0.48 + 1.2$; $E = 1.68$ volt.

GROUPING OF CELLS :

1. Electric Cells in Series:

When 'n' identical cells each of EMF 'E' and internal resistance 'r' are connected in series to an external resistance 'R', then

- total emf of the combination $= nE$
- effective internal resistance $= nr$
- Current through external resistance $i = \dfrac{nE}{R + nr}$
- If $R << nr$ then $i = \dfrac{E}{r}$ = current from one cell
- If $R >> nr$ then $i = \dfrac{nE}{R}$
- If two cells of different emf's are in series

$E_{eq} = E_1 + E_2$

$r_{eq} = r_1 + r_2$

$$i = \frac{E_1 + E_2}{r_1 + r_2 + R}$$

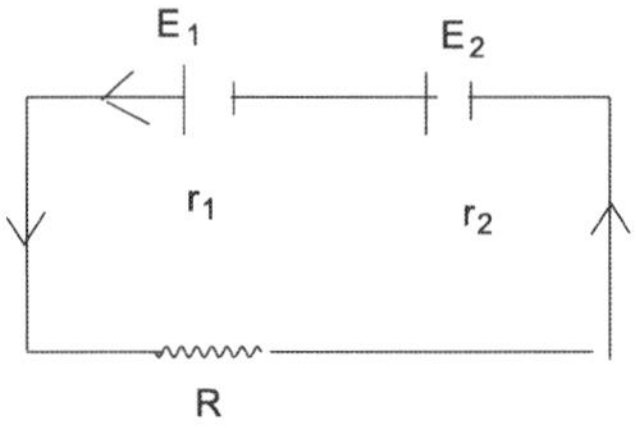

T.P.D across the first cell $V_1 = E_1 - ir_1$

T.P.D across the second cell $V_2 = E_2 - ir_2$

- If one of the cell is in reverse connection ($E_1 > E_2$) then $E_{eq} = E_1 - E_2$

$r_{eq} = r_1 + r_2$

$$i = \frac{E_1 - E_2}{r_1 + r_2 + R}$$

First cell is discharging then $V_1 = E_1 - ir_1$

Second cell is charging then $V_2 = E_2 + ir_2$

cell having less emf in charging state.

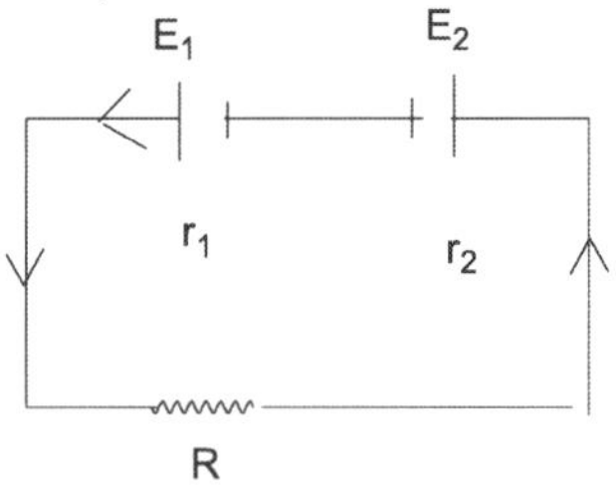

WRONGLY CONNECTED CELLS

- By mistake if 'm' cells out of 'n' cells are wrongly connected to the external resistance 'R'

(a) total emf of the combination $= (n - 2m)E$ (b) total internal resistance $= nr$

(c) total resistance $= R + nr$

(d) current through the circuit (i) $= \dfrac{(n - 2m)E}{R + nr}$

W.E - 34 : Two cells A and B with same e.m.f of 2 V each and with internal resistances $r_A = 3.5\Omega$ and $r_B = 0.5\Omega$ are connected in series with an external resistance $R = 3\Omega$. Find the terminal voltages across the two cells.

Solution: Current through the circuit

$i = \frac{\varepsilon}{(R+r)} = \frac{2+2}{(3+3.5+0.5)} = \frac{4}{7}$

i) $R = 3\Omega, r_A = 3.5\Omega, E = 2V$

Terminal voltages A, $V_A = E - ir$

$= 2 - \frac{4}{7} \times 3.5 = 0$ volt

ii) $r_B = 0.5\Omega, R = 3\Omega, E = 2V$

Terminal voltage at B, $V_B = E - ir$

$= 2 - \frac{4}{7} \times 0.5 = 1.714$ *volts.*

W.E - 35 : Two cells A and B each of 2 V are connected in series to an external resistance R=1 ohm. The internal resistance of A is r_A =1.9 ohm and B is r_B =0.9 ohm. Find the potential difference between the terminals of A.

Solution : Total current through the circuit $i = \frac{voltage}{Total\ resistance}$

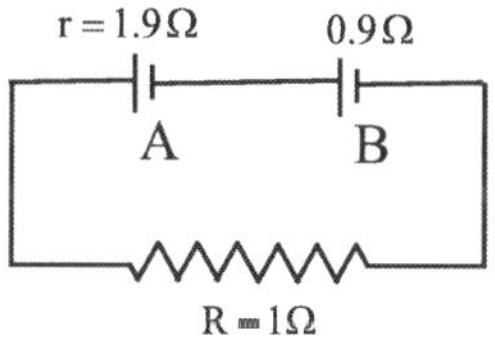

$= \frac{4}{(1+1.9+0.9)} = \frac{4}{3.8}A$.

potential difference at A, $V_A = \varepsilon - ir$, $= 2 - \frac{4}{3.8} \times 1.9 = 2 - 2 = 0.$

W.E - 36 : In the given circuit as shown below, calculate the magnitude and direction of the current

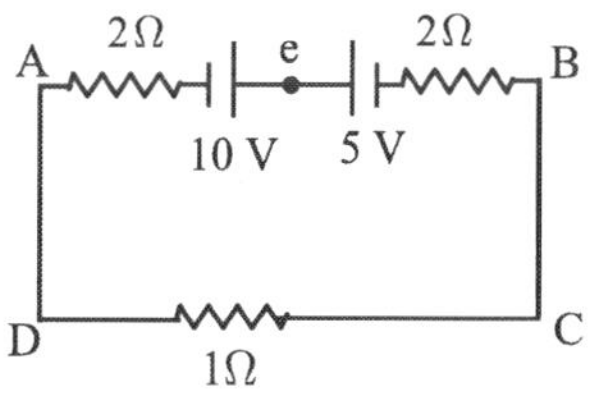

Solution : Effective resistance of the circuit is $R_{eff} = 2 + 2 + 1 = 5\Omega$

$\therefore$ total current in the circuit is $i = \frac{V_1 - V_2}{R_{eff}}$

$i = \frac{10-5}{5} = 1A$

Since the cell of larger emf decides the direction of flow of current, the direction of current in the circuit is **from A to B through e**

W.E - 37 : A voltmeter resistance 500Ω is used to measure the emf of a cell of internal resistance 4Ω. The percentage error in the reading of the voltmeter will be

Solution : $V = E - ir$

$\therefore$ Percentage error $=\frac{\Delta E}{E}\times 100=\frac{ir}{E}\times 100$

$$=\frac{\left(\frac{E}{R+r}\right)r}{E}\times 100=\left(\frac{r}{R+r}\right)\times 100$$

$$=\left(\frac{4}{500+4}\right)\times 100=0.8\%$$

ELECTRIC CELLS IN PARALLEL:

When 'n' identical cells each of EMF 'E' and internal resistance 'r' are connected in parallel to an external resistance 'R', then

- total emf of the combination = E
- effective internal resistance $=\frac{r}{n}$
- total resistance in the circuit = R + $\frac{r}{n}$
- **current through the external resistance** $i=\frac{E}{R+\frac{r}{n}}=\frac{nE}{nR+r}$
- If R >> $\frac{r}{n}$, then i = $\frac{E}{R}$ = current from one cell.
- If R << $\frac{r}{n}$, then i = $\frac{nE}{r}$
- If two cells of emf E_1 and E_2 having internal resistances r_1 and r_2 are connected in parallel to an external resistance 'R', then

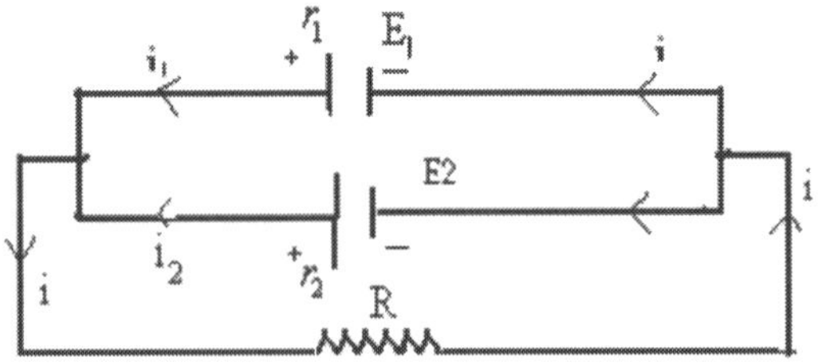

the effective emf, E = $\frac{E_1r_2+E_2r_1}{r_1+r_2}$

the effective internal resistance, $r_{eff}=\frac{r_1r_2}{r_1+r_2}$

Current through the circuit, $i=\frac{E}{r_{eff}+R}$

$i=i_1+i_2$

$i_1 = \frac{E_1 - iR}{r_1}$ and $i_2 = \frac{E_2 - iR}{r_2}$

Potential difference across R, i.e terminal potential of the cells is $V = iR = \frac{ER}{R + r_{eff}}$

- When the cell E_2 is reversed in polarity then we should use - E_2 in all the above equations.

MIXED GROUPING:

If n identical cell's are connected in a row and such m rows are connected in parallel as then

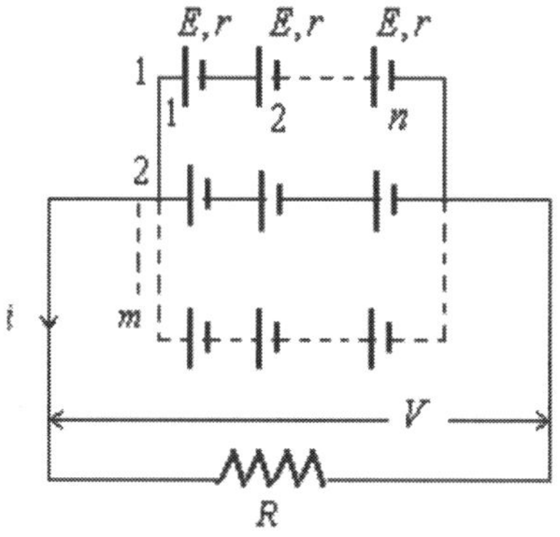

- Equivalent emf of the combination $\mathbf{E_{eq} = nE}$
- Equivalent internal resistance of the combination $\mathbf{r_{eq}} = \frac{nr}{m}$
- Main current flowing through the load $i = \frac{nE}{R + \frac{nr}{m}} = \frac{nmE}{mR + nr}$
- Condition for maximum power $R = \frac{nr}{m}$ and $P_{max} = (mn)\frac{E^2}{4r}$
- **Condition for maximum current**

$\frac{R}{n} + \frac{r}{m} = \text{minimum}$

$\frac{d}{dm}\left[\frac{mR}{N} + \frac{r}{m}\right] = 0$; $\left[n = \frac{N}{m}\right]$

$\frac{R}{N} - \frac{r}{m^2} = 0; i.e., \frac{R}{n} = \frac{r}{m}$ **(N = n x m)**

So in case of mixed grouping of cells, current in the circuit will be maximum when $\left(\frac{R}{n} = \frac{r}{m}\right)$

$I_{max} = \frac{nE}{2R} = \frac{mE}{2r}$

- Total number of cells = $m \times n$

KIRCHHOFF'S LAWS

When the circuit is complecated to find current kirchhoff's laws are formulated.

(i) **Kirchhoff s First Law (Junction Law or Current law) :** It states that the sum of the currents flowing into a junction is equal to the sum of the currents flowing out of the junciton .

OR

"The algebraic sum of currents at a junction is zero".

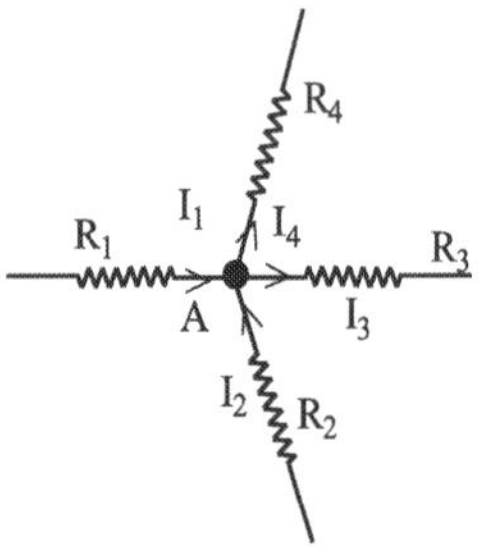

Distribution of current at a junction in the circuit

$I_1 + I_2 = I_3 + I_4$ or $I_1 + I_2 - I_3 - I_4 = 0$

If we take currents approaching point A in Fig as positive and that leaving the point as negative, then the above relation may be written as

$I_1 + I_2 + (-I_3) + (-I_4) = 0$

$\Sigma I = 0$

NOTE :

Thus, Kirchhoff's first law is accordance with **law of conservation of charge**, since no charge can accumulate at a junction.

(ii) **Kirchhoff s Second Law (Loop Law or Potential law) :** Kirchhoff's second law states that the algebraic sum of changes in potential around any closed loop is zero.

(Kirchhoff's second law) can be expressed as $\Sigma V = 0$.

In terms of potential drops and emfs, the law is expressed as $\Sigma(iR) + \Sigma E = 0$

Sign conventions:

(a) The change in potential in traversing a resistance in the direction of current is –IR while in the opposite direction +IR as shown in the figure.

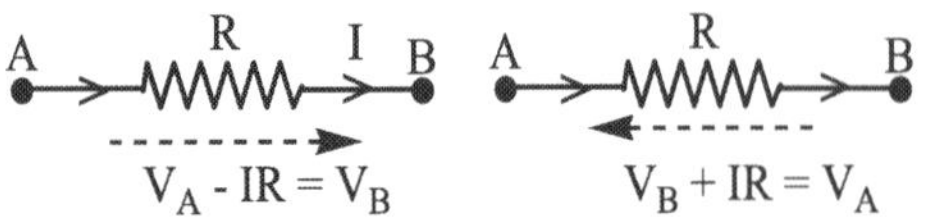

$V_A - IR = V_B$ $V_B + IR = V_A$

$V_A - V_B = IR$ $V_A - V_B = IR$

(b) The change in potential in traversing an emf source from negative to positive is +E while in the opposite direction –E irrespective of the direction of current in the circuit as shown in the figure.

$V_A - E = V_B$ $V_B + E = V_A$

$V_A - V_B = E$ $V_A - V_B = E$

Example 1:

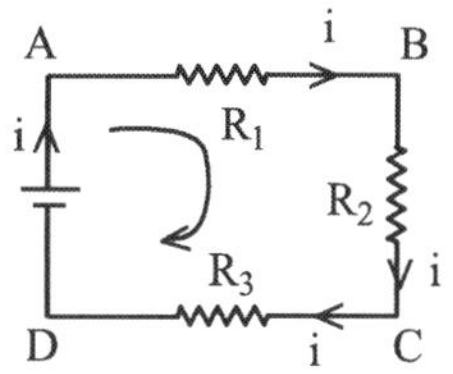

Apply the kirchhoff's second law to the loop ABCDA, then

$-iR_1 - iR_2 - iR_3 + E = 0$

$$\therefore i = \frac{E}{(R_1 + R_2 + R_3)}$$

Example :

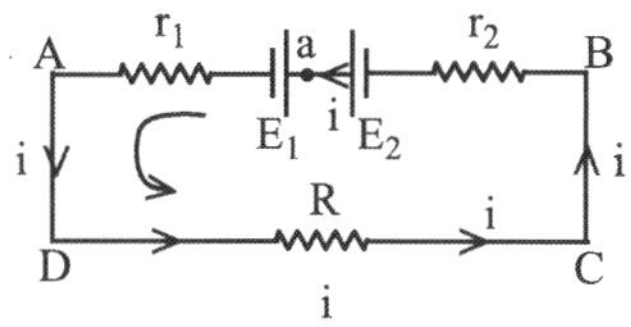

Apply the kirchhoff's second law to the loop ADCBA, then

$-iR - ir_2 + E_2 - E_1 - ir_1 = 0$

$$i(r_1 + R + r_2) = E_2 - E_1 \Rightarrow i = \frac{E_2 - E_1}{r_1 + r_2 + R}$$

NOTE :

1) This law represents "**conservation of energy**"

2) If there are n meshes in a circuit, the number of independent equations in accordance with loop rule will be (n – 1).

Application : This is the most general case of parallel grouping in which E and r of different cells are different and the positive terminals cells are connected as shown

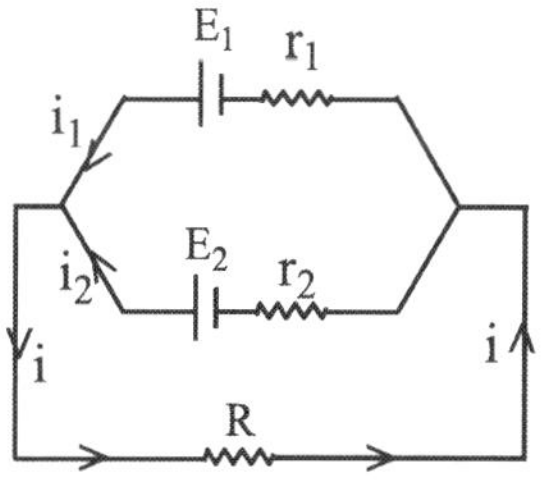

Kirchhoff's second law in diferent loops gives the following equations,

$E_1 - iR - i_1 r_1 = 0$

$$or \; i_1 = \frac{E_1}{r_1} - \frac{iR}{r_1} \ldots\ldots\ldots\ldots\ldots \textbf{(1)}$$

$E_2 - iR - i_2 r_2 = 0$

$$i_2 = \frac{E_2}{r_2} - \frac{iR}{r_2} \ldots\ldots\ldots\ldots\ldots \textbf{(2)}$$

Adding Eqs. (1), (2) we get

$$i_1 + i_2 = (E_1 / r_1) + (E_2 / r_2) - iR(1/r_1 + 1/r_2)$$

or $i[1+R(1/r_1+1/r_2)]=(E_1/r_1)+(E_2/r_2)$

$$\therefore i=\frac{(E_1/r_1)+(E_2/r_2)}{1+R(1/r_1+1/r_2)}$$

W.E - 38 :

Find the emf (V) and internal resistance (r) of a single battery which is equivalent to a parallel combination of two batteries of emfs V_1 and V_2 and internal resistances r_1 and r_2 respectively, with polarities as shown in figure

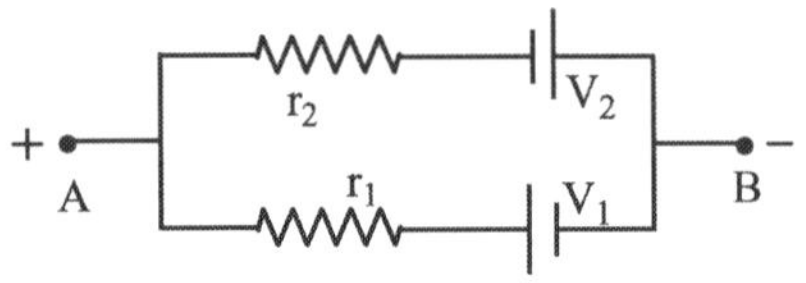

Solution:

EMF of battery is equal to potential difference across the terminals, when no current is drawn from battery (for external circuit) [Here, all the elements in the circuit are in series]

Current in internal circuit = i

$$\therefore i=\frac{\text{Net emf}}{\text{Total resistance}} \quad \text{or } i=\frac{V_1+V_2}{r_1+r_2}$$

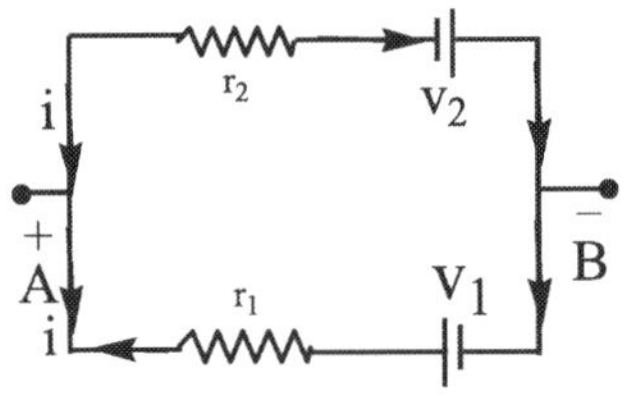

$\therefore V_A - V_B = V_1 - ir_1$ [$\because$ V_1 cell is discharging]

$$\text{or} \quad V_A - V_B = V_1 - \left(\frac{V_1+V_2}{r_1+r_2}\right)r_1$$

$$\text{or} \quad V_A - V_B = \frac{V_1r_2 - V_2r_1}{r_1+r_2}$$

$\therefore$ Equivalent emf of the battery $= V$

$$\therefore V=\frac{V_1r_2 - V_2r_1}{r_1+r_2}$$

(ii) Internal resistance of equivalent battery. r_1 and r_2 are in parallel.

$$\frac{1}{r}=\frac{1}{r_1}+\frac{1}{r_2} \quad \text{or} \quad r=\frac{r_1r_2}{r_1+r_2}$$

W.E - 39 : In the given circuit values are as follows

$\varepsilon_1 = 2V, \varepsilon_2 = 4V, R_1 = 1\Omega$ **and** $R_2 = R_3 = 1\Omega$. **Calculate the Currents through R_1, R_2 and R_3.**

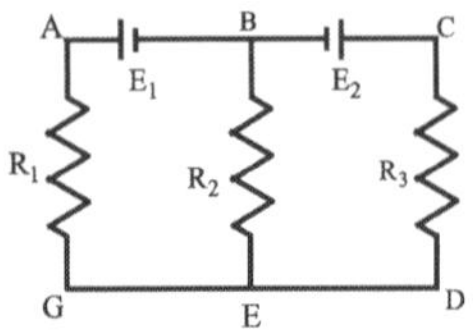

Solution:

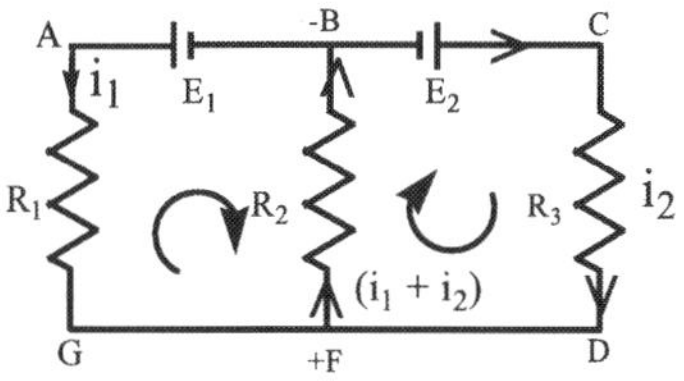

Let i_1, i_2 are currents across R_1 and R_3.

$(i_1 + i_2)$ is current across R_2.

Their direction are taken as shown

From Kirchhoff's second law for AGFBA loop

$$-i_1R_1 - (i_1 + i_2)R_2 + E_1 = 0$$

$$i_1 + i_1 + i_2 = 2$$

$$2i_1 + i_2 = 2 \rightarrow (1)$$

From Kirchhoff's second law for BCDEB loop

$$-i_2R_3 - (i_1 + i_2)R_2 + E_2 = 0$$

$$i_2 + i_1 + i_2 = 4$$

$$i_2 + 2i_2 = 4 \rightarrow (2)$$

Solving equation (1) and (2) we get $i_1 = 0A$, $i_2 = 2A$

Thus currents across R_1 is 0, while across R_3 and R_2 are 2A each.

W.E - 40 :

Solve for current values in figure.

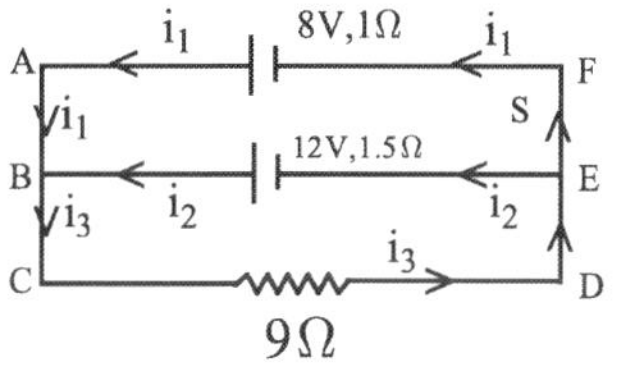

Solution:

Applying Kirchhoff's first law at the junction B we have $i_1 + i_2 = i_3$ .... (1)

Applying Kirchhoff's second law to loop ABEFA

$-12 + i_2 \times 1.5 - i_1 \times 1 + 8 = 0$

$i_1 - 1.5\, i_2 = -4.$ .... (2)

From loop BCDEB

$-(i_2 \times 1.5) - (i_3 \times 9) + 12 = 0$

$1.5\, i_2 + 9i_3 = 12$ .... (3)

on solving $i_1 = -1A$ and $i_3 = 1A$

WHEATSTONE BRIDGE

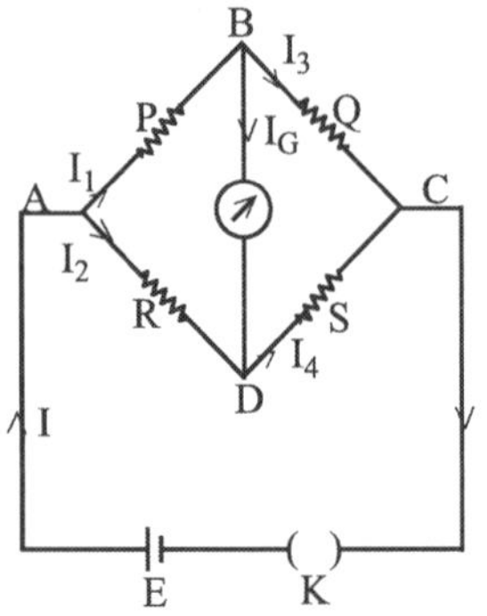

Condition for balancing of bridge :

Applying Kirchhoff's first law at junction B and D we get $I_1 - I_3 - I_G = 0$; and $I_2 + I_G - I_4 = 0$

Applying Kirchhoff's second law for closed loop ABDA, $-I_1P - I_GG + I_2R = 0$

Applying Kirchhoff's second law for closed loop BCDB , $-I_3Q + I_4S + I_GG = 0$

The values of P, Q, R, S are adjusted such that I_G becomes zero. At this stage the bridge is set to be in balance condition.

i.e., In balanced condition of bridge $I_G = 0$

⇒ In balanced condition the above equations respectively become

$I_1 = I_3$ ....(1)

and $I_2 = I_4$ ....(2)

$I_1P = I_2R$ ....(3)

$I_3Q = I_4S$ ....(4)

Dividing equation (3) by equation (4)

$$\frac{I_1P}{I_3Q} = \frac{I_2R}{I_4S}$$

Using eqns (1) and (2) we get

$$\frac{P}{Q} = \frac{R}{S} \qquad ....(5)$$

This is the balancing condition for Wheat stone bridge.

APPLICATIONS ON WHEATSTONE BRIDGE :

1. We can compare two unknown resistances R and S from $\frac{P}{Q} = \frac{R}{S}$
2. In place of resistances we can use capacitors to form a D.C. Wheatstone bridge with four capacitors of capacitances C_1, C_2, C_3 and C_4. The balancing condition will be $\frac{C_1}{C_2} = \frac{C_3}{C_4}$
3. It has been found that the bridge has the greatest sensitivity when the resistances are as nearly equal as possible.
   The bridge is most sensitive if P = Q = R = S.

4. Equivalent resistance of balanced bridge across the ends of battery when the bridge is balanced is given by $\frac{(P+Q)(R+S)}{P+Q+R+S}$

5. There are seven closed meshes in wheatstone's bridge

Application :

Direction of current in an unbalanced wheatstone's bridge :

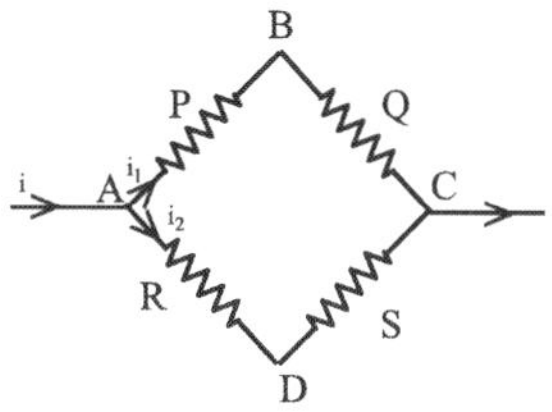

$$V_{AB} = V_A - V_B = i_1 P = i\frac{(R+S)P}{P+Q+R+S}$$

$$V_{AD} = V_A - V_D = i_2 R = i\frac{(P+Q)R}{P+Q+R+S}$$

$$(V_B - V_D) = \frac{i}{P+Q+R+S}[(P+Q)R-(R+S)P] = \frac{i}{P+Q+R+S}[QR-PS]$$

if $QR > PS,\ V_B > V_D \Rightarrow$ current flows from B to D

$QR < PS, V_B < V_D \Rightarrow$ current flows from D to B

$QR = PS,\ V_B = V_D \Rightarrow$ Balanced bridge

W.E - 41 : Determine the current in each branch of the network shown in fig.

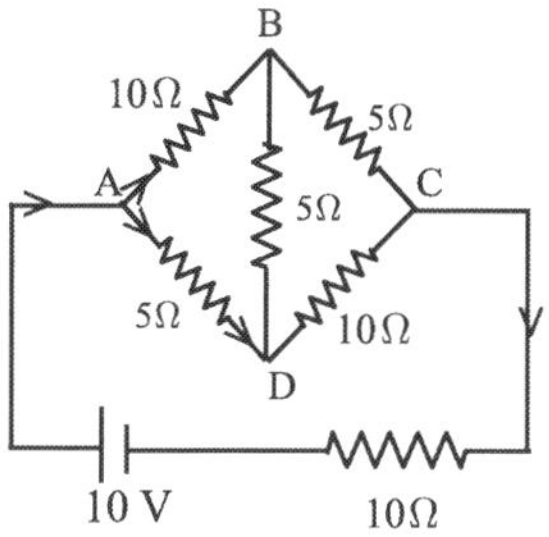

Sol:

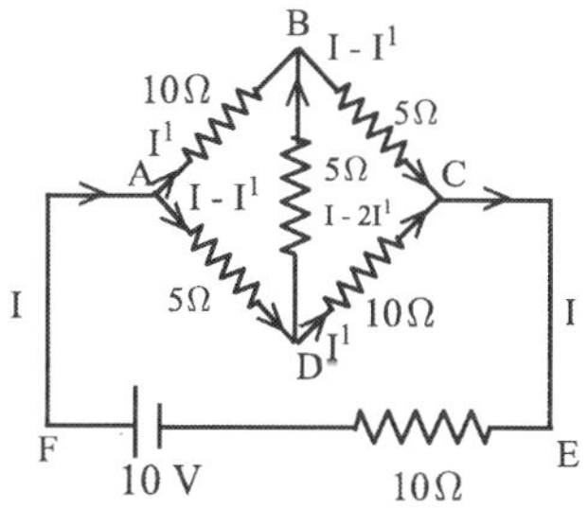

Apply KVL in loop ABDA

$-10I^1+5(I-2I^1)+5(I-I^1)=0$

$2I=5I'$ ....(i)

Apply KVL in ADCEFA loop

$-5(I-I^1)-10I^1+10-10I=0$

$5I^1+15I=10$ ....(2)

From equation (1) and (2)

$I=\frac{10}{17};\quad I^1=\frac{2I}{5}=\frac{4}{17}A$

Current in AB branch $=\frac{4}{17}$

$=I-I^1=\frac{10}{17}-\frac{4}{17}=\frac{6}{17}A$

Current in DB branch

$I-2I'=\frac{10}{17}-\frac{8}{17}=\frac{2}{17}A$

METRE BRIDGE:

➢ It works on the principle of Wheatstone Bridge $\left(\frac{P}{Q}=\frac{R}{S}\right)$

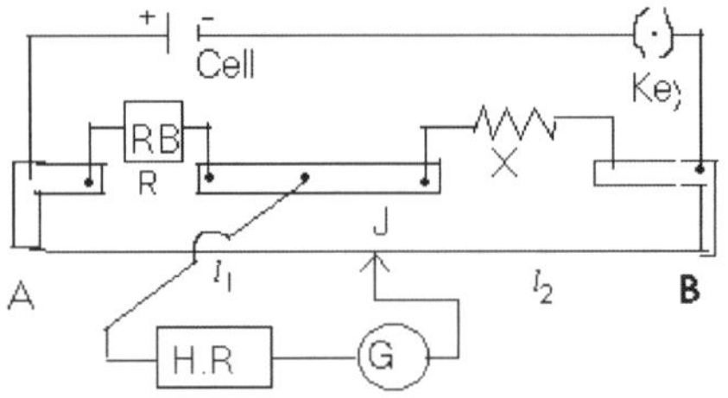

➢ When the Meter bridge is balanced then

$$\frac{R}{X}=\frac{l_1}{l_2}=\frac{l_1}{100-l_1}$$

Where l_1 *is* the balancing length from the left end.

NOTE :

1. If resistance in the left gap increases or resistance in the right gap decreases, balancing point shifts towards right side.
2. If resistance in the left gap decreases or resistance in the right gap increases, balancing point shifts towards left.
3. If a cm, b cm are the end corrections at A and B, then $\frac{R}{X}=\frac{l_1+a}{l_2+b}$
4. Meter bridge is more sensitive if $l_1=50$ cm
5. The resistance of copper strip is called end resistance

W.E - 42 : In the shown arrangement of the experiment of the meter bridge if AC corresponding to null deflection of galvanometer is x, what would be its value if the radius of the wire AB is doubled?

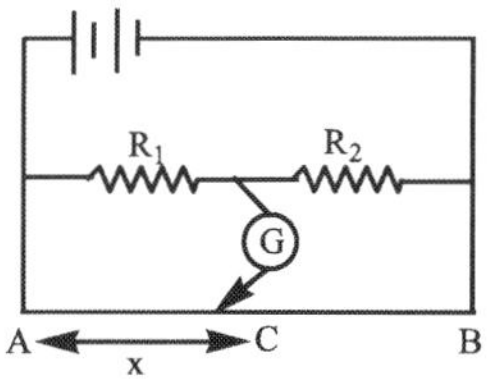

Solution: For null deflection of galvanometer in a metrebridge experiment,

$$\frac{R_1}{R_2}=\frac{R_{AC}}{R_{CB}} \text{ or } \frac{R_1}{R_2}=\frac{x}{(100-x)}$$

Since R_1 / R_2 remains constant, x / (100-x) also remains constant. The value of x remains as such.

$\therefore$ Length of AC = x

W.E - 43 :

A resistance of 2Ω is connected across one gap of a metre-bridge (the length of the wire is 100 cm) and an unknown resistance, greater than 2Ω, is connected across the other gap. When these resistances are interchanged, the balance point shifts by 20 cm. Neglecting any corrections, the unknown resistance is

Solution: Refer to the diagrum

Apply the conditions of the balanced Wheatstone's bridge for the two cases.

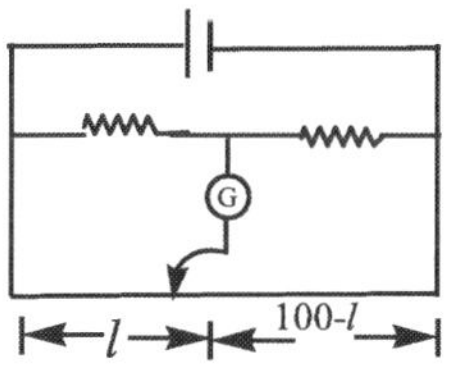

$$\frac{2}{x}=\frac{\ell}{100-\ell} \text{ ....(i)}$$

$$\frac{x}{2}=\frac{\ell+20}{80-\ell} \text{ ....(ii)}$$

Equations (i) and (ii) give $x=3\Omega$

POTENTIOMETER:

Potentiometer is an instrument which can measure accurately the emf of a source or the potential difference across any part of an electric circuit without drawing any current.

a) Principle : The principle of potentiometer states that when a constant current is passed through a wire of uniform area of cross-section, the potential drop across any portion of the wire is directly proportional to the length of that portion.

The principle of potentiometer require that

i) potentiometer wire should be of uniform area of cross-section and

ii) current through the wire should remain constant.

c) Theory of potentiometer :

The end of the potentiometer wire AB are connected to a standard cell of emf E or a source of emf E that supplies constant current. The current through the potentiometer wire can be varied by means of a series resistance R_s which is adjustable.

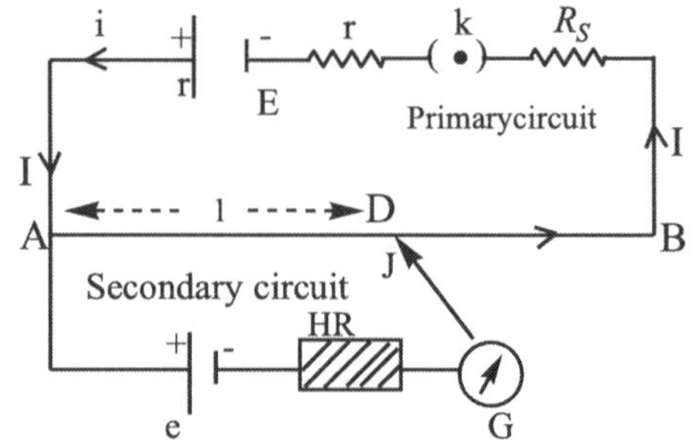

Let r be the internal resistance of the cell of emf E connected across the potentiometer wire of length L and resistance R. The current through the potentiometer wire is $I = \frac{E}{r+R+R_s}$

The potential of the wire decreases from the end A to the end B. The potential fall or potential drop across a length l of the potentiometer wire is

V = Current x Resistance of length l of the potentiometer wire $= I \times \left(\frac{R}{L}\right) l$

If the resistance per unit length of the wire , $\frac{R}{L}$ is denoted by ρ , the potential drop across the wire is

$V = I \times \rho \times \ell$

$\frac{V}{l}$ is called potential drop per unit length of the potentiometer wire or potential gradient of the wire. It is

given by $\frac{V}{l} = I\rho = \left(\frac{E}{r+R+R_S}\right)\frac{R}{L}$

Thus, the unknown voltage V is measured when no current is drawn from it.

1) When specific resistance (S) of potentiometer wire is given then potential gradiant $X = \frac{IS}{A} = \frac{IS}{\pi r^2}$

where A = area of cross - section of potentiometer wire r = Radius of potentiometer wire.

2) When two wires of length $L_1 \; and \; L_2$ and resistance $R_1 \; and \; R_2$ are joined together to form the potentiometer wire, then $\frac{x_1}{x_2} = \frac{R_1}{L_1}\frac{L_2}{R_2}$

Potential gradient depends on

a) Resistance per unit length of the potentiometer wire (ρ =R/L)

b) Radius of crosssection of the potentiometer wire, when the series resistance is included in the circuit and cell in the primary circuit is not ideal.

c) Current flowing through potentiometer wire.

d) emf of the cell in primary circuit

e) Series resistance in the primary circuit

f) Total length (L) and resistance (R) of the potentiometer wire.

g) If cell in primary circuit is ideal and in the absence of series resistance potential gradient **only depends on emf of cell** in primary circuit and length of potentiometer wire

TO DETERMINE THE INTERNAL RESISTANCE OF A PRIMARY CELL:

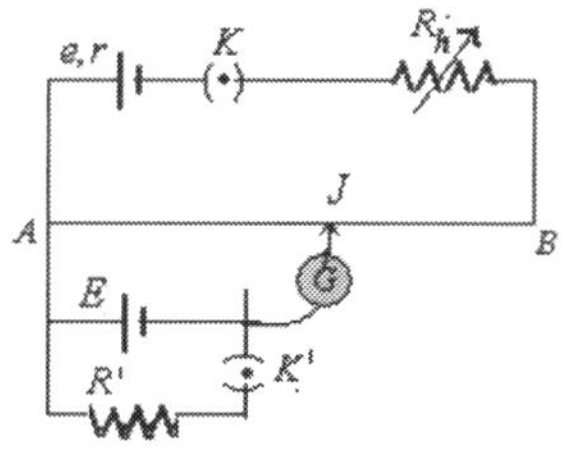

- Initially in secondary circuit key K' remains open and balancing length (l_1) is obtained. Since cell E is in open circuit so it's emf balances on length l_1

  i.e $E = xl_1$ .....(i)

- Now key K ' is closed so cell E comes in closed circuit. If the process of balancing is repeated again keeping constant then potential difference V balances on length l_2

  i.e $V = xl_2$ .....(ii)

- **By using formula internal resistance** $r = \left(\frac{E}{V} - 1\right).R'$

  Where E = emf of cell in secondary circuit
  V = Terminal voltage

  i.e p.d on R, $r = \left(\frac{l_1 - l_2}{l_2}\right)R'$

  $\because \frac{E}{V} = \frac{l_1}{l_2}$, $\frac{E}{V} - 1 = \frac{l_1}{l_2} - 1$

COMPARISON OF EMF'S OF TWO CELLS:

- Let l_1 and l_2 be the balancing length with the cell E_1 and E_2 respectively, then $E_1 = xl_1$ and $E_2 = xl_2 \Rightarrow \frac{E_1}{E_2} = \frac{l_1}{l_2}$

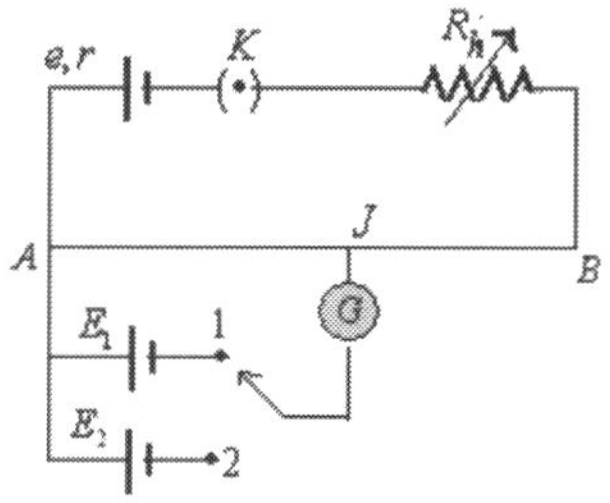

- Let $E_1 > E_2$ and both are connected in series. If balancing length is l_1 when cells assist each other and it is l_2 when they oppose each other as shown then:

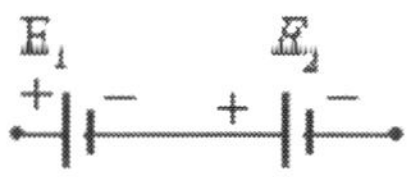

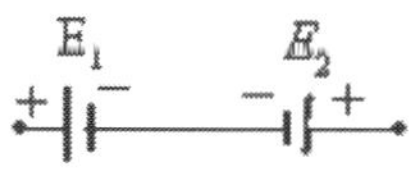

$$(E_1+E_2)=xl_1 \qquad (E_1-E_2)=xl_2$$

$$\Rightarrow \frac{E_1+E_2}{E_1-E_2}=\frac{l_1}{l_2} \text{ (or) } \frac{E_1}{E_2}=\frac{l_1+l_2}{l_1-l_2}$$

Comparison of resistances:

Let the balancing length for resistance R_1 (when XY is connected) be l_1 and let balancing length for resistance R_1+R_2 (when YZ is connected) be l_2. keeping X constant

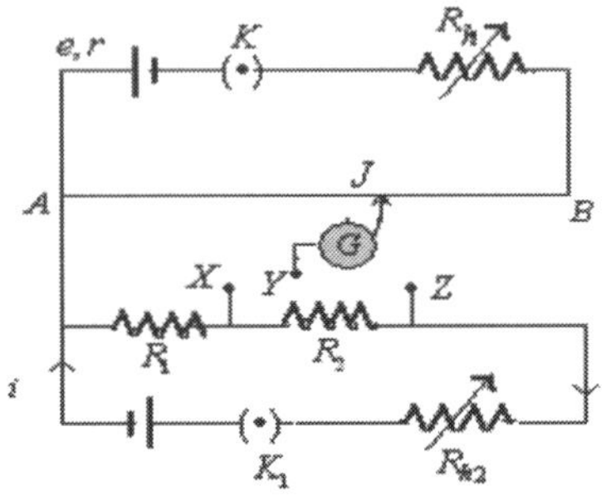

Then $iR_1=xl_1$ and

$$i(R_1+R_2)=xl_2 \Rightarrow \frac{R_2}{R_1}=\frac{l_2-l_1}{l_1}$$

To determine thermo emf:

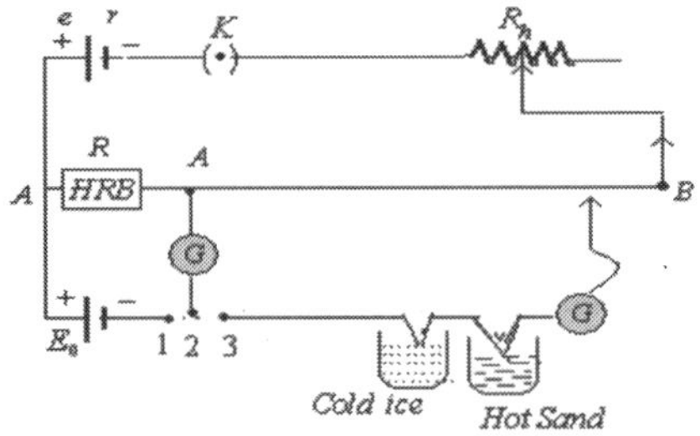

- The value of thermo-emf in a thermocouple for ordinary temperature difference is very low $(10^{-6}\,volt)$.

  For this the potential gradient x must be also very low $(10^{-4}V/m)$. Hence a high resistance (R) is connected in series with the potentiometer wire in order to reduce current in the primary circuit
- The potential difference across R must be equal to the emf of standard cell

  i.e $iR=E_0$ $\qquad \therefore i=\frac{E_0}{R}$
- **The small thermo emf produced in the** thermocouple $e=xl$
- $x=i\rho=\frac{iR^{|}}{L} \quad \therefore \; e=\frac{iR^{|}I}{L}$

  where L = Length of potentiometer wire, ρ =resistance per unit length, l = balancing length of e and $R^{|}$ = Resistance of potentiometer wire

SENSITIVITY OF POTENTIO METER

1. Sensitivity of potentio meter is estimated by its potential gradient.
2. Sensitivity is inversly proportional to potential gradient so lower the potential gradient higher will be the sensitivity.
3. The best instrument for accurate measurement of e.m.f. of a cell is potentiometer, because it does not draw any current from the cell.

CALIBRATION OF AMMETER:

Checking the correctness of ammeter readings with the help of potentiometer is called calibration of ammeter.

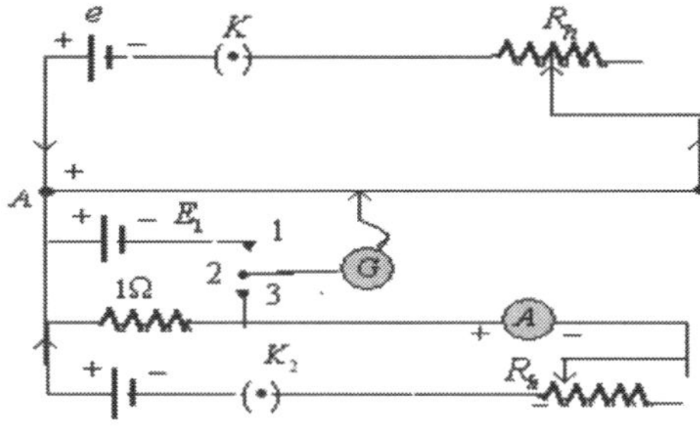

➢ For the calibration of an ammeter, 1Ω resistance coil is specifically used in the secondary circuit of the potentiometer, because the potential difference across 1Ω is equal to the current following through it i.e $V = i$

➢ If the balancing length for the emf

E_0 is l_0 then $E_0 = xl_0$ $\Rightarrow x = \dfrac{E_0}{l_0}$ (Process of standardisation)

➢ Let i' current flows through 1Ω resistance giving potential difference as $V' = i'(1) = xl_1$ where l_1 is the balancing length. so error can be found as $\Delta i = i - i' = i - xl_1 = i - \dfrac{E_0}{l_0} \times l_1$

Here i is ammeter reading

CALIBRATION OF VOLTMETER:

➢ Checking the correctness of voltmeter readings with the help of potentiometer is called calibration of voltmeter.

➢ If l_0 is balancing length for E_0 the emf of standard cell by connecting 1 and 2 of bi-directional key, then $x = E_0 / l_0$

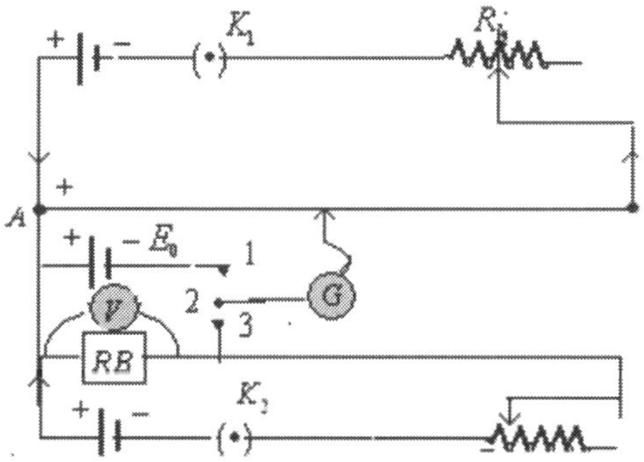

➢ The balancing length l_1 for unknown potential difference V' is given by (closing 2 and 3) $V^{|} = Xl_1 = \left(\frac{E_0}{l_0}\right)l_1$

If the voltmeter reading is V then the error will be $\left(V - V^{|}\right)$ which may be +ve, -ve or zero

W.E - 44 : The length of a potentiometer wire is 1m and its resistance is 4Ω. A current of 5 mA is flowing in it. An unknown source of e.m.f is balanced on 40 cm length of this wire, then find the e.m.f of the source.

Solution : $x = I\rho = I\frac{R}{L} = \frac{5\times 4}{1} = 20\,mV$

E = 1 x = 0. 40 x 20 = 8 mV

W.E - 45 : A cell of e.m.f 2 volt and internal resistance 1.5Ω is connected to the ends of 1m long wire. The resistance of wire is $0.5\Omega/m$. Find the value of potential gradient on the wire.

Solution : $X = \frac{IR}{L} = \left(\frac{E}{R+r}\right)\frac{R}{L} = \frac{2\times 0.5}{0.5+1.5}$ **=0. 5 V/m**

W.E - 46 : In a potentiometer experiment the balancing length with a cell is 560 cm. When an external resistance of 10Ω is connected in parallel to the cell, the balancing length changes by 60 cm. Find the internal resistance of the cell.

Solution : Balancing length $\ell_1 = 560$ cm

Change in balancing length $(\ell_1 - \ell_2) = 60$ cm

$560 - \ell_2 = 60$

$\therefore \ell_2 = 500$ cm

$r = R\left(\frac{\ell_1 - \ell_2}{\ell_2}\right) \Rightarrow r = 10\times\frac{60}{500} = \frac{6}{5} = 1.2\Omega$.

W.E - 47 : In a potentiometer experiment when a battery of e.m.f. 2V is included in the secondary circuit, the balance point is 500cm. Find the balancing length of the same end when a cadimum cell of e.m.f. 1.018V is connected to the secondary circuit.

Solution : $E \propto \ell$; $\frac{E_1}{E_2} = \frac{\ell_1}{\ell_2}$

$\ell_2 = \frac{E_2}{E_1}\times\ell_1 = \frac{1.018}{2}\times 500 = 254.5\,cm$.

CONCEPTUAL

ELECTRIC CURRENT AND DRIFT VELOCITY

1. **If n, e, τ, m, are representing electron density, charge, relaxation time and mass of an electron respectively then the resistance of wire of length l and cross sectional area A is given by**

1) $\frac{ml}{ne^2\tau A}$ 2) $\frac{2mA}{ne^2\tau}$ 3) $ne^2\tau A$ 4) $\frac{ne^2\tau A}{2m}$

2. **Among the following dependences of drift velocity v_d on electric field E, Ohm's Law obeyed is**

1) $v_d \alpha E$ 2) $v_d \alpha E^2$ 3) $v_d \alpha \sqrt{E}$ 4) v_d = constant

3. **A steady current is passing through a linear conductor of nonuniform cross-section. The net quantity of charge crossing any cross section per second is**
1) independent of area of cross-section
2) directly proportional to the length of the conductor
3) directly proportional to the area of cross section.
4) inversely proportional to the area of the conductor

4. **Given a current carrying wire of non-uniform cross section. Which of the following quantity or quantities are constant throughout the length of the wire?**
1) current, electric field and drift speed 2) drift speed only
3) current and drift speed 4) current only

5. **When electric field ($\vec{E}$) is applied on the ends of a conductor, the free electrons starts moving in direction**
1) similar to $\vec{E}$ 2) Opposite to $\vec{E}$ 3) Perpendicular to $\vec{E}$ 4) Cannot be predicted

6. **The drift speed of an electron in a metal is of the order of**
1) 10^{-13} m/s 2) 10^{-3} mm/s 3) 10^{-4} m/s 4) 10^{-30} m/s

7. **In metals and vacuum tubes charge carriers are**
1) electrons 2) protons 3) both 4) positrons

8. **The electric intensity E, current density j and conductivity σ are related as :**
1) $j = \sigma E$ 2) $j = E / \sigma$ 3) $jE = \sigma$ 4) $j = \sigma^2 E$

9. **Electric field (E) and current density (J) have relation**
1) $E \propto J^{-1}$ 2) $E \propto J$ 3) $E \propto \frac{1}{J^2}$ 4) $E^2 \propto \frac{1}{J}$

10. **Assertion : A current flows in a conductor only when there is an electric field within the conductor.**
Reason : The drift velocity of electron in presence of electric field decreases.
1) Both (A) and (R) are true and (R) is the correct explanation of A.
2) Both (A) and (R) are true but (R) is not the correct explanation of A.
3) (A) is true but (R) is false 4) (A) is false but (R) is true

OHM'S LAW & FACTORS EFFECTING RESISTANCE

11. **In an electric circuit containg a battery, the charge (assumed positive) inside the battery**
1) always goes form the positive terminal to the negative terminal
2) may move from the positive terminal to the negative terminal
3) always goes from the negative terminal to the positive terminal 4) does not move

12. **From the following the quantity which is analogous to temperature in electricity is**
1) potential 2) resistance 3) current 4) charge

13. **The flow of the electric current through a metallic conductor is**
1) only due to electrons 2) only due to +ve charges
3) due to both nuclei and electrons. 4) can not be predicted.

14. **For making standard resistance, wire of following material is used**
1) Nichrome 2) Copper 3) Silver 4) manganin

15. **Material used for heating coils is**
1) Nichrome 2) Copper 3) Silver 4) Manganin

16. **A piece of silver and another of silicon are heated from room temperature. The resistance of**
1) each of them increases 2) each of them decreases
3) Silver increases and Silicon decreases 4) Silver decreases and Silicon increases

17. **i-v graph for a metal at temperatures t_1, t_2, t_3 are as shown. The highest temperature is**

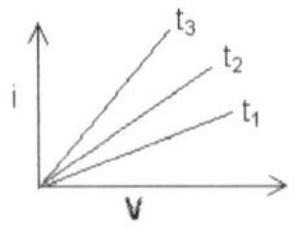

1) t_1 2) t_2 3) t_3 4) $t_1 = t_2 = t_3$

18. **A certain piece of copper is to be shaped into a conductor of minimum resistance. Its length and cross sectional area should be**

1) L and A 2) 2L and A/2 3) L/2 and 2A 4) 3L and A/3

19. **When light falls on semiconductors, their resistance**

1) decreases 2) increases 3) does not change 4) can't be predicted

20. **With the increase of temperature, the ratio of conductivity to resistivity of a metal conductor**

1) Decreases 2) Remains same 3) Increases
4) May increase or decrease

21. **The conductivity of a super conductor, in the super conducting state is**

1) Zero 2) Infinity 3) Depends on temp 4) Depends on free election

22. **When a piece of aluminium wire of finite length is drawn through a series of dies to reduce its diameter to half its original value, its resistance will become**

1) Two times 2) Four times 3) Eight times 4) Sixteen times

23. **Metals have**

1) Zero resistivity 2) High resistivity 3) Low resistivity 4) Infinite resistivity

24. **Consider a rectangular slab of length L, and area of cross-section A. A current I is passed through it, if the length is doubled the potential drop across the end faces**

1) Becomes half of the initial value 2) Becomes one-forth of the initial value
3) Becomes double the initial value 4) Remains Same

25. **A metallic block has no potential difference applied across it, then the mean velocity of free electrons is (T = absolute temperature of the block)**

1) Proportional to T 2) Proportional to $\sqrt{T}$
3) Zero 4) Finite but independent of temperature.

26. **The resistance of a metal increases with increasing temperature because**

1) The collisions of the conducting electrons with the electrons increases.
2) The collisions of the conducting electrons with the lattice consisting of the ions of the metal increases
3) The number of the conduction electrons decreases.
4) The number of conduction electrons increase.

27. **In the absence of applied potential, the electric current flowing through a metallic wire is zero because**

1) The average velocity of electron is zero
2) The electrons are drifted in random direction with a speed of the order of 10^{-2}cm/s.
3) The electrons move in random direction with a speed of the order close to that of velocity of light.
4) Electrons and ions move in opposite direction.

28. **A long constan wire is connected across the terminals of an ideal battery. if the wire is cut in to two equal pieces and one of them is now connected to the same battery, what will be the mobility of free electrons now in the wire compared to that in the first case?**

1) same as that of previous value 2) double that of previous value
3) half that of previous value 4) four times that of previous value

29. **Ohm's law is not applicable for**

1) insulators 2) semi conductors 3) vaccum tube 4) all the above

30. **V - I graphs for two materials is shown in the figure. The graphs are drawn at two different temperatures.**

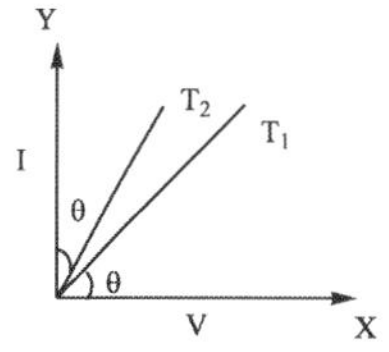

1) $T_1 - T_2 \propto \cot 2\theta$

2) $T_1 - T_2 \propto \sin 2\theta$

3) $T_1 - T_2 \propto \tan 2\theta$ 4) $T_1 - T_2 \propto \cos 2\theta$

31. **Wires of Nichrome and Copper of equal dimensions are connected in series in electrical circuit. Then**

1) More current will flow in copper wire 2) More current will flow in Nichrome wire

3) Copper wire will get heated more 4) Nichrome wire will get heated more

32. **At absolute zero silver wire behaves as**

1) Super conductor 2) Semi conductor 3) Perfect insulator 4) Semi insulator

33. **Fuse wire is a wire of**

1) low melting point and low value of α 2) high melting pointand high value of α

3) high melting point and low value of α 4) low melting point and high value of α

34. **Assertion : Material used in the construction of a standard resistance is constantan or manganin.**
Reason : Temperature coefficient of constantan is very small.

1) Both (A) and (R) are true and (R) is the correct explanation of A.

2) Both (A) and (R) are true but (R) is not the correct explanation of A.

3) (A) is true but (R) is false 4) (A) is false but (R) is true

35. **Assertion (A) : Bending of a conducting wire effects electrical resistance.**
Reason (R) : Resistance of a wire depends on resistivity of that material.

1) Both (A) and (R) are true and (R) is the correct explanation of A.

2) Both (A) and (R) are true but (R) is not the correct explanation of A.

3) (A) is true but (R) is false 4) (A) is false but (R) is true

36. **Assertion (A) : When the radius of a copper wire is doubled, its specific resistance gets increased.**
Reason (R):Specific resistance is independent of cross-section of material used

1) Both (A) and (R) are true and (R) is the correct explanation of A.

2) Both (A) and (R) are true but (R) is not the correct explanation of A.

3) (A) is true but (R) is false 4) (A) is false but (R) is true

THERMISTOR

37. **The thermistors are usually made of**

1) metals with low temperature coefficient of resistivity

2) metals with high temperature coefficient of resistivity.

3) metal oxides with high temperature coefficient of resistivity

4) semiconducting materials having low temperature coefficient of resistivity

38. **For a chosen non-zero value of voltage, there can be more than one value of current in**

1) copper wire 2) thermistor 3) zener diode 4) manganin wire

ELECTRIC POWER

39 **A heater coil is cut into two equal parts and only one part is used in the heater. Then the heat generated becomes**

1) become one fourth 2) halved 3) doubled 4) become four times

40. **Two lamps have resistance r and R, R being greater than r. If they are connected in parallel in an electric circuit, then**

1) the lamp with resistance R will shine more brightly
2) the lamp with resistance r will shine more brightly
3) the two lamps will shine equal brightly
4) the lamp with resistance R will not shine at all

41. **Two bulbs are fitted in a room in the domestic electric installation. If one of them glows brighter than the other, then**

1) the brighter bulb has smaller resistance
2) the brighter bulb has larger resistance
3) both the bulbs have the same resistance
4) nothing can be said about the resistance unless other factors are known

42. **Three identical bulbs P, Q and R are connected to a battery as shown in the figure. When the circuit is closed**

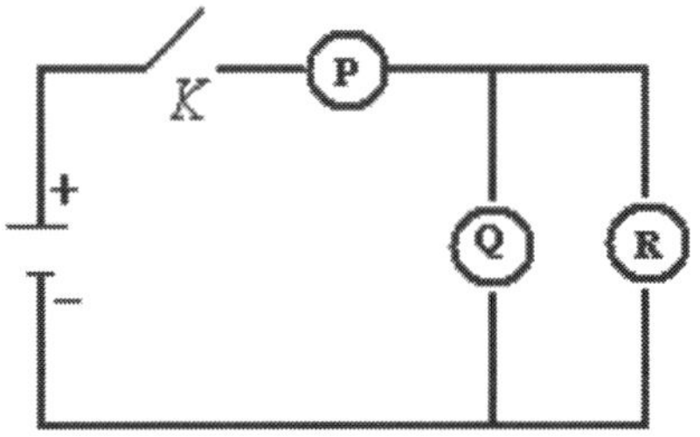

1) Q and R will be brighter than P
2) Q and R will be dimmer than P
3) All the bulbs will be equally bright
4) Q and R will not shine at all

43. **Figure shows three similar lamps L_1, L_2, L_3 connected across a power supply. If the lamp L_3 fuses. The light emitted by L_1 and L_2 will change as**

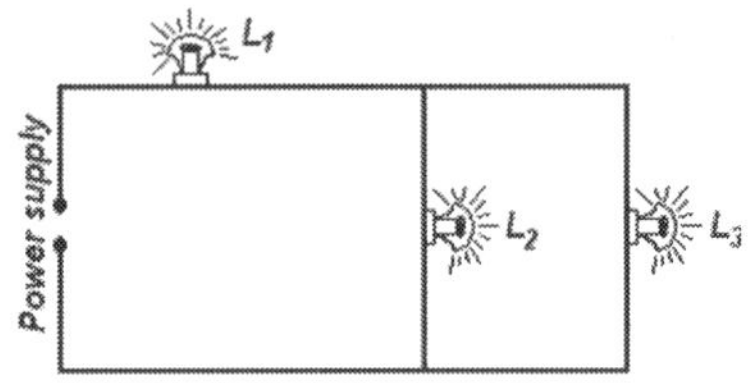

1) no change
2) brilliance of L_1 decreases and that of L_2 increases
3) brilliance of both L_1 and L_2 increases
4) brilliance of both L_1 and L_2 decreases

44. **The potential difference across a conductor is doubled, the rate of generation of heat will**

1) become one fourth 2) be halved 3) be doubled 4) become four times

45. **Two metallic wires of same material and same length have different diameters. When the wires are connected in parallel across an ideal battery the rate of heat produced in thinner wire is Q_1 and that in thicker wire is Q_2. The correct statement is**

1) $Q_1 = Q_2$ 2) $Q_1 < Q_2$ 3) $Q_1 > Q_2$ 4) It will depend on the emf of the battery

46. **There are two metalic wires of same material, same length but of different radii. When these are connected to an ideal battery in series, heat produced is H_1 but when connected in parallel, heat produced is H_2 for the same time. Then the correct statement is**

1) $H_1 = H_2$ 2) $H_1 < H_2$ 3) $H_1 > H_2$ 4) No relation

47. **Two electric bulbs rated P_1 watt and V volt, are connected in series, across V-volt supply. The total power consumed is**

1) $\frac{P_1 + P_2}{2}$ 2) $\sqrt{P_1 \cdot P_2}$ 3) $\frac{P_1 \cdot P_2}{P_1 + P_2}$ 4) $(P_1 + P_2)$

48. In above question, if the bulbs are connected in parallel, total power consumed is

1) $\frac{P_1+P_2}{2}$ 2) $\sqrt{P_1 \cdot P_2}$ 3) $\frac{P_1 \cdot P_2}{P_1+P_2}$ 4) (P_1+P_2)

49. Which of the following causes production of heat, when current is set up in a wire

1) Fall of electron from higher orbits to lower orbits 2)Inter atomic collisions

3)Inter electron collisions 4)Collisions of conduction electrons with atoms

50. A constant voltage is applied between the two ends of a metallic wire. If both the length and the radius of the wire are doubled, the rate of heat developed in the wire

1) will be doubled 2) will be halved 3) will remain the same 4) will be quadrupled

51. A resistor R_1 dissipates the power P when connected to a certain generator. If the resistor R_2 is put in series with R_1, the power dissipated by R_1

1) Decreases 2) Increases

3) Remains the same 4) Any of the above depending upon the relative values of R_1 and R_2

CELL-INTERNAL RESISTANCE EMF

52. Back emf of a cell is due to

1) Electrolytic polarization 2) Peltier effect

3) Magnetic effect of current 4) Internal resistance

53. The direction of current in a cell is

1) $(-)$ ve pole to $(+)$ ve pole during discharging

2) $(+)$ ve pole to $(-)$ ve pole during discharging

3) Always $(-)$ ve pole to $(+)$ ve pole 4) always flows from (+)ve ploe to (-) ve pole

54. When an electric cell drives current through load resistance, its Back emf,

1) Supports the original emf 2) Opposes the original emf

3) Supports if internal resistance is low 4) Opposes if load resistance is large

55. The terminal voltage of a cell is greater than its emf. when it is

1) being charged 2) an open circuit 3) being discharged 4) it never happens

56. What is constant in a battery (also called a source of emf) ?

1) current supplied by it 2) terminal potential difference

3) internal resistance 4) emf

57. From the following the standard cell is

1) Daniel cell 2) Cadmium cell 3) Leclanche cell 4) Lead accumulator

58. A cell is to convert

1) chemical energy into electrical energy 2) electrical energy into chemical energy

3) heat energy into potential energy 4) potential energy into heat energy

59. 'n' identical cells, each of internal resistance (r) are first connected in parallel and then connected in series across a resistance (R). If the current through R is the same in both cases, then

1) R = r/2 2) r = R/2 3) R = r 4) r = 0

60. The value of internal resistance of ideal cell is

1) Zero 2) infinite 3) 1 Ω 4) 2Ω

61. In a circuit two or more cells of the same emf are connected in parallel in order

1) Increases the pd across a resistance in the circuit

2) Decreases pd across a resistance in the circuit

3) Facilitate drawing more current from the battery system

4) Change the emf across the system of batteries

62. **The resistance of an open circuit is**

1) Infinity 2) Zero 3) Negative 4) cann't be predicted

63. **According to joule's law if potential difference across a conductor having a material of specific resistance ρ, remains constant, then heat produced in the conductor is directly proportional to**

1) ρ 2) ρ^2 3) $\frac{1}{\sqrt{\rho}}$ 4) $\frac{1}{\rho}$

64. **Internal resistance of a cell depends on**

1) concentration of electrolyte 2) distance between the electrodes

3) area of electrode 4) all the above

65. **When cells are arranged in series**

1) the current capacity decreases 2) The current capacity increases

3) the emf increases 4) the emf decreases

66. **To supply maximum current, cells should be arrange in**

1) series 2) parallel 3) Mixed grouping

4) depends on the internal and external resistance

67. **The terminal Pd of a cell is equal to its emf if**

1) external resistance is infinity 2) internal resistance is zero

3) both 1 and 2 4) internal resistance is 5 Ω

68. **The electric power transfered by a cell to an external resistance is maximum when the external resistance is equal to ...(r internal resistance)**

1) $\frac{r}{2}$ 2) 2r 3) r 4) r^2

69. **Which depolarizers are used to neutralizes hydrogen layer in cells**

1) Potassium dichromite 2) Manganese dioxide

3) 1 or 2 4) hydrogen peroxide

70. **Assertion : Series combination of cells is used when their internal resitance is much smaller than the external resistance.**

Reason : $I = \frac{n\varepsilon}{R + nr}$ **where the symbols have their standard meaning,in series connection**

1) Both (A) and (R) are true and (R) is the correct explanation of A.

2) Both (A) and (R) are true but (R) is not the correct explanation of A.

3) (A) is true but (R) is false 4) (A) is false but (R) is true

71. **Assertion (A) : To draw more current at low P.d; parallel connection of cells is preferred.**

Reason (R) : In parallel connection, current $i = \frac{nE}{r}$**, if** r **>> R.**

1) Both (A) and (R) are true and (R) is the correct explanation of A.

2) Both (A) and (R) are true but (R) is not the correct explanation of A.

3) (A) is true but (R) is false 4) (A) is false but (R) is true

KIRCHHOFF'S LAWS WHEATSTONE BRIDGE

72. **Kirchoff's law of meshes is in accordance with law of conservation of**

1) charge 2) current 3) energy 4) angular momentum

73. **Kirchoff's law of junctions is also called the law of conservation of**

1) energy 2) charge 3) momentum 4) angular momentum

74. **Wheatstones's bridge cannot be used for measurement of very ——— resistances.**
1) high 2) low 3) low(or) high 4) zero

75. **In a balanced Wheatstone's network, the resistances in the arms Q and S are interchanged. As a result of this :**
1) galvanometer and the cell must be interchanged to balance
2) galvanometer shows zero deflection
3) network is not balanced 4) network is still balanced

76. **If galvanometer and battery are interchanged in balanced wheatstone bridge, then**
1) the battery discharges 2) the bridge still balances
3) the balance point is changed 4) the galvanometer is damaged due to flow of high current

77. **Wheatstone bridge can be used**
1) To compare two unknown resistances. 2) to measure small strains produced in hardmetals
3) as the working principle of meterbridge 4) All the above

78. **In a wheatstone's bridge three resistances P,Q,R connected in three arms and the fourth arm is formed by two resitances S_1,S_2 connected in parallel.The condition for bridge to be balanced will be**

1) $\frac{P}{Q}=\frac{R}{S_1+S_2}$ 2) $\frac{P}{Q}=\frac{2R}{S_1+S_2}$ 3) $\frac{P}{Q}=\frac{R(S_1+S_2)}{S_1S_2}$ 4) $\frac{P}{Q}=\frac{R(S_1+S_2)}{2S_1S_2}$

79. **Assertion : At any junction of a network, algebraic sum of various currents is zero**
Reason : At steady state there is no accumulation of charge at the junction.
1) Both (A) and (R) are true and (R) is the correct explanation of A.
2) Both (A) and (R) are true but (R) is not the correct explanation of A.
3) (A) is true but (R) is false 4) (A) is false but (R) is true

METERBRIDGE

80. **Metal wire is connected in the left gap, semi conductor is connected in the right gap of meter bridge and balancing point is found. Both are heated so that change of resistances in them are same. Then the balancing point**
1) will not shift 2) shifts towards left
3) shifts towards right 4) depends on rise of temperature

81. **A metre bridge is balanced with known resistance in the right gap and a metal wire in the left gap. If the metal wire is heated the balance point.**
1) shifts towards left 2) shifts towards right
3) does not change 4) may shift towards left or right depending on the nature of the metal.

82. **In metre bridge experiment of resistances, the known and unknown resistances are interchanged . The error so removed is**
1) end correction 2) index error
3) due to temperature effect 4) random error

83. **In a metre-bridge experiment, when the resistances in the gaps are interchanged, the balance-point did not shift at all. The ratio of resistances must be**
1) Very large 2) Very small 3) Equal to unity 4) zero

84. **Assertion (A) : Meterbridge wire is made up of manganin**
Reason (R) : The temperature coeffiecient of resistance is very small for manganin
1) Both (A) and (R) are true and (R) is the correct explanation of A.
2) Both (A) and (R) are true but (R) is not the correct explanation of A.
3) (A) is true but (R) is false 4) (A) is false but (R) is true

POTENTIOMETER

85. A potentiometer is superior to voltmeter for measuring a potential because

1) voltmeter has high resistance
2) resistance of potentiometer wire is quite low
3) potentiometer does not draw any current from the unknown source of emf. to be measured.
4) sensitivity of potentiometer is higher than that of a voltmeter.

86. In comparing emf's of 2 cells with the help of potentiometer, at the balance point, the current flowing through the wire is taken from

1) Any one of these cells
2) both of these cells
3) Battery in the primary circuit
4) From an unknown source

87. A potentiometer wire is connected across the ideal battery now, the radius of potentiometer wire is doubled without changing its length. The value of potential gradient

1) increases 4 times
2) increases two times
3) Does not change
4) becomes half

88. In a potentiometer of ten wires, the balance point is obtained on the sixth wire. To shift the balance point to eighth wire, we should

1) increase resistance in the primary circuit.
2) decrease resistance in the primary circuit.
3) decrease resistance in series with the cell whose emf. has to be measured.
4) increase resistance in series with the cell whose emf. has to be measured.

89. If the emf of the cell in the primary circuit is doubled, with out changing the cell in the secondary circuit, the balancing length is

1) Doubled
2) Halved
3) Uncharged
4) Zero

90. The potential gradients on the potentiometer wire are V_1 and V_2 with an ideal cell and a real cell of same emf in the primary circuit then

1) $V_1 = V_2$
2) $V_1 > V_2$
3) $V_1 < V_2$
4) $V_1 \le V_2$

91. If the current in the primary circuit is decreased, then balancing length is obtained at

1) Lower length
2) Higher length
3) Same length
4) 1/3rd length

92. Temperature coefficient of resistance 'α' and resistivity 'ρ' of a potentiometer wire must be

1) high and low
2) low and high
3) low and low
4) high and high

93. A series high resistance is preferable than shunt resistance in the galvanometer circuit of potentiometer. Because

1) shunt resistances are costly
2) shunt resistance damages the galvanometer
3) series resistance reduces the current through galvanometer in an unbalanced circuit
4) high resistances are easily available

94. The sensitivity of potentiometer wire can be increased by

1) decreasing the length of potentiometer wire
2) increasing potential gradient on its wire
3) increasing emf of battery in the primary circuit
4) decreasing the potential gradient on its wire

95. A cell of emf 'E' and internal resistance 'r' connected in the secondary gets balanced against length 'ℓ' of potentiometer wire. If a resistance 'R' is connected in parallel with the cell, then the new balancing length for the cell will be

1) $\left(\frac{R}{R-r}\right)l$
2) $\left(\frac{R-r}{R}\right)l$
3) $\left(\frac{R}{r}\right)$
4) $\left(\frac{R}{R+r}\right)l$

96. Potentiometer is an ideal instrument, because

1) no current is drawn from the source of unknown emf
2) current is drawn from the source of unknown emf
3) it gives deflection even at null point
4) it has variable potential gradient

97. On increasing the resistance of the primary circuit of potentiometer, its potential gradient will

1) become more
2) become less
3) not change
4) become infinite

98. **If the value of potential gradient on potentiometer wire is decreased, then the new null point will be obtained at**

1) lower length 2) higher length 3) same length 4) nothing can be said

99. **A cell of negligible internal resistance is connected to a potentiometer wire and potential gradient is found. Keeping the length as constant, if the radius of potentiometer wire is increased four times, the potential gradient will become (no series resistance in primary)**

1) 4 times 2) 2 times 3) half 4) constant

100. **For the working of potentiometer, the emf of cell in the primary circuit (E) compared to the emf of the cell in the secondary circuit (E^1) is**

1) $E > E^1$ 2) $E < E^1$ 3) Both the above 4) none of the above

101. **The balancing lengths of potentiometer wire are l_1 and l_2 when two cells of emf E_1 and E_2 are connected in the secondary circuit in series first to help each other and next to oppose each other $\frac{E_1}{E_2}$ is equal to ($E_1 > E_2$).**

1) $\frac{l_1}{l_2}$ 2) $\frac{l_1 - l_2}{l_1 + l_2}$ 3) $\frac{l_1 + l_2}{l_1 - l_2}$ 4) $\frac{l_2}{l_1}$

102. **At the moment when the potentiometer is balanced,**

1) Current flows only in the primary circuit 2) Current flows only in the secondary circuit

3) Current flows both in primary and secondary circuits 4) current does not flow in any circuit

103. **The quantity that cannot be measured by a potentiometer is ....**

1) Resistance 2) emf 3) current in the wire 4) Inductance

104. **Assertion : Potentiometer is much better than a voltmeter for measuring emf of cell**

Reason : A potentiometer draws no current while measuring emf of a cell

1) Both (A) and (R) are true and (R) is the correct explanation of A.

2) Both (A) and (R) are true but (R) is not the correct explanation of A.

3) (A) is true but (R) is false 4) (A) is false but (R) is true

105. **A : The emf of the cell in secondary circuit must be less than emf of cell in primary circuit in potentiometer.**

R : Balancing length cannot be more than length of potentiometer wire.

1) Both (A) and (R) are true and (R) is the correct explanation of A.

2) Both (A) and (R) are true but (R) is not the correct explanation of A.

3) (A) is true but (R) is false 4) (A) is false but (R) is true

KEY CONCEPTUAL

1) 1	2) 1	3) 1	4) 4	5) 2	6) 3	7) 1	8) 1	9) 2	10) 3
11) 2	12) 1	13) 1	14) 4	15) 1	16) 3	17) 1	18) 3	19) 1	20) 1
21) 2	22) 4	23) 1	24) 3	25) 3	26) 2	27) 1	28) 1	29) 4	30) 1
31) 4	32) 1	33) 4	34) 1	35) 2	36) 4	37) 3	38) 2	39) 3	40) 2
41) 1	42) 2	43) 2	44) 4	45) 2	46) 2	47) 3	48) 4	49) 4	50) 1
51) 1	52) 1	53) 1	54) 2	55) 1	56) 4	57) 2	58) 1	59) 3	60) 1
61) 3	62) 1	63) 4	64) 4	65) 3	66) 4	67) 3	68) 3	69) 3	70) 1
71) 1	72) 3	73) 2	74) 2	75) 3	76) 2	77) 4	78) 3	79) 1	80) 3
81) 2	82) 1	83) 3	84) 1	85) 3	86)3	87)3	88) 1	89) 2	90) 2
91) 2	92) 2	93) 3	94) 4	95) 4	96) 1	97) 2	98) 2	99) 4	100) 1
101) 3	102) 1	103) 4	104) 1	105) 1					

LEVEL - I (CW)

ELECTRIC CURRENT & DRIFT VELOCITY

1. **If the electron in a Hydrogen atom makes 6.25×10^{15} revolutions in one second, the current is**
   1) 1.12 mA 2) 1 mA 3) 1.25 mA 4) 1.5 mA

2. **The current through a wire connected to a condencer varies with time as $i = (2t+1)A$**
   The charge transport to the condenser from $t=0$ to $t=5s$ is
   1) 5C 2) 55C 3) 30C 4) 60C

3. **A copper wire of cross-sectional area 2.0 mm^2, resistivity = 1.7×10^{-8} Ωm, carries a current of 1 A. The electric field in the copper wire is**
   1) 8.5×10^{-5} V/m 2) 8.5×10^{-4} V/m 3) 8.5×10^{-3} V/m 4) 8.5×10^{-2} V/m

OHM'S LAW AND COMBINATION OF RESISTANCES

4. **Using three wires of resistances 1 ohm, 2ohm and 3 ohm, then no.of different values of resistances that possible are**
   1) 6 2) 4 3) 10 4) 8

5. **Six resistances of each 12 ohm are connected as shown in the fig. The effective resistance between the terminals A and B is**

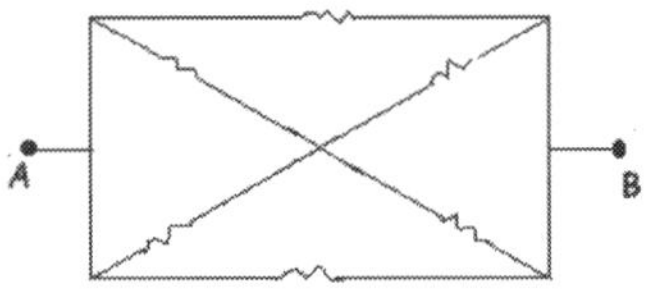

   1) 8 Ω 2) 6 Ω 3) 4 Ω 4) 12 Ω

6. **Current 'i' coming from the battery and ammeter reading are**

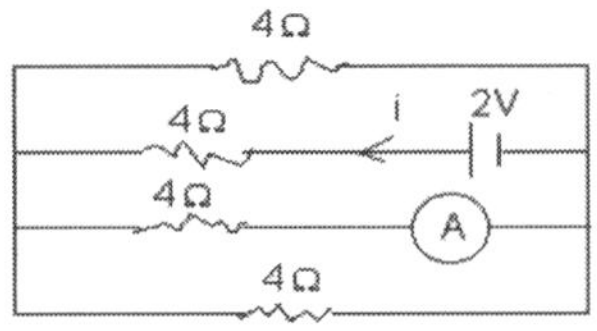

   1) $\frac{3}{8}A, \frac{1}{8}A$ 2) $\frac{1}{8}A, \frac{1}{8}A$ 3) $2A, \frac{2}{3}A$ 4) $2A, \frac{1}{8}A$

7. **In the circuit shown, the reading of the voltmeter and the ammeter are**

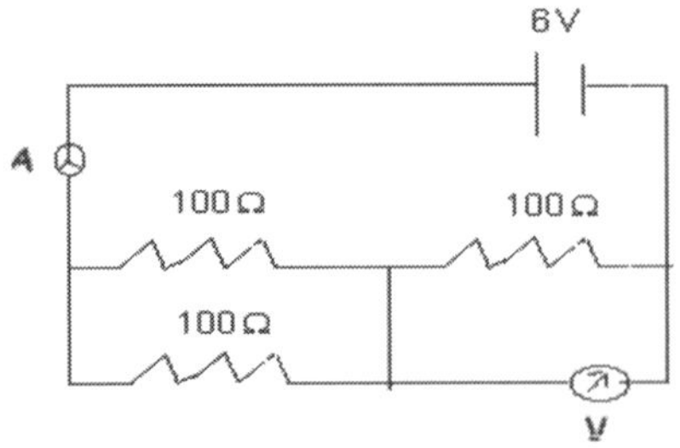

   1) 4V, 0.2A 2) 2V, 0.4A 3) 3V, 0.6A 4) 4V, 0.04A

8. **The resistance of a wire of 100 cm length is 10 Ω. Now, it is cut into 10 equal parts and all of them are twisted to form a single bundle. Its resistance is**
   1) 1 Ω 2) 0.5 Ω 3) 5 Ω 4) 0.1 Ω

9. **A metallic wire of resistance 20 ohm stretched until its length is doubled. Its resistance is**
   1) 20 Ω 2) 40 Ω 3) 80 Ω 4) 60 Ω

10. A wire of resistance 20Ω is bent in the form of a square. The resistance between the ends of diagonal is

1) 10Ω 2) 5Ω 3) 20Ω 4) 15Ω

11. Resistance of each 10Ω are connected as shown in the fig. The effective resistance betweeen A and G is

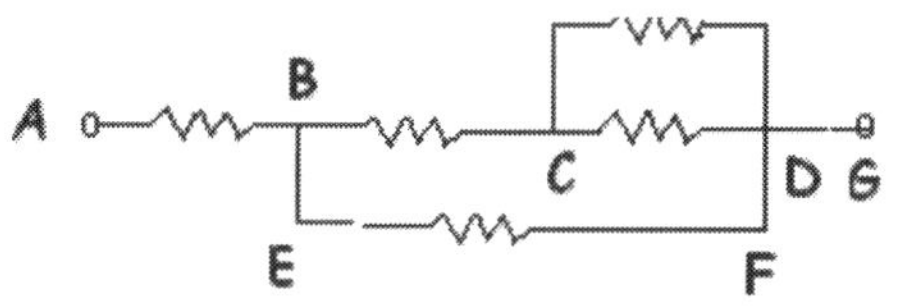

1) 16Ω 2) 20Ω 3) 12Ω 4) 8Ω

12. Which arrangement of four identical resistances should be used to draw maximum energy from a cell of voltage V

1) 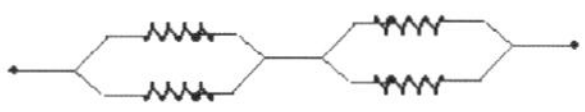

2) 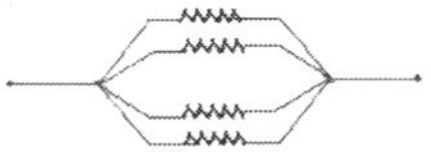

3) 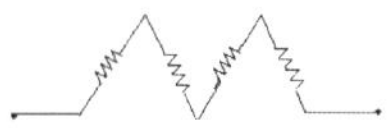

4) 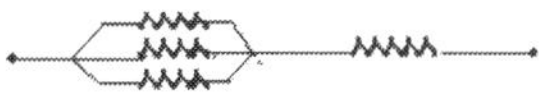

13. If four resistances are connected as shown in the fig. between A and B the effective resistance is

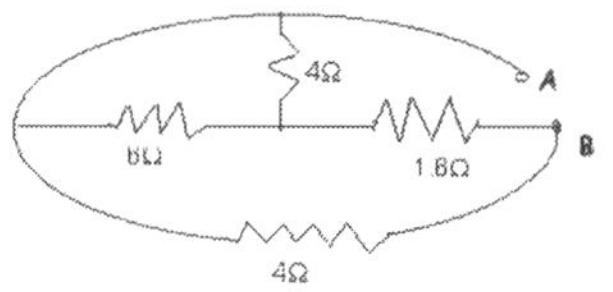

1) 4Ω 2) 8Ω 3) 2.4 Ω 4) 2Ω

14. A letter 'A' is constructed as a uniform wire of resistance 1 ohm/cm. The sides of the letter are 20 cm long and the cross piece in the middle is 10cm long while the vertex angle is 60^0. The resistance of the letter between the two ends of the legs is

1) 40/3 Ω 2) 80/3Ω 3) 40 Ω 4) 10Ω

15. Find the value of colour coded resistance shown is fig

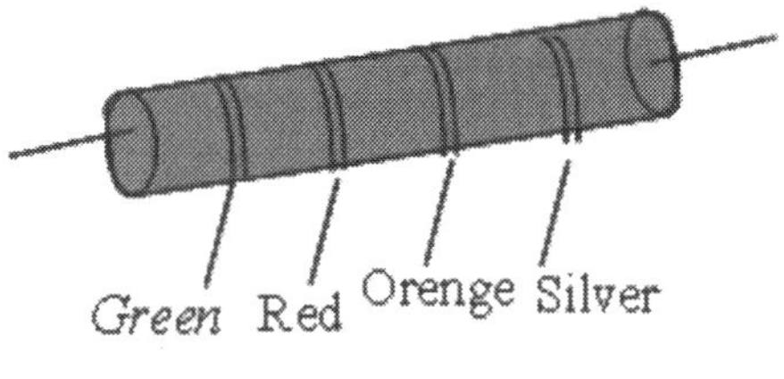

1) $520 \pm 10\%$ 2) $5200 \pm 1\%$ 3) $52000 \pm 10\%$ 4) $52000 \pm 1\%$

16. The resistance of a wire is 2Ω. If it is drawn in such a way that it experiences a longitudinal strain 200%. Its new resistance is

1) 4Ω 2) 8Ω 3) 16Ω 4) 18Ω

17. 'n' conducting wires of same dimensions but having resistivities 1, 2, 3,...n are connected in series. The equivalent resistivity of the combination is

1) $\frac{n(n+1)}{2}$ 2) $\frac{n+1}{2}$ 3) $\frac{n+1}{2n}$ 4) $\frac{2n}{n+1}$

18. An Aluminium (α = 4 x $10^{-3}K^{-1}$) resistance R_1 and a carbon (α = -0.5 x $10^{-3}K^{-1}$) resistance R_2 are connected in series to have a resultant resistance of 36 Ω at all temperatures. The values of R_1 and R_2 in Ω respectively are :

1) 32, 4 2) 16, 20 3) 4, 32 4) 20, 16

19. The temperature coefficient of a wire is 0.00125°C^{-1}. At 300 K its resistance is one ohm. The resistance of the wire will be 2Ω at

1) 1154 K 2) 1100 K 3) 1400 K 4) 1127 K

20. The electrical resistance of a mercury column in a cylindrical container is 'R'. The mercury is poured into another cylindrical container with half the radius of cross-section. The resistance of the mercury column is

1) R 2) 2R 3) 16R 4) 5R

21. Four conductors of same resistance connected to form a square. If the resistance between diagonally opposite corners is 8 ohm, the resistance between any two adjacent corners is

1) 32 ohm 2) 8 ohm 3) 1/6 ohm 4) 6 ohm

22. The resistivity of a material is S ohm meter. The resistance between opposite faces of a solid cube of edge 10 cm is (in ohm)

1) S/2 2) S/10 3) 100S 4) 10S

23. Four wires made of same material have different lengths and radii, the wire having more resistance in the following case is

1) $\ell = 100cm, r = 1mm$ 2) $\ell = 50cm, r = 2mm$

3) $\ell = 100cm, r = \frac{1}{2}mm$ 4) $\ell = 50cm, r = \frac{1}{2}mm$

24. Two different wires have specific resistivities, lengths, area of cross-sections are in the raio 3:4, 2:9 and 8:27. Then the ratio of resistance of two wires is

1) $\frac{16}{9}$ 2) $\frac{9}{16}$ 3) $\frac{8}{27}$ 4) $\frac{27}{8}$

25. Two wires made of same material have their length are in the ratio 1:2 and their masses in the ratio 3 : 16. The ratio of resistance of two wires is

1) $\frac{3}{4}$ 2) 1:2 3) 2:1 4) 4:3

26. A wire of resistance 18 ohm is drawn until its radius reduce $\frac{1}{2}$ th of its original radius then resistance of the wire is

1) 188 Ω 2) 72 Ω 3) 288 Ω 4) 388 Ω

27. A piece of wire of resistance 4Ω is bent through 180° at its midpoint and the two halves are twisted together. Then the resistance is

1) 8 Ω 2) 1 Ω 3) 2 Ω 4) 5 Ω

28. If three wires of equal resistance are given then number of combinations they cany be made to give different resistance is

1) 4 2) 3 3) 5 4) 2

29. The effective resistance between A and B in the given circuit is

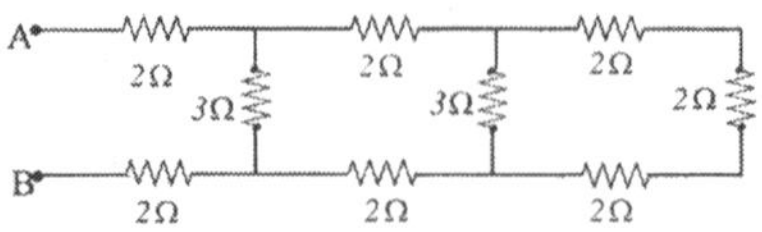

1) 20 Ω 2) 7 Ω 3) 3 Ω 4) 6 Ω

30. How many cells each marked $(6V - 12A)$ should be connected in mixed grouping so that it may be marked $(24V - 24A)$

1) 4 2) 8 3) 12 4) 6

31. The effective resistance in series combination of two equal resistance is 's'. When they are joined in parallel the total resistance is p. If s = np then the minimum possible value of 'n' is

1) 4 2) 1 3) 2 4) 3

ELECTRIC POWER & JOULES LEW

32. A 25 watt, 220 volt bulb and a 100 watt, 220 volt bulb are connected in series across 440 volt line

1) only 100 watt bulb will fuse 2) only 25 watt bulb will fuse
3) none of the bulb will fuse 4) both bulbs will fuse

33. There are 5 tube-lights each of 40W in a house. These are used on an average for 5 hours per day. In addition, there is an immersion heater of 1500W used on an average for 1 hour per day. The number of units of electricity are consumed ina month is

1) 25 units 2) 50 units 3) 75 units 4) 100 units

34. Three equal resistors connected in series across a source emf together dissipate 10 watt. If the same resistors are connected in parallel across the same emf the power dissipate will be

1) 10 watt 2) 30 watt 3) 10/3 watt 4) 90 watt

35. Time taken by a 836 W heater to heat one litre of water from $10^0 C \, to \, 40^0 C$ is

1) 50 s 2) 100 s 3) 150 s 4) 200 s

36. A lamp of 600W-240V is connected to 220V mains. Its resistance is

1) 96 Ω 2) 84 Ω 3) 90 Ω 4) 64 Ω

37. A 200W - 200V lamp is connected to 250V mains. It power consumption is

1) 300 W 2) 312.5W 3) 292 W 4) 250 W

38. If the current in a heater increases by 10% , the percentage change in the power consumption

1) 19% 2) 21% 3) 25% 4) 17%

39. The power of a heating coil is P. It is cut into two equal parts. The power of one of them across same mains is

1) 2P 2) 3P 3) P/2 4) 4P

40. In a house there are four bulbs each of 50W and 5 fans each of 60W. If they are used at the rate of 6 hours a day, the electrical energy consumed in a month of 30 days is

1) 64 KWH 2) 54 KWH 3) 72 KWH 4) 42 KWH

41. An electric kettle has two coils. When one coil is switched on it takes 15 minutes and the other takes 30 minutes to boil certain mass of water. The ratio of times taken by them, when connected in series and in parallel to boil the same mass of water is

1) 9 :2 2) 2:9 3) 4:5 4) 5:4

42. A resistance coil of $60\,\Omega$ is immersed in 42kg of water. A current of 7A is passed through it. The rise in temperature of water per minute is

1) 4^0C 2) 8^0C 3) 13^0C 4) 12^0C

43. What is the required resistance of the heater coil of an immersion heater that will increase the temperature of 1.50 kg of water from 10^0C to 50^0C in 10 minutes while operating at $240V$?

1) 25Ω 2) 12.5Ω 3) 250Ω 4) 125Ω

44. A 5^0C rise in the temperature is observed in a conductor by passing some current. When the current is doubled, then rise in temperature will be equal to

1) 5^0C 2) 10^0C 3) 20^0C 4) 40^0C

CELLS AND COMBINATION OF CELLS

45. In the following diagram, the pd across 6V cell is

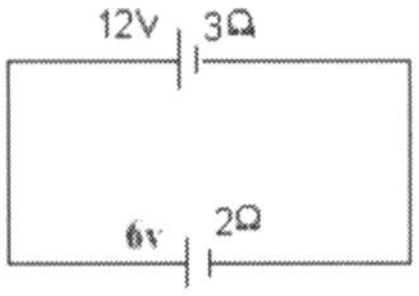

1) 6V 2) 5.6V 3) 8.2V 4) 8.4V

46. While connecting 6 cells in a battery in series, in a tape recorder, by mistake one cell is connected with reverse polarity. If the effective resistance of load is 24 ohm and internal resistance of each cell is one ohm and emf 1.5V, the current devlivered by the battery is

1) 0.1A 2) 0.2A 3) 0.3A 4) 0.4A

47. A 10m long wire of resistance 15 ohm is connected in series with a battery of emf 2V (no internal resistance) and a resistance of 5 ohm. The potential gradient along the wire is

1) 0.15 Vm^{-1} 2) 0.45V m^{-1} 3) 1.5Vm^{-1} 4)4.5Vm^{-1}

48. When a resistance of 2 ohm is placed across a battery the current is 1A and when the resistance across the terminals is 17 ohm, the current is 0.25A. the emf of the battery is

1) 4.5 V 2) 5 V 3) 3 V 4) 6 V

49. A battery has six cells in series. Each has an emf 1.5V and internal resistance 1 ohm. If an external load of 24Ω is connected to it. The potential drop across the load is

1) 7.2V 2) 0.3V 3) 6.8V 4) 0.4V

50. 12 cells of each emf 2V are connected in series among them, if 3 cells are connected wrongly. Then the effective emf. of the combination is

1) 18 V 2) 12 V 3) 24 V 4) 6 V

51. When a battery connected across a resistor of 16Ω, the voltage across the resistor is 12V.When the same battery is connected across a resistor of 10Ω, voltage across it is 11V. The internal resistance of the battery in ohms is

1) 10/7 2) 20/7 3) 25/7 4) 30/7

KIRCHOFF'S LAWS, WHEATSTONE BRIDGE

52. Six resistors of each 2 ohm are connected as shown in the figure. The resultant resistance between A and B is.

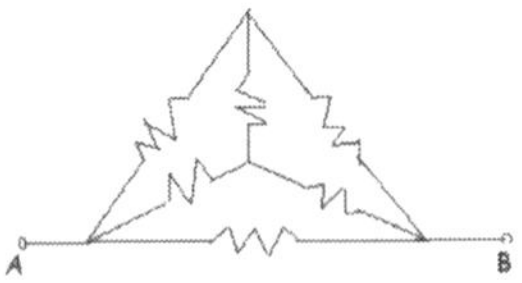

1) 4Ω 2) 2Ω 3) 1Ω 4) 10Ω

53. In the given circuit current through the galvanometer is

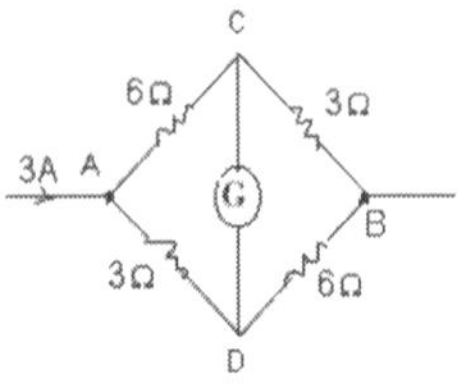

1) Zero 2) Flows from C to D

3) Flows from D to C 4) In sufficient information

Chapter - 5

(Part - 2)
CURRENT ELECTRICITY

54. The potential difference between A & B in the given branch of a circuit is

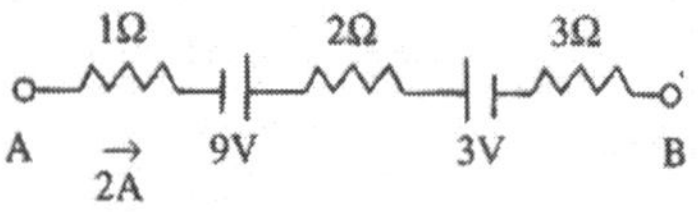

1) 6V 2) 12V 3) 9V 4) 0V

55. The resistance between A and B is

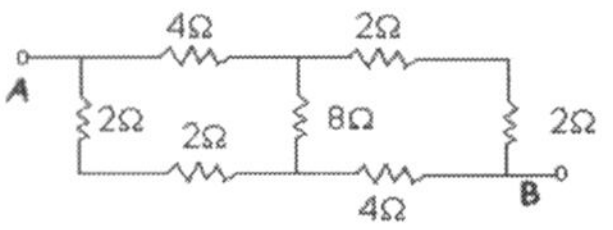

1) 8 Ω 2) 4 Ω 3) 3.75 Ω 4) 2 Ω

56. The resistance between A and B is

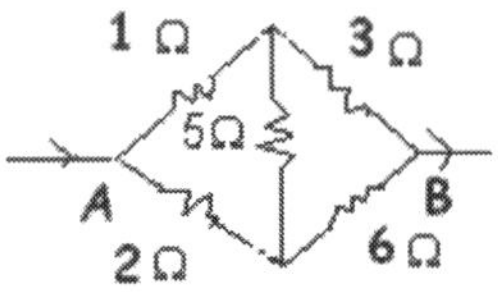

1) $\frac{288}{56}\Omega$ 2) 12 Ω 3) $\frac{8}{3}\Omega$ 4) $\frac{9}{4}\Omega$

57. The value of E of the given circuit is

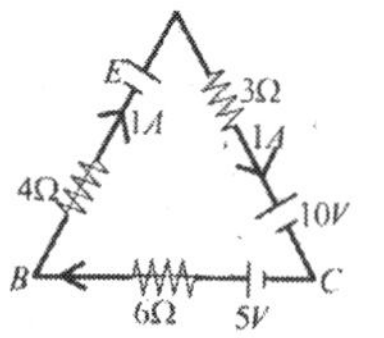

1) 10 V 2) 12 V 3) 14 V 4) 0 V

58. In the circuit shown in the figure, the value of Resistance X, when potential difference between the poins B and D is zero will I

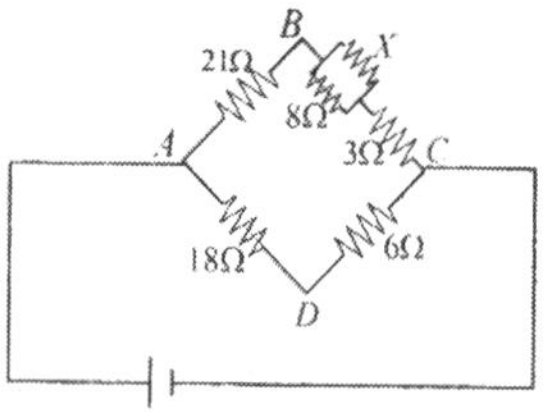

1) 9 Ω 2) 8 Ω 3) 6 Ω 4) 4 Ω

METRE BRIDGE

59. When an unknown resistance and a resistance of 4 Ω are connected in the left and right gaps of a Meterbridge, the balance point is obtained at 50cm. The shift in the balance point if a 4 Ω resistance is now connected in parallel to the resistance in the right gap is

1) 66.7cm 2) 16.7 cm 3) 34.6 cm 4) 14.6 cm

60. In a meter bridge, the gaps are closed by resistances 2 and 3 ohms. The value of shunt to be added to 3 ohm resistor to shift the balancing point by 22.5 cm is

1) 1 Ω 2) 2 Ω 3) 2.5 Ω 4) 5 Ω

61. Two equal resistance are connected in the gaps of a metre bridge. If the resistance in the left gap is increased by 10%, the balancing point shift

1) 10 % to right 2) 10% to left 3) 9.6% to right 4) 4.8% to right

POTENTIO METER

62. **A potentiometer having a wire of 4m length is connected to the terminals of a battery with a steady voltage. A leclanche cell has a null point at 1m. If the length of the potentiometer wire is increased by 1m, The position of the null point is**

1) 1.5m 2) 1.25m 3) 10.05m 4) 1.31m

63. **The emf of a battery A is balanced by a length of 80cm on a potentio meter wire. The emf of a standard cell 1v is balanced by 50cm. The emf of A is**

1) 2 V 2) 1.4 V 3) 1.5 V 4) 1.6 V

64. **When 6 identical cells of no internal resistance are connected in series in the seondary circuit of a potentio meter, the balancing length is 'l', balancing length becomes 1/3 when some cells are connected wrongly, the number of cells conected wrognly are**

1) 1 2) 3 3) 2 4) 4

65. **In a potentiometer experiment, the balancing length with a cell is 560cm. When an external resistance of 10ohms is connected in parallel to the cell the balancing length changs by 60cm. The internal resistance of the cell in ohm is**

1) 3.6 2) 2.4 3) 1.2 4) 0.6

66. **The resistivity of a potentio meter wire is, if the are of cross section of the wire is $4cm^2$. The current flowing in the circuit is 1A, the poetntial gradient is 7.5 v/m**

1) $3 \times 10^{-3}\ \Omega$ 2) $2 \times 10^{-6}\ \Omega$ - m 3) $4 \times 10^{-6}\ \Omega$ - m 4) $5 \times 10^{-4}\ \Omega$ - m

67. **A potentiometer wire of 10m legnth and 20 Ohm resistance is connected in series with a resistance R ohms and a battery of emf 2V, negligible internal resistance, Potential gradient on the wire is 0.16 millivolt / centimetre then R is ...ohms**

1) 50 Ω 2) 60 Ω 3) 230 Ω 4) 46 Ω

KEY LEVEL -I

1) 2	2) 3	3) 3	4) 4	5) 3	6) 1	7) 4	8) 4	9)3	10) 2
11) 1	12) 2	13) 4	14) 2	15) 3	16) 4	17) 2	18) 3	19) 4	20) 3
21) 4	22) 4	23) 3	24) 2	25) 4	26) 3	27) 2	28) 1	29) 4	30) 1
31) 1	32) 2	33) 3	34) 4	35) 3	36) 1	37) 2	38) 2	39) 1	40) 2
41) 1	42) 3	43) 4	44) 3	45) 4	46) 2	47) 1	48) 2	49) 1	50) 2
51) 2	52) 3	53) 3	54) 1	55) 2	56) 3	57) 4	58) 2	59) 2	60) 2
61) 4	62) 2	63) 4	64) 3	65) 3	66) 1	67) 3			

LEVEL - I (CW) HINTS

1. i = qf

2. $q = \int_0^5 idt$

3. $E = \frac{i\rho}{A}$

4. no of combinations $x = 2^n$

5. combination of resistors

6. combination of resistors

7. combination of resistors

8. $R_{eff} = \frac{R}{n^2}$

9. $R = \frac{\rho l}{A}$, $V = A.l$

10. $R^1 = \frac{20}{4} = 5$

$R_1 = 10\Omega, R_2 = 10$

$R_P = \frac{10}{2} = 5$

11. Solving for effective resistence by series and parallel combination
12. combination of resistors
13. combination of resistors
14. combination of resistors
15. $R = 52 \times 10^3 \pm 10\%$
16. $R \propto l^2$
17. $R = R_1 + R_2 + ... R_n$

$\frac{\rho(nl)}{A} = 1\frac{l}{A} + 2\frac{l}{A} + ... + n\frac{l}{A}$

$\rho n = 1 + 2 + 3 + ... + n$

$\rho n = \frac{n(n+1)}{2} = \frac{n+1}{2}$ S

18. $R_1 \alpha_1 = R_2 \alpha_2$ and $R_1 + R_2 = 36$ ohm.
19. $\alpha = \frac{R_2 - R_1}{R_1 t_2 - R_2 t_1}$
20. $R = \frac{\rho l}{A}$, $V = A.l$
21. concept
22. $R = \frac{\rho l}{A}$
23. $R\alpha \frac{l}{r^2}$. Check the options
24. $R = \frac{\rho l}{A}$
25. $R \propto \frac{l^2}{m}$
26. $R \propto \frac{1}{r^4}$
27. $R^1 = \frac{R_1 R_2}{R_1 + R_2}$
28. 2^{n-1}

29. combination of resistors

30. Number of rows $= \dfrac{\text{required current}}{\text{Given current}}$

$= \dfrac{24A}{12A} = 2 = m$

Number of cells in each row

$= \dfrac{\text{requried potential}}{\text{Given potential}}$

$= \dfrac{24}{6} = 4 = n$

$\therefore$ Total no of cells $= n \times m$

$= 2 \times 4 = 8$

31. $R_S = n.R_P$

32. concept

33. $P = \dfrac{E}{t}$, 1 K.W.H = 1 unit

34. $P = \dfrac{V^2}{R} \Rightarrow \dfrac{P_S}{P_P} = \dfrac{R_P}{R_S}$

35. $W = JQ \Rightarrow P \times t = Jms\Delta T$

36. $R = \dfrac{(V')^2}{P'}$

37. $P = \dfrac{V^2}{R}$

38. $P \propto i^2$

39. $P = \dfrac{V^2}{R}$; $R = \dfrac{\rho l}{A}$

40. 1 unit $= \dfrac{no.watts \times no.of\ hours}{1000}$

41. $t_s = t_1 + t_2$

$t_p = \dfrac{t_1 t_2}{t_1 + t_2}$

42. $JQ = i^2Rt$, $mS\Delta t = i^2Rt$

43. use Joule's law

. ms α T $= i^2Rt \Rightarrow \Delta t \alpha i^2$

44. $\dfrac{\Delta t_1}{\Delta t_2} = \dfrac{i_1^2}{i_2^2} \Rightarrow \Delta t_2 = 20^\circ C$

45. $V = E + ir$

46. $i = \dfrac{(n-2m)E}{(R+nr)}$

47. Potential gradient $= \dfrac{E}{r+R+R_S} \times \dfrac{R}{L_P}$

$= \dfrac{2}{0+15+5} \times \dfrac{15}{10} = 0.15$

48. $i = \dfrac{E}{R+r}$

49. $V = \left(\dfrac{nE}{R+nr}\right)R$

50. $E_{eq} = (N-2m)E$

51. $r = \left(\dfrac{E-V_1}{V_1}\right)R_1 = \left(\dfrac{E-V_2}{V_2}\right)R_2$. Solve for E and substitute for r

52. K.V.L.
53. K.V.L.
54. K.V.L.
55. K.V.L.
56. K.V.L.
57. from K V L

58. $\dfrac{P}{Q} = \dfrac{R}{S}$

59. $\dfrac{x}{y} = \dfrac{50}{50}$ ---- (1)

$\dfrac{4}{2} = \dfrac{l}{(100-l)}$ ----- (2)

$l - 50 = 16.7$

60. $\dfrac{2}{3} = \dfrac{l}{100-l} \Rightarrow l = 40\ cm$

$\dfrac{2}{\dfrac{3r}{3+r}} = \dfrac{62.5}{100-62.5}$

61. $\dfrac{X}{R} = \dfrac{l}{100-l}$

62. $l \alpha L \Rightarrow \dfrac{l_1}{l_2} = \dfrac{L_1}{L_2}$

63. $v = i\rho l$

64. $E' = (n-2m)E$

65. $r = R\left[\frac{\ell_1 - l_2}{\ell_2}\right]$

66. $P.G = \frac{iR}{l}$

Resistivity $\rho = \frac{RA}{L}$

67. $v = i\rho l$

LEVEL - I (HW)

ELECTRIC CURRENT & DRIFT VELOCITY

1. **The current passing through a conductor is 5 ampere. The charge that passes through that conductor in 5 minute is**

   1) 1200C 2) 300 C 3) 1000C 4)1500C

2. **In a hydrogen atom, an electron is revolving with an angular frequency 6.28 rad/s around the nucleus. Then the equivalent electric current is ..... $\times 10^{-19}$ A**

   1) 0.16 2) 1.6 3) 0.016 4) 16

3. **A current of 1.6 A is flowing in a conductor. The number of electrons flowing per second through the conductor is**

   1) 10^9 2) 10^{19} 3) 10^{16} 4) 10^{31}

4. **If an electron revolves in the circular path of radius $0.5A^0$ at a frequency of 5×10^{15} cycles/sec. The equivalent electric current is**

   1) 0.4 mA 2) 0.8 mA 3) 1.2 mA 4) 1.6 mA

5. **A current flows in a wire of circular cross section with the free electrons travelling with drift velocity $\vec{V}$. If an equal current flows in a wire of twice the radius, new drift velocity is**

   1) $\vec{V}$ 2) $\frac{\vec{V}}{2}$ 3) $\frac{\vec{V}}{4}$ 4) $2\vec{V}$

OHM'S LAW AND COMBINATION OF RESISTANCES

6. **Three resistances each of 3Ω are connected as shown in fig. The resultant resistance between A and F is**

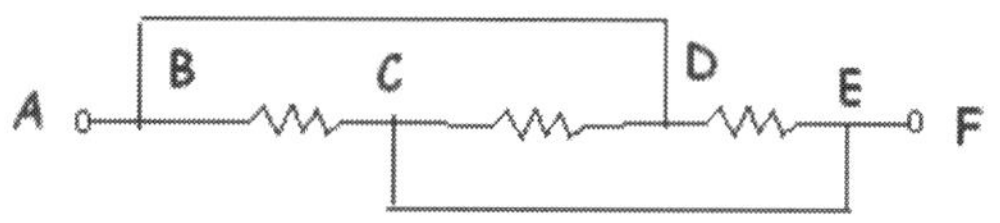

   1) 9Ω 2) 2Ω 3) 4Ω 4) 1Ω

7. **Two wires made of same material have lengths in the ratio 1 : 2 and their volumes in the same ratio. The ratio of their resistances is**

   1) 4 : 1 2) 2 : 1 3) 1 : 2 4) 1 : 4

8. **Two wires made of same material have their electrical resistances in the ratio 1 : 4. If their lengths are in the ratio 1 : 2, the ratio of their masses is**

   1) 1: 1 2) 1 : 8 3) 8 : 1 4) 2 : 1

9. There are five equal resistors. The minimum resistance possible by their combination is 2 ohm. The maximum possible resistance we can make with them is

1) 25 ohm 2) 50 ohm 3) 100 ohm 4) 150 ohm

10. An electric current is passed through a circuit containing two wires of the same material, connected in parallel. If the lengths and radii of the wires are in the ratio 4/3 and 2/3, then the ratio of the currents passing through the wires will be

1) 3 2) 1/3 3) 8/9 4) 2

11. A current of 1 A is passed through two resistances $1\,\Omega$ and $2\,\Omega$ connected in parallel. The current flowing through $2\,\Omega$ resistor will be

1) 1/3 A 2) 1 A 3) 2/3 A 4) 3 A

12. The colour coded resistance of corbon resistance is (Initial three bands are red and fourth band is silver)

1) $222\,\Omega \pm 10\%$ 2) $2200\,\Omega \pm 10\%$ 3) $333\,\Omega \pm 5\%$ 4) $33000\,\Omega \pm 10\%$

13. The resistance of a wire is 10 ohm. The resistance of a wire whose length is twice and the radius is half, if it is made of same material is

1) $20\,\Omega$ 2) $5\,\Omega$ 3) $80\,\Omega$ 4) $40\,\Omega$

14. The resultant resistance of two resistors when connected in series is 48 ohm. The ratio of their resistances is 3 : 1. The value of each resistance is

1) $20\,\Omega$ $28\,\Omega$ 2) $32\,\Omega, 16\,\Omega$ 3) $36\,\Omega, 12\,\Omega$ 4) $24\,\Omega, 24\,\Omega$

15. The resistance of a bulb filament is 100Ω at a temperature of $100^0 C$. If its temperature coefficient of resistance be 0.005 per ^0C, its resistance will become 200Ω at temperature of :

1) $300^0 C$ 2) $400^0 C$ 3) $500^0 C$ 4) $200^0 C$

16. The current 'i' in the circuit given aside is

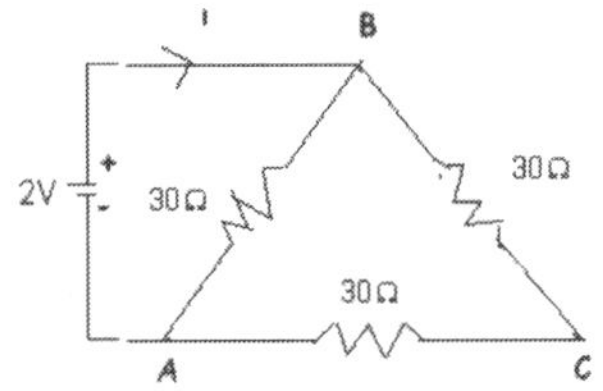

1) 0.1 A 2) 0.2A 3) 1.0A 4) 2.0 A

17. The combined resistance of two conductors in series is $1\,\Omega$. If the conductance of one conductor is 1.1 siemen, the conductance of the other conductor in siemen is

1) 10 2) 11 3) 1 4) 1.1

18. Four conductors of resistnace 16Ω each are connected to form a square. The equivalent resistance across two adjacent corners is (in ohm)

1) 6 2) 18 3) 12 4) 16

19. When two resistances are connected in parallel then the equivalent resistance is $\frac{6}{5}\Omega$. When one of the resistance is removed then the effective resistance is 2Ω. The resistance of the wire removed will be

1) 3 ohm 2) 2 ohm 3) $\frac{3}{5} ohm$ 4) $\frac{6}{5} ohm$

20. A material 'B' has twice the specific resistance of 'A'. A circular wire made of 'B' has twice the diameter of a wire made of 'A'. Then for the two wires to have the same resistace, the ratio l_B/l_A of their respective lengths must

1) 1 2) 1/2 3) 1/4 4) 2/1

22. **If a wire of resistance 'R' is melted and recasted in to half of its length, then the new resistance of the wire will be**

1) $\frac{R}{4}$ 2) $\frac{R}{2}$ 3) R 4) 2R

23. **When a wire is drawn until its radius decreases by 3%. Then percentage of increase in resistance is**

1) 10% 2) 9% 3) 6% 4) 12%

24. **When three wires of unequal resistances are given the number of combinations they can be made to give different resistances is**

1) 6 2) 4 3) 2 4) 8

25. **The resistance of a coi is 4.2Ω at 100°C and the temperature coefficient of resistance of its material is 0.004/°C. Its resistance at 0°C is**

1) 6.5 Ω 2) 5 Ω 3) 3 Ω 4) 2.5 Ω

26. **You are given several identical resistors each of value 10Ω and each capable of carrying a maximum current of 1A. It is required to make a suitable combination of these to resistances to produce a resistance of 5Ω which can carry a current of 4A. The minimum number of resistors required for this job is**

1) 4 2) 8 3) 10 4) 20

27. **A wire of resistance 50Ω is cut into six equal parts and they ae bundled together side by side to form a thicker wire. The resistance of the bundle is**

1) $\frac{18}{25}\Omega$ 2) $\frac{9}{12.5}\Omega$ 3) $\frac{25}{9}\Omega$ 4) $\frac{25}{18}\Omega$

28. **Three conductors of resistance 12Ω each are connected to form an equilateral triangle. The resistance between any two vertices is**

1) 4 Ω 2) 2 Ω 3) 6 Ω 4) 8 Ω

29. **When three equal resistance are connected in parallel, the effective resistance is $\frac{1}{3}\Omega$. If all are connected in series, the effective resistance is**

1) 9 Ω 2) 3 Ω 3) 6 Ω 4) 12 Ω

30. **A technician has only two resistance coils. By using them in series or in paralle he is able to obtain the resistance 3,4,12 and 16 ohms. The resistance of two coils are**

1) 6, 10 2) 4, 12 3) 7, 9 4) 4, 16

31. **The effective resistance between A&B in the given circuit is**

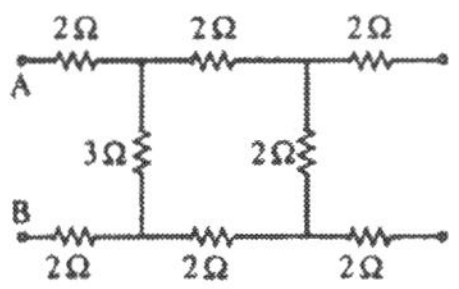

1) 7 Ω 2) 2 Ω 3) 6 Ω 4) 5 Ω

32. **The effective resistance between A and B is 3Ω then the value of R is**

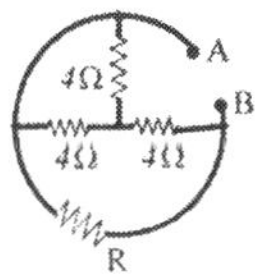

1) 2 Ω 2) 4 Ω 3) 6 Ω 4) 8 Ω

33. The effective resistance between A and B in the given circuit is

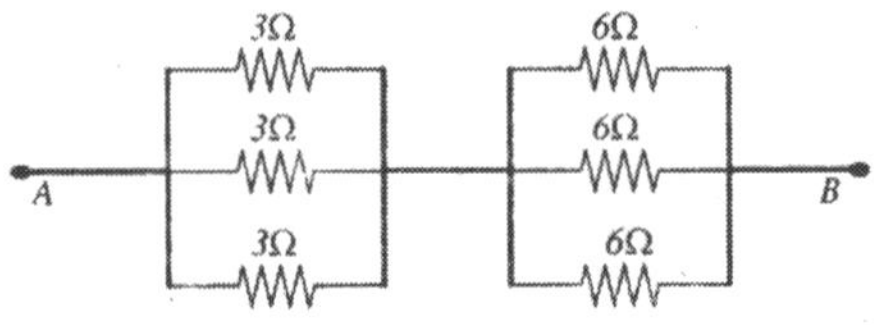

1) 2 Ω 2) 4 Ω 3) 3 Ω 4) 6 Ω

ELECTRIC POWER & JOULES LAW

34. An electric bulb is rated 220 volt and 100 watt. Power consumed by it when operated on 110 volt is

1) 50 watt 2) 75 watt 3) 90 watt 4)25 watt

35. A heater coil is cut in to two parts of equal length and only one of them is used in the heater. The ratio of the heat produced by this half-coil to that by the original coil is

1) 2 : 1 2) 1 : 2 3) 1 : 4 4) 4 : 1

36. If the electric current in a lamp decreases by 5% then the power output decreases by

1) 20% 2) 10% 3) 5% 4) 2.5%

37. Two electric bulbs whose resistances are in the ratio of 1 : 2 are connected in parallel to a constant voltage source. The powers dissipated in them have the ratio

1) 1 : 2 2) 1 : 1 3) 2 : 1 4) 1 : 4

38. A bulb rated 60 W -120V is connected to 80V mains. What is the current through the bulb

1) $\frac{1}{3}A$ 2) $\frac{2}{3}A$ 3) $\frac{5}{3}A$ 4) $\frac{3}{5}A$

39. An electric bulb has the following specifications 100 watt, 220 volt. The resistance of bulb

1) 384 Ω 2) 484 Ω 3) 344 Ω 4) 584 Ω

41. A 200W and 100W bulbs, both meant for operation at 220V, are connected in series to 220V. The power consumption by the combination is

1) 46 W 2) 66 W 3) 56 W 4) 75 W

42. Five bulbs, each rated at 40 W-220 V are used for 5 hours daily on 20V line. How may units of electric energy is consumed in a month of 30 days?

1) 20 units 2) 25 units 3) 15 units 4) 30 units

43. An electric Kettle has two heating coils. When one of them is switched on water in it boils in 6 minutes and when other is switched on water boils in 4 minutes. In what time will the water boil if both coil are switched on simultaneously

1) 1.6 min 2) 2.8 min 3) 2.4 min 4) 3 min

JOULE'S LAW

45. A 10 V storage battery of negligible internal resistance is connected across a 50Ω resistor. How much heat energy is produced in the resistor in 1 hour

1) 7200J 2) 6200J 3) 5200J 4) 4200J

CELLS AND COMBINATION OF CELLS

46. A cell of emf 6V is being charged by 1A current. If the internal resistance of the cell is 1 ohm, the potential difference across the terminals of the cell is

1) 5V 2) 7V 3) 6V 4) 8V

47. When two identical cells are connected either in series or in parallel across 2 ohm resistor they send the same current through it. The internal resistance of each cell is

1) 2 ohm 2) 1.2 ohm 3) 12 ohm 4) 21 ohm

48. The emf of a Daniel cell is 1.08V. When the terminals of the cells are connected to a resistance of 3 Ω, the potential difference across the terminlas is found to be 0.6V. Then the internal resistance of the cell is

1) 1.8 Ω 2) 2.4 Ω 3) 3.24 Ω 4) 0.2 Ω

49. Four cells each of emf 2V and internal resistance 1 ohm are connected in parallel with an external resistance of 6 ohm. The current in the external resistance is

1) 0.32 A 2) 0.16 A 3) 0.2 A 4) 0.6 A

50. A student is asked to connected four cells of emf of 1 V and internal resistance 0.5 ohm in series with an external resistance of 1 ohm. But one cell is wrongly connected by him with its terminal reversed, the current in the circuit is

1) $\frac{1}{3}A$ 2) $\frac{2}{3}A$ 3) $\frac{3}{4}A$ 4) $\frac{4}{3}A$

51. Two cells of emf 1.25V, 0.75V and each of internal resistance 1Ω are connected in parallel. The effective emf will be

1) 1 V 2) 1.25 V 3) 2 V 4) 0.5 V

52. The emf of a cell is 2V. When the terminals of the cell is connected to a resistance 4Ω. The potential difference across the terminals, if internal resistance of cell is 1Ω is

1) $\frac{3}{5}V$ 2) $\frac{8}{5}V$ 3) $\frac{6}{5}V$ 4) $\frac{5}{8}V$

53. If the external resistance is equal to internal resistance of a cell of emf E. The current across the circuit is

1) $\frac{E}{r}$ 2) $\frac{r}{E}$ 3) $\frac{r}{2E}$ 4) $\frac{E}{2r}$

54. Two cells each of emf 10V and each 1Ω internal resistance are used to send a current through a wire of 2Ω resistance. The cells are arranged in parallel. Then the current through the circuit

1) 2A 2) 4A 3) 3A 4) 5A

KIRCHOFF'S LAWS, WHEATSTONE BRIDGE

56. The figure below shows current in a part of electric circuit. The current i is

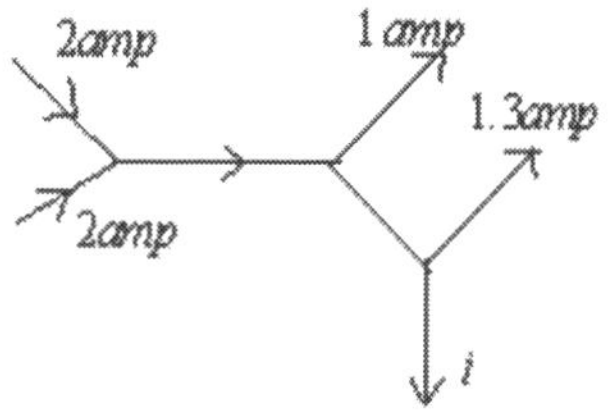

1) 1.7*amp* 2) 3.7 *amp* 3) 1.3 *amp* 4) 1 *amp*

57. Current in the main circuit shown is

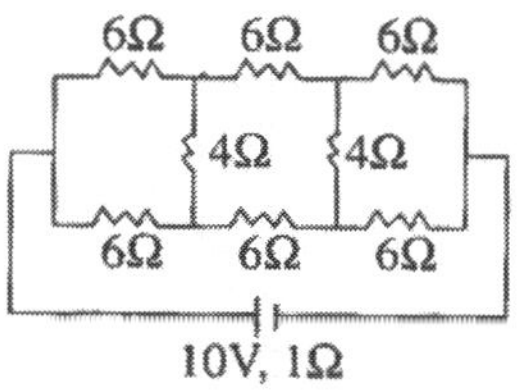

1) 1.5 A 2) 2 A 3) 0.6 A 4) 1 A

59. **Find 'i' for the given loop.**

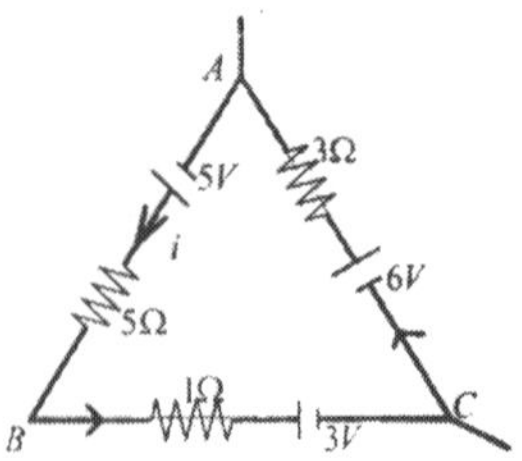

1) $\frac{6}{5}A$ 2) $\frac{8}{9}A$ 3) $\frac{1}{2}A$ 4) 1A

60. **The potential difference between points A and B is**

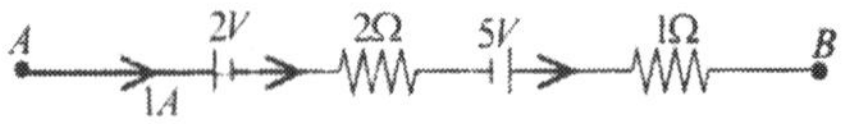

1) 0 V 2) 10 V 3) 4 V 4) 5 V

61. **In wheat stone bridge, P and Q are approximately equal. When R is 500Ω, the bridge is balanced. On interchanging P and Q, the values of R is 505Ω for balanching . The value of 'S' is**

1) 500.5 Ω 2) 501.5 Ω 3) 502.5 Ω 4) 503.5 Ω

62. **To balance the bridge in the circuit, the values of R is**

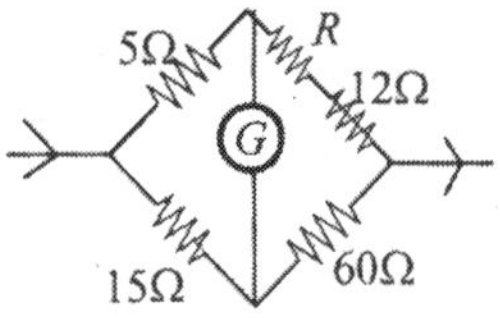

1) 8 Ω 2) 4 Ω 3) 20 Ω 4) 12 Ω

METRE BRIDGE

63. **The point in a Metre bridge is at 35.6 cm. If the resistances in the gaps are interchanged,the new balance point is**

1) 64.4 cm 2) 56 cm 3) 41.2 cm 4) 56.7 cm

64. **In a metre bridge expt, when the resistances in the gaps are interchanged the balance point is increases by 10cm. The ratio of the resistances is**

1) $\frac{15}{5}$ 2) $\frac{12}{8}$ 3) $\frac{11}{9}$ 4) $\frac{10}{9}$

65. **When an unknown resistance and a resistance 6Ω are connected in the left and right gaps of a meter bridge, the balance point is obtained at 50cm. If 3Ω resistance is connected in parallel to resistance in right gap, the balance point is**

1) decrease by 25 cm 2) increase by 25 cm
3) decrease by 16.7 cm 4) increase by 16.7 cm

66. **When un known resistance and a resistance of 5Ω are used in left and right gaps of meter bridge the balance point is 50cm. The balanceing point if 5Ω resistance is now connected in seriece to the resistor in right gap**

1) 20 cm 2) 33.3 cm 3) 60 cm 4) 60 cm

67. **In a meter bridge experiement two unkonwn resistances X and y are connected to left and right gaps of a meter bridge and the balancing point is obtained at 20cm from right (X>Y) the new position of the null point from left if one decides balance a resistance of 4X against Y.**

1) 114 cm 2) 80 cm 3) 53.3 cm 4) 70 cm

POTENTIO METER

68. In a potentiometer the balance length with standard cadmium cell is 509 cm. The emf of a cell which when connected in the place of the standard cell gave a balance length of 750 cm is (emf of standard cell is 1.018V)

1) 1.5V 2) 0.5V 3) 1.08V 4) 1.2V

69. Two cells of emf's E_1 and E_2 when placed in series produce null deflection at a distance of 204 cm in a potentio meter. When one cell is reversed they produce null deflection at 36 cm if E_1 1.4v then E_2=

1) 0.98 V 2) 2.47 cm 3) 0.098 V 4) 98.8 V

70. When 6 identical cells of no internal resistance are connected in series in the second arycircuit of a poetntio meter, the balancing length is *l*. If two of them are wrongly connected the balancing length becomes

1) $\frac{l}{4}$ 2) $\frac{l}{3}$ 3) l 4) $\frac{2l}{3}$

71. In an experiment to determine the internal resistance of a cell with potentiometer, the balancing length is 165cm. When a reistance of 5 ohm is joined in parallel with the cell the balancing length is 150cm. The internal resistance of cell is

1) 2.2Ω 2) $1.1\ \Omega$ 3) 3.3Ω 4) $0.5\ \Omega$

72. The resistivity of a potentio meter wire is $40 \times 10^{-8}\ \Omega - m$ and its area of cross section is $8 \times 10^{-6}\ m^2$. If 0.2A current is flowing through the wire, the potential gradient will be

1) 10^{-2} V/m 2) 10^{-1} V/m 3) 3.2×10^{-2} V/m 4) 1 V/m

73. The emf of a cell is Ev, and its its internal resistance is 1Ω. A resistance of 4Ω is joined to battery in parallel. This is connected in secondary circuit of poetntio meter. The balancing length is 160cm. If 1V cell balances for 100cm of potentio meter wire, the emf of cell E is

1) 1 V 2) 3 V 3) 2 V 4) 4 V

LEVEL -I KEY

1) 4	2) 2	3) 2	4) 2	5) 3	6) 4	7) 4	8) 1	9) 2	10) 2
11) 1	12) 2	13) 3	14) 3	15) 2	16) 1	17) 2	18) 3	19) 1	20) 4
21)	22) 1	23) 4	24) 4	25) 3	26) 2	27) 4	28) 4	29) 2	30) 2
31) 3	32) 3	33) 3	34) 4	35) 1	36) 2	37) 3	38) 1	39) 2	40)
41) 2	42) 4	43) 3	44)	45) 1	46) 2	47) 1	48)2	49) 1	50) 2
51) 1	52) 2	53) 4	54) 2	55)	56) 1	57) 4	58)	59) 2	60) 1
61) 3	62) 1	63) 1	64) 3	65) 2	66) 2	67) 3	68) 1	69) 1	70) 2
71) 4	72) 1	73) 3							

HINTS LEVEL - I

1. $i = \frac{q}{t}$

2. $i = qf$

3. $i = \frac{ne}{t} \Rightarrow n = \frac{it}{e}$

4. $i = q\,f$

5. $V_d \propto \frac{1}{r^2}$

6. The 3 resistances are parallel

7. $R \propto l^2$ $\because$ V constant

8. $R \propto \frac{l^2}{m}$

9. $\frac{R}{5} = 2 \quad R_{max} = 5R \;\; R_{min} = \frac{R}{5}$.

10. $\frac{i_1}{i_2} = \frac{r_1^2}{r_2^2} \times \frac{l_2}{l_1}$

11. $i_2 = \frac{iR_1}{R_1 + R_2}$

12. $R = 22 \times 10^2 \pm 10\%$

13. $\frac{R_1}{R_2} = \frac{l_1}{l_2} \times \frac{r_2^2}{r_1^2}$

14. $R_S = R_1 + R_2, R_P = \frac{R_1 R_2}{R_1 + R_2}$ Solving for R_1 & R_2

15. $\alpha = \frac{R_2 - R_1}{R_1 t_2 - R_2 t_1}$

16. use ohm's law

17. $R_1 + R_2 = 1$ ohm.

$\frac{1}{R_1} = 1.1 \;\; => \;\; R_1 = \frac{10}{11}$

$R_2 = 1 - R_1$

$\therefore \frac{1}{R_2} = \frac{1}{1 - R_1}$

18. concepts

19. $\frac{R_1 R_2}{R_1 + R_2} = \frac{6}{5}$. If R_2 is removed $R_1 = 2\Omega$

$\frac{2R_2}{2 + R_2} = \frac{6}{5} \Rightarrow 5R_2 = 6 + 3R_2 \Rightarrow R_2 = 3\Omega$

20. $R = \frac{\rho\ell}{A} \Rightarrow \ell = \frac{RA}{\rho} \Rightarrow l \propto \frac{r^2}{\rho}$

21.

22. $R \propto \ell^2$

23. $R \propto \frac{1}{r^4}$

24. 2^n

25. $R_t = R_0(1 + \alpha t)$

26. $R_P = \frac{R}{n}$; $i_p = mi \Rightarrow$ m × n = ...

27. $R' = \frac{R}{n^2}$

28. $R' = \frac{2R}{3}$

29. $R_P = \frac{R}{n}$ and $R_S = nR$

30. $R_1 = \dfrac{R_S + \sqrt{R_s^2 - 4R_sR_p}}{2}$

$R_2 = \dfrac{R_s - \sqrt{R_s^2 - 4R_sR_p}}{2}$

31. diagram
32. diagram
33. diagram

34. $P = \dfrac{V^2}{R}$

35. $W = JQ \Rightarrow Q = \dfrac{V^2}{RJ} \Rightarrow \dfrac{Q_1}{Q_2} = \dfrac{R_2}{R_1}$

36. $P = i^2R \Rightarrow P\alpha i^2 \Rightarrow \dfrac{\Delta P}{P}\times 100 = 2\times\dfrac{\Delta I}{I}\times 100$

37. $P = \dfrac{V^2}{R}$

38. $R = \dfrac{V'^2}{P'}$ and $i = \dfrac{V}{R}$

39. $R = \dfrac{V^2}{P}$

41. $P = \dfrac{P_1P_2}{P_1 \mid P_2}$

42. $\dfrac{no.of\ watts \times no.of\ hours}{1000}$

43. $t_s = t_1 + t_2$

$t_p \dfrac{t_1t_2}{t_1 + t_2}$

45. $\mathrm{Q} = \dfrac{\mathrm{V}^2}{\mathrm{R}}\mathrm{t} \Rightarrow \mathrm{Q} = 7200\mathrm{J}$

46. $V = E + ir$

47. $i_S = i_P$; $\dfrac{nE}{(R+nr)} = \dfrac{E}{\left(R+\dfrac{r}{n}\right)}$

48. $r = \left(\dfrac{E-V}{V}\right)R$

49. $i = \dfrac{E}{R+\dfrac{r}{n}} = \dfrac{2}{6+\dfrac{1}{4}} = \dfrac{2\times4}{25} = \dfrac{8}{25}$

50. $i = \dfrac{(N-2n)E}{R+Nr} = \dfrac{(4-2)\times1}{1+4\times0.5} = \dfrac{2\times1}{3}$

51. $\mathrm{E_{eff}} = \dfrac{\mathrm{E_1r_2 + E_2r_1}}{\mathrm{r_1 + r_2}}$

52. $V = E - ir$

53. $i = \dfrac{E}{R+r}$

54. $i = \dfrac{E}{\frac{r}{n} + R}$

56. K.C.L

57. K.C.L

59. K.C.L

60. K.C.L

61. $\dfrac{P}{Q} = \dfrac{R}{S}$

62. $\dfrac{P}{Q} = \dfrac{R}{S}$

63. $\dfrac{P}{Q} = \dfrac{35.6}{64.4}, \dfrac{Q}{R} = \dfrac{64.4}{35.6}$

64. $\dfrac{X}{R} = \dfrac{l}{100-l}$

65. $\dfrac{X}{R} = \dfrac{l}{100-l}$

66. $\dfrac{X}{R} = \dfrac{l}{100-l}$

67. $\dfrac{x}{y} = \dfrac{l}{100-l}$

68. $\dfrac{E_1}{E_2} = \dfrac{l_1}{l_2} 3$

69. $\dfrac{E_1}{E_2} = \dfrac{\ell_1 + \ell_2}{\ell_1 - \ell_2}$

70. concept

71. $r = R\left[\dfrac{\ell_1 - l_2}{\ell_2}\right]$

72. $P.G = \dfrac{i\rho}{A}$

73. $r = R\left[\dfrac{\ell_1 - l_2}{\ell_2}\right]$

LEVEL - II (CW)

ELECTRIC CURRENT AND DRIFT VELOCITY

1. **In a hydrogen tube it is observed that through a given cross - section 3.13 x 10^{15} electrons per sec, moving from right to left and 3.12 x 10^{15} protons per sec are moving from left to right. The electric current in the discharge tube and its direction is**

1) 1 mA towards left 2) 1 mA towards right 3) 1.5 mA towards right 4) 2 mA towards left

2. **An electron of mass m, moves around the nucleus in a circular orbit of radius 'r' under the action of centripetal force 'F'. The equivalent electric current is**

1) $\dfrac{e}{2\pi}\sqrt{\dfrac{F}{mr}}$ 2) $\dfrac{e}{\pi}\sqrt{\dfrac{Fr}{m}}$ 3) $\dfrac{e}{2\pi}\sqrt{\dfrac{Fm}{r}}$ 4) $\dfrac{e}{\pi}\sqrt{\dfrac{Fm}{r}}$

3. **The current in a conductor varies with time 't' as $I = 3t + 4t^2$. Where I in amp and t in sec. The electric charge flows through the section of the conductor between t = 1s and t=3s**

1) $\dfrac{100}{3}C$ 2) $\dfrac{127}{4}C$ 3) $\dfrac{140}{3}C$ 4) $\dfrac{150}{3}C$

4. **A conductor has a non-uniform section as shown in the figure. A steady current is flowing through it. Then the drift speed of the electrons (M-2012)**

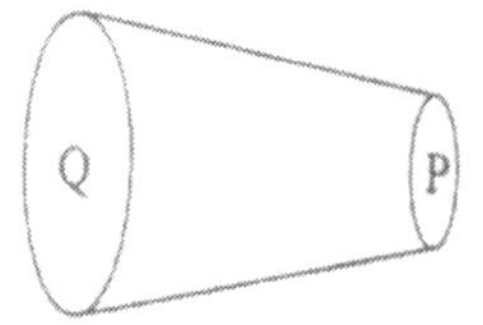

1) is constant throughout the wire 2) varies unpredictably

3) decreases from P & Q 4) increases from P to Q

5. **A current of 16 A is made to pass through a conductor in which the number of density of free electrons is 4 × 10^{28} m^{-3} and its area of cross section is 10^{-5} m^2. The average drift velocity of free electrons in the conductor is**

1) 1.6 × 10^{-4} ms^{-1} 2) 2.5 × 10^{-4} ms^{-1} 3) 6.4 × 10^{-4} ms^{-1} 4) 3.2 × 10^{-4} ms^{-1}

OHM's LAW AND COMBINATION OF RESISTANCES

6. A current of 3A flows in a circuit shown in the figure. The potential difference between A and B is

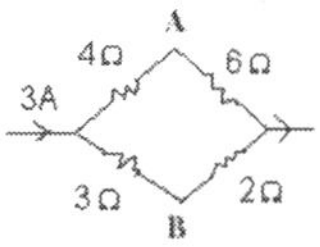

1) 4V 2) 3V 3) 2V 4) 5V

7. The resistance of the network between the terminals A and B is

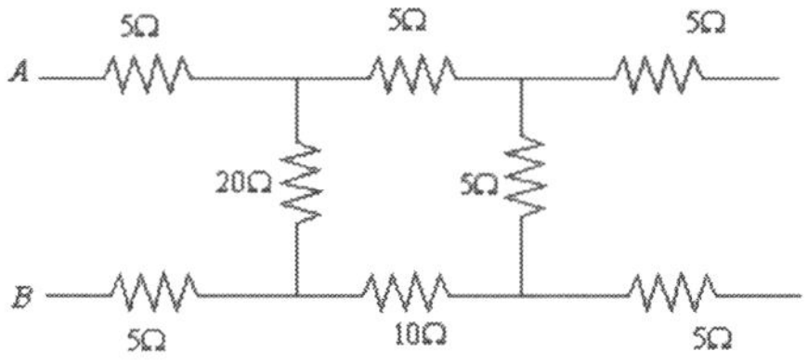

1) 30Ω 2) 20Ω 3) 50Ω 4) 60Ω

8. In the figure, the value of resistance to be connected between C and D so that the resistance of the entire circuit between A and B does not change with the number of elementary sets used is

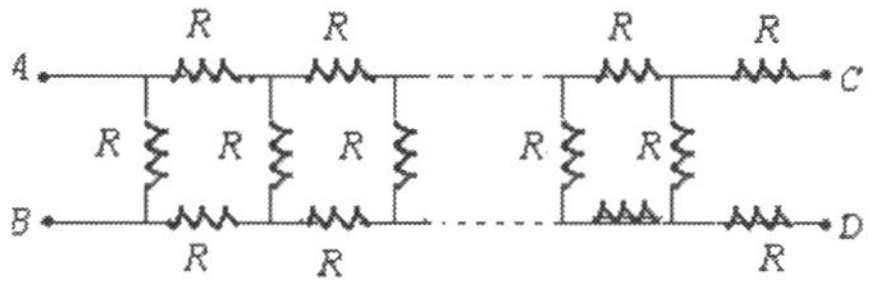

1) R 2) $R\left(\sqrt{3}-1\right)$ 3) 3R 4) $R\left(\sqrt{3}+1\right)$

9. The effective resistance across the points A and I is

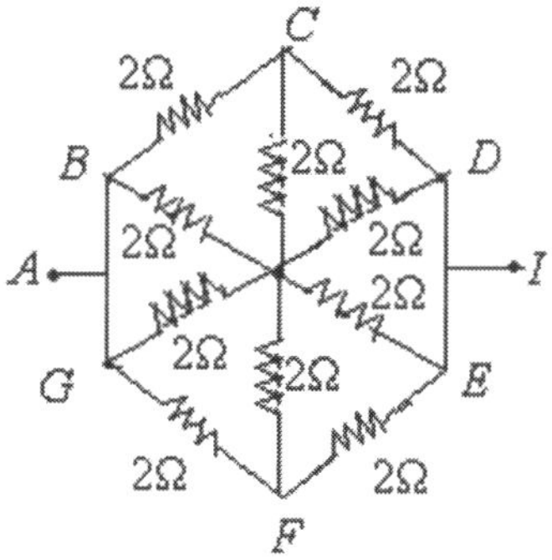

1) 2Ω 2) 1Ω 3) 0.5Ω 4) 5Ω

10. In the circuit shown below, the cell has an emf of 10V and internal resistance of 1 *ohm*. the other resistances are shown in the figure. The potential difference $V_A - V_B$ is

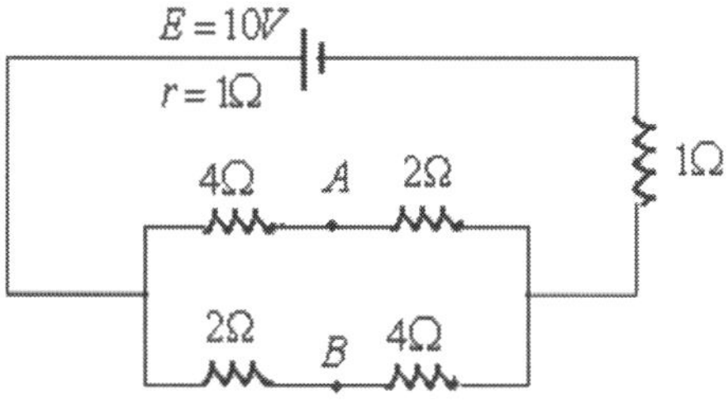

1) 6V 2) 4V 3) 2V 4) -2 V

11. **A uniform wire of resistance $20\,\Omega$ having resistance $1\,\Omega/\mathrm{m}$ is bent in the form of circle as shown in fig. If the equivalent resistance between M and N is $1.8\,\Omega$, then the length of the shorter section is**

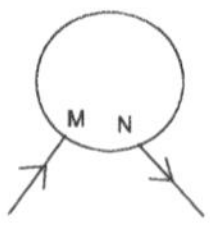

1) 2 m 2) 5 m 3) 1.8 m 4) 18 m

12. **If the voltmeter reads 0.2 V and the ammeter reads 0.101A, the resistance of the voltmeter is (in ohm)**

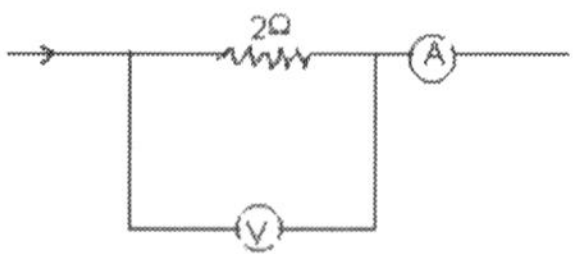

1) 500 2) 1000 3) 200 4) 400

13. **In the given circuit Ammeter reading is same when both switches S_1, S_2 are closed or opened. The value of resistance R is**

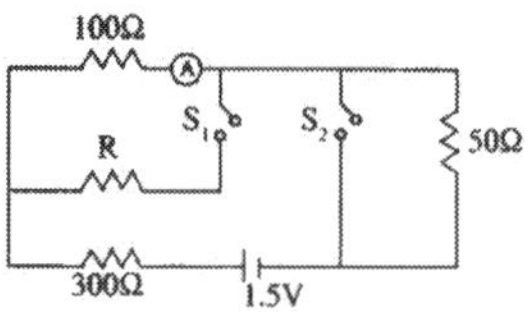

1) $200\,\Omega$ 2) $100\,\Omega$ 3) $400\,\Omega$ 4) $600\,\Omega$

14. **In the following diagram ammeter reading is 4A, voltmeter reading is 20V, the value of R is**

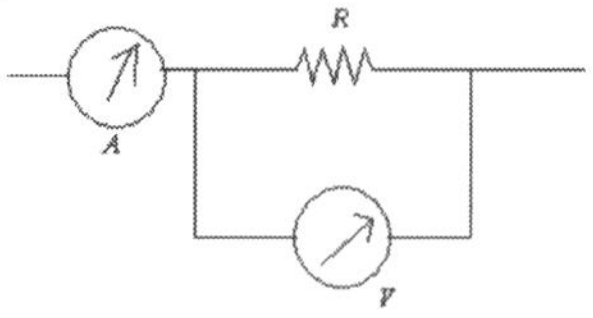

1) $> 5\Omega$ 2) $< 5\Omega$ 3) $= 5\Omega$ 4) $\le 5\Omega$

15. **Twelve resistances each of resistance R are connected in the circuit as shown in fig. Net resistance between points A and C would be**

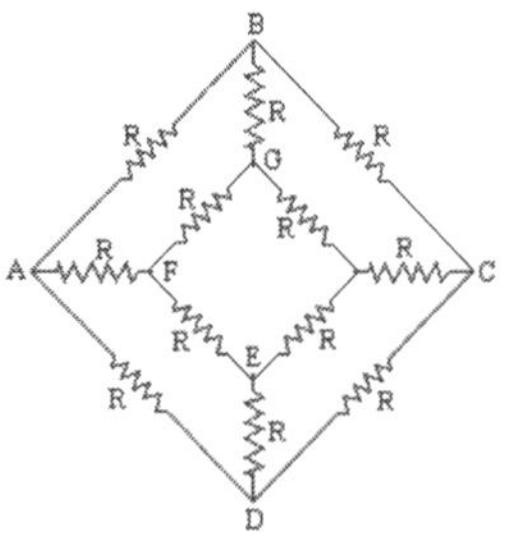

1) $\frac{5R}{3}$ 2) $\frac{7R}{6}$ 3) R 4) $\frac{3R}{4}$

16. **A resistance is made by connecting two wires (series) of same material of radii 2 mm and 5 mm and length 8 cm and 5 cm. A potential difference of 22V is applied to them. The potential difference on the longer wire is**

1) 15 V 2) 18 V 3) 16 V 4) 20 V

17. A 220 V and 800 W electric kettle and three 220V and 100W bulbs are connected in parallel. On connecting this combination with 200 V supply, the total current in the circuit will be

1) 0.15 A 2) 5A 3) 5.5A 4) 4.55A

18. What is the equivalent resistance of the circuit

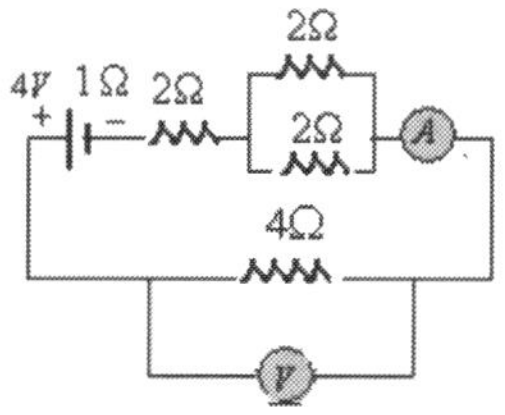

1) 6Ω 2) 7Ω 3) 8Ω 4) 9Ω

19. The temperature coefficient of resistance of platinum is $\alpha = 3.92\times10^{-3}K^{-1}$ at 20^0C. Find the temperature at which the increase in the resistance of platinum wire is 10% of its value at 20^0C

1) 40.5^0C 2) 45.5^0C 3) 48.5^0C 4) 43.5^0C

20. Four identical resistance are joined as shown in fig. The equivalent resistance between points A and B is R_1 and that between A and C is R_2. Then ratio of $\frac{R_1}{R_2}$ is

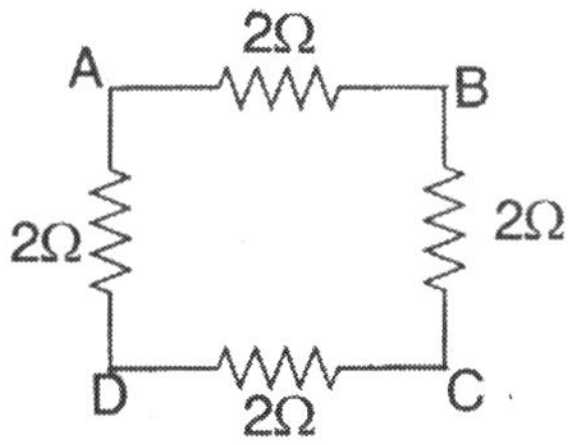

1) 1:5 2) 3:4
3) 2:5 4) 1:2

21. If the galvanometer reading is zero in the given circuit, the current passing through resistance 250 Ω is

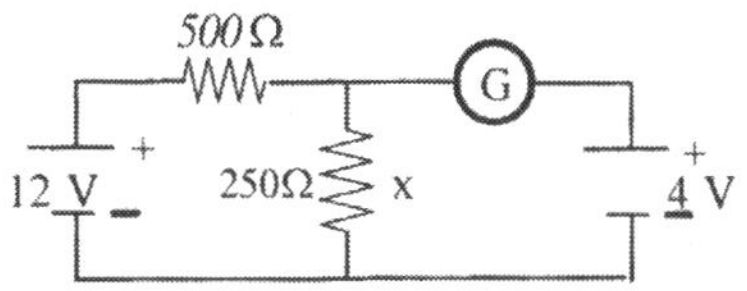

1) 0.016 A 2) 0.16 A 3) 0.032A 4) 0.042 A

22. The effective resistance between A and B is the given circuit is

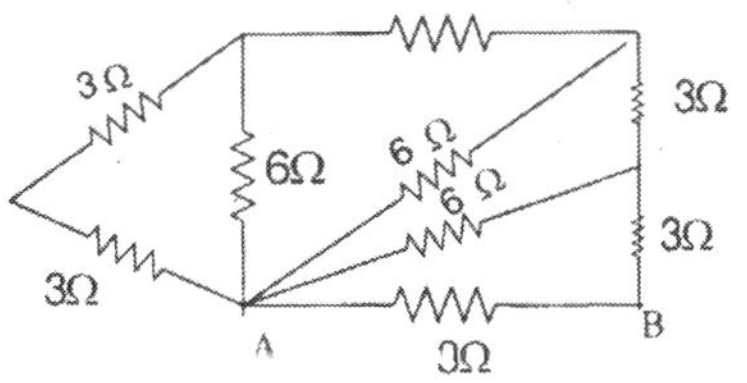

1) 3 Ω 2) 2 Ω 3) 4 Ω 4) 6 Ω

23. The equivalent resistance between points A and B of an infinite network of resistance each of 1Ω connected as shown is

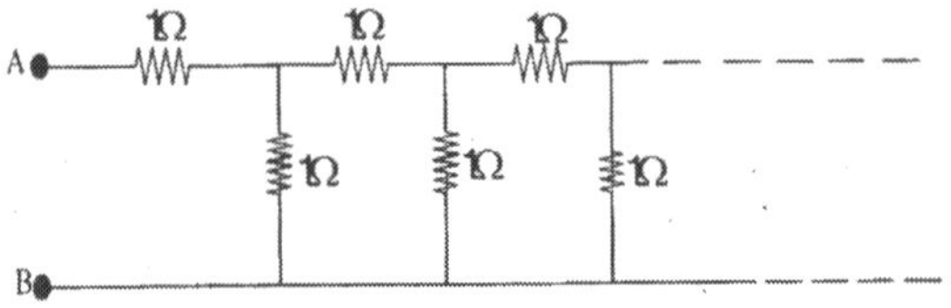

1) $\frac{1+\sqrt{5}}{2}$ 2) $\frac{2+\sqrt{5}}{4}$ 3) $\frac{3+\sqrt{5}}{2}$ 4) $\frac{1+\sqrt{7}}{3}$

24. Equivalent resistance across A and B in the given circuit is

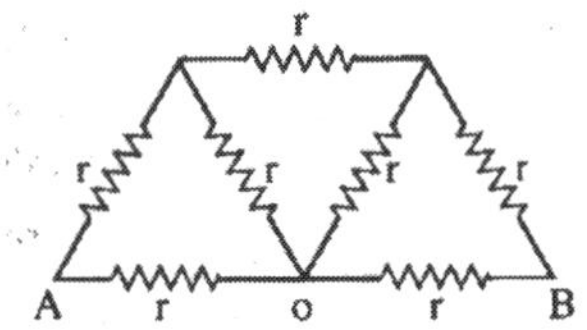

1) $\frac{2r}{3}$ 2) $\frac{5r}{3}$ 3) $\frac{7r}{3}$ 4) 6r

25. Two resistances of 400Ω and 800Ω are connected in series with 6V battery of negligible internal resistance. A voltmeter of resistance $10{,}000\Omega$ is used to measure the p.d. across 400Ω. The error in the measurement of p.d. in volts approximately

1) 0.05 V 2) 0.5 V 3) 0.75 V 4) 5 V

26. Copper and carbon wires are connected in series and the combined resistor is kept at 0°C. Assuming the combined resistane does not vary with temperature, the ratio of the resistances of carbon and copper wires at 0°C is (Temperature coefficients of resisti-vity of copper and carbon respectively are 4×10^{-3}/°C and -0.5×10^{-3}/°C)

1) 2 2) 4 3) 8 4) 6

27. Three resistances of equal values are arranged in four different configurations as shown below. Power dissipation in the increasing order is

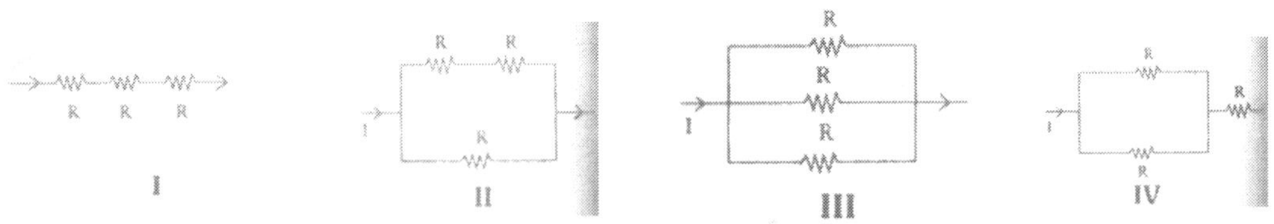

1) (III) < (II) < (IV) < (I) 2) (II) < (III) < (IV) < (I)
3) (I) < (IV) < (III) < (II) 4) (I) < (III) < (II) < (IV)

28. If 400 Ω of resistance is made by adding four 100 Ω resistances of tolerance 5%, then the tolerance of the combination is

1) 5% 2) 10% 3) 15% 4) 20%

ELECTRIC POWER

29. Two wires A and B with lengths in the ratio of 3 : 1, diameters in the ratio of 1:2 and resistivities in the ratio of 1:20 are joined in parallel with a source of emf. 2V. Ratio of the R_1 / R_2 is:

1) 5 : 2 2) 2 : 5 3) 5 : 3 4) 3 : 5

30. An electric heater operating at 220 volts boils 5 litre of water in 5 minutes. If it is used on 110 volts, it will boil the same amount of water in

1) 10 minutes 2) 20 minutes 3) 15 minutes 4) 25 minutes

31. **Three electric bulbs of 40W, 60W and 100W have the tungsten wire of the same diameter. Then the longer wire is used by**

1) 60W 2) 100W 3) 40W 4) All use the same length

32. **A fuse wire with radius of 0.2 mm blows off with a current of 5 Amp. The fuse wire of same material but of radius 0.3 mm will blow off with a current of**

1) $5\times\frac{3}{2}amp$ 2) $\frac{5\sqrt{3}}{2}amp$ 3) $5\sqrt{\frac{27}{8}}amp$ 4) 5 amp

33. **In a large building, there are 15 bulbs of 40 W, 5 bulbs of 100 W, 5 fans of 80 W and 1 heater of 1 kW. The voltage of electric mains is 220 V. The minimum capacity of the main fuse of the building will be:**

1) 8 A 2) 10 A 3) 12 A 4) 14 A

34. **The supply voltage to room is 120 V. The resistance of the lead wires is 6Ω. A 60 W bulb is already switched on. What is the decrease of voltage across the bulb, when a 240 W heater is switched on in parallel to the bulb?**

1) zero 2) 2.9 Volt 3) 13.3 Volt 4) 10.04 Volt

INTERNAL RESISTANCE AND EMF

35. **In the circuit shown here, cells A and B have emf 10 V each and the internal resistance is 5Ω for A and 3Ω for B. For what value of R will the potential difference across the cell A will be zero?**

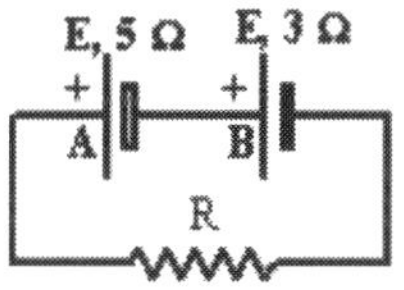

1) zero 2) 1 ohm 3) 2 ohm 4) 3 ohm

36. **In the circuit of fig. with steady current, the potential drop across the capacitor is**

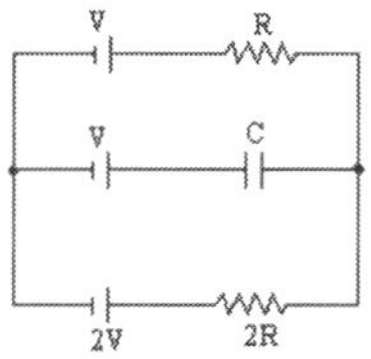

1) V 2) $\frac{V}{2}$ 3) $\frac{V}{3}$ 4) $\frac{2V}{3}$

37. **In the circuit, the galvanometer G shows zero deflection. If the batteries A and B have negligible internal resistance, the value of the resistor R will be :**

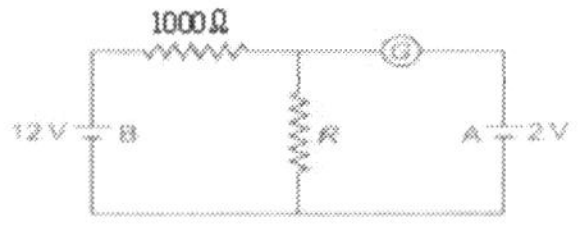

1) 100Ω 2) 200Ω 3) 500Ω 4) 1000Ω

38. **Twenty four cells each of emf 1.5V and internal resistance 0.5 ohms are to be connected to a 3 ohm resistance. For maximum current through this resistance the number of rows and number of columns that you connect these cells is.**

1) 12 cells in series 2 rows in parallel 2) 8 cells in series 3 rows in parallel
3) 4 cells in series 6 rows in parallel 4) 6 cells in series 4 rows in parallel

39. A battery of four cells in series each having an emf of 1.5V and internal resistance 1 Ω are connected in series with an ammeter, a coil of resistance 2 Ω and a filament lamp. If the ammeter reads 0.5A, the resistance of the filment lamp is

1) 4 Ω 2) 6 Ω 3) 2 Ω 4) 12 Ω

40. A 5V battery with internal resistance 2Ω and a 2V battery with internal resistance 1Ω are connected in parallel with unlike polarities connected together .This combination is connected to 10Ω resistor The current in the 10Ω resistor is

1) 0.27 A 2) 0.05 A 3) 0.25 A 4) 0.3 A

41. A voltmeter resistance 500Ω is used to measure the emf of a cell of internal resistance 4Ω. The percentage error in the reading of the voltmeter will be

1) 0.4% 2) 0.6% 3) 0.8% 4) 1.2%

42. When two identical cells are connected either in series or in parallel across a 4 ohm resistor, they send the same current through it. The internal resistance of the cell in ohm is

1) 4 Ω 2) 2 Ω 3) 1 Ω 4) 7 Ω

43. Two cells with same e.m.f. 'E' and different internal resistances r_1 and r_2 are connected in series to an external resistance 'R'. The value of R so that the p.d. across the first cell be zero is

1) $r_2 - r_1$ 2) $r_1 - r_2$ 3) $r_1 + r_2$ 4) $\frac{r_1 - r_2}{2}$

44. Two conductors have the same resistance at 0°C but their temperature coefficients of resistance are α_1 and α_2. The respective temperature coefficients of their series and parallel combinations are nearly

1) $\frac{\alpha_1+\alpha_2}{2}, \alpha_1+\alpha_2$ 2) $\alpha_1+\alpha_2, \frac{\alpha_1+\alpha_2}{2}$ 3) $\alpha_1+\alpha_2, \frac{\alpha_1\alpha_2}{\alpha_1+\alpha_2}$ 4) $\frac{\alpha_1+\alpha_2}{2}, \frac{\alpha_1+\alpha_2}{2}$

KIRCHOFF'S LAWS AND WHEAT SHONE'S BRIDGE

45. The electric current i in the circuit shown is

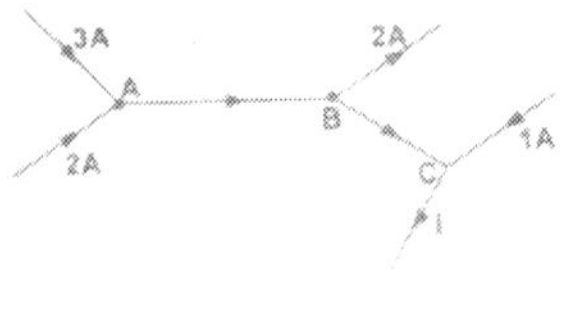

1) 6 A 2) 2 A 3) 3 A 4) 4 A

46. In the circuit shown in the figure, the current 'I' is

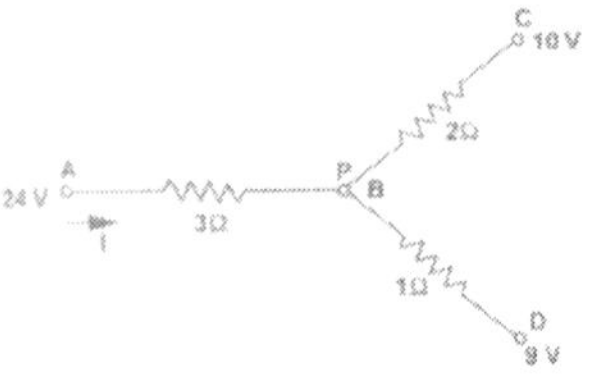

1) 6 A 2) 2 A 3) 4 A 4) 7 A

47. Four resistors A, B, C and D form a wheatstone's bridge. The bridge is balanced when C = 100 Ω. If A and B are inter-changed, the bridge balances for C = 121 Ω. The vlaue of D is

1) 10 Ω 2) 100 Ω 3) 110 Ω 4) 120 Ω

48. In the circuit shown below, the ammeter reading is zero. Then the value of the resistance R is

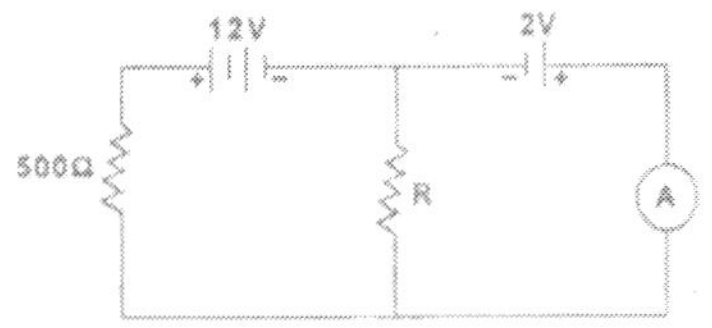

1) 50Ω 2) 100Ω 3) 200Ω 4) 400Ω

METRE BRIDGE

49. Two unknown resistrance X and Y are connected to left and right gaps of a meter bridge and the balancing point is obtained at 80 cm from left. When a 10Ω resistance is connected in parallel to 'x', the balance point is 50 cm from left. The values of X and Y respectively are

1) 40Ω, 9Ω 2) 30Ω, 7.5Ω 3) 20Ω, 6Ω 4) 10 Ω, 3Ω

50. In the meter bridge experiment, the length AB of the wire is 1 m. The resistors X and Y have values 5Ω and 2Ω respectively. When a shunt resistance S is connected to X, the balancing point is found to be 0.625 m from A. Then, the resistance of the shunt is

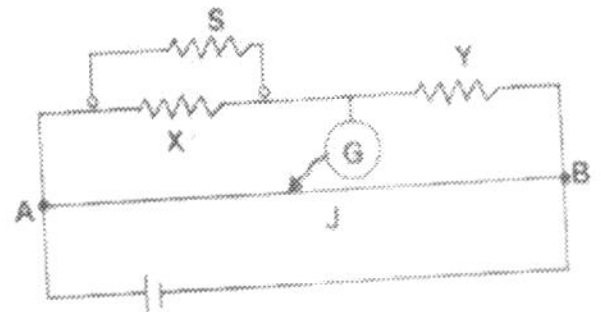

1) 5Ω 2) 10Ω 3) 7.5Ω 4) 12.5Ω

POTENTIO METER

51. The potential gradient along the length of a uniform wire is 10 volt/m B and C are two points at 30 cm and 60 cm in a scale fitted along the wire. The pd between B and C is

1) 3V 2) 0.4V 3) 7V 4) 4V

52. In the determination of the internal resis-tance of a cell using a potentiometer, when the cell is shunted by a resistance "R" and connected in the secondary circuit, the balance length is found to be L_1. On doubling the shunt resistance, the balance length is found to increase to L_2. The value of the internal resistance is

1) $\frac{2R(L_2 - L_1)}{(L_1 - 2L_2)}$ 2) $\frac{2R(L_2 - L_1)}{(2L_1 - L_2)}$ 3) $\frac{R(L_2 - L_1)}{(L_1 - 2L_2)}$ 4) $\frac{R(L_2 - L_1)}{(2L_1 - L_2)}$

53. Figure shows a potentiometer circuit for comparision of two resistances. The balance point with a standard resistor $R = 10.0\,\Omega$ is found to be 58.3 cm, while that with the unknown resistance X is 68.5 cm. The value of X is

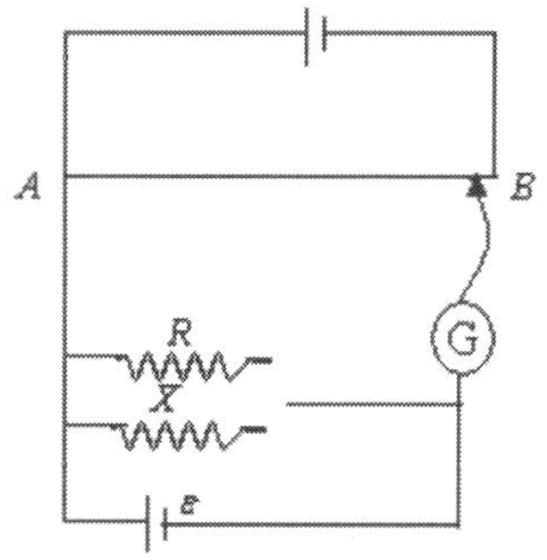

1) 11.75Ω 2) 12.55Ω 3) 9.55Ω 4) 12.75Ω

54. **In a experiment for calibration of voltmeter a standard cell of emf 1.5 V is balanced at 300 cm length of potentiometer wire. The P.D across a resistance in the circuit is balancedat 1.25 m. If a voltmeter is connected across the same resistance it reads 0.65 V. The error in the volt meter is**

1) 0.5 V 2) 0.025V 3) 0.05 V 4) 0.25V

55. **The current in the primay circuit of a potentiometer is 0.2 A. The specific resistance and cross-section of the potentiometer wire are 4 × 10^{-7} m², respectively. The potential gradient will be equal to**

1) 1 V/m 2) 0.5 V/m 3) 0.1 V/m 4) 0.2 V/m

KEY LEVEL - II

1) 2	2) 1	3) 3	4) 3	5) 2	6) 3	7) 2	8) 2	9) 2	10) 4
11) 1	12) 3	13) 4	14) 1	15) 4	16) 4	17) 4	18) 3	19) 2	20) 2
21) 1	22) 2	23) 1	24) 2	25) 1	26) 3	27) 1	28) 1	29) 3	30) 2
31) 3	32) 3	33) 3	34) 4	35) 3	36) 3	37) 2	38) 1	39) 2	40) 3
41) 3	42) 1	43) 2	44) 4	45) 4	46) 3	47) 3	48) 2	49) 2	50) 2
51) 1	52) 2	53) 1	54) 2	55) 3					

LEVEL - II (CW) HINTS

1. $i = \frac{(n_e + n_p)e}{t}$

2. $i = \frac{q}{t} = \frac{ev}{2\pi r}$

3. $q = \int_{t_1}^{t_2} I.dt$

4. $v_d \propto \frac{1}{A}$

5. $v_d = \frac{i}{neA}$

6. V = iR

7. End resistors are not considered

8. Resitance between AB is independent of number of sets used. Let x be the R_{eff} between AB

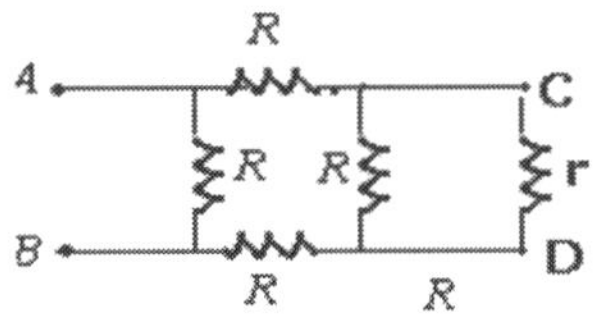

$$A = R \,||\, \left[2R + \frac{Rr}{R+r}\right] \quad \rightarrow (1)$$

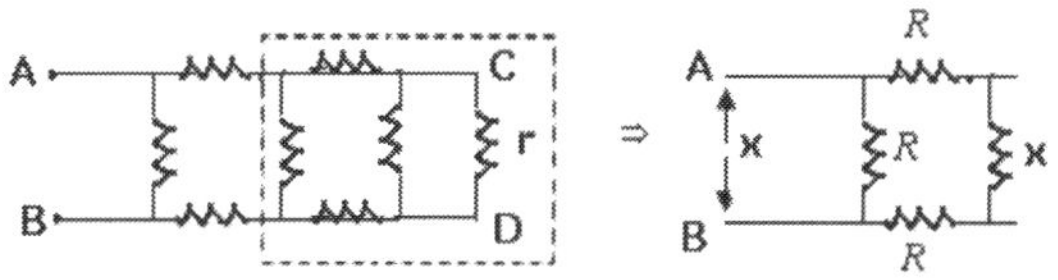

$x = R \| [2R + x]$ $\rightarrow (2)$ Solve for r

9. The equivalent circuit is shown

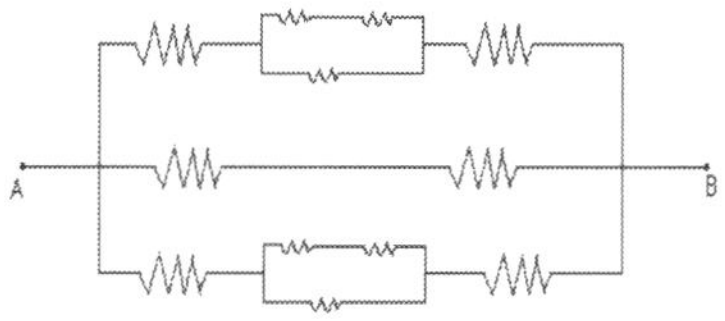

10. Apply Ohm's law.

11. Let the resistance of shorter part MN be x.

Then resistance of longer part is (20 – x) Ω

$$R_{eq} = \frac{(20-x)x}{20-x+x} = 1.8\,\Omega$$

Solving we get x = 2 Ω

So length of shorter part = 2m

12. $\left[\frac{2x}{2+x}\right](0.101) - 0.2$ solve for 'x'.

13. When S_1 and S_2 are opened, $i = \frac{1.5}{450}$ ..... (1)

When S_1 and S_2 are closed,

$$i = \frac{1.5[100+R]}{400R + 30{,}000} \quad .... (2)$$

14. $iR = 20$, here $i < 4A \Rightarrow R > 5\Omega$

15. When a cell is connected across A and C, no current flows in the arms BG and ED due to symmetry in the arrangement. Then equivalent circuit will be as shown in Fig.(a) and (b). The effective resistance between A and C is

$$\frac{1}{R_{eff}} = \frac{1}{2R} + \frac{1}{3R} + \frac{1}{2R}$$

$$= \frac{3+2+3}{6R} = \frac{8}{6R} = \frac{4}{3R} \quad \text{or } R_{eff} = \frac{3R}{4}.$$

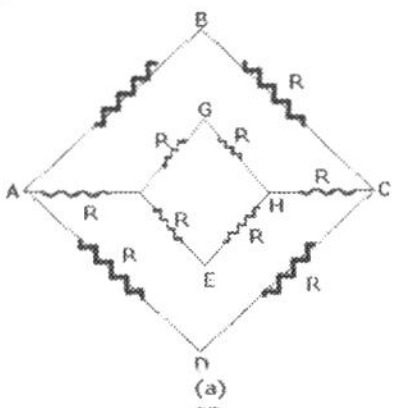

(a)

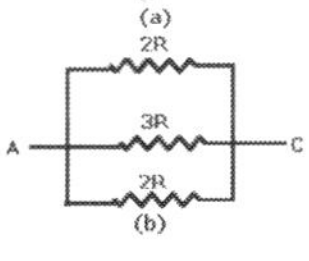

(b)

16. Apply Ohm's law.

17. $R = \frac{V^2}{P}$

find R_{eq} then $i = \frac{V}{R_{eq}}$

18. Combination of resistors

19. $\Delta t = \frac{R_2 - R_1}{R_1 \alpha}$

20. Combination of resistors

21. $i = \frac{V}{R_{total}}$

22. Combination of resistors

23. Combination of resistors

24. Combination of resistors

25. error = $V_1 = V_1^1$

26. $R_1\alpha_1 = R_2\alpha_2$

27. $P = i^2 R$

28. $4R = 400 \pm 20$

29. $H = \frac{V^2}{R} \times t$

$R = \frac{\rho l}{\pi r^2}$

30. $H = \frac{V^2}{R} t$

$\therefore \frac{t_2}{t_1} = \frac{V_1^{\ 2}}{V_2^{\ 2}}$

31. $P = \frac{V^2}{R} = \frac{V^2 A}{\rho l} \Rightarrow P \alpha \frac{1}{l}$

32. $i^2 \alpha\, r^3$

33. $p = v \times i$

34. $P = \frac{V^2}{R}$

35. $i = \frac{e_1 - e_2}{R + r_1 + r_2}, v = e_1 - ir_1 = 0$ 7

36. $I = \dfrac{2V - V}{3R} = \dfrac{V}{3R}$

Apply loop law in upper loop

$IR + V - V - V_C = 0$

or $V_C = IR = \dfrac{V}{3R}R$ or $V_C = \dfrac{V}{3}$.

37. i = E / r + R

38. $i = \dfrac{mnE}{mR + nr}$ & $m \times n = 24$

39. $i = \dfrac{nE}{R + nr}$

40. $I_s = \dfrac{2E}{2r + R}$, *and* $I_P = \dfrac{2E}{r + 2R}$

r = R

41. % error = $\dfrac{\Delta E}{E} \times 100 = \dfrac{ir}{E} \times 100$

42. when $i_s = i_p$

r = R

43. $i = \dfrac{E_1 + E_2}{R + r_1 + r_2}$

$V_1 = E_1 - i r_1$

44. $R_1 = R_0(1 + \alpha_1(\Delta t))$

$R_2 = R_0(1 + \alpha_2(\Delta t))$

Series $R = R_1 + R_2$

Parallel $\dfrac{1}{R} = \dfrac{1}{R_1} + \dfrac{1}{R_2}$

45. Applying kirchoff's first law
46. Using kirchoff's law
47. Using wheatstone bridge principle
48. Applying kirchoff's law

49. $\dfrac{X}{Y} = \dfrac{80}{100 - 80}$ ------(1)

$\dfrac{\dfrac{10x}{10 + x}}{Y} = \dfrac{50}{50}$ -----(2)

50. $\frac{X}{R}=\frac{l}{100-l}$

51. Potential gradient $=\frac{V}{l}$

52. $\left(\frac{E}{R+r}\right)R=K.l$

53. $\frac{E_1}{E_2}=\frac{l_1}{l_2};\frac{IR}{IX}=\frac{l_1}{l_2}$

54. $\frac{V_1}{V_2}=\frac{l_1}{l_2}\Rightarrow V_2=0.625V$

error $=0.65-0.625$

55. $V=i\rho l$

<u>LEVEL - II (HW)</u>

ELECTRIC CURRENT AND DRIFT VELOCITY

1. **An electron of mass 9 x 10^{-31}kg moves around a nucleus in a circular orbit of radius 2A^0 under the action of centripetal force 3.2N. Then the equivalent electric current is**

1) $\frac{32}{3\pi}$ **2)** $\frac{3\pi}{32}$ **3)** $\frac{16}{3\pi}$ **4)** $\frac{3\pi}{16}$

2. **The current in a conductor varies with time 't' as I = 2- 0.02t ampers. The electric charge that passes from t = 0 to t = 100 sec is**

1) 50 C **2) 100 C** **3) 25 C** **4) 75 C**

OHM's LAW AND COMBINATION OF RESISTANCES

3. **Four resistances 10Ω, 5Ω,7Ω and 3Ω are connected so that they form the sides of a rectangle AB, BC, CD and DA respectively. Another resistance of 10Ω is connected across the diagonal AC. The equivalent resistance between A and B is**

1) 2Ω 2) 5Ω 3) 7Ω 4) 10Ω

4. **A 3Ω resistor and a 6Ω resistor are connected in parallel and the combination is connected in series to a battery of 5V and a 3Ω resistor. The potential difference across the 6Ω resistor**

1) 2V 2) 4V 3) 3V 4) 1V

5. **You are given a wire of length 100 cm and linear resistance of 1 ohm/cm. If it is cut into two parts, so that when they are in parallel, the effective resistance is 24 ohm. The lengths of the two parts are**

1) 30cm & 70cm 2) 60cm & 40cm 3) 70cm & 30cm 4) 20cm & 80cm

6. **The resistance of a platinum wire of a platinum resistance thermometer at the ice point is 5Ω and at steam point is 5.4Ω. When the thermometer is inserted in a hot bath, the resistance of the platinum wire is 6.2Ω. Find the temperature of the hot bath.**

1) 3000^oC 2) 30^oC 3) 300^oC 4) $300K$

7. Three unequal resistors in parallel are equivalent to a resistance 1 ohm. If two of them are in the ratio 1 : 2 and if no resistance value is fractional, the largest of the three resistance in ohm is

1) 4 2) 6 3) 8 4) 12

8. A carbon filament has resistance of 120Ω at 0^oC what must be the resistance of a copper filament connected in series with carbon so that combination has same resistance at all temperatures $(\alpha_{carbon} = -5\times10^{-4}/^oC, \alpha_{copper} = 4\times10^{-3}/^oC)$

1) 120Ω 2) 15Ω 3) 60Ω 4) 210Ω

9. The equivalent resistance across XY in fig.

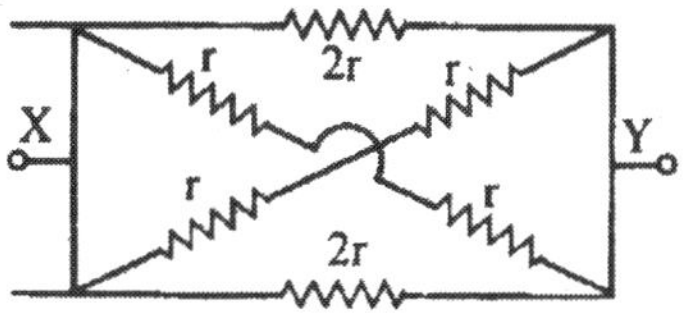

1) r 2) 2r 3) 4r 4) $\frac{r}{2}$

10. If the resistance of a circuit having 12V source is increased by 4Ω, the current drops by 0.5A. What is the original resistance of circuit

1) 4Ω 2) 8Ω 3) 16Ω 4) $\frac{1}{16}\Omega$

11. An electric current is passed through a circuit containing two wires of the same material connected in parallel. If the lengths and radii of the wire are in the ratio 4/3 and 2/3, then the ratio of the currents passing through the wires will be

1) $\frac{1}{3}$ 2) $\frac{3}{1}$ 3) $\frac{4}{3}$ 4) $\frac{3}{4}$

12. When 'n' wires which are identical are connected in series, the effective resistance exceeds that when they are in parallel by $\frac{X}{Y}\Omega$. Then the resistance of each wire is

1) $\frac{xn}{y(n^2-1)}$ 2) $\frac{yn}{x(n^2-1)}$ 3) $\frac{xn}{y(n-1)}$ 4) $\frac{yn}{x(n-1)}$

13. The equivalent resistance across A and B is

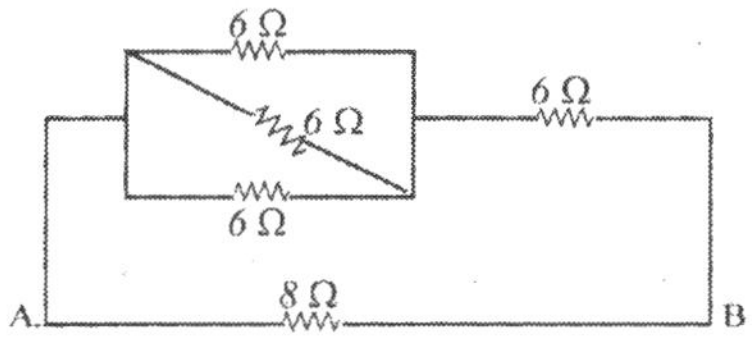

1) 2Ω 2) 4Ω 3) 8Ω 4) 12Ω

14. An ammeter A is connected as shown in the diagram. Ammeter reading is

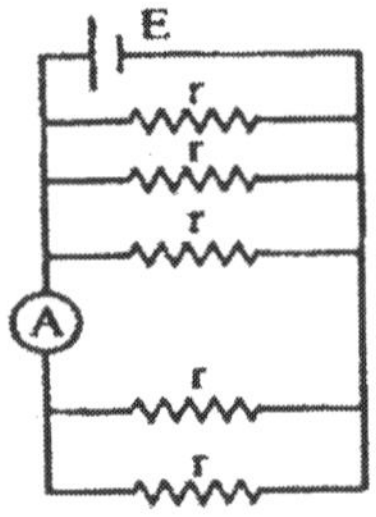

1) $\frac{E}{r}$ 2) $\frac{2E}{r}$ 3) $\frac{r}{2E}$ 4) $\frac{E}{2r}$

15. A current of 7A flows thorugh the circuit as shown in the figure the potential difference across points B and D is

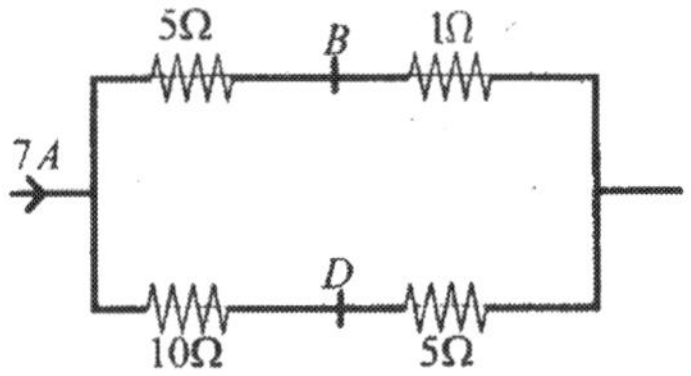

1) 5 V 2) 3 V 3) 10 V 4) 7 V

16. If a wire is stretched to make it 0.1% longer, its rsistance will :

1) increase by 0.2% 2) decrease by 0.2%

3) decrease by 0.05% 4) increase by 0.05%

ELECTRIC POWER

17. Two resistances R_1 and R_2 when connected in series consume power equal to 25W. When connected in parallel they consume 100W. The ratio of power of each is

1) 1/4 2) 1/3 3) 1/2 4) 1

18. Two electric bulbs marked 500 W, 220 V are put in series with 110V line. The power dissipated in each of the bulb is

1) $\frac{125}{4}$W 2) $\frac{25}{4}$W 3) $\frac{225}{4}$W 4) $\frac{325}{4}$W

19. A conductor of resistance 3 ohm is stretched uniformly till its length is doubled. The wire now is bent in the form of an equilateral triangle. The effective resistance between the ends of any side of the triangle in ohms is

1) $\frac{9}{2}$ 2) $\frac{8}{3}$ 3) 2 4) 1

20. The energy in kilowatt hour is consumed in operating ten 50W bulbs for 10 hours per day in a month of 30 days is

1) 1500 2) 15000 3) 15 4) 150

21. Two electric bulbs rated 25 W - 220 V and 100 W – 220 V are connected in series to a 440 V supply. Which of the bulbs will fuse?

1) Both 2) 100 W 3) 25 W 4) Neither

INTERNAL RESISTANCE AND EMF

22. Two batteries of different emf and internal resistances connected in series with each other and with an external load resistor. The current is 3.0 A. When the polarity of one battery is reversed, the current becomes 1.0 A. The ratio of the emf of the two batteries is

1) 2.5 : 1 2) 2 : 1 3) 3 : 2 4) 1 : 1

23. The pd across terminals of a cell is found to be 29 volt and 28 volt respectively when it delivers a current of 1 ampere and 2 ampere respectively. The emf and internal resistance of a cell are respectively

1) 30V, 2Ω   2) 30V,1Ω   3) 29V, 1Ω   4) 28V,2Ω

24. The current in a circuit containing a battery connected to 2 Ω resistance is 0.9 A. When a resistance of 7 Ω connected to the same battery, the current observed in the circuit is 0.3A. Then the internal resistance of the battery is

1) 0.1 Ω   2) 0.5 Ω   3) 1 Ω   4) Zero

25. The potential difference across the terminals of a battery is 10 V when there is a current of 3A in the battery from the negative to the positive terminal. When the current is 2A in the reverse direction, the potential difference becomes 15V. The internal resistance of the battery is

1) 1Ω   2) 0.4Ω   3) 0.6Ω   4) 0.8Ω

26. Two cells of emf 3V and 5V and internal resistance r_1 and r_2 respectively are in series with an external resistance R. If the p.d. across 1st cell is zero, then R is

1) $\frac{5r_1-3r_2}{3}$ 2) $\frac{2r_1-3r_2}{4}$ 3) $\frac{3r_1-5r_2}{3}$ 4) $\frac{4r_1-5r_2}{3}$

27. A battery when connected by a resistance of 16Ω gives a terminal voltage of 12V. and when connected by a resistance of 10Ω gives a terminal voltage of 11V. Then the emf of the battery and its internal resistance.

1) 12.8 V   2) 13.7 V   3) 10.7 V   4) 9 V

28. When a resistor of 11 ohm is connected in series with an electric cell the current flowing in it is 0.5A. Instead, when a resistor of 5Ω is connected to the same electric cell in series, the current increases by 0.4A. The internal resistance of cell is

1) 1.5Ω   2) 2Ω   3) 2.5Ω   4) 6Ω

29. Two cells of emf 4V and 8V are connected to two resister 4Ω and 6Ω as shown. If 8V cell is short circuited. Then current through resistance 4Ω and 6Ω

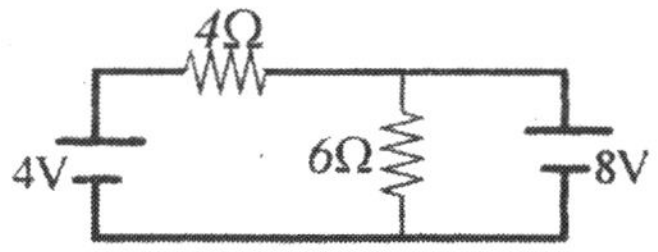

1) 2 A   2) 1 A   3) 2.5 A   4) 3 A

30. If in the circuit shown below, the internal resistnce of the battery is 1Ω and V_P and V_Q are the potentials at P and Q respectively, the potential difference between the points P and Q is

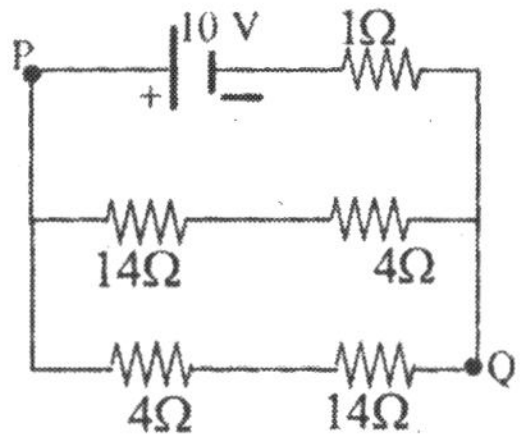

1) 9 V   2) 11 V   3) 7 V   4) 6 V

31. **Voltmeter reading in the given circuit is (voltmeter is ideal)**

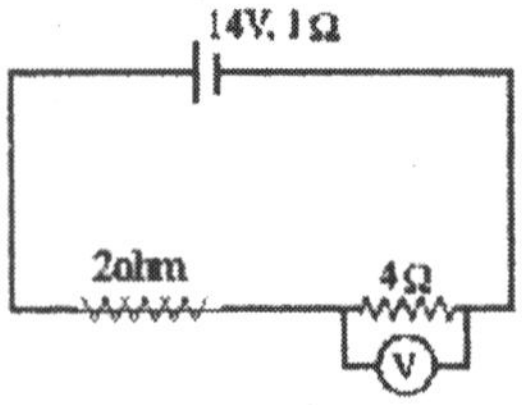

1) 6 V 2) 8 V 3) 10 V 4) 14 V

32. **For a cell the graph between the p.d(v) across the terminals of the cells and the current (I) drawn from the cell as shown. The emf and inernal resistance is**

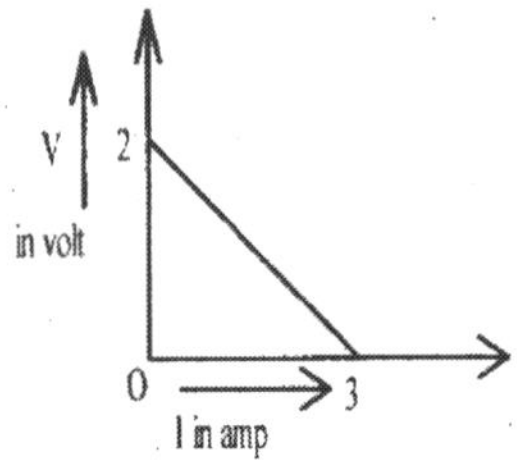

1) $\frac{3}{2}\Omega$ 2) $\frac{1}{3}\Omega$ 3) 3Ω 4) $\frac{2}{3}\Omega$

33. **The minimum number of cells in mixed grouping required to produced a maximum current of 1A through external resistance of 20Ω given the emf of each cell is 2V and internal resistance 1Ω is**

1) 25 2) 20 3) 16 4) 30

34. **A battery of emf 'E' and internal resistance 'r' is connected to a resistor of restance 'r_1' and Q Joules of heat is produced in a certain time 't'. When the same battery is connectted to another resistor of resistance r_2 the same quantity of heat is produced in the same time 't'. Then, the value of 'r' is**

1) $\frac{r_1^2}{r_2}$ 2) $\frac{r_2^2}{r_1}$ 3) $\frac{1}{2}(r_1+r_2)$ 4) $\sqrt{r_1 r_2}$

35. **The emf of a cell E is 15 V as shown in the figure with an internal resistance of 0.5 Ω. Then the value of the current drawn from the cell is**

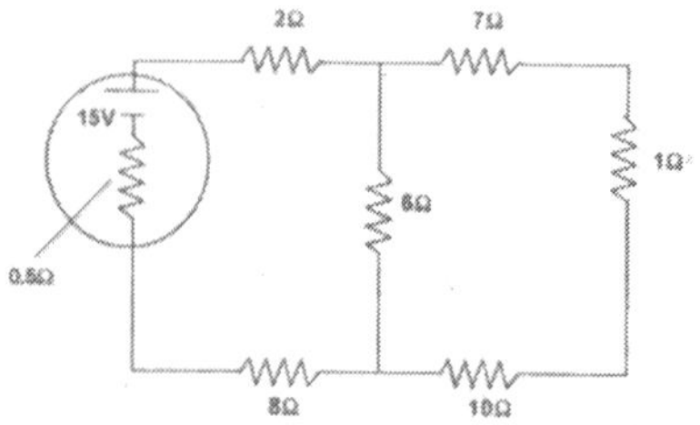

1) 1 A 2) 3 A 3) 2 A 4) 5 A

KIRCHOFF'S LAWS AND WHEAT SHONE'S BRIDGE

36. **The current i drawn from the 5 volt source will be**

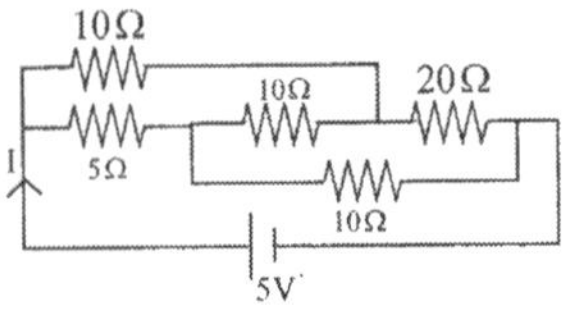

1) 0.5 2) 2 A 3) 1.5 A 4) 3 A

37. **In the given circuit which is a part of a closed cirucit the current i_1, i_2 are respcetively**

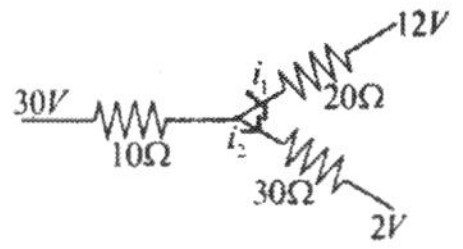

1) 0.4 A 2) 0.6 A 3) 1.6 A 4) 2 A

38. **A 5V battery with internal resistance 2 Ω and a 2V battery with internal resistance 1Ω are connected to a 10Ω resistor as shown in the figure**

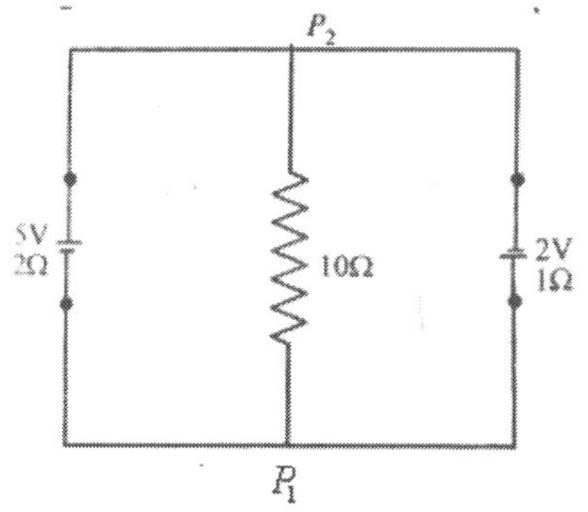

The current in the 10Ω resistor is

1) 0.27 A P_2 to P_1 2) 0.03 A P_1 to P_2 3) 0.03 A P_2 to P_1 4) 0.27 A P_1 to P_2

METRE BRIDGE

39. **When a conductor is connected in the left gap and known resistance in the right gap the balancing length is 50cm. If the wire is stretched so that its length increased by 20% , New balancing length is**

1) 40.98 cm 2) 38.23 cm 3) 42.56 cm 4) 48.21 cm

40. **In a meter bridge experiment when a resistance wire is connected in the left gap, the balance point is found at the 30th cm. When the wire is replaced by another wire, the balance point is found at the 60th cm. find the balance point when the two wires connectd in series and in parallel in the left gap successively**

1) 20 cm 2) 25 cm 3) 23 cm 4) 30 cm

41. **Shown in the figure below is a meter-bridge set up with null deflection in the galvanometer.**

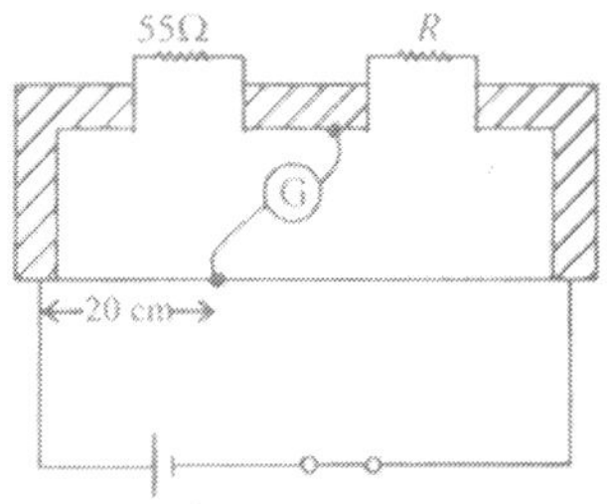

The vlaue of the unknown resistor R is

1) 13.75Ω 2) 220Ω 3) 110Ω 4) 55Ω

POTENTIO METER

42. **A potentiometer wire 10 m long has a resistance of 40Ω. It is connected in series with a resistanced box and a 2 v storage cell. If the potential gradient along the wire is 0.01V/m the resistance unplugged in the box is**

1) 760Ω 2) 260Ω 3) 1060Ω 4) 960Ω

43. **The ratio of potential gradients is 1 : 2, the resistance of two potentiometer wires of same length are $2\Omega \& 4\Omega$ respectively. The current flowing through them are in the ratio**

1) 1 : 2 2) 2 : 1 3) 1 : 3 4) 1 : 1

44. **The length of potentiometer wire is 100 cm and the emf of its standard cell is E volts. It is employed to measure the emf of a battery whose internal resistance is 0.5 ohm. If the balance point is obtained at l = 30 cm from positive end, the emf of the battery is**

1) $\frac{30E}{100.5}$ 2) $\frac{30E}{100-0.5}$ 3) $\frac{30E}{100}$ 4) $\frac{100E}{30}$

45. **1 ohm resistance is in series with an Ammeter which is balanced by 75 cm of potentiometer wire. A standard cell of 1.02 V is balanced by 50 cm. The ammeter shows a reading of 1.5 A. The error in the ammeter reading is**

1) 0.002 A 2) 0.03 A 3) 1.01 A 4) no error

KEY LEVEL - II (HW)

1) 1	2) 2	3) 2	4) 1	5) 2	6) 3	7) 2	8) 2	9) 4	10) 2
11) 1	12) 1	13) 2	14) 2	15) 1	16) 1	17) 4	18) 1	19) 2	20) 4
21) 3	22) 2	23) 2	24) 2	25) 1	26) 1	27) 2	28) 3	29) 2	30) 1
31) 2	32) 4	33) 2	34) 4	35) 1	36) 1	37) 2	38) 3	39) 1	40) 2
41) 2	42) 1	43) 4	44) 3	45) 2					

LEVEL - II (HW) HINTS

1. $i=\frac{q}{t}=\frac{qv}{2\pi r}$ $\left[\because F=\frac{mv^2}{r}\right]$

2. $q=\int_{t_1}^{t_2} i\,dt$

3. Combination of resistors

4. Combination of resistors

5. Combination of resistors

6. $R_2=R_1\left(1+\alpha\Delta T\right)$

7. $\frac{1}{R_1}+\frac{1}{R_2}+\frac{1}{R_3}=1$

$R_2:R_3=1:2$

8. $R_1\alpha_1=R_2\alpha_2$

9. Combination of resistors

10. $V = iR$

$12 = (i - 0.5)(R + 4)$

11. V - iR

V = constant

$i_1 R_1 = i_2 R_2$

$\frac{i_1}{i_2} = \frac{R_2}{R_1} \quad \because R \propto \frac{l}{A}$

12. $R_s = R_p + \left(\frac{X}{Y}\right)$

13. Combination of resistors

14. $i = \frac{5E}{r} - \frac{3E}{r} =$ --------

15. $i_1 = \frac{i \times R_2}{R_1 + R_2}$

$i_2 = \frac{i \times R_1}{R_1 + R_1}$

16. $R \propto l^2$

17. $P_s = \frac{P_1 P_2}{P_1 + P_2}$; $P_P = P_1 + P_2$

18. $R_1 = R_2 = \frac{V^2}{P} = R$

$i = \frac{V}{2R}$

$\therefore P = i^2 R$

19. $R \propto L^2, R_{eff} = \frac{R_1 R_2}{R_1 + R_2}$

20. $P \propto i^2$

21. $P = Vi$

$R = \frac{V^2}{P}$

22. $3 = \frac{E_1 + E_2}{r_1 + r_2 + R}$ -----(1)

$1 = \frac{E_1 - E_2}{r_1 + r_2 + R}$ -----(2) ; $\frac{(1)}{(2)}$

23. $r = \left(\frac{E-V}{V}\right)R$

24. $\frac{E}{2+r} = 0.9 - - - - - (1)$

$\frac{E}{7+r} = 0.3 - - - - - (2)$

$\frac{(1)}{(2)} \; \frac{7+r}{2+r} = 3$

25. $E - 3r = 10 \; ....(i)$

$E + 2r = 15 \; ....(ii)$

solving (i) & (ii) r = 1

26. $V = E - ir$

27. $E = \frac{V}{R}(R+r) = \text{constant}$

28. E = i(R + r)

29. **V = iR**

30. $i = \frac{V}{R}$

31. V = iR

32. $\mathbf{i} = \frac{E}{r}$

33. No. of cells = m × n

$i_{max} = mR = nr$

34. $Q = \frac{V^2}{r_1} = \frac{E^2 r_1}{(r+r_1)^2}$

$Q = \frac{V^2}{r_2} = \frac{E^2 . r_2}{(r+r_2)^2}$

35. using wheastone bridge

36. using kirchoff's laws

37. Applying kirchoff's law

38. $\frac{X}{R} = \frac{l}{100-l}$

39. $\frac{X}{R} = \frac{l}{100-l}$

40. $\frac{X}{R}=\frac{l}{100-l}$

41. $i=\frac{E}{R}$

42. $K=\frac{E}{(R+R_s)}\frac{R}{L}$

43. $\frac{i_1}{i_2}=\frac{V_1}{V_2}\frac{R_2}{R_1}$

44. $V\alpha\, l\,,\frac{E'}{E}=\frac{30}{100}$

45. $\frac{V_1}{V_2}=\frac{l_1}{l_2}=\frac{75}{50}\Rightarrow V_1=1.53V$

$I_1=\frac{V_1}{1}=1.53A$, error = 1.53-1.50=0.03

LEVEL - III

ELECTRIC CURRENT AND DRIFT VELOCITY

1. **The electron of hydrogen atom is considered to be revolving around the proton in circular orbit of radius $\frac{\hbar^2}{me^2}$ with velocity $\frac{e^2}{\hbar}$, where $\hbar=\frac{h}{2\pi}$. The current I is**

1) $\frac{4\pi^2me^2}{h^2}$ 2) $\frac{4\pi^2me^2}{h^3}$ 3) $\frac{4\pi^2m^2e^2}{h^3}$ 4) $\frac{4\pi^2me^5}{h^3}$

2. **In a straight conductor of uniform cross-section charge q is flowing for time t. Let s be the specific charge of an electron. The momentum of all the free electrons per unit length of the conductor, due to their drift velocity only is**

1) $\frac{q}{ts}$ 2) $\left(\frac{q}{ts}\right)^2$ 3) $\sqrt{\frac{q}{ts}}$ 4) qts

3. **Potential difference of 100 V is applied to the ends of a copper wire one metre long. Find the ratio of average drift velocity and thermal velocity of electrons at 27^oC. (Consider there is one conduction electron per atom. The density of copper is 9.0×10^3; Atomic mass of copper is 63.5 g. $N_A=6.0\times10^{23}$ per gram-mole, conductivity of copper is $5.81\times10^7\,\Omega^{-1}$. $K=1.38\times10^{-23}\,JK^{-1}$)**

1) 3.67×10^{-6} 2) 4.3×10^{-6} 3) 6×10^{-5} 4) 5.6×10^{-6}

OHM'S LAW AND COMBINATION OF RESISTANCES

4. The sides of rectangular block are 2cm, 3cm and 4cm. The ratio of the maximum to minimum resistance between its parallel faces is

1) 3 2) 4 3) 2 4) 1

5. Find the equivalent resistance across AB:

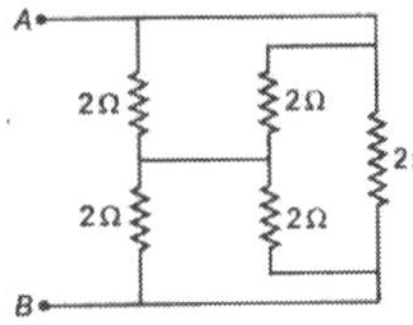

1) 1Ω 2) 2Ω 3) 3Ω 4) 4Ω

6. Two wires of the same material have length 6cm and 10cm and radii 0.5 mm and 1.5 mm respectively. They are connected in series across a battery of 16V. The p.d. across the shorter wire is

1) 5V 2) 13.5 V 3) 27 V 4) 10 V

7. Three ammeters P,Q and R with internal resistances r, 1.5r ,3r respectively . Q and R parallel and this combination is in series with P , The whole combination concted between X and Y . When the battery connected between X and Y , the ratio of the readings of P,Q and R is

1) 2:1:1 2) 3:2:1 3) 3:1:2 4) 1:1:1

8. The potential difference between the points A and B is

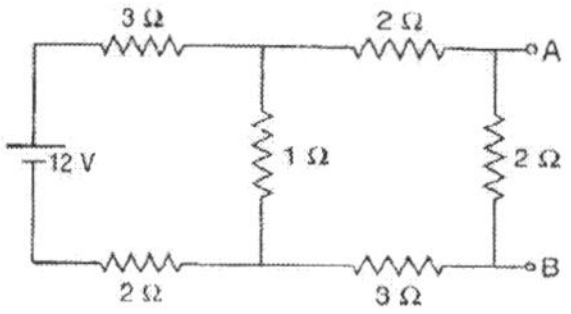

1) 1.50 V 2) 2.50 V 3) 1.00 V 4)0.50 V

9. The resistance of a semicircle shown in fig. between its two end faces is (Given that radial thickness = 3 cm, axial thickness = 4 cm, inner radius = 6 cm and resistivity = $4\times10^{-6}\Omega cm$)

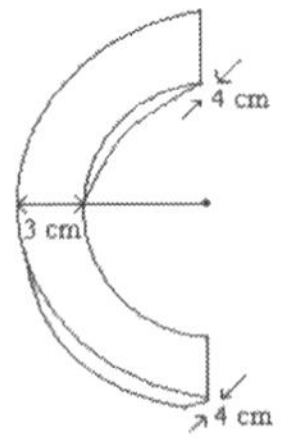

1) $24.15\times10^{-6}\Omega$ 2) $7.85\times10^{-7}\Omega$ 3) $7.85\times10^{-6}\Omega$ 4) $7.85\times10^{-5}\Omega$

10. ABCD is a square where each side is a uniform wire of resistance 1Ω. A point E lies on CD such that if a uniform wire of resistance 1Ω is connected across AE and constant potential difference is applied across A and C, then B and E are equipotential .

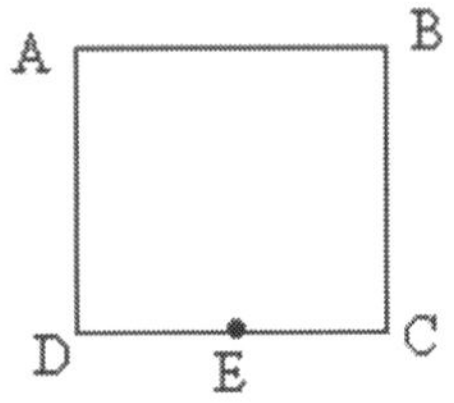

1) $\frac{CE}{ED}=1$ 2) $\frac{CE}{ED}=\frac{1}{\sqrt{2}}$ 3) $\frac{CE}{ED}=\frac{1}{2}$ 4) $\frac{CE}{ED}=\sqrt{2}$

11. A heating element using nichrome connected to a 230 V supply draws an initial current of 3.2 A which settles after a few seconds to a steady value of 2.8 A. What is the steady temperature of the heating element if the room temperature is $27.0^{o}C$? Temperature coefficient of resistance of nichrome averaged over the temperature range involved is $1.70\times10^{-4}C^{-1}$?

1) $680^{0}C$ 2) $867^{0}C$ 3) $920^{0}C$ 4) $750^{0}C$

CELLS, KIRCHOFF'S LAW 'S , WHEAT STONE BRIDGE

12. A group of N cells where e.m.f. varies directly with the internal resistance as per the equation $E_N = 1.5\ r_N$ are connected as shown in the figure. The current I in the circuit is:

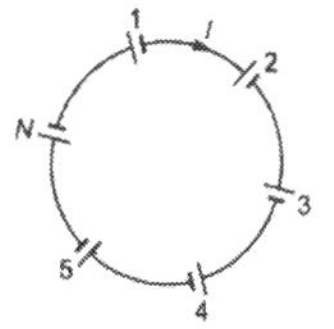

1) 0.51 A 2) 5.1 A 3) 0.15 A 4) 1.5 A

13. Cell A has emf $2E$ and internal resistance $4r$. Cell B has emf E and internal resistance r. The negative of A is connected to the positive of B and a load resistance of R is connected across the battery formed. If the terminal potential difference across A is zero, then R is equal to

1) $3r$ 2) $2r$ 3) r 4) $5r$

14. For the circuit shown in the figure, potential difference between points A and B is 16V. Find the current passing through 2Ω

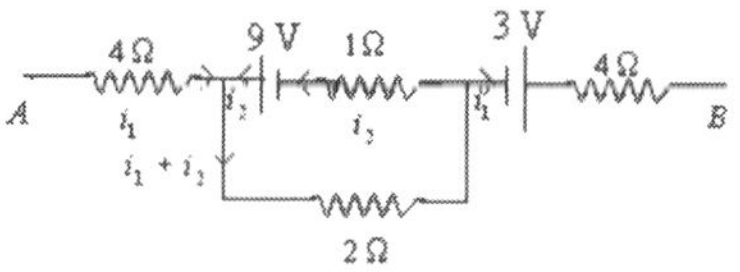

1) 3.5A 2) 3A 3) 4.5A 4) 5.5A

15. The minimum number of cells in mixed group ing required to produce a maximum current of 1.5 A through an external resistance of 30Ω,given the emf of each cell is 1.5 V and internal resistance is 1Ω is

1) 30 2) 120 3) 40 4) 60

16. The p.d between the terminals A & B is

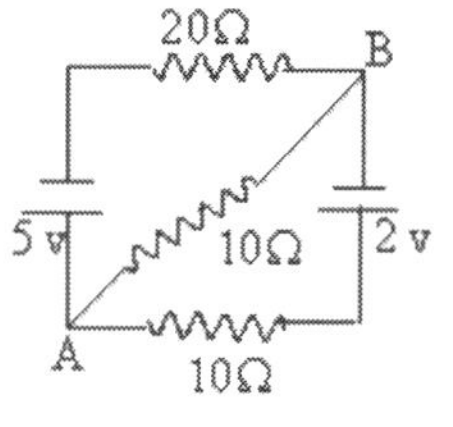

1) 2V 2) 3V 3) 3.6 V 4) 1.8 V

17. The energy stored in the capcitor is

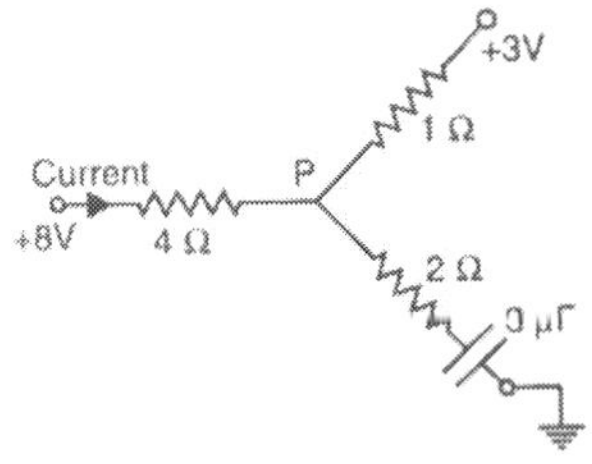

1) $12\mu J$ 2) $24\mu J$ 3) $36\mu J$ 4) $48\mu J$

18. **In the circuit shown in figure, the potentials of B,C and D are :**

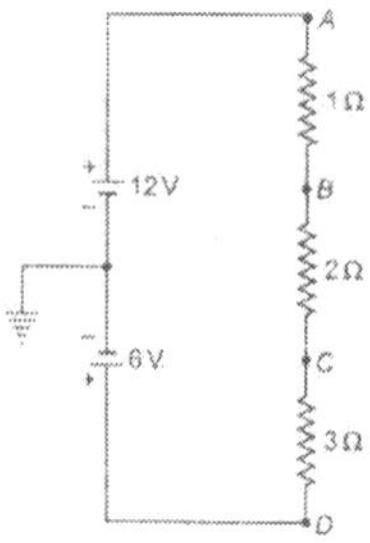

1) $V_B = 6V; V_C = 9V; V_D = 11V$ 2) $V_B = 11V; V_C = 9V; V_D = 6V$

3) $V_B = 9V; V_C = 11V; V_D = 6V$ 4) $V_B = 9V; V_C = 6V; V_D = 11V$

19. **A current of 0.10 A flows through the 25Ω resistor represented in the diagram to the right. The current through the 80Ω resistor is:**

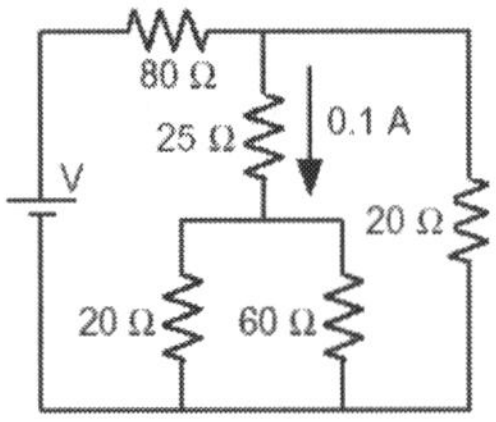

1) 0.10 A 2) 0.20 A 3) 0.30 A 4) 0.40 A

20. **In Wheat stone's bridge shown in the adjoining figure galvanometer gives no deflection on pressing the key, the balance condition for the bridge is :**

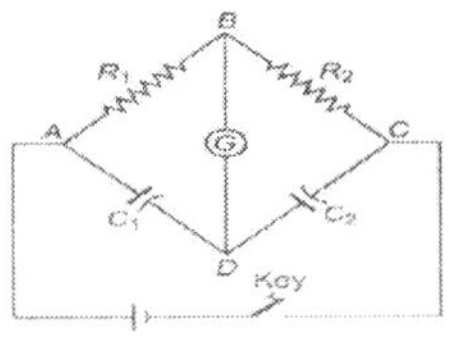

1) $\frac{R_1}{R_2} = \frac{C_1}{C_2}$ 2) $\frac{R_1}{R_2} = \frac{C_2}{C_1}$

3) $\frac{R_1}{R_1 + R_2} = \frac{C_1}{C_1 - C_2}$ 4) $\frac{R_1}{R_1 - R_2} = \frac{C_1}{C_1 + C_2}$

21. **In the steady state, the energy stored in the capacitor is :**

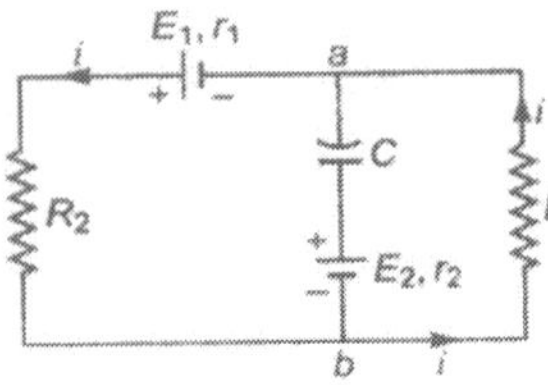

1) $\frac{1}{2}C(E_1 + E_2)^2$ 2) $\frac{1}{2}C(E_1 - E_2)^2$

3) $\frac{1}{2}C\left(\frac{E_1R_1 + E_2R_2}{r_1 + r_2 + R_1 + R_2}\right)^2$ 4) $\frac{1}{2}C\left(E_2 + \frac{E_1R_1}{r_1 + R_1 + R_2}\right)^2$

22. **A part of circuit in steady state along with the currents flowing in the branches, the value of resistances is shown in figure. Calculate the energy stored in the capacitor.**

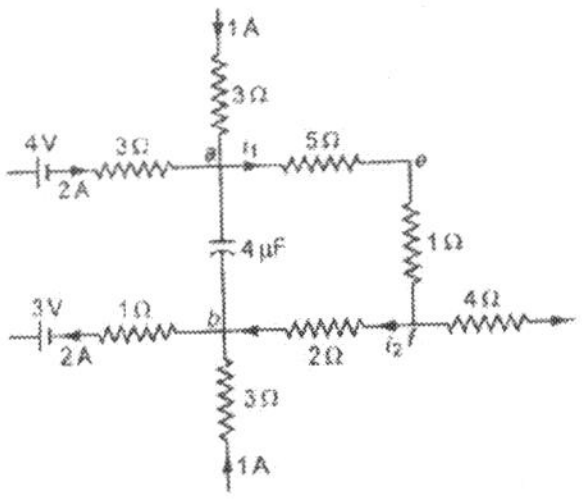

1) $8\times10^{-1}J$ 2) $8\times10^{-2}J$ 3) $8\times10^{-3}J$ 4) $8\times10^{-4}J$

23. **Equivalent resistance across A and B in the given circuit if r = 10 Ω, R = 20 Ω is**

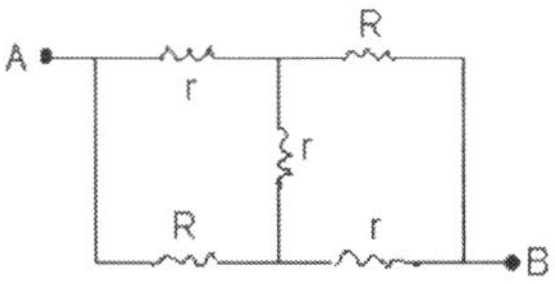

1) 7 Ω 2) 14 Ω 3) 35 Ω 4) 20/3 Ω

24. **For a cell, the graph between the p.d.(V) across the terminals of the cell and the current I drawn from the cell is shown in the fig. the emf and the internal resistance of the cell is E and r repectively.**

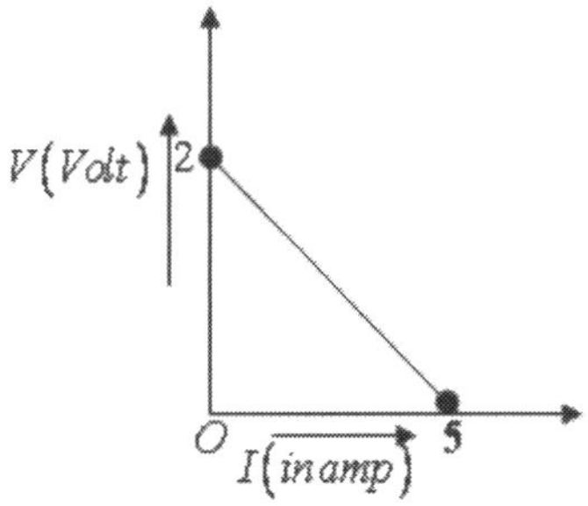

1) E = 2V, r = 0.5Ω 2) E = 2V, r = 0.4Ω

3) E > 2V, r = 0.5Ω 4) E > 2V, r = 0.4Ω

25. **The charge developed on 4 μF condenser is**

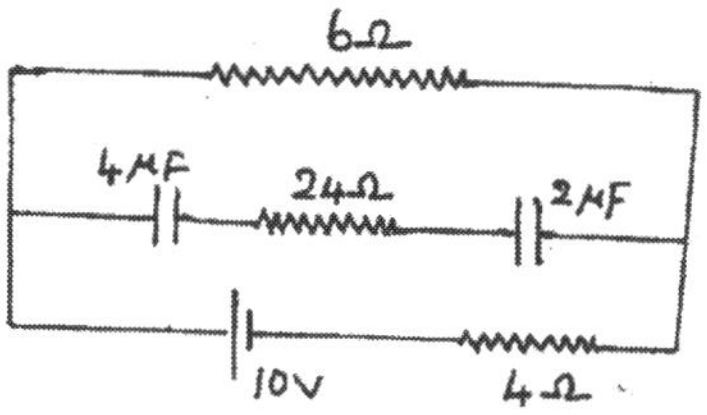

1) 18 μC 2) 4 μC 3) 8 μC 4) Zero

26. **Find out the value of current through 2Ω resistance for the given circuit.**

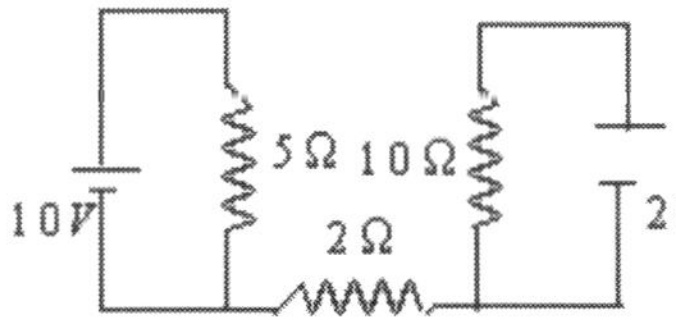

1) 0 2) 1.6 A 3) 2.4 A 4) 3A

ELECTRIC POWER, JOULE'S LAW

27. Same mass of copper is drawn into 2 wires of 1mm thick and 3mm thick. Two wires are connected in series and current is passed. Heat produced in the wires is the ratio of

1) 3 : 1 2) 9 : 1 3) 81 : 1 4) 1 : 81

28. Masses of three are in the ratio 1:3:5. Their lengths are in the ratio 5:3:1. When they are connected in series to an external source, the amounts of heats produced in them are in the ratio

1) 125 : 15 : 1 2) 1 : 15 : 125 3) 5 : 3 : 1 4) 1 : 3 : 5

29. A heater coil rated at 1000W is connected to a 110V mains. How much time will take to melt 625 gm of ice at $0^0 C$. (for ice L = 80 cal/gm)

1) 100s 2) 150s 3) 200s 4) 210s

30. In the following circuit, 5Ω resistor develops 45 J/s due to current flowing through it. The power developed across 12Ω resistor is

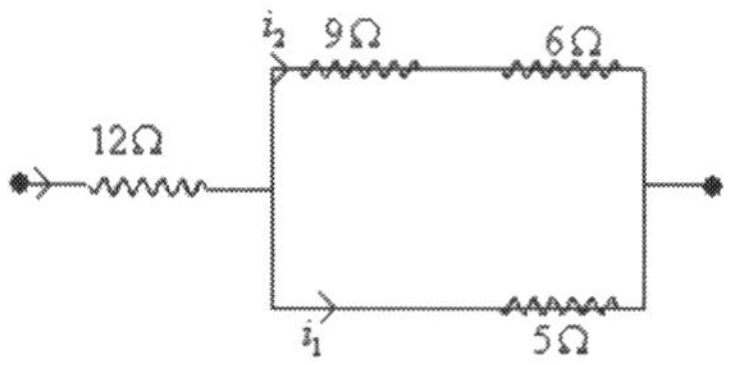

1) 16 W 2) 192 W 3) 36 W 4) 64 W

31. Two wires 'A' and 'B' of the same material have their lengths in the ratio 1 : 2 and radii in the ratio 2 : 1. The two wires are connected in parallel across a battery. The ratio of the heat produced in 'A' to the heat produced in 'B' for the same time is

1) 1 : 2 2) 2 : 1 3) 1 : 8 4) 8 : 1

32. An electric motor operating on 50 volt D.C. supply draws a current of 10 amp. If the efficiency of motor is 40%, then the resistance of the winding of the motor is

1) 1.5Ω 2) 3Ω 3) 4.5Ω 4) 6Ω

33. The resistance of a 240 V – 200 W electric bulb when hot is 10 times the resistance when cold. The resistance at room temperature and the temperature coefficient of the filament are (given working temperature of the filament is $2000°C$)

1) $28.8\Omega, 4.5\times10^{-3}/°C$ 2) $14.4\Omega, 4.5\times10^{-3}/°C$

3) $28.8\Omega, 3.5\times10^{-3}/°C$ 4) $14.4\Omega, 3.5\times10^{-3}/°C$

34. A wire of length L and 3 identical cells of negligible internal resistances are connected in series. Due to the current, the temperature of the wire is raisied by ΔT in a time t. A number N of similar cells is now connected in series with a wire of the same material and cross-section but the length 2L The temperature of the wire is raised by the same amount ΔT in the same t. The value of N is :

1) 3 2) 2 3) 6 4) 4

35. Three bulbs with their power and working voltage are connected as shown in the circuit diagram to a 12 V battery. The total power consumed by the bulbs is (ignore the internal resistance of the battery shown)

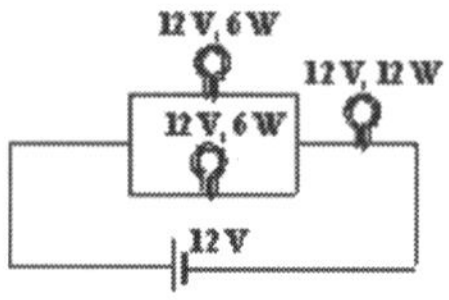

1) 24 W 2) 12 W 3) 6 W 4) 15 W

36. A cell of emf 12 V and internal resistance 6Ω is connected in parallel with another cell of emf 6 V and internal resistance 3Ω, such that the positive of the first cell joins the positive of the second cell and similarly the negative of first cell joins the negative of the second cell. A bulb of filament resistance 14Ω is connected across the combination. The power delivered to be bulb is

1) 4.0 W 2) 3.5 W 3) 8.5 W 4) 2.5 W

37. A cell develops the same power across two resistances $R_1 \& R_2$ separately. The internal resistance of the cell is

1) $\sqrt{R_1R_2}$ 2) $\sqrt{2R_1R_2}$ 3) R_1+R_2 4) R_1-R_2

METRE BRIDGE

38. A metallic conductor at $10^\circ C$ connected in the left gap of meter bridge gives balancing length 40 cm. When the conductor is at $60^\circ C$, the balancing point shifts by --- cm, (temperature coefficeint of resistance of the material of the wire is $(1/220)/^0C$)

1) 4.8 2) 10 3) 15 4) 7

39. When a conducting wire is connected in the right gap and known resistance in the left gap, the balancing length is 60cm. The balancing length becomes 42.4 cm when the wire is stretched so that its length increases by

1) 10% 2) 20% 3) 25% 4) 42.7%

40. '*n*' identical resistors are taken. '*n*/2' resistors are connected in series and the remaining are connected in parallel. The series connected group is kept in the left gap of a meter bridge and the parallel connected group in the right gap. The distance of the balance point from the left end of the wire is

1) $\frac{100n^2}{n^2+4}$ 2) $\frac{100n^2}{n^2+1}$ 3) $\frac{400}{n^2+4}$ 4) $\frac{400}{n^2+1}$

41. In a metere bridge, the balance length from left end (standard resistance of 1Ω is in the right gap) is found to be 20 cm, the length of resistance wire in left gap is $\frac{1}{2}m$ and radius is 2mm its specific resistance is

1) $\pi\times10^{-6}ohm-m$ 2) $2\pi\times10^{-6}ohm-m$ 3) $\frac{\pi}{2}\times10^{-6}ohm-m$ 4) $3\pi\times10^{-6}ohm-m$

POTENTIO METER

42. In an experiment with potentiometer to mea- sure the internal resistance of a cell, when the cell is shunted by 5Ω, the null point is obtained at 2m. When cell is shunted by 20Ω the null point is obtained at 3m. The internal resistance of cell is

1) 2Ω 2) 4Ω 3) 6Ω 4) 8Ω

43. A potentiometer wire of length 100cm has a resistance 5ohm. It is connected in series with a resistance and a cell of emf 2v and of negli gible internal resistance. A source of emf 5mv balanced by 10 cm length of potentiometer wire. The value of external reistance is _____

1) 540Ω 2) 195Ω 3) 190Ω 4) 990Ω

44. 1 ohm resistance is in series with an Ammeter which is balanced by 75 cm of potentiometerwire. A standard cell of 1.02V is balanced by 50 cm. The Ammeter shows a reading of 1.5A. The error in the Ammeter reading is

1) 0.002A 2) 0.03A 3) 1.01A 4) no error

45. **An ideal battery of emf 2V and a series resistance R are connected in the primary circuit of a potentio meter of length 1m and resistance 5Ω. The value of R to give a potential difference of 5mV across the 10cm of potentiometer wire is**

1) 180Ω 2) 190Ω 3) 195Ω 4) 200Ω

46. **In an experiment for calibration of voltmeter, a standard cell of emf 1.5V is balanced at 300cm length of potentiometer wire. The P.D.across a resistance in the circuit is balanced at 1.25m. If a voltmeter is connected across the same resistance, it reads 0.65V. The error in the volt meter is**

1) 0.05V 2) 0.025V 3) 0.5V 4) 0.25V

47. **In the circuit shown in fig., the potential difference between the points C and D is balanced against 40 cm length of poten-tiometer wire of total length 100 cm. In order to balance the potential difference between the points D and E. The jockey be pressed on potentiometer wire at a distance of**

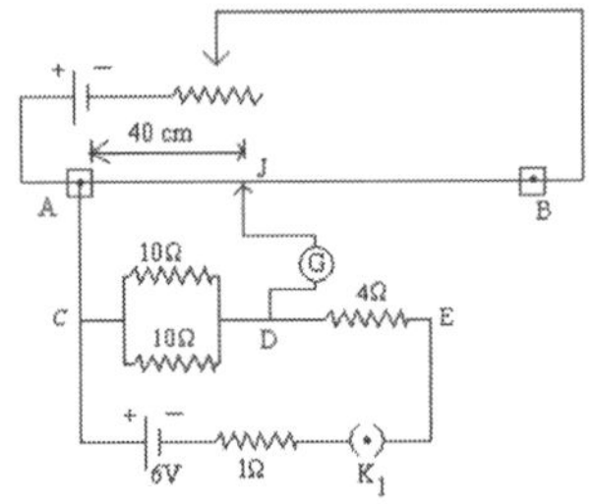

1) 16 cm 2) 32 cm 3) 56 cm 4) 80 cm

KEY LEVEL-III

1) 4	2) 1	3) 1	4) 2	5) 1	6) 2	7) 2	8) 4	9) 3	10) 4
11) 2	12) 2	13) 4	14) 3	15)2	16) 2	17) 2	18) 4	19) 2	20)3
21) 3	22) 2	23) 4	24) 2	25) 2	26) 3	27) 1	28) 3	29)1	30) 4
31) 2	32)4	33) 2	34)1	35) 3	36) 3	37)2	38) 1	39)1	40) 4
41) 1	42) 2	43)2	44) 2	45) 2	46) 3	47) 2			

HINTS LEVEL - III

1. $I = \frac{e}{t} = \frac{e}{2\pi r / v} = \frac{ev}{2\pi r}$

2. $I = nAev_d$ or $v_d = \frac{I}{nAe} = \frac{q/t}{nAe}$

No. of free electrons per unit length of conductor

$N = nA \times 1$

$\therefore$ Momentum of all the free electrons is

$p = Nmv_d$

3. $\therefore v_d = \frac{\sigma E}{ne}$ $\quad v_{rms} = \sqrt{\frac{3K_B T}{m_e}}$

4. $R_{max} = \frac{\rho a}{bc}, R_{min} = \frac{\rho c}{ab}$

$a > b > c = 4 > 3 > 2$

$$\frac{R_{max}}{R_{main}} = \frac{Pa}{bc} \times \frac{ab}{Pc} = \frac{a^2}{c^2} = \frac{4^2}{2^2} = 4.$$

5. Apply series and parallel combinations

6. $l_1 = 6cm \ , l_2 = 10cm,$

$r_1 = 0.5 \times 10^{-3}, r_2 = 1.5 \times 10^{-3}$

In series combination i = constant

$$\frac{V_1}{V_2} = \frac{R_1}{R_2} = \frac{\frac{Pl_1}{A_1}}{\frac{Pl_2}{A_2}} = \frac{l_1}{l_2} \times \frac{A_2}{A_1} \quad V_1 + V_2 = 16V$$

Solving for $V_1 = 13.5V$

7. $i = \frac{V}{R}$

8. 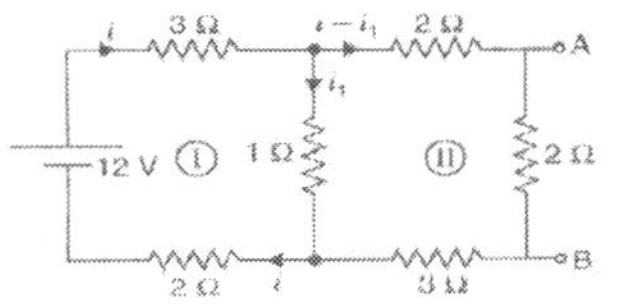

For the first loop $12 = 5i + i_1$

For the second loop $0 = 7(i - i_1) - i_1$

or $8i_1 = 7i \ or, \ i_1 = (7/8)i$

Therefore, we obtain $12 = 5i + \frac{7}{8}i = \frac{47i}{8}$

or, $i = \frac{12 \times 8}{47} A = 2.04A$

$i_1 = \frac{7}{8} \times 2.04A = 1.79A$

Thus, the p.d across A and B is

$V_A - V_B = (i - i_1) \times 2 = 0.25 \times 2 = 0.50V$

9. Here, A = 4 cm × 3 cm = 12 cm^2

$l = \pi r = \pi(6 + 3/2) = \pi \times 7.5$ cm

$$R = \frac{\rho l}{A} = \frac{4 \times 10^{-6} \times \pi \times 7.5}{12}$$

$= 7.85 \times 10^{-6}$ ohm.

10. 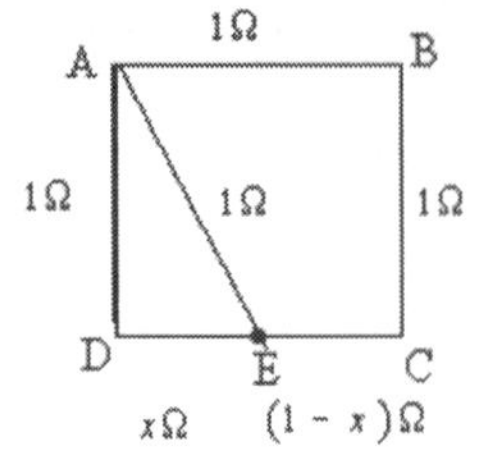

Equivalent resistance between A and E is

$$y=\frac{(x+1)}{x+2}$$

For B and E to be at equal potential , we get

$$\frac{R_{AE}}{R_{AB}}=\frac{R_{EC}}{R_{BC}}\Rightarrow\frac{x+1}{(x+2)1}=\frac{1-x}{1}$$

Solving $x=\sqrt{2}-1$

Now $\frac{CE}{ED}=\frac{1-x}{x}=\sqrt{2}$

11. $R_{27}=\frac{V}{I}=\frac{230}{3.2}$, $R_\theta=\frac{230}{2.8}$ $\quad\quad \frac{R_{27}}{R_\theta}=\frac{R_0(1+\alpha\times 27)}{R_0(1+\alpha\times\theta)}$

12. $i=\frac{E_N}{r_N}=1.5$

13. $i=\frac{e_1+e_2}{R+r_1+r_2}, V_1=E_1-ir_1=0$

14. $V_A-V_B=16$

$4i_1+2(i_1+i_2)-3+4i_1=16....(1)$

$9-i_2-2(i_1+i_2)=0$

Solving eqs (1) and (2) $i_1=1.5\,A\,and\,i_2=2A$

15. $i_{max}=\frac{mE}{2r}=\frac{nE}{2R}$

n = number of cells in each row.

m = number of rows.

$1.5=\frac{1.5\times n}{2\times 30}$

n = 60

$1.5=\frac{1.5\times m}{2\times 1}$

m = 2

$\therefore$ total no of cells; $=n\times m=2\times 60=120$

16. $$i=\frac{\frac{E_1}{r_1}+\frac{E_2}{r_2}}{1+R\left(\frac{1}{r_1}+\frac{1}{r_2}\right)}=\frac{\frac{5}{20}+\frac{2}{10}}{1+10\left(\frac{1}{20}+\frac{1}{10}\right)}$$

$V=iR=i\times10=1.8v$

17. 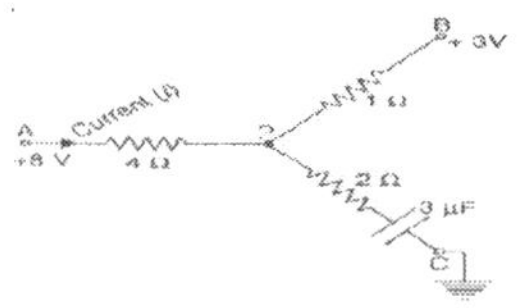

We have

$V_A-V_B=i\times(4+1)$

$8-3=i\times5$

$5=i\times5$

$i=1A$

$V_A-V_p=4\times1$

$8-V_P=4;\ V_P=4\,volt$

Now $V_C=0$. So, the energy stored in the capacitor is $\xi=\frac{1}{2}\times3\times16=24\mu J$

18. Potential at O is zero being earthed.
Applying Kirchhoff's second law

$i(1+2+3)=12-6\ or\ i=1A$

$V_A-V_D=(1+2+3)\times1=6V$

$V_A-V_B=1\times1=1V$

$V_A-V_C=(1+2)\times1=3V$

Also, $V_A-V_O=12V$ or $V_A=12V$

Thus, $V_D=12-6=6V$,

$V_B=12-1=11V$, $V_C=12-3=9V$

19. $i=\frac{V}{r}$

20. At balance, the potentials of point B and D are same and there will be no current in the arm BD. Thus,

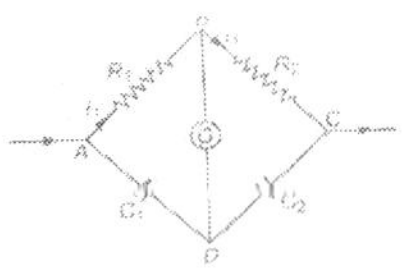

$$V_A-V_B=V_A-V_D$$

$$i_1R_1 = \frac{q}{C_1} \qquad .... \text{(i)}$$

where q is the charge on both the capacitor plates connected in series.

Quite similarly $V_B - V_C = V_D - V_C$

or $$i_1R_2 = \frac{q}{C_2} \qquad .... \text{(ii)}$$

Dividing eqs. (i) and (ii), we get $\frac{R_1}{R_2} = \frac{C_2}{C_1}$

21. When the capacitor plate acquire full cahrge q_0, there will be no current in the capacitor arm. Applying Kirchhoff's second law to the current carrying circuit

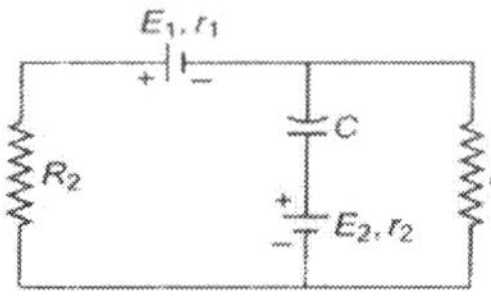

$i(R_1 + R_2) = E_1 - ir_1$ or $\quad i = \frac{E_1}{r_1 + R_1 + R_2}$

Now $V_a - V_b = -iR_1 = -\frac{E_1R_1}{r_1 + R_1 + R_2}$

and $V_C = \frac{q_0}{C} = E_2 + iR_1 = E_2 + \frac{E_1R_1}{r_1 + R_1 + R_2}$

Now energy stored in the capacitor

$$U = \frac{1}{2}CV_c^2 \qquad = \frac{1}{2}C\left[E_2 + \frac{E_1R_1}{r_1 + R_1 + R_2}\right]^2$$

22. When the capacitor plates get fully charged, there will be no current in branch ab, Remember capacitance acts as the open circuit since capacitance offers infinite resistance to d.c.The capacitance simply collects the charge. Applying Kirchooff's first law to the junctions a and b, we find $i_1 = 3A \; and \; i_2 = 1A$. Now applying Kirchhoff's second law to the closed mash aefba, we get $\quad 3\times5 + 3\times1 + 1\times2 = V_a - V_b$

$$V_a - V_b = 20V$$

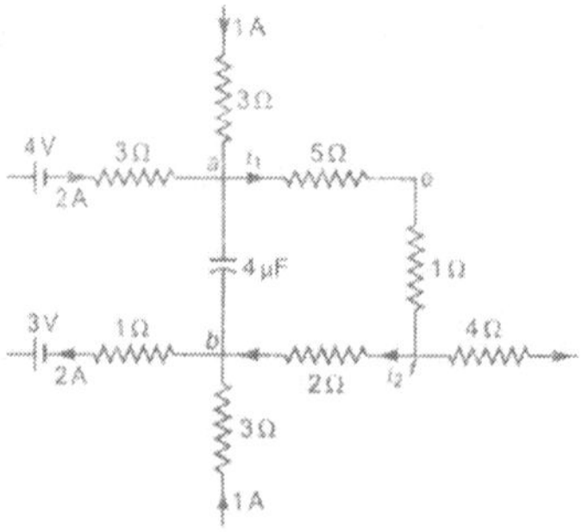

Energy stored in the capacitor

$$U=\frac{1}{2}C(V_a-V_b)^2=\frac{1}{2}\times 4\times 10^{-6}\times (20)^2$$
$$=8\times 10^{-4}J$$

23. $R^1 = r\left[\dfrac{3R+r}{3r+R}\right]$

24. $V=E-ir$

$i=0,\ V=E=2V$

$V=0,\ r=\dfrac{E}{i}=0.4\Omega$

25. current i = 1 amp

P.D across $6\,\Omega=6\,Volt$

P.D across $4\,\Omega=2\,Volt$

$\therefore$ Charges on $4\mu F=8\,\mu C$

26. $2\,\Omega$ resistor is in open circut so current is 0

27. $JQ=i^2Rt$

$Q\alpha R\alpha\dfrac{1}{A^2}$

$\dfrac{Q_1}{Q_2}=\dfrac{r_2^4}{r_1^4}=\dfrac{3^4}{1^4}=81.1$

28. $m_1:m_2:m_3=1=3:5$

$l_1:l_2:l_3=5:3:1$

$Q=i^2Rt \quad R=\dfrac{pdl^2}{m}$

$Q\,\alpha\,R=\dfrac{pdl^2}{m}$

$\dfrac{Q_1}{Q_2}=\dfrac{l_1^2}{l_2^2}\times\dfrac{m_2}{m_1}\Rightarrow$

$Q_1:Q_2:Q_3=125:15:1$

29. $JQ=P\times t$

$J\times mL=P\times t$

$1\times 625\times 10^{-3}\times\dfrac{80\times 4.2}{10^{-3}}=1000\times t$

$t=210s$

30. $P = i^2R = 192W$

31. $Q = \frac{V^2}{R}$

$\frac{Q_1}{Q_2} = \frac{R_2}{R_1} = \frac{l_2}{l_1} \times \frac{r_1^2}{r_2^2}$.

$\frac{Q_1}{Q_2} = \frac{2}{1} \times \frac{(2)^2}{1^2} = \frac{8}{1}$

32. Input power $P = VI = 50 \times 10$

Power dissipated as heat = $I^2 R = 100 R$

$$\text{Efficiency} = \frac{\textit{Out put power}}{\textit{Input power}}$$

$$\frac{40}{100} = \frac{500 - 100R}{500}$$

33. Resistance of the hot bulb $R_2 = \frac{V^2}{P} = \frac{240 \times 240}{200}$

Resistance of the bulb at room temperature

$R_1 = \frac{R_2}{10} = \frac{288}{10}$ $\alpha = \frac{R_2 - R_1}{R_1 t}$

34. In the first case, three indentical cells are connected in seires with a wire of length L. Let the terminal potential difference of each cell is V and resistance of the wire is R. Then heat developed in the wire in

time t is $H = \frac{(3V)^2}{R} t = ms\Delta T$

where m is the mass of the wire, s-the specific heat of its material and ΔT is the rise in its temperature. When N such indentical cells are connected in series, the effective terminal otential is NV volt and if the length of the wire is doubled, its resistance and mass also doubled. Then heat develope in the wire is

$$H^` = \frac{(NV)^2}{2R} t = (2m)s.\Delta T$$

Dividing both the equations, we get

$$\frac{N^2}{2 \times 9} = 2 \Rightarrow N = 6$$

35 . $P = \frac{V^2}{R}$

36. $i = \frac{E_{eff}}{R_{eff}}, p = i^2 R$

37. $P_1 = i^2 R_1 = \left(\frac{E}{R_1 + r}\right)^2 R_1$

$$P_2 = i^2 R_2 = \left(\frac{E}{R_2 + r}\right)^2 R_2$$

38. $\frac{X}{R} = \frac{40}{60} = \frac{2}{3}$; $\frac{X_0(1+\alpha t_1)}{R} = \frac{2}{3}$

$$\frac{X_0(1+\alpha t_2)}{R} = \frac{l}{100-l}$$

$$\frac{1+\alpha t_1}{1+\alpha t_2} = \frac{2}{3} \times \left[\frac{100-l}{l}\right]$$

$$l = 44.8$$

Balancing point shifts by =44.8-40=4.8 .

39. $\frac{X}{R} = \frac{60}{40}$;

$\frac{X}{R^{'}} = \frac{42.4}{57.6}$;

$$\frac{R^{'}}{R} = \frac{60}{40} \times \frac{57.6}{42.4}$$

$R \alpha l^2$

$$\frac{l^1 - l}{l} \times 100 = \left(\frac{\sqrt{R^1} - \sqrt{R}}{\sqrt{R}}\right) \times 100$$

40. $\frac{X}{100-X} = \frac{nr/2}{2r/n}$

41. $X = \frac{sl}{A}$ $\quad \frac{X}{R} = \frac{20}{80}$

$$\frac{1}{4} = \frac{S \times \frac{1}{2}}{\pi \times (2 \times 10^{-3})^2} \quad X = \frac{R \times 2}{8}$$

$$X = \frac{1}{4} \quad S = 2\pi \times 10^{-6} \Omega - m$$

42. $\frac{V_1}{V_2} = \frac{l_1}{l_2}$

$$V_1 = \left[\frac{E}{R_1 + r}\right] R_1, V_2 = \left[\frac{E}{R_2 + r}\right] R_2$$

$$\frac{R_1(R_2+r)}{R_2(R_1+r)}=\frac{l_1}{l_2}\Rightarrow r=4\Omega$$

43. $E' = iP\,l = \left[\frac{E}{R+R_3}\right]\frac{R}{L}\cdot l$

$$5\times10^{-3}=\left[\frac{2}{5+R_3}\right]\frac{5}{100}\times10$$

$$R_S=195\Omega$$

44. $1.02V \rightarrow 50cm$

$? \qquad \rightarrow 75cm$

$V=\frac{75\times1.02}{50}=1.53;$

$\text{error}=1.53-1.5=0.03$

45. $5\times10^{-3}=\frac{V}{L}l=\frac{iR}{L}l$

$$5\times10^{-3}=\left(\frac{2}{R+5}\right)\frac{5}{1}\times10\times10^{-2}$$

46. $300\times10^{-3}m\Rightarrow1.5V$

$1.25m\rightarrow?$

$V=0.625V$

Error in ammeter reading

$=0.625-0.65=0.025v.$

47. $\frac{V_1}{V_2}=\frac{iR_1}{iR_2}=\frac{l_1}{l_2} \qquad \therefore\frac{5}{4}=\frac{40}{l_2}$

LEVEL - IV

Instructions for Assertion & Reason Type questions:

1) Both (A) and (R) are true and (R) is the correct explanation of A.

2) Both (A) and (R) are true but (R) is not the correct explanation of A.

3) (A) is true but (R) is false **4) (A) is false but (R) is true**

1. **Assertion : Terminal voltage of a cell is greater than emf of cell, during charging of the cell.**
Reason : The emf of a cell is always greater than its terminal voltage.

2. **Assertion : In metrebridge experiment, a high resistance is connected in series with the galvanometer.**
Reason : As resistance increases, current through the circuit increases

3. **Assertion (A) : In a metrebridge ; copper wire is connected in the left gap and silicon is connected in the right gap, when the temp of both wires increase, balancing point shifts to right.**
Reason (R) : Temperature coeficient of copper is -Ve and that of silicon is +Ve.

4. Assertion : If a current flows through a wire of non-uniform cross-section, potential difference per unit length of the wire in the direction of current is same at different points.

Reason : V = iR and current in the wire is same throughout.

5. Assertion : Voltmeter is much better than a potentiometer for measuring emf of cell.

Reason : A potentiometer draws no current while measuring emf of a cell.

6. Assertion : The equivalent resistance between the points X and Y in the figure, is 10Ω.

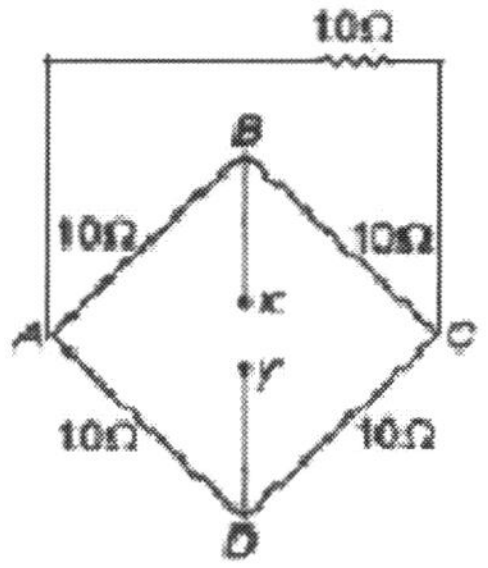

Reason : According to wheatstone bridge points A and C have the same potential.

7. Assertion : The drift velocity of electrons in a metallic wire will decrease, if the temperature of the wire is increased.

Reason : On increasing temperature, conductivity of metallic wire decreases.

8. Assertion : The electric bulb glows immediately when switch is on.

Reason : The drift velocity of electrons in a metallic wire is very high.

9. Assertion : If the length of the conductor is doubled, the drift velocity will become half of the original value (keeping potential difference unchanged)

Reason : At constant potential difference, drift velocity is inversely proportional to the length of the conductor.

10. Assertion : If the current of a lamp decreases by 20%, the percentage decrease in the illumination of the lamp is 40%

Reason : Illumination of the lamp is directly proportional to square of the current through lamp.

11. Assertion : However long a fuse wire may be, the safe current that can be allowed is the same.

Reason : The safe current that can be allowed to pass through a fuse wire depends on the radius of the wire.

12. Assertion : The resistance of an ideal voltmeter should be infinite.

Reason : The potential difference measured by a voltmeter across a resistor is always less than the actual potential difference across the resistor.

13. Assertion : Current is passed through a metallic wire, heating it red. When cold water is poured on half of its portion, then rest of the half portion becomes more hot.

Reason : Resistance decreases due to decrease in temperature and then current through wire increases.

MATCHING TYPE QUESTIONS

14. Match list - I with List - II

List - I	List- II
a) Ohm's law	e) conservation of charge
b) Joule's Law	f) conservation of energy
c) Kirchhoff's I Law	g) $v = Ri$
d) Kirchhoff's II Law	h) $H = i^2Rt$

1) a-h, b-g, c-e, d-f 2) a-g, b-h, c-e, d-f 3) a-h, b-f, c-e, d-g 4) a-e, b-f, c-g, d-h

15. Match list - I with List - II

List - I	List - II
a) Potentimeter	e) For measuring current
b) Metrebridge	f)For measuring internal resistance
c) Ammeter	g)For measuring specific resistance of wire
d) Voltmeter	h) For measuring potential difference

1) a-f, b-g, c-e, d-h 2) a-g, b-e, c-f, d-h 3) a-h, b-e, c-f, d-g 4) a-h, b-f, c-e, d-g

16. Match list - I with List - II

List - I	List - II
a) Thermistor	e) High $+ve'\alpha'$
b) Carbon	f) α almost zero
c) Nichrome	g) either positive or negative $'\alpha'$
d) Constantan and manganin	h) Negative $'\alpha'$

1) a-g, b-h, c-e, d-f 2) a-h, b-g, c-e, d-f 3) a-e, b-f, c-g, d-h 4) a-e, b-g, c-h, d-f

17. Match list - I with List - II

List - I	List - II
a) Resistivity	e) Volt
b) Conductivity	f) Siemen
c) emf	g) ohm - metre
d) conductance	h) $mho-metre^{-1}$

1) a-e, b-f, c-g, d-h 2) a-f, b-e, c-g, d-h 3) a-g, b-h, c-e, d-f 4) a-h, b-g, c-e, d-f

18. Three wires of same material are connected in parallel to a source of emf. The length ratio of the wires is 1 : 2 : 3 and the ratio of their area of cross section is 2 : 4 : 1.

Table – 1	Table - 2
(a) Resistance ratio	(p) 6 : 6 : 1
(b) Current ratio	(q) 1 : 6 : 6
(c) Power ratio	(r) 1 : 1 : 6
	(s) None

1) a-r,b-q,c-p 2) a-p,b-q,c-r 3) a - r; b - p, c - p 4) a-q,b-p,c-r

19. In the figure shown, each resistance is R.

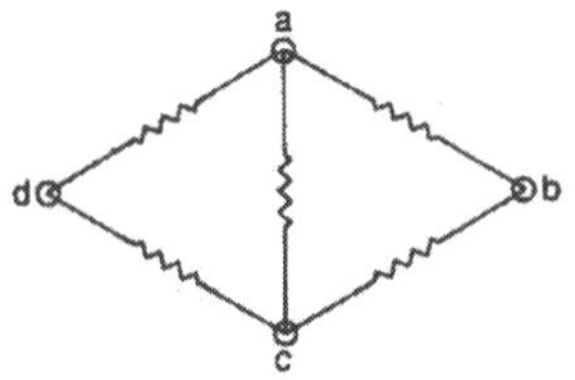

Table – 1	Table - 2
(a) Resistance between a and b	(p) $\frac{R}{2}$
(b) Resistance between a and c	(q) $\frac{5}{8}R$
(c) Resistance between b and d	(r) R
	(s) None

1) a-q,b-p,c-r 2) a - r; b - p; c - p 3) a-p,b-q,c-s 4) a-s,b-r,c-p

20. Six batteries of increasing emf and increasing internal resistance are as shown in figure.

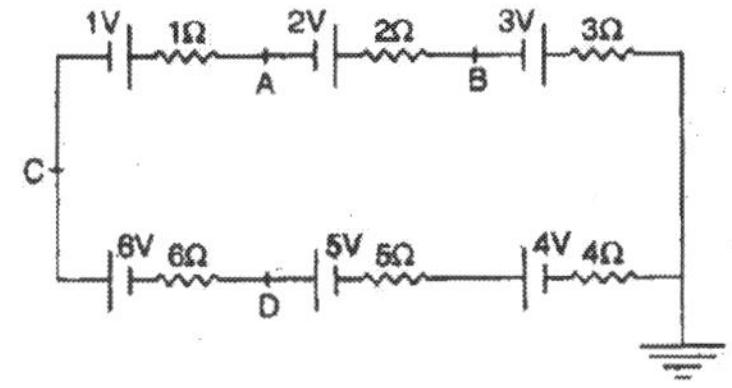

Table – 1	Table - 2
(a) Potential of point A	(p) Zero
(b) Potential of point B	(q) 2 V
(c) Potential of point C	(r) 4 V
(d) Potential of point D	(s) 6 V
	(t) none

1) a - p; b - p; c - p,d-p 2) a-p,b-s,c-q 3) a-q,b-q,c-s 4) a-r,b-r,c-p

21. In the potentiometer arrangement shown in figure, null point is obtained at length l.

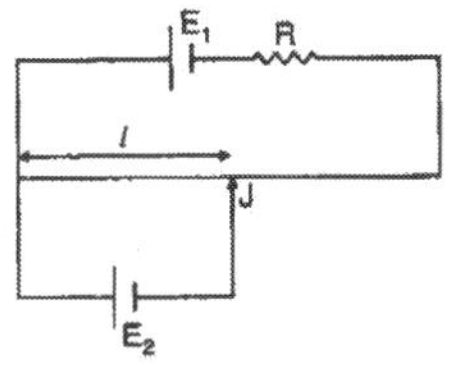

Table – 1	Table - 2
(a) If E_1 is increased	(p) l should increase
(b) If R is increased	(q) l should decrease
(c) If E_2 is increased	(r) l should remain the same to again get the null point

1) a-q,b-q,c-p 2) a-r,b-r,c-q 3) a-p,b-p,c-r 4) a - q; b - p; c - p

22. In the circuit shown in figure, if a resistance R is connected in parallel with R_2

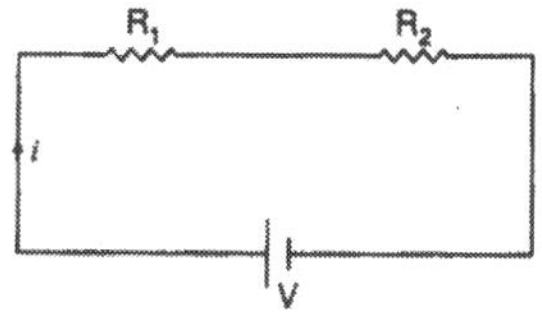

Table – 1	Table - 2
(a) Main current i	(p) will increase
(b) Power across R_1	(q) will decrease
(c) Power across R_2	(r) will remain same

1) a-p,b-p,c-p 2) a-q,b-q,c-q 3) a - p, b - p, c - q 4) a-r,b-r,c-r

23. In the circuit shown,

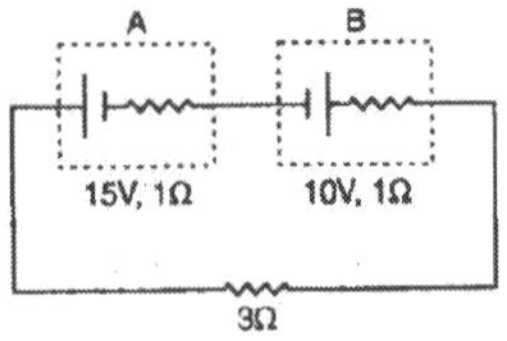

Table – 1	Table - 2
(a) Potential difference across battery A	(p) A
(b) Potential difference across battery B	(q) B
(c) Power is supplied by battery	(r) 14 V
(d) Power is consumed by battery	(s) 9 V
	(t) 11V

1) a-p; b-p;c-q;d-r 2) a-q;b-q;c-r;d-s 3) a-r, b-t, c-p, d-q 4) a-r;b-r;c-s;d-t

24. Current is flowing through a wire of non-uniform cross section. Cross section of wire A is less than the cross section of wire at B.

Table – 1	Table - 2
(a) current at A	(p) is zero
(b) drift velocity of electrons at A	(q) is more than at B
(c) electric field in the wire at A	(r) is less than at B
(d) current density at A	(s) is equal to that at B

1) a-p;b-q;c-q;d-p 2) a-s, b-q, c-q, d-q 3) a-q;b-r;c-p;d-q 4) a-r;b-s;c-r;c-s

25. In the circuit shown in figure, $R_1 = R_2 = R_3 = R$.

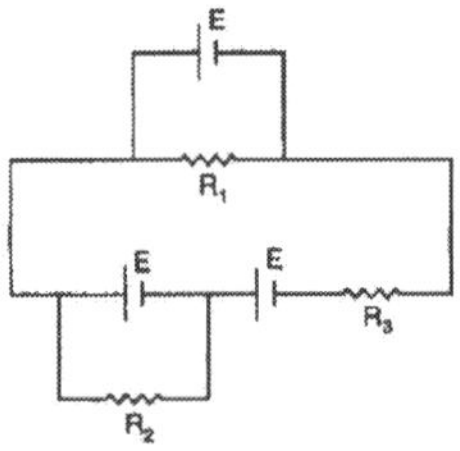

Table – 1	Table - 2
(a) current through R_1	(p) E/R
(b) current through R_2	(q) 2E/R
(c) current through R_3	(r) E/2R
	(s) Zero

1) a-p, b-p, c-p 2) a-p;b-q;c-s 3) a-q;b-r;c-r 4) a-r;b-s;c-p

26. Matrix Matching

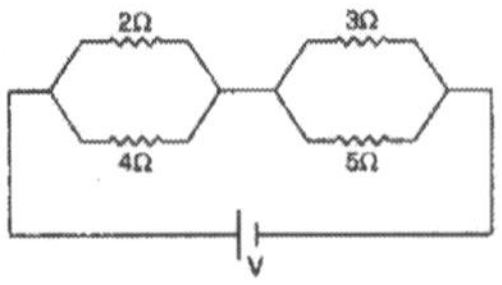

Table – 1	Table - 2
(a) Minimum current will flow through	(p) 2Ω
(b) Maximum current will flow through	(q) 4Ω
(c) Maximum power will be generated across	(r) 3Ω
(d) Minimum power will be generated across	(s) 5Ω

1) a-p;b-p;c-q;d-r 2) a-r;b-r;c-p;d-s 3) a - q, b - p, c - r, d - q 4) a-s;b-s;c-r;d-q

27. **Statement (A) : Thermistor can have only negative temperature coefficeints of resistances.**

Statement (B) : Thermistors with negative temperature coefficients of resistance are used as resistance thermometers, to measure low temperatures of the order of 10 K.

1) both A and B are true
2) both A and B are false
3) A is true and B is false
4) A is false, but B is true

28. **Statement (A) : Resistivity of insulators is about 10^{22} times the resistivity of metallic conductors.**

Statment (B) : Metals like silver, copper and aluminium have very high values of resistivity.

1) A and B are true
2) A and B are false
3) A is true, B is false
4) A is false, B is true

29. **Statement (A) : Series combination of cells is preferred when external resistance is large compared to internal resistance of cell.**

Statement (B) : Parallel combination of cells is preferred when external resistance is small compared to the internal resistance of each cell.

1) A and B are true 2) A and B are false 3) A is true, B is false 4) A is false, B is true

30. **Statement (A) : The difference between a new torch light cell and an old one is due to increase in internal resistance.**

Statement (B) : At 0 kelvin specific resistance of prefect insulator is infinity.

1) Both A and B are true
2) A is true, B is false
3) A is false, B is true
4) Both A and B are false

31. **Statement-1: The temperature dependence of resistance is usually given as $R = R_0(1+\alpha\Delta t)$. The resistance of a wire changes from 100Ω to 150Ω when its temperature is increased from 27°C to 227°C. This implies that $\alpha = 2.5\times10^{-3}/^0C$**

Statement 2 : - $R = R_0(1+\alpha\Delta t)$ is valid only when the change in the temperature ΔT is small and $\Delta R = (R-R_0) << R_0$

1) Statement -1 is true, statement -2 is true, Statement-2 is the correct explanation of statement -1.
2) Statement -1 is true, statement -2 is true,statement -2 is not the correct explanation of statement -1
3) Statement - 1 is false, Statement - 2 is true
4) Statement - 1 is true, Statement - 2 is false

32. **In the above circuit, *C* denotes the balance position on the potentiometer wire *AB*. Which of the following procedures can shift *C* towards the end *B*?**

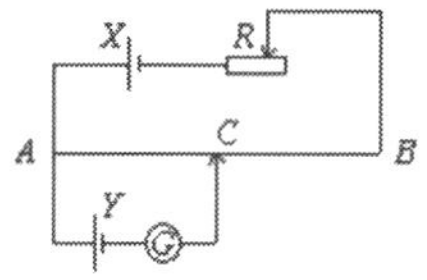

a) replacing the driving cell *X* by one with a smaller EMF

b) adding a resistance in series with the galvanometer *G*

c) increasing the resistance of the rheostat *R*

1) a, b and c 2) a and b only 3) b and c only 4) a and (c) only

33. **The potentiometer circuit shown is used to find the internal resistance of the cell *E*. At balance, the galvanometer pointer does not deflect, and NO current flows through**

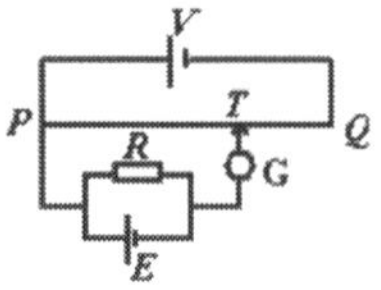

A) the potentiometer wire PQ **B) the resistor R**

C) the galvanometer G

1) A, B and C 2) A and B only 3) B and C only 4) C only

34. **The two ends of a uniform conductor are joined to a cell of emf E and some internal resistance. Starting from the mid point P of the conductor, we move in the direction of current and return to P. The potential V at every point on the path is plotted against the distance covered (x). One of the following best represent the resulting curve?**

1) 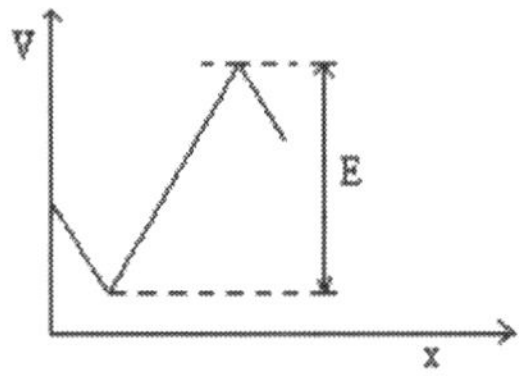

2) 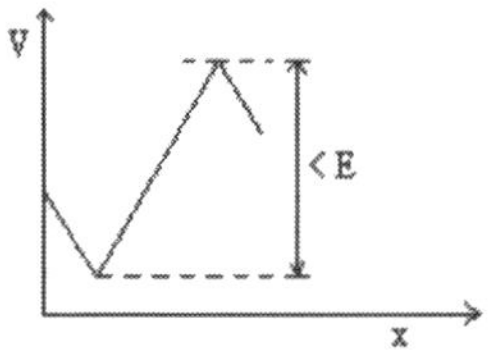

3) 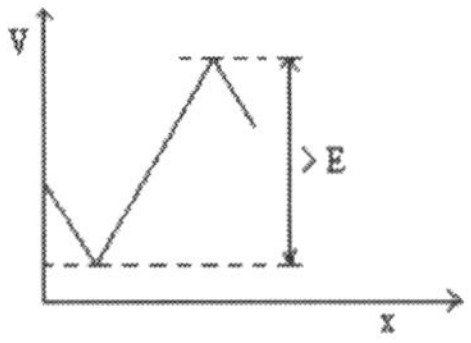

4) 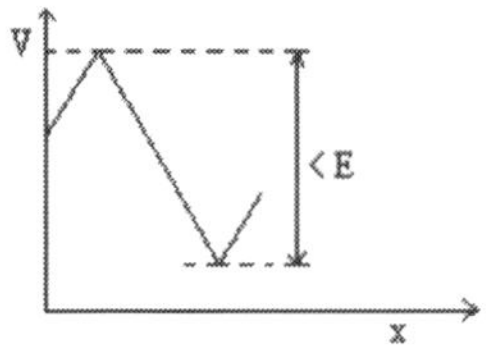

35. **In the circuit shown L1, L2, L3, and L4 are** identical light bulbs. There are six **voltmeters connected to the circuit as shown. Assume that the voltmeters are ideal. If L3 were to** burn out, opening the circuit, which **voltmeter(s) would read zero volts?**

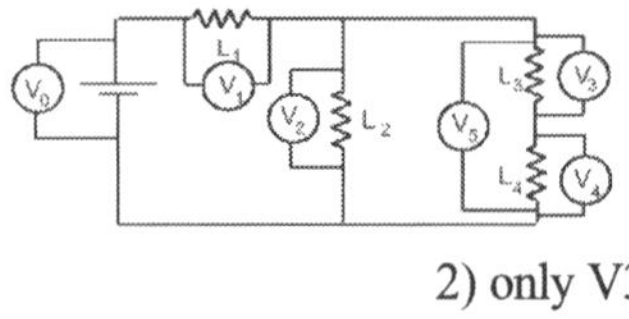

1) none would read zero 2) only V3

3) only V4 4) only V3, V4, and V5

KEY LEVEL-IV

1) 3	2) 3	3) 3	4) 4	5) 4	6) 1	7) 2	8) 3	9) 1	10) 4
11) 1	12) 2	13) 1	14) 2	15) 1	16) 1	17) 3	18) 3	19) 1	20) 1
21) 4	22) 3	23) 3	24) 2	25) 1	26) 3	27) 4	28) 3	29) 1	30) 1
31) 3	32) 4	33) 4	34) 2	35) 3					

Chapter - 6

(Part - 1) MOVING CHARGES AND MAGNETISM

MOVING CHARGES AND MAGNETISM

SYNOPSIS

1. INTRODUCTION :

i) A current carrying wire produces a magnetic field of its own. This was first observed by Oersted.

ii) When current is flowing through a conductor, only magnetic field is produced around it, which is non conservative.

iii) The direction of magnetic lines of force due to straight current carrying conductor will be concentric circles around the conductor in a plane which is always perpendicular to the length of the conductor.

2. The direction of magnetic field can be found by using :

i) AMPERE'S SWIMMING RULE :

Imagine a person swimming along a current carrying wire in the direction of the current facing a magnetic needle below the wire, then the magnetic north pole of the needle deflects towards his left hand.

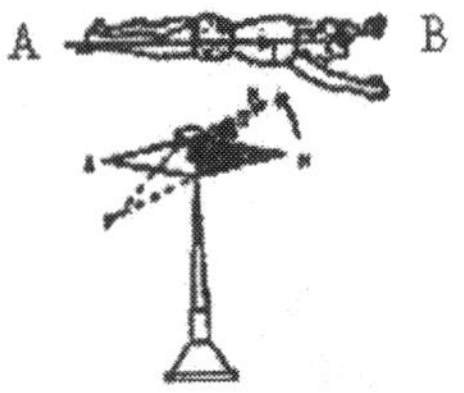

ii) AMPERE'S RIGHT HAND THUMB RULE :

When a straight conductor carrying current is held in the right hand such that the thumb is pointing along the direction of current, then the direction in which fingers curl round it gives the direction of magnetic lines of force.

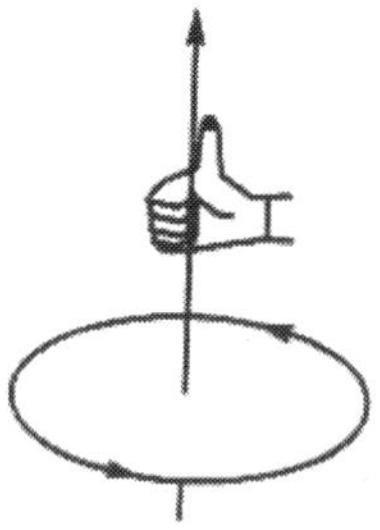

. The direction of magnetic field for current carrying conductor is as given below.

⊗ indicates $\overline{B}$ into the plane of paper

⊙ indicates $\overline{B}$ out of the plane of paper

iii) **MAXWELL'S CORK SCREW RULE :**

Imagine a right handed cork screw advancing in the direction of current, then the direction of rotation of the screw head gives the direction of magnetic lines of force,

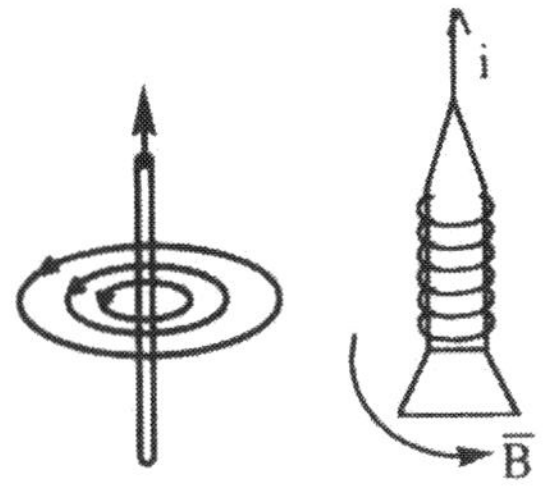

3.1 AMPERE'S CIRCUITAL LAW

STATEMENT :

The line integral of the magnetic induction field (B) along any closed path in air (or) vacuum is equal to μ_0 times the net current across the area bounded by this path.

$$\oint \vec{B}.\vec{dl} = \mu_0 i_{net}$$

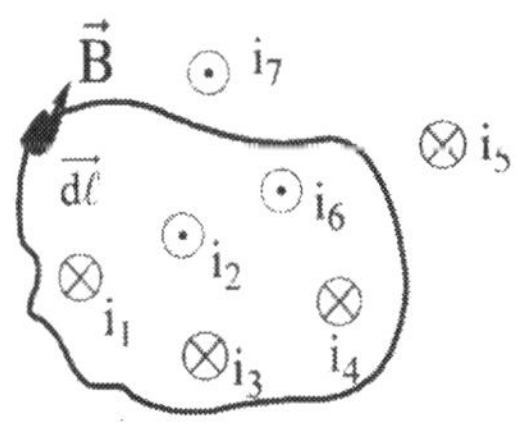

Consider a closed plane curve as shown in figure. $\vec{dl}$ is a small length element on the curve. Let $\vec{B}$ be the resultant magnetic field at the position of $\vec{dl}$. If the scalar product $\vec{B}.\vec{dl}$ is integrated by varying $\vec{dl}$ on the closed curve it is called line integral of $\vec{B}$ along the curve and it is represented by $\oint \vec{B}.\vec{dl}$

The rule for deciding whether an enclosed current is positive or negative : The fingers of the right hand are to be taken in the direction of integration around the path. If a current pierces the membrane stretched across the area in the direction of the thumb, then it is positive current. If the current pierces the membrane in the opposite direction, then it is negative.

For the above closed path

$$\oint B.dl = \mu_0 (i_1 - i_2 + i_3 + i_4 - i_6)$$

Points to remember regrading Ampere's Law

a) The line integral does not depend on the shape of the closed path or on the position of the current carrying wire in the loop.

b) If a conductor carrying current is outside the closed path, the line integral of B due to that conductor is zero i.e., we need not consider the currents that do not pierce the area of the closed path.

c) Ampere's circuital law is always true no matter how distorted the path or how complicated may be the magnetic field. In most cases even though Ampere's circuital law is true it is inconvenient because it is impossible to perform the path integral. However in few special symmetric cases it is easy to perform path integral using ampere's law.

d) Ampere's circuital law is applicable for conductors carrying steady current.

e) Ampere's circuital law is analogous to Gauss law.

f) Ampere's circuital law is not independent of Biot-Savart's law. It can be derived from Biot-Savart's law. Its relation with Biot-Savart's Law is similar to the relation between Gauss Law and Coulomb's Law in electrostatics.

WE -1:

Eight wires cut the page perpendicular to the points shown. Each wire carries current i_0. Odd currents are out of the page and even current into the page. Find the line integral $\oint \vec{B}.\vec{dl}$ along the loop.

Solution : According to Ampere's, circuital law

$$\oint \vec{B}.\vec{dl} = \mu_0 i_{enclosed}$$

$$\oint \vec{B}.\vec{dl} = \mu_0 \left[i_8 - i_5 + i_2 + i_4\right]$$

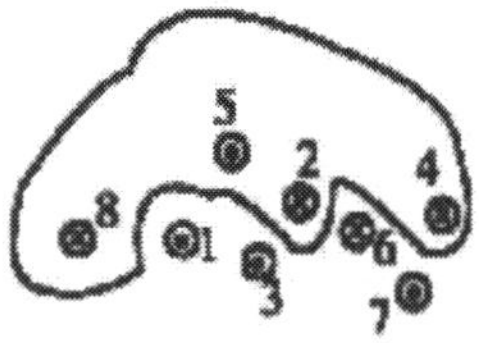

since, all the wires carry same current of i_0, we have $\oint \vec{B}.\vec{dl} = 2\mu_0 i_0$.

3.2 INTENSITY OF MAGNETIC INDUCTION (B) NEAR A LONG STRAIGHT CONDUCTOR :

Consider an infinitely long wire carrying current i as shown in figure. P is a point at a perpendicular distance r from the conductor. The magnetic induction field produced by the conductor is radially symmetric i.e., magnetic lines of force are concentric circles centred at the conductor. The tangent drawn to the line of force at any point gives the direction of magnetic induction field $\vec{B}$ at that point. $\vec{dl}$ is a small element on the circle of radius r and angle between $\vec{B}$ and $\vec{dl}$ is 0^0 every where on this path. From Ampere's circuital law

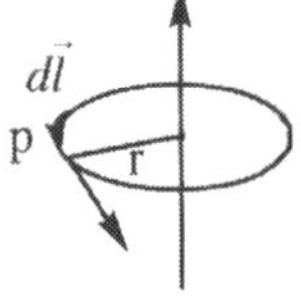

$$\oint \vec{B}.\vec{dl} = \mu_0 i$$

$$\oint B dl \cos 0^0 = \mu_0 i$$

$$B\oint dl = \mu_0 i$$

$$B(2\pi r) = \mu_0 i \Rightarrow B = \frac{\mu_0 i}{2\pi r}$$

Here r must be much less than the length of conductor.

Magnetic induction at any point along the axis of ocnductor is zero.

4.1 MAGNETIC FIELD DUE TO A CURRENT ELEMENT - BIOT-SAVART LAW :

All magnetic fields are due to currents (or moving charges) and due to intrinsic magnetic moments of particles. Here, the relation between current and the magnetic field produced by the current is given by the Biot-Savart's law.

Biot and Savart conducted several experiments and established the relation between magnetic induction $\left(\vec{B}\right)$ and current(i).

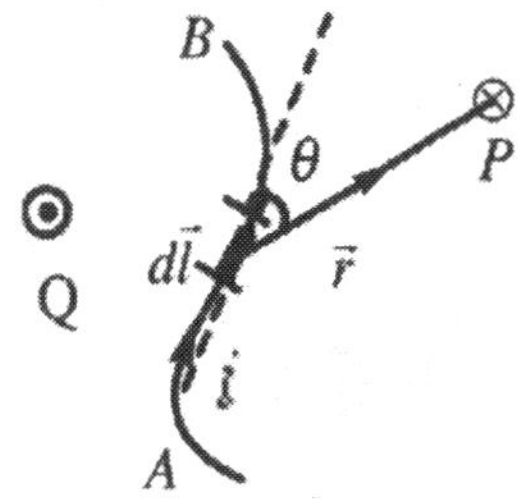

The above figure shows a finite conductor AB carrying current 'i'. Consider an infinitesimal element $\overrightarrow{dl}$ of the conductor. The magnetic field $\overrightarrow{dB}$ due to this element is to be determined at point 'P' which is at a distance 'r' from it. Let θ be the angle between $i\overrightarrow{dl}$ and the radius vector $\vec{r}$.

According to Biot-Savart's law, the magnitude of magnetic induction dB.

a) is directly proportional to the current(i) flowing through the element i.e.,

$dB\alpha i \rightarrow (i)$

b) is directly proportional to the length (dl) of the element i.e.,

$dB\alpha dl \rightarrow (ii)$

c) is directly proportional to the sine of the angle (θ) between length of the element and the line joining the element to the point P.

$dB\alpha \sin\theta \rightarrow (iii)$

d) is inversely proportional to the square of the distance (r) of the point from the element.

$$dB\alpha \frac{1}{r^2} \rightarrow (iv)$$

If the conductor is in vacuum (or) air then

$$dB = \frac{\mu_0}{4\pi}\frac{idl\sin\theta}{r^2}$$

Here $\frac{\mu_0}{4\pi}$ is the proportionality constant and μ_0 is called as permeability of free space or air.

The value of μ_0 is $4\pi \times 10^{-7}\, tesla - m/A$

The above equation gives the magnitude of the magnetic field produced due to small current element at a distance 'r' from it.

If current flows in the direction as shown in the figure, the direction of dB at P is directed perpendicular to the plane of the paper in the inward direction.

In vector form the above equation can be written

$$\overrightarrow{dB} = \frac{\mu_0}{4\pi}\frac{i\overrightarrow{dl}\times\bar{r}}{r^3}$$

The resultant field at P due to the entire conductor can be obtained by integrating the above equation.

$$B = \int_A^B \frac{\mu_0 i}{4\pi}\frac{\overrightarrow{dl}\times\vec{r}}{r^3}$$

4.2 MAGNETIC FIELD DUE TO A STRAIGHT CURRENT CARRYING WIRE :

Consider a straight conductor carrying current 'i'. Let 'P' be a point at a perpendicular distance 'd' from the conductor.

Let 'dy' be a small current element at a distance 'r' from 'P'.

According to Biot-Savart's law, the magnetic induction at P due to the small element is

$$dB = \frac{\mu_0}{4\pi}\frac{idy\sin\theta}{r^2}$$

As every element of the wire contributes to $\vec{B}$ in the same direction, the magnetic induction due to the entire conductor is

$$B = \int dB = \frac{\mu_0 i}{4\pi}\int\frac{dy.\sin\theta}{r^2}$$

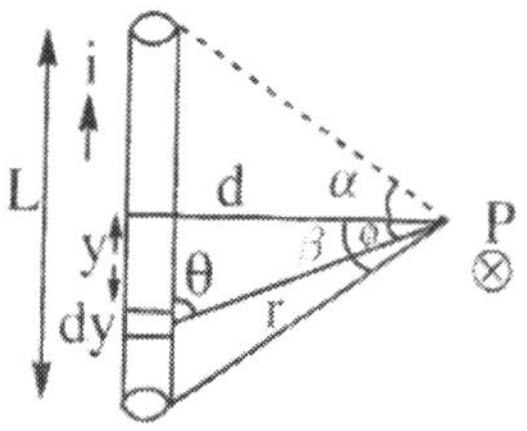

$$\tan\varnothing = y/d$$

$$y = d\tan\phi \Rightarrow dy = d(\sec^2\phi)d\phi$$

$$\frac{r}{d} = \sec\phi \qquad r = d\sec\phi$$

$$B = \frac{\mu_0 i}{4\pi}\int\frac{d(\sec^2\phi).d\phi\sin(90^0-\phi)}{d^2\sec^2\phi}$$

$$[\because \theta = (90-\phi)]$$

$$B = \frac{\mu_0 i}{4\pi}\int_{-\beta}^{\alpha}\frac{d(\sec^2\phi)d\phi\cos\phi}{d^2\sec^2\phi}$$

$$B = \frac{\mu_0 i}{4\pi d}\int_{-\beta}^{\alpha}\cos\phi d\phi$$

($-\beta$ is taken because the angle is measured anti clockwise)

$$B = \frac{\mu_0 i}{4\pi d}(\sin\alpha + \sin\beta)$$

Similary B is given as

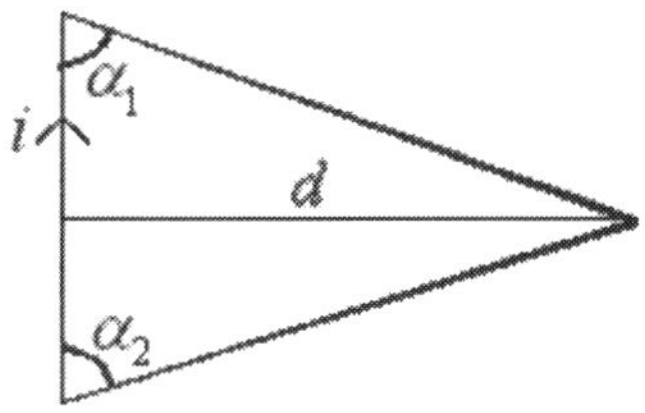

$$B = \frac{\mu_0 i}{4\pi d}[\cos\alpha_1 + \cos\alpha_2]$$

4.3 Special Cases :

i) If the point is along the length of the wire (but not on it then as $\vec{dl}$ and $\vec{r}$ will be either parallel (or) antiparallel i.e. $\theta = 0\ (or)\ \pi$

so $\vec{dl} \times \vec{r} = 0$ and hence $\vec{B} = \int_A^B \vec{dB} = \vec{0}$

ii) If a point is at a perpendicular distance d from the wire then the magnetic field B varies inversely with distance d.

$$B\alpha \frac{1}{d}$$

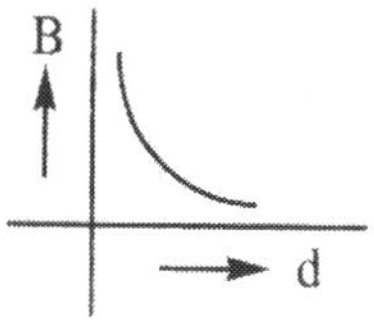

iii) If the wire is of finite length 'L' and the point is on its perpendicular bisector, at a distance 'd' from the wire, i.e $\alpha = \beta$

$$B = \frac{\mu_0}{4\pi}\frac{2i}{d}\sin\alpha \text{ with } \sin\alpha = \frac{L}{\sqrt{L^2 + 4d^2}}$$

iv) If wire is of infinite length and the point P lies at a distance 'd' from the wire which is at a large distance from its ends as shown in figure, $\alpha = \beta = \pi / 2$

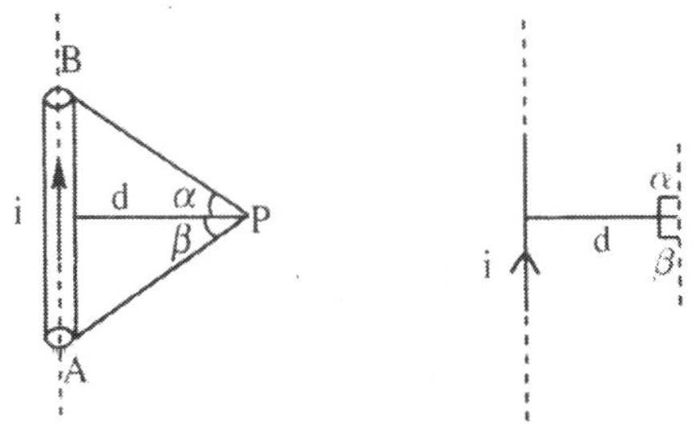

$$B = \frac{\mu_0}{4\pi}\frac{i}{d}(2) = \frac{\mu_0}{4\pi}\frac{2i}{d} = \frac{\mu_0 i}{2\pi d}$$

v) At a point away from the conductor and near the edge of conductor

$\alpha = 90^0, \qquad \beta = 0^0$

$$B = \frac{\mu_0}{4\pi}\frac{i}{d}$$

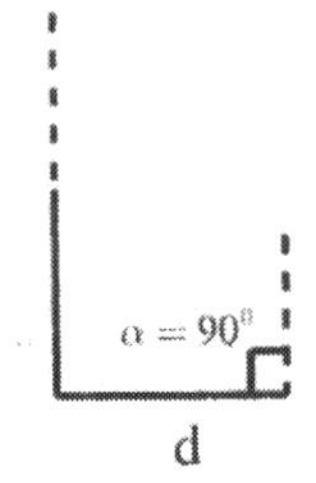

vi)a) Magnetic induction at the centre of current carrying wire bent in the form of square of side 'a' is

$$B_{net} = 4B_{side}$$

$$B_{net} = 4\frac{\mu_0}{4\pi} \times \frac{i}{a/2}(\sin 45^0 + \sin 45^0)$$

$$B = 8\sqrt{2}\left(\frac{\mu_0 i}{4\pi a}\right) \otimes$$

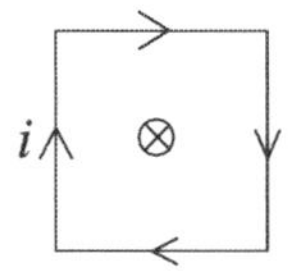

b) Magnetic induction at the centroid of current carrying wire bent in the form of equilateral triangle of side 'a' is

$$B_{net} = 3B_{eachside}$$

$$B_{net} = 3\frac{\mu_0}{4\pi} \times \frac{i}{r}(\sin 60^0 + \sin 60^0)$$

$$\left(where \;\; r = \frac{a}{2\sqrt{3}}\right)$$

$$B = 18\frac{\mu_0 i}{4\pi a} \otimes$$

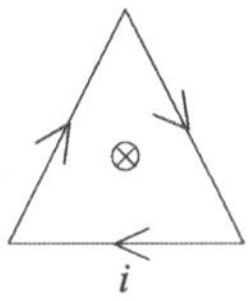

c) Magnetic induction at the centre of current carrying wire bent in the form of hexagon of side 'a' is given by

$$B_{net} = 6B_{eachside}$$

Here $\alpha = \beta = 30^0$

$$B = 4\sqrt{3}\frac{\mu_0 i}{4\pi a} \otimes$$

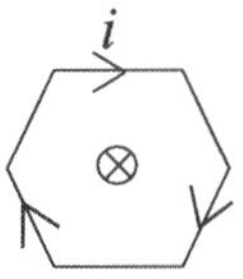

4.4 THE MAGNETIC FIELD DUE TO A LONG STRAIGHT CURRENT CARRYING CONDUCTOR.

a) Taking a circular ampere loop centered to the wire of radius r<R. To find B inside the conductor using ampere's circuital law (ACL), we have

$\oint \vec{B}.\vec{dl} = \mu_0 i'$,

Here $i' = J.\pi r^2$

Where $J = \dfrac{i}{\pi R^2}$

$\Rightarrow \oint Bdl\cos 0 = \mu_0 J\pi r^2$

or $B\oint dl = \mu_0 J\pi r^2$

or $B2\pi r = \mu_0 J\pi r^2$

(or) $B = \dfrac{\mu_0 J}{2} r; \quad r \le R$

$B \alpha r$

b) At a point outside the wire (r>R) $\oint B.dl\cos 0^0 = \mu_0 i'$, Where $i' = i$ because the amperian encloses total current or $B\oint dl = \mu_0 i$ (or) $B2\pi r = \mu_0 i$

$\Rightarrow B = \dfrac{\mu_0 i}{2\pi r}; \quad r \ge R$

$B \alpha \dfrac{1}{r}$

c) B varies linearly inside the conductor and hyperbolically outside the conductor.

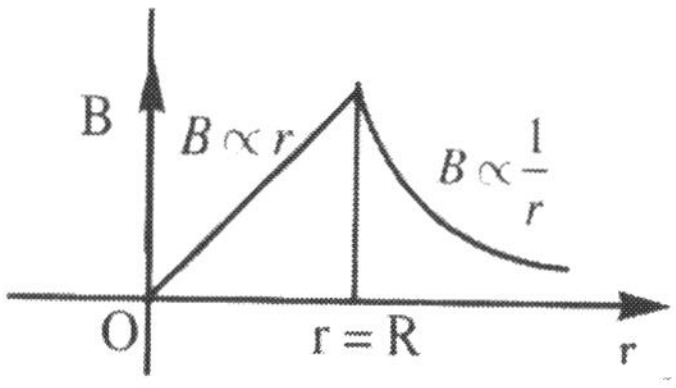

Magnetic induction is maximum at the periphery of the wire

d) The variation of $\vec{B}$ as the function of radial distance r due to a hollow cylinder carrying a current i_0.

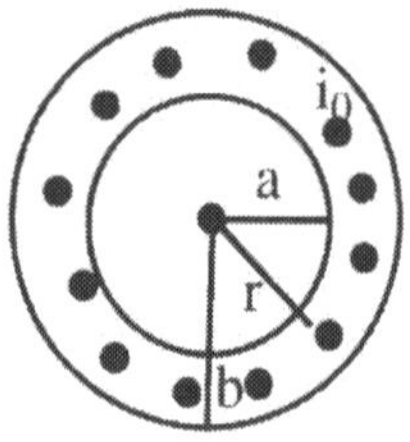

Taking a circular amperian loop of radius r(>a) and applying ACL,

$\int \vec{B}.\vec{dl} = \mu_0 i$; $\quad B2\pi r = \mu_0 i,$

Where $i = \frac{i_0}{\pi(b^2 - a^2)}.\pi(r^2 - a^2)$

$= \frac{i_0(r^2 - a^2)}{b^2 - a^2}$

then $B = \frac{\mu_0 i_0 (r^2 - a^2)}{2\pi(b^2 - a^2)r} \quad a \le r \le b$

B=0 for $r \le a$ (as because i=0)

for $r > b \qquad B = \frac{\mu_0 i_0}{2\pi r}$

e) **For thin hollow cylinder**

i) $B_{inside} = 0 \qquad$ ii) $B_{surface} = \frac{\mu_0 i}{2\pi R}\ (r = R)$

iii) $B_{outside} = \frac{\mu_0 i}{2\pi r}\ (r > R)$

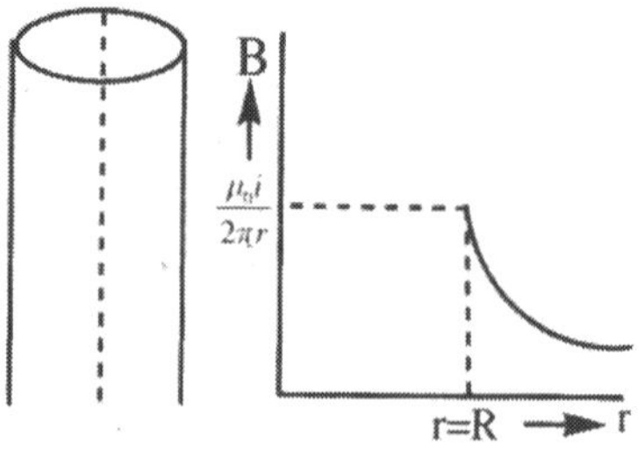

f) **Work done to move a unit north pole through a small distance dl' along the tangent at a distance 'r' away from current carrying conductor**

$\Rightarrow dw = \vec{F}.\vec{dl}$

$\vec{F} = m\vec{B} = \vec{B} (\because m = 1)$

but $dw = \vec{F}.\vec{dl} \Rightarrow dw = \vec{B}.\vec{dl}$

Total work done in moving it once around the conductor. $W = \oint dw$

$W = \oint \vec{B}.\vec{dl}$

But from Ampere's circuital law

$\oint \vec{B}.\vec{dl} = \mu_0 i \qquad \Rightarrow W = \mu_0 i$

If a pole of strength 'm' is rotated for 'n' times around the current carrying conductor, then the work done is

$W = \mu_0 i \times nm$

Here $W \neq 0$, the magnetic field produced by crrent carrying conductor is a non-conservative field.

WE -2:

Find the magnetic induction due to a straight conductor of length 16cm carrying current of 5A at a distance of 6cm from the midpoint of conductor.

Solution :

$$B = \frac{\mu_0}{4\pi}\frac{I}{r}(\sin\theta + \sin\theta)$$

but $\sin\theta = \frac{8}{10} = \frac{4}{5}$

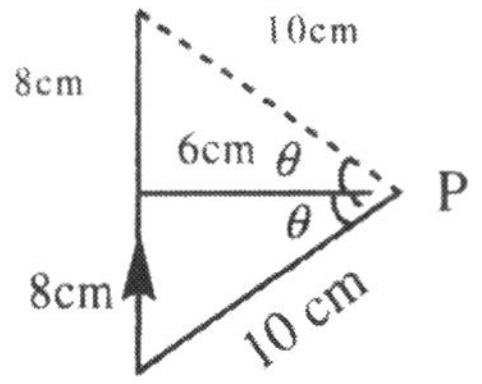

$$B = 10^{-7} \times \frac{5}{6\times10^{-2}} \times 2 \times \frac{4}{5}$$

$$= \frac{40}{3}\mu T$$

WE -3:

If a straight conductor of length 40cm bent in the form of a square and the current 2A is allowed to pass through square, then find the magnetic induction at the centre of the square loop

Solution :

$B_{net} = 4B_{side}$

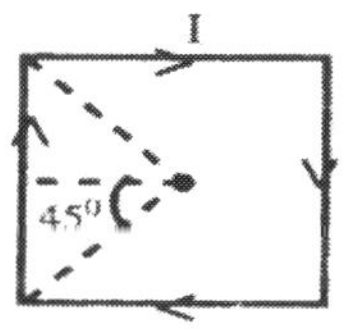

$$B_{net} = 4\frac{\mu_0}{4\pi} \times \frac{I}{L/2}(\sin 45^0 + \sin 45^0)$$

$$= 4X \frac{\mu_0}{4\pi} \times \frac{I}{L/2} (\sqrt{2}) = \frac{\mu_0}{4\pi} \frac{8\sqrt{2}I}{L}$$

$$= 10^{-7} \times 8\sqrt{2} \times 2 \times 10 = 16\sqrt{2}\mu T$$

WE -4:

If a thin uniform wire of length 1m is bent into an equilateral traiangle and crries a current of $\sqrt{3}A$ in anitclockwise direction, find the net magnetic induction at the centroid

Solution :

$$B_{net} = 3B_{eachside}$$

$$B_{net} = 3\frac{\mu_0}{4\pi} \times \frac{I}{r} (\sin 60^0 + \sin 60^0)$$

$$\left(\because r = \frac{a}{2\sqrt{3}}\right) \quad = 3 \times \frac{\mu_0}{4\pi} \frac{I}{r} (2\sin 60^0) \quad = 3 \times \frac{\mu_0}{4\pi} \frac{I(2\sqrt{3})}{a} 2 \times \frac{\sqrt{3}}{2} = 18 \frac{\mu_0}{4\pi} \frac{I}{a}$$

$$B = 18 \times 10^{-7} \times \frac{\sqrt{3}}{1/3} = 54\sqrt{3} \times 10^{-7} T$$

WE -5:

A large straight current carrying conductor is bent in the form of L shape. Find $\vec{B}$ at P.

Solution : Let us divide the conductor into two semi infinite segments 1 and 2. Then, induction at P is

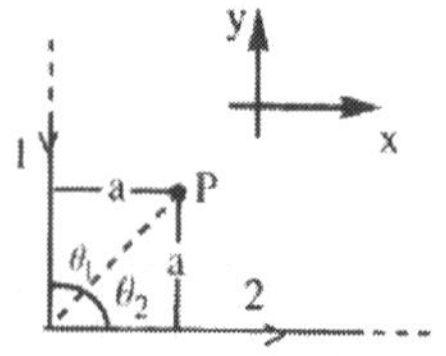

$$\vec{B} = \vec{B}_1 + \vec{B}_2 \qquad ..i$$

$$\vec{B}_1 = \frac{\mu_0 i}{4\pi a} (\sin(90^0 - \theta_1) + \sin 90^0)\hat{k} \qquad ..ii$$

$$\vec{B}_2 = \frac{\mu_0 i}{4\pi a} (\sin(90^0 - \theta_2) + \sin 90^0)\hat{k} \qquad ..iii$$

then $\vec{B} = \frac{\mu_0 i}{4\pi a} (\cos\theta_1 + \cos\theta_2 + 2)\hat{k}$, where $\cos\theta_1 = \cos\theta_2 = \frac{1}{\sqrt{2}}$

Hence, $\vec{B} = (2+\sqrt{2})\frac{\mu_0 i\hat{k}}{4\pi a}$

WE -6:

Infinite number of straight wires each carrying current I are equally placed as shown in the figure. Adjacent wires have current in opposite direction. Find net magnetic field at point P ?

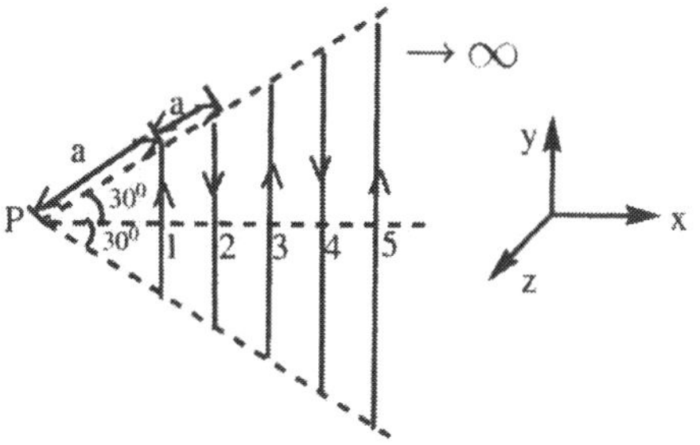

Solution :

$$B_{net} = \frac{\mu_0 I}{4\pi}(\sin 30^0 + \sin 30^0)\hat{k}\left[\frac{1}{d} - \frac{1}{2d} + \frac{1}{3d} - \frac{1}{4d} ..... + \infty\right]$$

$$\left(Where\ d = a\cos 30 = \frac{\sqrt{3}a}{2}\right)$$

$$\therefore B_{net} = \frac{\mu_0 I}{2\sqrt{3}\pi a}\hat{k}\left[1 - \frac{1}{2} + \frac{1}{3} - \frac{1}{4} + ...\infty\right]$$

$$= \frac{\mu_0 I}{2\sqrt{3}\pi a} In2\hat{k} = \frac{\mu_0 I}{4\pi}\frac{In4}{\sqrt{3}a}\hat{k}$$

WE -7:

Find the magnetic field at P due to the arrangement shown

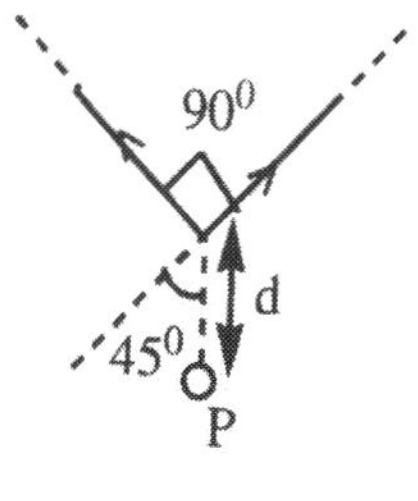

Solution : $B_{net} = 2 \times \frac{\mu_0 I}{4\pi r}\left(\sin\frac{\pi}{2} - \sin\frac{\pi}{4}\right)$

here $r = \frac{d}{\sqrt{2}}$

$$B_{net} = \frac{\mu_0 I}{\sqrt{2}\pi d}\left(1 - \frac{1}{\sqrt{2}}\right)$$

WE -8:

A pair of stationary and infinitely long bent wires are placed in the x-y plane as shown in figure. The wires carry current of 10 ampere each as shown. The segment L and M are along the x-axis. The segment P and Q are parallel to the Y-axis such that OS=OR=0.02m. Find the magnitude and direction of the magnetic induction at the origion O.

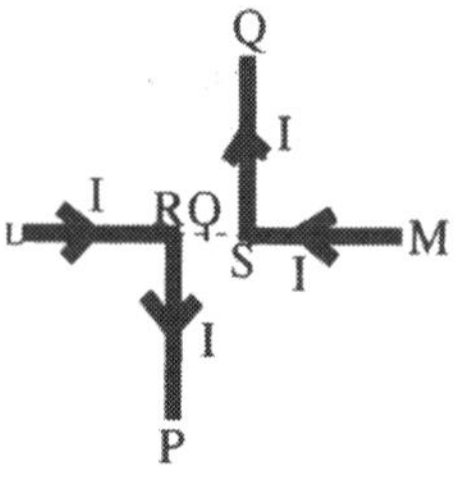

Solution : Since point O is along the length of segment L and M the field at O due to these two segments will be zero

$\therefore$ Magnetic field at O is due to QS and RP.

$$\therefore B_{SQ} = \frac{\mu_0}{4\pi} \times \frac{I}{OS} = 10^{-7} \times \frac{10}{0.02} \odot$$

$$B_{RP} = \frac{\mu_0}{4\pi} \times \frac{I}{OR} = 10^{-7} \times \frac{10}{0.02} \odot$$

$$\therefore B_0 = B_{SQ} + B_{RP} = 10^{-7} \times \frac{10}{0.02} \times 2 = 10^{-4} T \odot$$

WE -9:

An equilateral triangle of side length *l* is formed from a piece of wire of uniform resistance. The current I is as shown in figure. Find the magnitude of the magnetic field at its centre O.

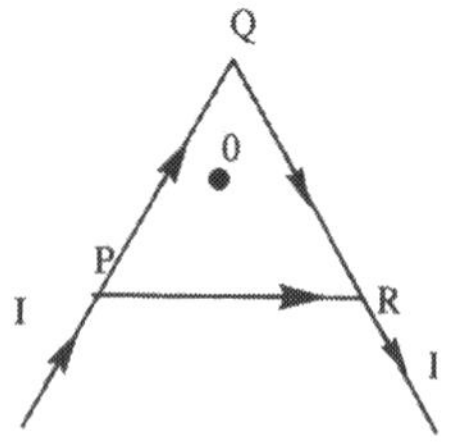

Solution :

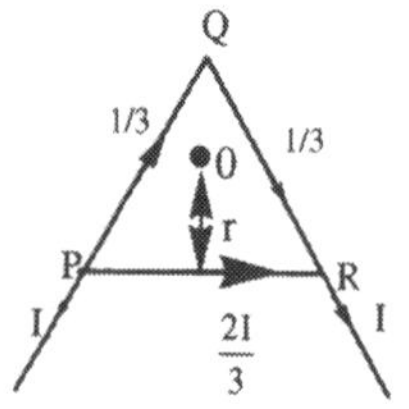

The magnetic field induction at O due to current through PR is

$$B_1 = \frac{\mu_0}{4\pi} \frac{2l/3}{r} \left[\sin 30^0 + \sin 30^0 \right]$$

$$= \frac{\mu_0}{4\pi} \frac{2l}{3r} \odot \qquad \text{(directed outside)}$$

The magnetic field induction at O due to current through PQR is

$$B_2 = 2\times\frac{\mu_0(l/3)}{4\pi r}\left[\sin 30^0 + \sin 30^0\right] = \frac{\mu_0}{4\pi}\frac{2l}{3r}\otimes \quad \text{(directed inside)}$$

$\therefore$ Resultant magnetic induction at O

$\Rightarrow B_1 - B_2 = 0$

WE -10:

A non planar loop of conducting wire carrying a current I is placed as shown in the figure each of the straight sections of the loop is of length 2a. Find the direction of magnetic field due to this loop at the point P (a,0,a)

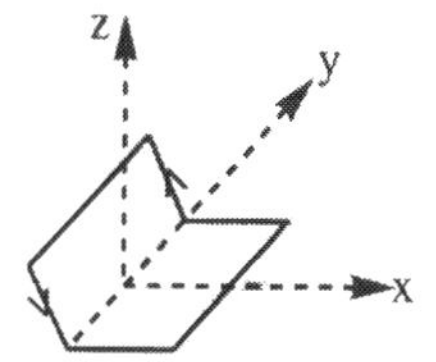

Sol : The magnetic field at P (a,0,a) due to the loop is equal to the vector sum of the magnetic fields produced by loops ABCDA and AFEBA as shown in the figure

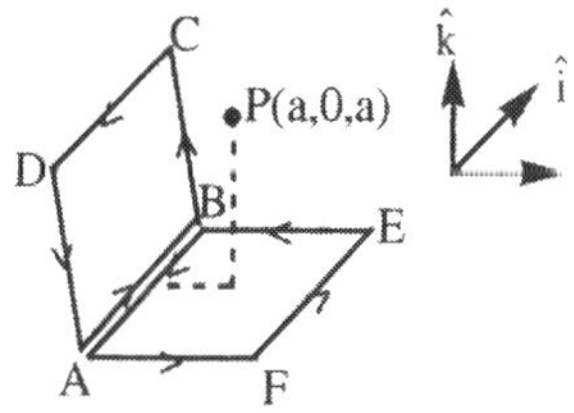

Magnetic field due to ABCDA will be along $\hat{i}$ and due to loop AFEBA, along $\hat{k}$. Magnitude of magnetic field due to both the loops will be equal.

There fore, direction of resultant magnetic field at P will be $\frac{1}{\sqrt{2}}(\hat{i}+\hat{k})$.

5.1. MAGNETIC FIELD AT THE CENTRE OF A CIRCULAR COIL CARRYING CURRENT

Consider a circular coil of radius R carrying a current i in clockwise direction. Consider any small element dl of the wire. The magnetic field at the centre O due to the current element $id\vec{l}$ is

$$dB = \frac{\mu_0}{4\pi}\frac{id\vec{l}\times\vec{R}}{R^3}$$

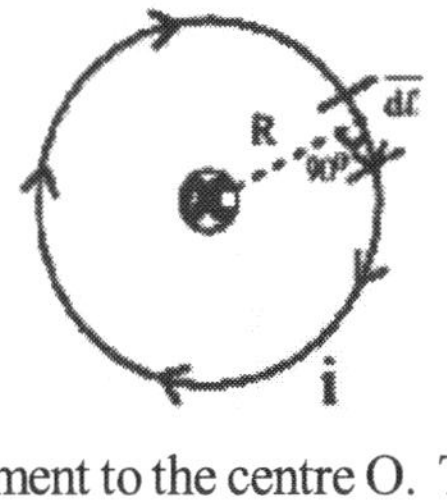

Where $\vec{R}$ is the vector joining the element to the centre O. The direction of this field is perpendicular to the plane of the diagram and is going into it.

The magnitude of the magnetic field is

$$dB = \frac{\mu_0}{4\pi}\frac{idl}{R^2}$$

As the fields due to all such elements have the same direction, the net field is also in this direction. It can, therefore, be obtained by integrating equation (i) under proper limits. Thus,

$$B = \int dB = \int \frac{\mu_0 i}{4\pi R^2} dl$$

If the coil has N turns $\int dl = 2\pi RN$

$$= B = \frac{\mu_0 i}{4\pi R^2}\int dl = \frac{\mu_0 i}{4\pi R^2} \times 2\pi RN = \frac{\mu_0 iN}{2R} \otimes$$

If the current is in clock wise direction, then the magnetic field produced is normally inwards and the face of the coil behaves as south pole.

If the current is in anti clock wise direction, then the magnetic field produced is normally outwards and the face of the coil behaves as north pole.

5.2 FIELD AT AN AXIAL POINT OF A CIRCULAR LOOP :

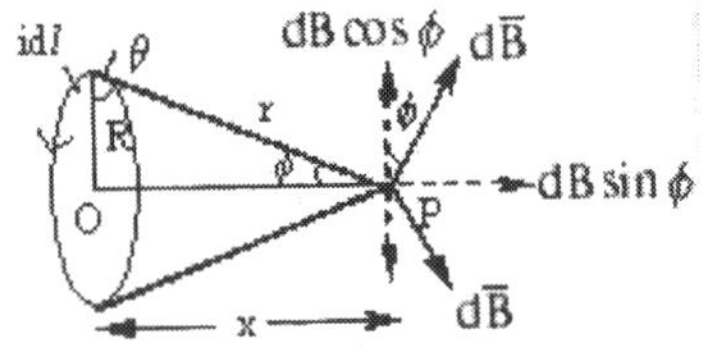

Consider a circular poop of radius R, carrying current in in yz plane with centre at origin O. Let P be a point on the axis of the loop at a distance 'x' from the centre 'O' of the loop.

Consider a conducting element dl of loop. According to Biot-Savart's law, the megnitude of magnetic field due to the current element is

$$\left|\overrightarrow{dB}\right| = \frac{\mu_0}{4\pi}\frac{\left|i\overrightarrow{dl} \times \vec{r}\right|}{r^3} \text{ where } r = \sqrt{x^2 + R^2}$$

Here the element dl is in yz plane where as the displacement vector $\bar{r}$ from $\overline{dl}$ to the point p is in xy plane. So $\left|i\overrightarrow{dl} \times \vec{r}\right| = idl \times r$

$$\left|\overrightarrow{dB}\right| = \frac{\mu_0}{4\pi}\frac{idl \times r}{r^3} = \frac{\mu_0}{4\pi}\frac{idl}{r^2}$$

The direction of $\overline{dB}$ is perpendicular to the plane formed by $\bar{r}$ and $\overline{dl}$.

In case of a point P on the axis of circular coil, for every current element 'idl' there is a symmetrically situated opposite element. The componenet of the field **dB** perpendicular to the axis cancel each other while component of the field **dB** along the axis add up and contributes to the net magnetic field.

i.e., $B = \int dB \sin\phi = \frac{\mu_0}{4\pi}\int \frac{idl \sin\theta}{r^2} \sin\phi$

Here angle θ between the element $\vec{dl}$ and $\vec{r}$ is $\pi/2$ every where and r is same for all elements and also $\sin\phi = (R/r)$ so,

$$B = \frac{\mu_0}{4\pi}\int \frac{idl \sin\theta}{r^2} \sin\phi = \frac{\mu_0}{4\pi}\int \frac{idl \sin 90^0}{r^2}\frac{R}{r}$$

$$= \frac{\mu_0}{4\pi}\frac{iR}{r^3}\int dl$$

for a loop $\int dl = 2\pi R$ and as $r^3 = (x^2 + R^2)^{3/2}$

$$B = \frac{\mu_0}{4\pi}\frac{2\pi i R^2}{(x^2 + R^2)^{3/2}} = \frac{\mu_0 i R^2}{2(x^2 + R^2)^{3/2}}$$

The direction of magnetic field B is along the axis of the loop.

i) The magnetic field B varies non linearly with distance **x** from centre as shown in figure.

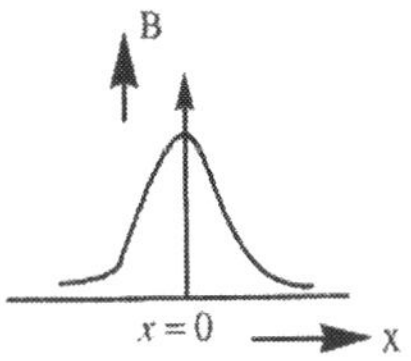

For a coil having N turns, $\int dl = 2\pi RN$

so, $B = \frac{\mu_0 N i R^2}{2(x^2 + R^2)^{3/2}}$.

It is maximum when $x^2 = 0$, i.e., at the centre of the coil whose value is given by

$$B = \frac{\mu_0}{4\pi}\frac{2\pi NI}{R} = \frac{\mu_0 Ni}{2R}$$

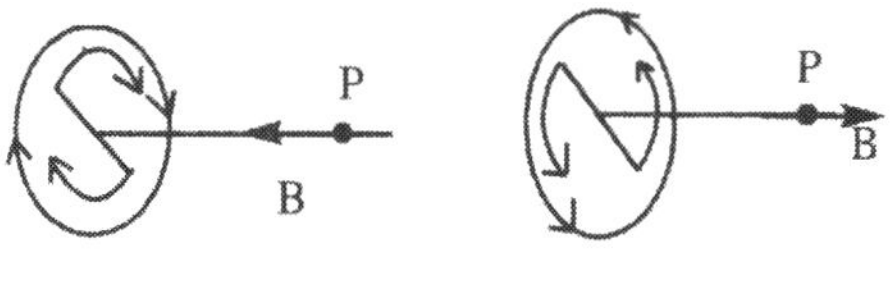

ii) if $x >> R$

$$B = \frac{\mu_0}{4\pi}\frac{2\pi NIR^2}{x^3} = \frac{\mu_0}{4\pi}\frac{2NIA}{x^3}$$

Where $A = \pi R^2$, area of the coil.

5.3 CIRCULAR CURRENT LOOPAS MAGNETIC DIPOLE :

From the above expression $B = \frac{\mu_0}{4\pi}\frac{2NIA}{x^3}$

Comparing with $B = \frac{\mu_o}{4\pi}\frac{2M}{x^3}$

a) Magnetic moment of the circular current carrying coil is M = NiA;

b) M is independent of shape of the coil

$\therefore$ Current loop behaves like a magnetic dipole with poles on either side of its face and it is known as "magnetic shell".

c) SI unit of magnetic moment (M) is $A-m^2$ and dimensional formula is IL^2.

d) Magnetic moment of a curent loop is a vector perpendicular to the plane of the loop and the direction is given by right hand thumb rule.

5.4 MAGNETIC DIPOLE MOMENT OF A REVOLVING ELECTRON :

Consider an electron revolving in a circular path of radius r around a nucleus with uniform speed v. The current in the orbit is

$$i = \frac{e}{T} = \frac{e}{2\pi r / v} = \frac{ev}{2\pi r}$$

Magnetic dipole moment of a revolving electron is $\mu = iA = \frac{ev}{2\pi r} \times \pi r^2 = \frac{evr}{2}$

Magnetic dipole moment of a revolving electron in the first orbit of hydrogen atom is called Bohr magneton (μ).

From Bohr second postulates, for an electron revolving in first orbit of hydrogen atom.

$$m_e vr = \frac{h}{2\pi}(n=1)$$

Where h = Planck's constant, m_e = mass of electron

$$\mu = \frac{evr}{2} = \frac{e}{2}\frac{h}{2\pi m_e} = \frac{eh}{4\pi m_e}$$

$$(\mu)_{\min} = \frac{e}{4\pi m_e}h$$

$$= \frac{1.60\times10^{-19}\times6.63\times10^{-34}}{4\times3.14\times9.11\times10^{-31}}$$

$$= 9.27\times10^{-24}\,Am^2$$

This value is called the . **Bohr magneton.**

5.5 Special cases :

i) For an arc shaped conductor carrying current subtending an angle θ at the centre.

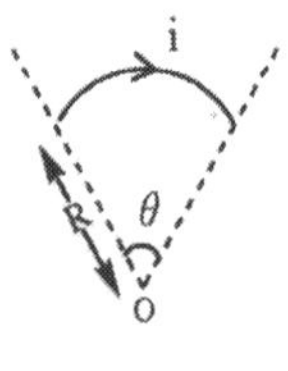

$$B = \frac{\mu_0 i}{2R}\frac{\theta}{2\pi} \otimes$$

$\therefore$ Magnetic induction at the centre $B = \frac{\mu_0 i\theta}{4\pi R} \otimes$

ii) For a quadrant circular wire carrying current.

$\theta = 90^0$

Magnetic induction at the centre $B = \frac{\mu_0 i}{8R} \otimes$

iii) If B_0 is magnetic induction at the centre of a circular current carrying coil of radius R having N turns and B_A is magnetic induction at a point on the axis of it at a distance x from centre then $B_A = \frac{B_0}{\left(1+\frac{x^2}{R^2}\right)^{3/2}}$

Proof: $B_0 = \frac{\mu_0 Ni}{2R}$ and $B_A = \frac{\mu_0 NiR^2}{2(R^2+x^2)^{3/2}}$

$$\Rightarrow B_A = \frac{\mu_0 Ni}{2R\left(1+\frac{x^2}{R^2}\right)^{3/2}} \Rightarrow B_A = \frac{B_0}{\left(1+\frac{x^2}{R^2}\right)^{3/2}}$$

$$B_A = B_0\left[1-\frac{3x^2}{2R^2}\right]$$

iv) If a particle of charge q moves in a circular path of radius r with a velocity v, then the magnetic induction at the centre of circular loop

$$B = \frac{\mu_0 i}{2r} = \frac{\mu_0}{2r} \times \frac{q\text{v}}{2\pi r} = \frac{\mu_0}{4\pi}\frac{q\text{v}}{r^2}$$

If f is the frequency of rotation

$$B = \frac{\mu_0}{2r} \times qf$$

If ω is the angular velocity, then

$$B = \frac{\mu_0}{2r} \times \frac{q\omega}{2\pi} = \frac{\mu_0}{4\pi}\frac{q\omega}{r}$$

v) A charge 'q' is moving with a velocity of 'v'. Then the expression of magnetic induction due to this charge at a position vector $\vec{r}$ from the charge is

Biot - Savart Law for a current element is

$$d\vec{B} = \frac{\mu_0 i\,\vec{dl} \times \vec{r}}{4\pi r^3}$$

If a charged particle of charge q and undergoes a displacement $\vec{dl}$ during a time dt put $i = \frac{q}{dt}$.

or $i\vec{dl}\times\vec{r}=\frac{q\vec{dl}}{dt}\times\vec{r}$

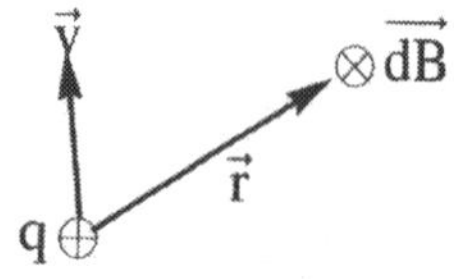

Putting $\frac{dl}{dt}=v$

$i\vec{dl}\times\vec{r}=q(\bar{v}\times\bar{r})$

Using the above equations, $d\vec{B}=\frac{\mu_0 q(\vec{v}\times\vec{r})}{4\pi r^3}$.

vi) a)When a wire of length 'l' carrying current 'i' is bent in a circular loop of 'n' turns then the magnetic induction at the centre of the loop is $B=\frac{\mu_0 ni}{2r}=\frac{\mu_0\pi n^2 i}{l}$ $(\because n\times 2\pi r=l)$

b) The same wire of length 'l' carrying current 'i' is first bent into a circular coil with n_1 turns and then into another circular coil with n_2 turns. If B_1, B_2 are magnetic inductions at their centres in the two cases, then

c) $\frac{B_1}{B_2}=\left(\frac{n_1}{n_2}\right)^2$

d) If r_1 and r_2 are radii of turns of the coil in the above case, then ratio of magnetic induction is $\frac{B_1}{B_2}=\left(\frac{r_2}{r_1}\right)^2$

e) If two circular coils are connected in series, then the ratio of magnetic induction at their centres is

$\frac{B_1}{B_2}=\left(\frac{n_1}{n_2}\right)\left(\frac{r_2}{r_1}\right)$

f) If the two coils are made up of same wire and connected in parallel, then the ratio of the magnetic induction at their centres is $\frac{B_1}{B_2}=\left(\frac{r_2}{r_1}\right)^2$.

vii)a) For semi circular wire carrying current.

$\theta=180^0$

Magnetic induction at the centre $B=\frac{\mu_0 i}{4R}\otimes$

b) To a circular wire, two straight wires are attached as shown. When current is passed through it the magnetic field at the centre is zero.

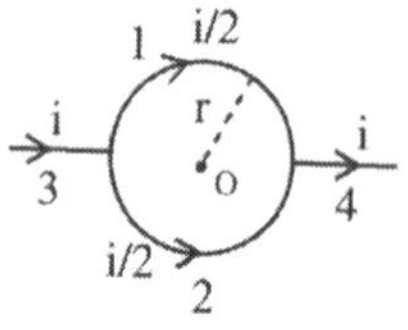

$$B_1 = \frac{\mu_0\left(2\right)}{4R} \otimes$$

$$B_3 = B_4 = 0$$

$$B_2 = \frac{\mu_0\left(\frac{i}{2}\right)}{4R} \odot \quad \therefore B_{net}\, at\, O = Zero$$

c) To a circular wire, two straight wires are attached as shown. When current is passed through it the magnetic field at the centre.

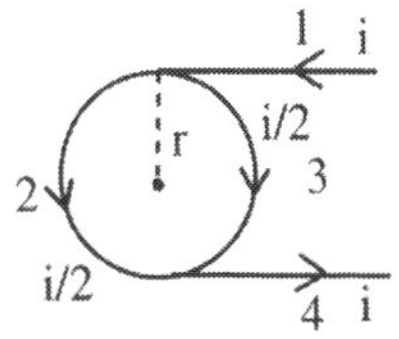

$$B_1 = \frac{\mu_0 i}{4\pi r} \odot$$

$$B_2 = \frac{\mu_0\left(\frac{i}{2}\right)}{4R} \odot$$

$$B_3 = \frac{\mu_0\left(\frac{i}{2}\right)}{4R} \otimes$$

$$B_4 = \frac{\mu_0 i}{4\pi r} \odot \qquad \vec{B}_{net} = \vec{B}_1 + \vec{B}_2 + \vec{B}_3 + \vec{B}_4$$

$$B_{net} = \frac{\mu_0 i}{2\pi r} \odot$$

d) The upper and lower halves of the ring have resistances R_1 and R_2. Two straight wires are connected to it as shown. The magnetic induction at the centre of the ring is

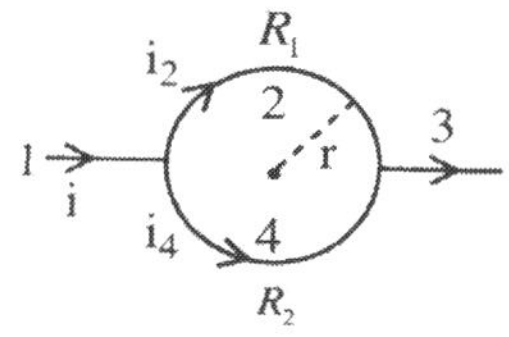

$$B_1 = B_3 = 0$$

$$B_2 = \frac{\mu_0 i_2}{4r} \otimes$$

$$B_4 = \frac{\mu_0 i_4}{4r} \odot$$

Since R_1 and R_2 are parallel to each other

$$i_2R_1 = i_4R_2; \qquad i_2 = \frac{i}{R_1 + R_2} \times R_2$$

$$i_4 = \frac{i}{R_1 + R_2} \times R_1$$

$$\vec{B}_{net} = \vec{B}_1 + \vec{B}_2 + \vec{B}_3 + \vec{B}_4$$

$$\vec{B}_{net} = \frac{\mu_0}{4r}(i_2 \sim i_4)$$

e) A straight current carrying conductor is held vertically in earth's magnetic field. It carries current in the upward direction, then the direction of magnetic filed (B) due to it

a) due north of the conductor is towards west $B_{net} = \sqrt{B^2 + B_H^2}$.

b) due west of the conductor is towards south $B_{net} = B - B_H$

c) due south of the conductor is towards east $B_{net} = \sqrt{B^2 + B_H^2}$.

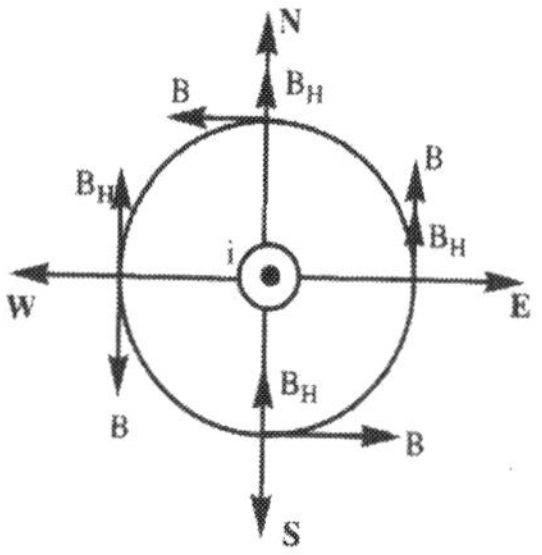

WE-11 :

A 2A current is flowing through a circular coil of radius 10cm containing 100 turns. Find the magnetic flux density at the centre of the coil.

Sol: $B = N\frac{\mu_0 i}{2r}$

$$= 100 \times \frac{2\pi \times 10^{-7} \times 2}{10 \times 10^{-2}}$$

$$= 1.26 \times 10^{-3} Wb/m^2$$

WE-12.

Three rings, each having equal radius R, are placed mutuallly perpendicular to each other and each having its centre at the origin of coordinate system. If current I is flowing through each ring then find the magnitude of the magnetic filed at the common centre.

Sol: 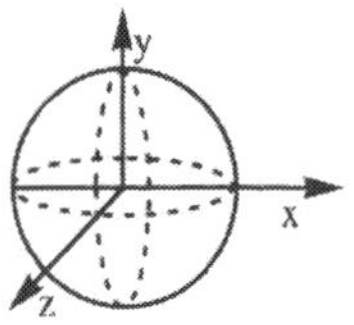

B due to the ring lying in XY-plane is $B_{xy} = \frac{\mu_0 I}{2R}$ along Z-axis.

B due to the ring lying in YZ-plane is $B_{yz} = \frac{\mu_0 I}{2R}$ along X-axis and

B due to the ring lying in XZ-plane is $B_{xz} = \frac{\mu_0 I}{2R}$ along Y-axis.

$\therefore \vec{B}_{net} = \frac{\mu_0 I}{2R}(\hat{i} + \hat{j} + \hat{k})$

$\Rightarrow B_{net} = \sqrt{3}\frac{\mu_0 I}{2R}$

WE-13

Two wires are wrapped over a wooden cylinder to form two co-oxial loops carrying currents i_1 and i_2. If $i_2 = 8i_1$ then find the value of x for B=0 at the origin 0.

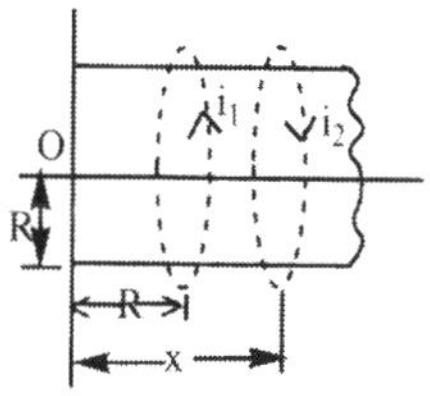

Sol: Magnetic induction at '0' due to 1st loop

$\vec{B}_1 = \frac{\mu_0 i_1 R^2}{2(R^2 + R^2)^{3/2}}$ to left Magnetic induction at '0' due to 2nd loop.

$\vec{B}_2 = \frac{\mu_0 i_2 R^2}{2(R^2 + R^2)^{3/2}}$ to right

$\vec{B}_1 + \vec{B}_2 = 0$

$\Rightarrow \frac{i_1}{(2R^2)^{3/2}} - \frac{i_1}{(R^2 + x^2)^{3/2}}$ and $i_2 = 8i_1$

$\Rightarrow x = \sqrt{7}R$

WE -14:

Two wires wrapped over a conical frame form the loops I and 2. If they produce no net magnetic filed at the apex P, Find the value of $\frac{i_1}{i_2}$.

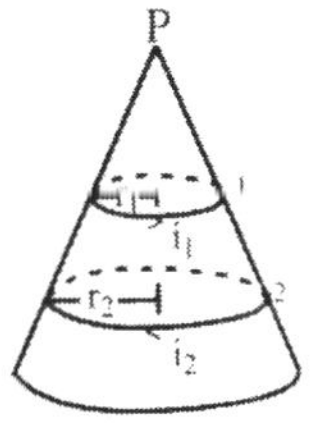

Sol : Magnetic induction due to a loop at apex,

$$B = \frac{\mu_0 i r^2}{2(r^2 + x^2)^{3/2}}$$

But $r^2 + x^2 = l^2 \Rightarrow (r^2 + x^2)^{3/2} = l^3$ where 'l' is slant length $B = \frac{\mu_0 i}{2r}\left(\frac{r}{l}\right)^3$

But $\frac{r}{l} = \sin\theta$ where θ is apex angle, same for both the loops

$\vec{B}_1 + \vec{B}_2 = 0$ (given)

$\Rightarrow B_1 = B_2 \Rightarrow \frac{i_1}{i_2} = \frac{r_1}{r_2}$

WE -15:

A thin insulated wire form a spiral of N=100 turns carrying a current of i=8mA. The inner and outer radii are equal to a=5cm and b=10cm. Find the magnetic field at the centre of the coil.

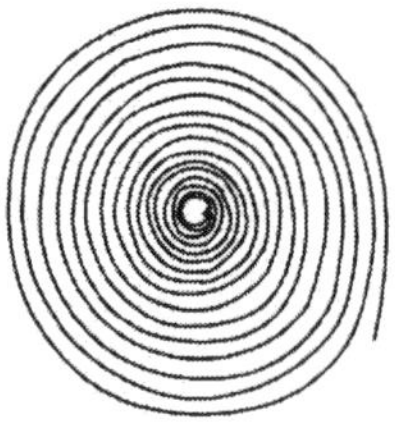

Sol: Let n = no. of turns per unit length along the radial of spiral. Consider a ring of radii x and x + dx.

No.of turns in the ring = ndx.

$$n = \frac{N}{(b - a)}$$

Magnetic field at the centre due to the ring is

$$dB = \frac{\mu_0 (ndx) i}{2x}$$

So net field

$$B = \int dB = \int_a^b \frac{\mu_0 n i dx}{2x} = \frac{\mu_0 n i}{2}\int_a^b \frac{dx}{x}$$

$$\text{or } B = \frac{\mu_0 n i}{2}\ln\frac{b}{a} \text{ or } B = \frac{\mu_0 N i}{2(b - a)}\ln\frac{b}{a}$$

$$= \frac{4\pi \times 10^{-7} \times 100 \times 8 \times 10^{-3}}{2(10 - 5) \times 10^{-2}}\ln\frac{10}{5}.$$

$$B = 6.9 \times 10^{-6} T$$

WE -16:

A plastic disc of radius 'R' has a chanrge 'q' uniformly distributed over its surface. If the disc is rotated with a frequency 'f' about its axis, then the magnetic induction at the centre of the disc is given by

Sol: $dB = \frac{m_0 di}{2x}, dq = \frac{q}{pR^2}(2px)dx$

$di = (dq)f = \frac{2qxdx}{R^2}f$

$dB = \frac{m_0 2qxdx}{2xR^2}\ \Omega\ B = \text{“}_0^R \frac{m_0 2q.dx}{2R^2}(f)$

$B = \frac{m_0 qf}{R^2}(R)\ \Omega\ B = \frac{m_0 qf}{R}$

WE -17:

A charge of 1C is placed at one end of a non conducting rod of length 0.6m. The rod is rotated in a vertical plane about a horizontal axis passing through the other end of the rod with angular frequency $10^4 p$ rad/s. Find the magnetic field at a point on the axis of rotation at a distance of 0.8m from the centre of the path.

Sol: $B = \frac{m_0 ir^2}{2(r^2+x^2)^{3/2}},\ i = \frac{qw}{2p}$

$B = \frac{m_0}{4p}\frac{qwr^2}{(r^2+x^2)^{3/2}}$

WE -18:

Two circular coils made of same material having radii 20 cm & 30 cm have turns 100 & 50 respectively. If they are connected

a) in series

b) in parallel

c) separately across a source of emf find the ratio of magnetic inductions at the centre of circles in each case

Sol: a) $B = \frac{m_0 ni}{2r}$

coils are in series Ω i is same in both

$B a \frac{n}{r}$

$\frac{B_1}{B_2} = \frac{100}{50}\text{ì}\frac{30}{20} = 3:1$

b) coils are parallel Ω potential difference is same $i a \frac{1}{R}$

Where $R = \frac{r(\text{ıp}\ r)}{A}$

Where A is area of cross section of wire which is same for both

$\Rightarrow R\alpha nr; \quad i\alpha\frac{1}{nr}$

but $B_0=\frac{\mu_0 ni}{2r}\Rightarrow B\alpha\frac{n}{r}\times\frac{1}{nr}\Rightarrow B\alpha\frac{1}{r^2}$

$\therefore\frac{B_1}{B_2}=\left(\frac{30}{20}\right)^2=\frac{9}{4}$

c) For the coils, potential difference is same

$i\alpha\frac{1}{R}$ where $R=\frac{\rho(n\pi r)}{A}$; $R\alpha nr$

$I\alpha\frac{1}{nr}\Rightarrow B_0\alpha\frac{1}{r^2}$

$\therefore\frac{B_1}{B_2}=\frac{9}{4}$

WE -19:

Two circular coils are made from a uniform wire the ratio of radii of circular coils are 2:3 & no.of turns is 3:4. If they are connected in parallel across a battery.

A : Find ratio of magnetic inductions at their centres

B : Find the ratio magnetic moments of 2 coils.

Sol: When connected in parallel

$a)B\alpha\frac{1}{r^2};\frac{B_1}{B_2}=\left(\frac{r_2}{r_1}\right)^2=\left(\frac{3}{2}\right)^2=\frac{9}{4}$

$b)M=ni\,A_{coil}$ but $i=\frac{V}{R}=\frac{V}{\rho l}A_{wire}$

$\Rightarrow M=\frac{Va_{wire}}{\rho(2\pi r_{coil})}(\pi r^2{}_{coil});\ M=\frac{VA_{wire}}{\rho X2}r_{coil}$

$\frac{M_1}{M_2}=\frac{r_1}{r_2}=\frac{2}{3}.$

WE -20:

Figure shows a square current carrying loop ABCD of side 2m and current $i=\frac{1}{2}A$. The magnetic moment $\overrightarrow{M}$ of the loop is

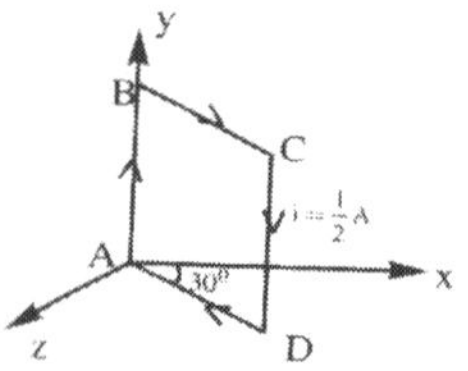

Sol: $DA = 2\cos 30^0 \hat{i} - 2\sin 30^0 \hat{k} (= \sqrt{3}\hat{i} - \hat{k})$

$\overrightarrow{AB} = 2\hat{j} \therefore \overrightarrow{M} = i(\overrightarrow{DA} \times \overrightarrow{AB})$

$= \frac{1}{2}\left[(-\sqrt{3}\hat{i} - \hat{k}) \times (2\hat{j})\right]$

$= -\sqrt{3}k + \hat{i} = (i - \sqrt{3}\hat{k})A - m^2.$

WE -21:

If two charged particles each of charge q mass m are connected to the ends of a rigid massless rod and is rotated about an axis passing through the centre and $\perp$ to length. Then find the ratio of magnetic moment to angular momentum.

Sol: $M = niA = 2 \times \frac{q}{t}\pi\left(\frac{l}{2}\right)^2 = 2 \times \frac{q\omega}{2\pi}\frac{\pi l^2}{4} = \frac{q\omega l^2}{4}$

$L = 2(mr^2\omega) = 2\left(m\frac{l^2}{4}\omega\right) = \frac{ml^2\omega}{2};$

$\frac{M}{L} = \frac{q}{2m}.$

WE -22:

Find the magnetic dipole moment of the spiral of total number of turns N, carrying current i having inner and outer radii a and b respectively.

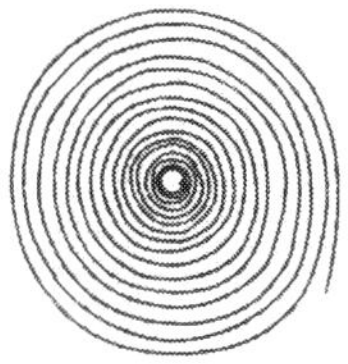

Sol:

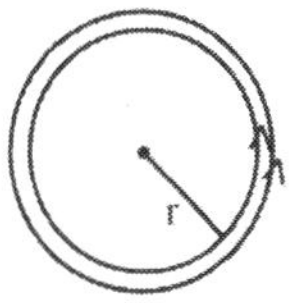

Let us take a thin coil of thickness dr. Then the number of turns of the coil is

$dN = \frac{N}{b-a}.dr$

the dipole moment of the coil is

$M = (dN)(i)(A) = \left(\frac{Ndr}{b-a}\right)(i)(\pi r^2) = \frac{\pi Ni}{b-a}\int_a^b r^2 dr$

$M = \frac{\pi i N}{3}(a^2 + ab + b^2).$

WE -23:

Consider a non conducting plate of radius a and mass m which has a charge q distributed uniformly over it, The plate is rotated about its own axis with an angular speed ω. Show that the magnetic moment M and the angular momentum L of the plate are related as $\frac{M}{L} = \frac{q}{2m}$.

Sol: If σ is the surface charge density, then $q = \sigma \pi a^2$

Current $i = \sigma \omega r \ \ dr$

The magnetic moment of the element ring

$$dM=(idA) = \sigma\omega\, dr(\pi r^2) = \pi\sigma\omega r^3 dr$$

and $$M=\pi\sigma\omega\int_0^a r^2\, dr = \frac{\pi\sigma\omega^4}{4} = M(\pi a^2\sigma)\frac{\omega a^2}{4} = \frac{q\omega a^2}{4}$$

The angular momentum of the disc about its axis $L = \frac{ma^2}{2}\omega$

The ratio $$\frac{M}{L} = \frac{4}{\frac{\omega m a^2}{2}} = \frac{q}{2m}$$

6. TANGENT GALVANOMETER

i) Tangent galvanometer works on the principle of Tangent law i.e., $B = B_H Tan\theta$

Here B = Magnetic induction at the centre of the current carrying coil $= \frac{\mu_0 ni}{2r}$

ii) It is a moving magnet type galvanometer

iii) During experiement, plane of the coil should be along the magnetic meridian [to fulfill the requirement of tangent law]

iii) current measured by Tangent galvanometer is $i = \left(\frac{2rB_H}{\mu_0 n}\right)Tan\theta = KTan\theta$

r = Radius of coil, K = reduction factor

n = number of turns of coil

iv) SI unit of reduction factor is ampere

v) Reading is more accurate when $\theta = 45^0$ since relative error $\frac{di}{i}\alpha\frac{1}{\sin 2\theta}$ and it is minimum for 45^0

vi) Sensitivity is maximum when $\theta = 0^0$ since $\frac{d\theta,}{di}\alpha \cos 2\theta$, which is maximum for $\theta = 0^0$

vii) Reduction factor K depends on horizontal component of earth's magnetic field.

viii) T.G gives different readings at different places for same current.

ix) T.G cannot be used at magnetic poles, since $B_H = 0$ at magnetic poles.

x) T.G is used to measure the current of the order of $10^{-6} A$.

WE -24:

A magnetic needle is arranged at the centre of a current carrying coil having 50 turns with radius of coil 20cm arranged along magnetic meridian. When a current of 0.5mA is allowed to pass through the coil the deflection is observed to be 30^0. Find the horizontal component of earth's magnetic field

Sol. $B = B_H \tan\theta \qquad \frac{\mu_0 ni}{2r\tan\theta} = B_H$

$$B_H = \frac{4\pi\times10^{-7}\times50\times5\times10^{-4}\times\sqrt{3}}{2\left(10^{-1}\right)(1)}$$

$= 5\sqrt{3}\pi\times10^{-8}T = 26.35\times10^{-8}T = 2.635\times10^{-7}T$

7. SOLENOID AND TOROID:

i) SOLENOID

a) If many turns of an insulated wire are wound around a cylinder the resulting coil is called a solenoid.

b) Field at a point on the axis of a solenoid is $B = \frac{\mu_0 ni}{2}(\sin\alpha + \sin\beta)$

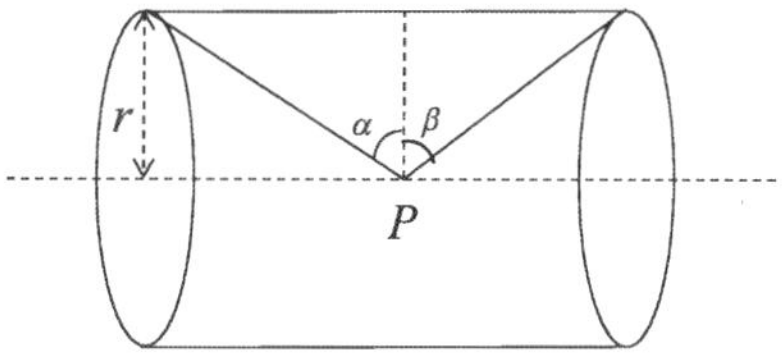

Where 'n' is number of turns per unit length of solenoid.

c) For a long solenoid magnetic induction on the axis is $B = \mu_0 ni \qquad (\because \alpha = \beta \cong 90^0)$

d) At one end of a long solenoid magnetic induction is $B = \frac{\mu_0 ni}{2} \quad (\because \alpha = 0^0, \beta \cong 90^0)$

ii) TOROID

a) If a solenoid is bent in a circular shape and the ends are joined then it is known as toroid. Let N be the total number of turns in the toroid of radius 'r' then number of turns per unit length $n = \frac{N}{2\pi r}$

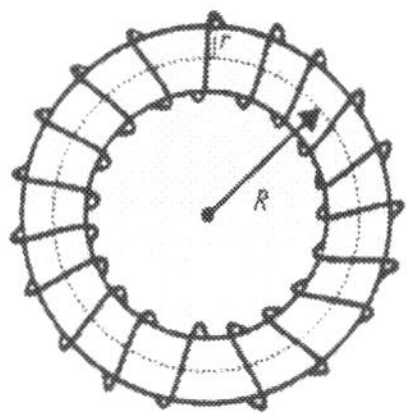

Applying Ampere's law

$$\oint \vec{B}.\vec{dl} = \mu_0 Ni$$

$$B = \frac{\mu_0 Ni}{2\pi r} = \mu_0 ni$$

WE -25:

A solenoid of length 8cm has 100 turns in it. If radius of coil is 3cm and if it is carrying a current of 2A, find the magnetic induction at a point 4cm from the end on the axis of the solenoid.

Solution : $B = \frac{\mu_0 ni}{2}(\sin\alpha + \sin\beta) = \frac{4\pi\times10^{-7}\times100\times2}{2}\times2\times\frac{4}{5} = 64\pi\mu T$

WE -26:

A solenoid 60cm long and of radius 4.0cm has 3 layers of windings of 300 turns each. A 2.0cm long wire of mass 2.5g lies inside the solenoid (near its centre) normal to its axis, both the wire and the axis of the solenoid are in the horizontal plane. The wire is connected through two leads parallel to the axis of the solenoid to an external battery which supplies a current of 6.0A in the wire. What value of current (with appropriate sense of circulation) in the windings of the solenoid can support the weight of the wire ? g=9.8 ms^{-2}.

Solution : $mg = Bi_{wire}l$

but $B = \mu_0 ni_{solenoid}$

$\Rightarrow mg = \mu_0 ni_{solenoid}\times i_{wire} Xl$

$i_{solenoid} = \frac{mg}{\mu_0 ni_{wire}l} = 108A$.

WE -27:

A toroid of non ferromagnetic has core of inner radius 25cm and outer radius 26cm. It has 3500 turns & carries a current of 11A, then find the magnetic field at a point

i) In the internal cavity of toroid

ii) At the midpoint of the windings

iii) At a point which is at a distance of 30cm from the centre of toroid

Solution : i) B = 0.

ii) $B = \frac{\mu_0}{2\pi}\frac{ni}{r} = 2\times10^{-7}\times\frac{3500\times11}{51\times10^{-2}}\times2$

$= \frac{88}{3}\times10^{-3} = 29.3\times10^{-3}T$

iii) B = 0

Based on magnetism for solenoid and toroid.

WE -28:

A solenoid of 2m long & 3cm diameter has 5 layers of winding of 500 turns per metre length in each layer & carries a current of 5A. Find intensity of magnetic field at the centre of the solenoid.

Solution : For long solenoid at the centre

$B = \mu_0 ni$

$H = \frac{B}{\mu_0} = ni = (500\times2)5\times5$

$= 2.5\times10^4\frac{A}{m}$.

8. FORCE ACTING ON A CHARGED PARTICLE MOVING IN A UNIFORM MAGNETIC FIELD :

i) If charge +q is moving with velocity $\vec{v}$, making an angle θ with the direction of field. force acting on the charge is, $\vec{F} = q(\vec{v} \times \vec{B})$

Magnitude of force is F=Bqv $Sin\theta$, direction of $\vec{F}$ is perpendicular to plane containing both $\vec{v}$ and $\vec{B}$.

ii) If $\theta = 0^0$ or 180^0, then the force acting on the particle is zero. And the particle keeps moving in the same path. i.e, **undeviated**.

iii) If the charged particle enters normal to the magnetic field, the force acting on it is maximum. ie $F_{max} = Bqv$

iv) This force acts right angles to $\vec{B}$ and $\vec{v}$. It acts as centripetal force and the path of particle will be **circular**.

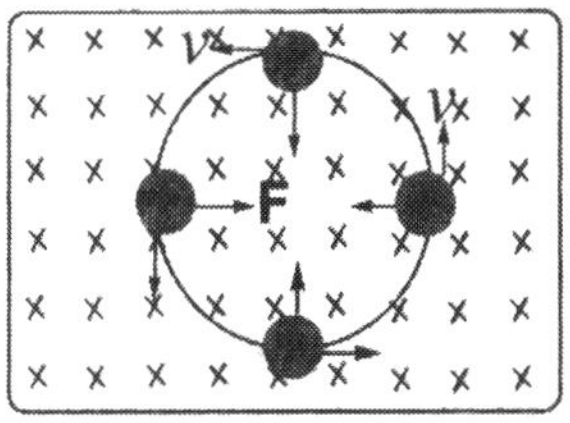

Then the radius of the circular path is given by

$$r = \frac{mv}{Bq} \Rightarrow r = \frac{P}{Bq} \text{ (from } Bqv = \frac{mv^2}{r})$$

Where p = momentum.

v) $r = \frac{\sqrt{2mK}}{qB}$ where K is kinetic energy of the particle.

vi) If charged particle is accelerated through a potential difference of V volts before it enters into the magnetic field normally then $r = \frac{\sqrt{2mqV}}{qB}$.

vii) Speed, kinetic energy remains constant, but velocity, acceleration, momentum and force are variable since their directions are continuously changing.

viii) The time period of rotation is

$$T = \frac{2\pi r}{v} \qquad \therefore T = \frac{2\pi m}{qB}$$

Angular frequency of rotation is $\omega = \frac{Bq}{m}$

$\therefore$ T and ω are independent of v and r of charged particle.

ix) When the particle enters the magnetic field at angle θ with $\vec{B}$, (such that $\theta \neq 0^0, \theta \neq 90^0, \theta \neq 180^0$), then the path followed by the particle will be **helical.**

x) Radius of circular path of the helix is given by $r = \frac{mv\sin\theta}{qB}$.

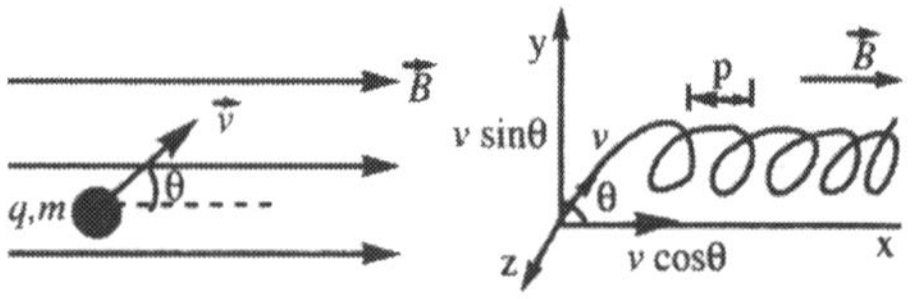

xi) Time period of rotation is $T = \frac{2\pi m}{qB}$

xii) Distance travelled by the particle along magnetic field in one complete rotation or pitch of helix is given by $P = (v\cos\theta)T$

$$P = \frac{2\pi m v\cos\theta}{qB}$$

xiii) Work done by the magnetic field on the charged particle is zero.

WE -29:

A magnetic field of $\left(4.0\times10^{-3}\hat{k}\right)T$ exerts a force $\left(4.0\hat{i}+3.0\hat{j}\right)\times10^{-10}N$ on a particle having a charge $10^{-9}C$ and moving in the x-y plane. Find the velocity of the particle.

Solution: Magnetic force $\overrightarrow{F_m} = \left(4.0\hat{i}+3.0\hat{j}\right)\times10^{-10}N$

Let velocity of the particle in x-y plane be.

$\vec{v} = v_x\hat{i} + v_y\hat{j}$ Then from the relation $\overrightarrow{F_m} = q\left(\vec{v}\times\vec{B}\right)$

We have $\left(4.0\hat{i}+3.0\hat{j}\right)\times10^{-10} = 10^{-9}\left[\left(v_x\hat{i}+v_y\hat{j}\right)\times\left(4\times10^{-3}\hat{k}\right)\right]$

$= \left(4v_y+10^{-12}\hat{i} - 4v_x\times10^{-12}\hat{j}\right)$

comparing the coefficient of $\hat{i}$ *and* $\hat{j}$ we have,

$4\times10^{-10} = 4v_y\times10^{-12}$

$\therefore v_y = 10^2 m/s = 100m/s$ and $3.0\times10^{-10} = -4v_x\times10^{-12}$

$\therefore v_x = -75m/s$

$\therefore \vec{v} = -75\hat{i}+100\hat{j}$

WE -30:

If a particle of charge $1\mu C$ is projected into a magnetic field $\vec{B} = \left(2\hat{i}+y\hat{j}-z\hat{k}\right)$T with a velocity $\vec{V} = \left(4\hat{i}+2\hat{j}-6\hat{k}\right)ms^{-1}$, then it passes undeviated. If it is now projected with a velocity $\vec{V} = \hat{i}+\hat{j}$, then find the force experienced by it

Solution: Charged particle moves in a magnetic field undeviated when $\vec{V}$ is parallel or anti parallel to $\vec{B}$

$$\frac{V_x}{B_x} = \frac{V_y}{B_y} = \frac{V_z}{B_z} = k; \frac{4}{2} = \frac{2}{y} = \frac{-6}{-z}$$

$$y = 1 \; z = 3$$

$$\therefore \vec{B} = \left(2\hat{i} + \hat{j} - 3\hat{k}\right) \quad \vec{F} = q\left(\vec{V} \times \vec{B}\right)$$

$$\vec{F} = 10^{-6}\begin{vmatrix} \hat{i} & \hat{j} & \hat{k} \\ 1 & 1 & 0 \\ 2 & 1 & -3 \end{vmatrix}$$

$$\vec{F} = 10^{-6}\left[i(-3) - \hat{j}(-3) + \hat{k}(-1)\right]$$

$$\left|\vec{F}\right| = 10^{-6}\left|\left(-3\hat{i} + 3\hat{j} - \hat{k}\right)\right| N = \sqrt{19}\mu N$$

9. DEVIATION OF CHARGED PARTICLE IN UNIFORM MAGNETIC FIELD

Case 1: Suppose a charged particle enters perpendicular to the uniform magnetic field if the magnetic field extends to a distance 'x' which is less than or equal to radius of the path.

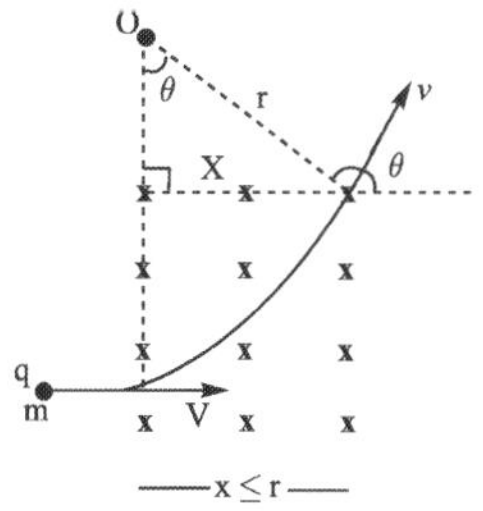

In this case, $r = \frac{mv}{Bq}$

Angle of deviation 'θ' can be determined by using the formula $\boxed{\sin\theta = \frac{x}{r}} = \frac{xqB}{mv}$

$$\therefore \theta = \sin^{-1}\left(\frac{xqB}{mv}\right)$$

The above relation can be used only when $x \leq r$.

Case 2: For x > r,

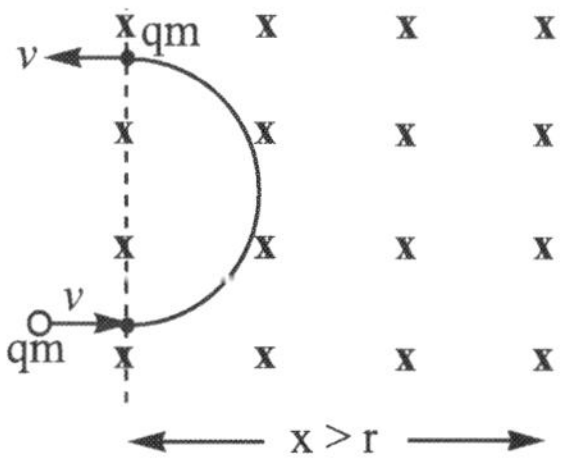

In this case, $r = \frac{mv}{Bq}$,

In this case, deviation $\theta = 180^0$.

Note: If particle moves for a time 't' in the field, then in such a case,

$\theta = \omega t$.

$$\boxed{\theta = \frac{Bq}{m}t}$$

WE -31:

An α - particle is accelerated by a potential difference of 10^4 V. Find the change in its direction of motion, if it enters normally in a region of thickness 0.1 m having transverse magnetic induction of 0.1 tesla. (Given : mass of α -particle 6.4×10^{-27} kg).

Sol: The situation is shown in Fig.

When a charged particle with charge q is accelerated through a potential difference V volt, then

$$\frac{1}{2}mv^2 = qV \text{ .....(i)}$$

or $$v = \sqrt{\left(\frac{2qV}{m}\right)} \text{ ....(i)}$$

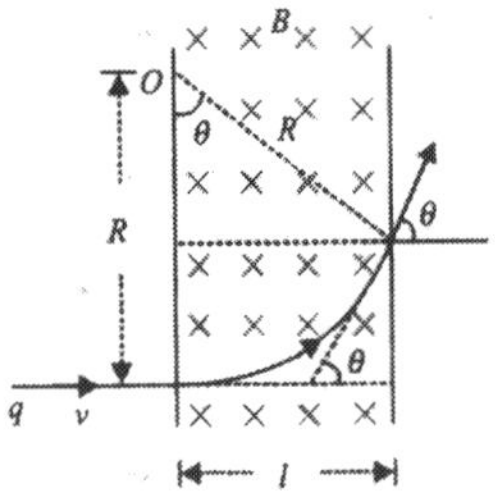

α - particle in magnetic field moves in a circle of radius R which is given by

$$R = \frac{mv}{qB} \text{ or } R = \frac{1}{B}\sqrt{\left(\frac{2mV}{q}\right)} \text{ ....(ii)}$$

The change in direction of α -particle (θ) from figure is given by

$$Sin\theta = \frac{l}{R} = lB\sqrt{\left(\frac{q}{2mV}\right)}$$

Here $l = 0.1m, B = 0.1$ tesla, $V = 10^4$ volt

$q = 2e = 2\times1.6X10^{-19} = 3.2\times10^{-19}C$ and $m = 6.4\times10^{-27}kg$

$$\therefore \sin\theta = 0.1\times0.1\times\sqrt{\left(\frac{3.2\times10^{-19}}{2\times6.4\times10^{-27}\times10^4}\right)} = \frac{1}{2} \text{ or } \theta = 30^0.$$

WE -32:

The magnetic field is confined in a square region. A positive charged particle of charge q and mass m is projected as shown in fig. Find the limiting velocities of the particle so that it may come out of face (1),(2),(3) and (4).

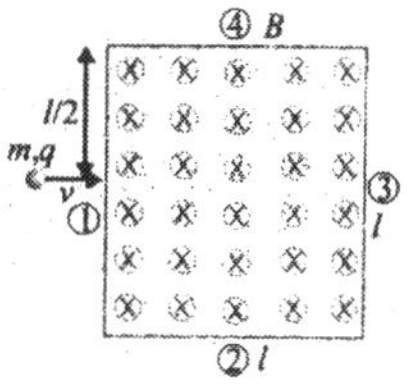

Sol : For the positive charge coming out from face (1), the radius of the path in magnetic field should be less than or equal to $l/4$. For limiting case $(2r = \frac{l}{2})$.

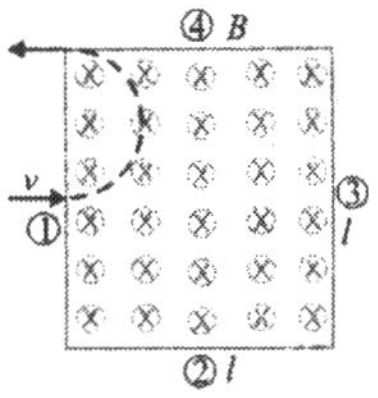

$$r_{max} = \frac{l}{4} = \frac{mv}{qB} \Rightarrow v_{max} = \frac{qBl}{4m}$$

Hence, if the velocity is $< \frac{qBl}{4m}$, the charge particle comes out of face (1).

We can observe from right palm rule that the particle cannot come out from face (1)

For a positive charge coming out of face (4) let particle come out at point N from ΔOMN

$(ON)^2 = (OM)^2 + (MN)^2$

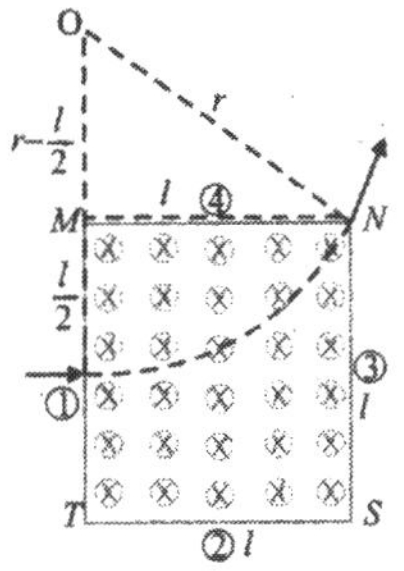

$$r^2 = \left(r - \frac{1}{2}\right)^2 + t^2 \Rightarrow r = \frac{5}{4}l$$

If the particle comes out from face (4), $r < \frac{5}{4}l \Rightarrow \frac{mv}{qB} < \frac{5}{4}l$ (or) $v < \frac{5}{4}\frac{qBl}{m}$. If velocity $v > \frac{5}{4}\frac{qBl}{m}$, the particle will come out from face (3).

WE -33:

A particle of mass m and change +q enters a region of magnetic filed with a velocity v, as shown in fig.

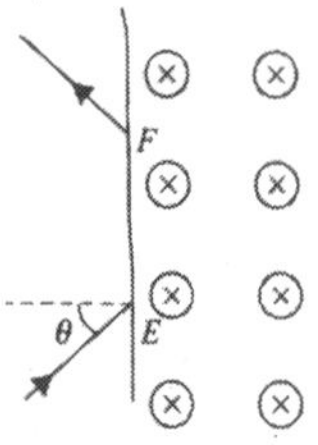

a) Find the angle subtended by the circular arc described by it in the magnetic field.

b) How long does the particle stay inside the magnetic filed ?

c) If the particle enters at E, what is the intercept EF ?

Sol: a) The particle circulates under the influence of magnetic field. As the magnetic field is uniform, the charge comes out symmetrically. The angle subtended at the centre is $(180-2\theta)$

b) The length of the arc traced by the particle, $l = R(\pi - 2\theta)$

Time spent in the field, $t = \frac{l}{v} = \frac{R(\pi-2\theta)}{v}$ and $R = \frac{mv}{Bq}$

which gives $t = \frac{m}{Bq}(\pi - 2\theta)$

As time period: $T = \frac{2\pi n}{Bq}$, hence

$t = \frac{T}{2\pi}(\pi - 2\theta)$

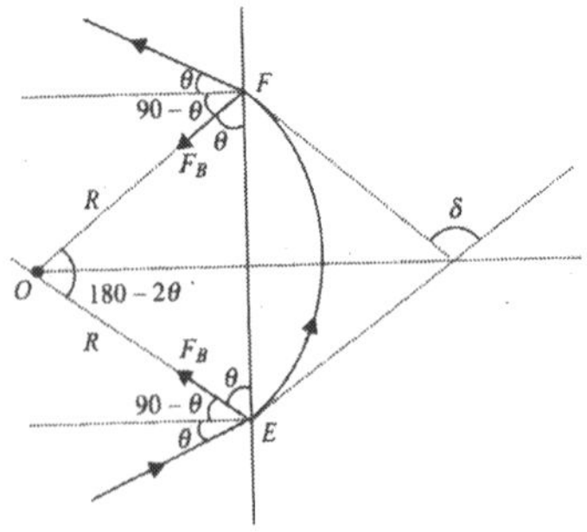

We can generalize this reslut. If $\varnothing$ is the angle subtended by the arc traced by the charged particle in the magnetic field, the time spent is $t = T\left(\frac{\phi}{2\pi}\right)$

c. Intercept EF = $2R\cos\theta$.

10. FLEMING'S LEFT HAND RULE :

Stretch the fore finger, central finger and thumb of left hand in mutually perpendicular directions, such that if fore finger indicates direction of magnetic field, Central finger indicates direction of current, then thumb indicates direction of force on conductor.

11. FORCE ON A CURRENT CARRYING CONDUCTOR KEPT IN UNIFORM MAGNETIC FIELD.

i) A conductor carrying current i is placed in a uniform magnetic field of induction B at an angle θ with the field direction. The force acting on it is given by

$\overline{F} = i(\bar{l} \times \overline{B})$. $\qquad |\overline{F}| = BilSin\theta$

ii) If B and l are parallel or anti-parallel F = 0

iii) If B and l are perpendicular, then $F_{Max} = Bil$.

iv) Direction of force can be found using Fleming's left hand rule.

12. LORENTZ FORCE :

i) When a charge enters a region where both electric and magnetic fields exists simulataneously, force acting on it is called Lorentz force and is given by $\overline{F} = \overline{F}_e + \overline{F}_m = q\left[\overline{E} + (\overline{V} \times \overline{B})\right]$.

ii) Cyclotron:

a) The cyclotron is a machine to accelerate charged particles or ions to high energies using both electric and magnetic fields in combination.

b) Cyclotron uses the fact that the frequency of revolution of the charged particle in a magnetic field is independent of its energy.

c) Centripetal force is provided by the magnetic force $\frac{mv^2}{r} = Bqv$

d) Radius of circular path is $r = \frac{mv}{Bq}$

e) Time period of charged particle is $T = \frac{2\pi r}{v}$

$T = \frac{2\pi m}{Bq}$ $f = \frac{1}{T} = \frac{Bq}{2\pi m}$ = cyclotron frequency.

f) K.E of charged particles is

K.E=1/2 $mv^2 = \frac{1}{2}m\left(\frac{Bqr}{m}\right)^2 = \frac{B^2q^2r^2}{2m}$

iii) Special Cases :

a). The force acting on a curved wire joining points a and b as shown in the figure is the same as that on a straight wire joining these points. It is given by $\vec{F} = i\vec{L} \times \vec{B}$ where $\overline{L} = ab$

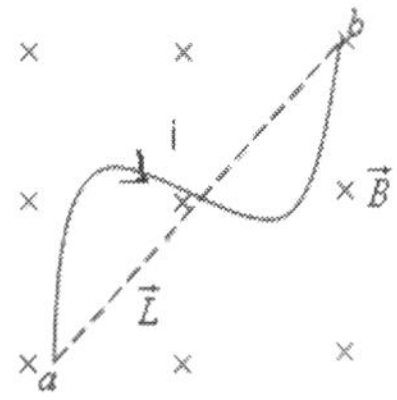

b) The force experienced by a semi circular wire of radius 'r' when it is carrying a current 'i' and is placed in a uniform external magnetic field of induction B as shown in the figure is given by F=BI(2r).

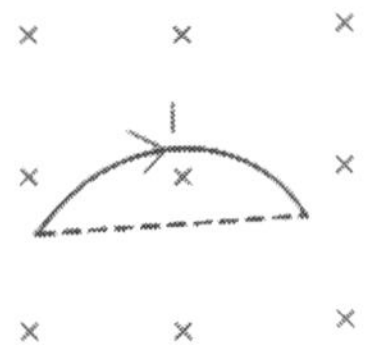

Force directed

Vertically upwards

c) The force on the wire shown $F = Bil\sin\frac{\theta}{2}$ towards left

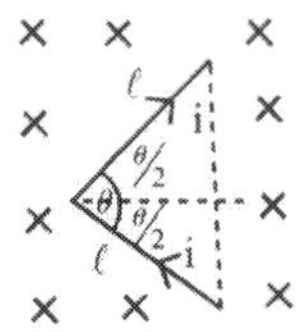

$= l_{eff} = 2l\sin\frac{\theta}{2}$

d) The force on a closed loop of any shape carrying current in a uniform magnetic field is always zero.

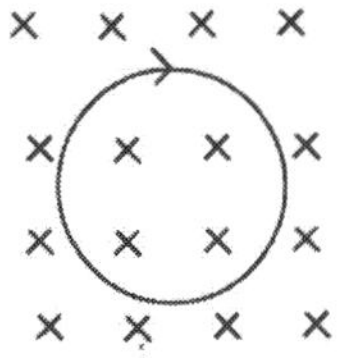

since $l_{eff} = 0$

e) The net force experienced by a closed current loop and current completes the loop in a uniform field is zero.

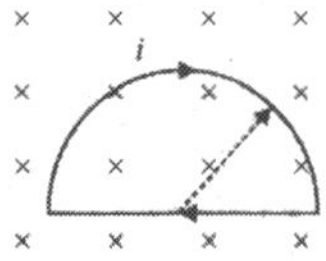

f) In case of a closed loop but current does not complete the loop the net force is not zero.

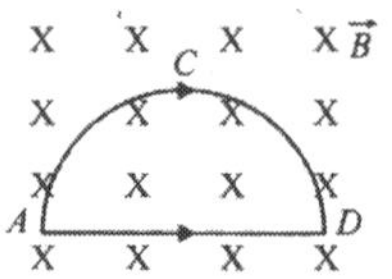

$\vec{F}_{ACD} = \vec{F}_{AD}$

$\therefore \vec{F}_{loop} = \vec{F}_{ACD} + \vec{F}_{AD} = 2\vec{F}_{AD}$

$\therefore \left|\vec{F}_{loop}\right| = 2\left|\vec{F}_{AD}\right|$

WE -34:

Find the force experienced by the wire carrying a current 2A if the ends P and Q of the wire have coordinates (1, 2, -3) m and (-2, -5, 1) m respectively when it is placed in a magnetic field $\vec{B}=\left(\hat{i}+\hat{j}+\hat{k}\right)T$

Sol. The force acting on the wire is

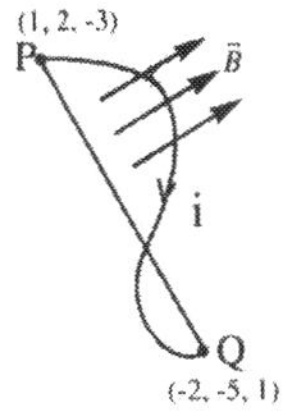

$$\vec{F}=i\overrightarrow{r_{21}}\times\vec{B}\ =i\left(\vec{r_2}-\vec{r_1}\right)\times\vec{B}$$

$$=2\left(-3\hat{i}-7\hat{j}+4\hat{k}\right)\times\left(\hat{i}+\hat{j}+\hat{k}\right)$$

$$=2\left(-11\hat{i}+7\hat{j}+4\hat{k}\right)N$$

WE -35:

In Fig. a semicircular wire loop is placed in uniform magnetic field B=1.0 T. The plane of the loop is perpendicular to the magnetic field. Current $i=2A$ flows in the loop in the direction shown. Find the magnitude of the magnetic force in both the cases (a) and (b). The radius of the loop is 1m.

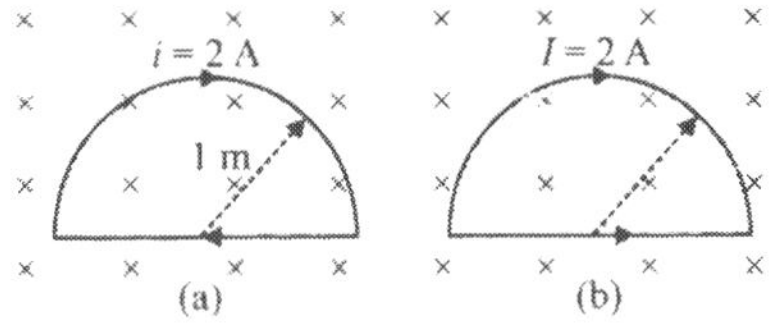

Sol: It forms a closed loop and the current completes the loop. Therefore, net force on the loop in uniform field should be zero. From the figure, net force on the loop in uniform field shoud be zero.

In case (b) although it forms a closed loop, but current does not complete the loop. Hence, net force is not zero.

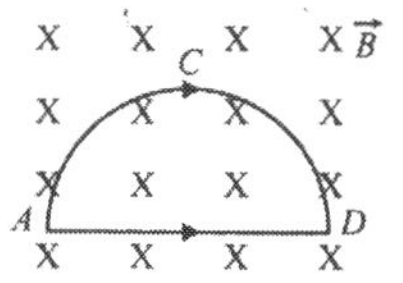

$$\vec{F}_{ACD}=\vec{F}_{AD}$$

$$\therefore\ \vec{F}_{loop}=\vec{F}_{ACD}+\vec{F}_{AD}=2\vec{F}_{AD}$$

$$\therefore\ \left|\vec{F}_{loop}\right|=2\left|\vec{F}_{AD}\right|$$

$$=2ilB\sin\theta(l\ =2r\ =2.0m)$$

$$=(2)(2)(2)(1)\sin 90^0\ =8N$$

WE -36:

A rough inclined plane inclined at angle of 37^0 with horizontal has a metallic wire of length 20cm with its length $\perp^r$ to length of inclined plane $(\mu = 0.1)$ When a current of its passignthrough the wire and a magnetic field is applied normal to the plane upwards, the wire starts moving up with uniform velocity for B = 0.5T. Then find the magnitude of current i, (mass of the wire = 50g)

Sol.

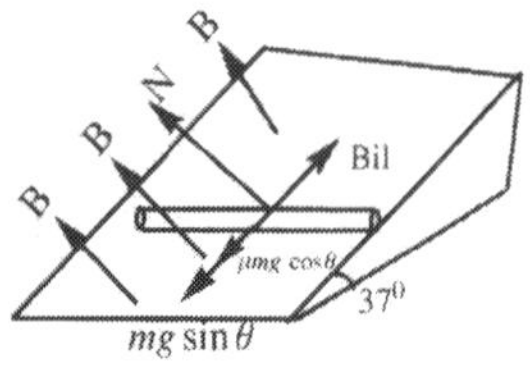

When the wire is in equilibrium

$$Bil = mg\sin\theta + f$$

$$Bil = mg(\sin\theta + \mu\cos\theta)$$

$$Bil = 5\times10^{-2}\times10\left(\frac{3}{5}+0.1\times\frac{4}{5}\right)i$$

$$= \frac{10^{-1}\times3.4}{10^{-1}} = 3.4A$$

WE -37:

A wire PQ of mass 10g at rest on two parallel metal rails. The separation between the rails is 4.9cm. A magnetic field of 0.80 tesla is applied perpendicular to the plane of the rails, directed in wards. The resistance of the circuit is slowly decreased. When the resistance decreases to below 20 ohm, the wire PQ begins to slide on the rails. Calculate the coefficient of friction between the wire and the rails.

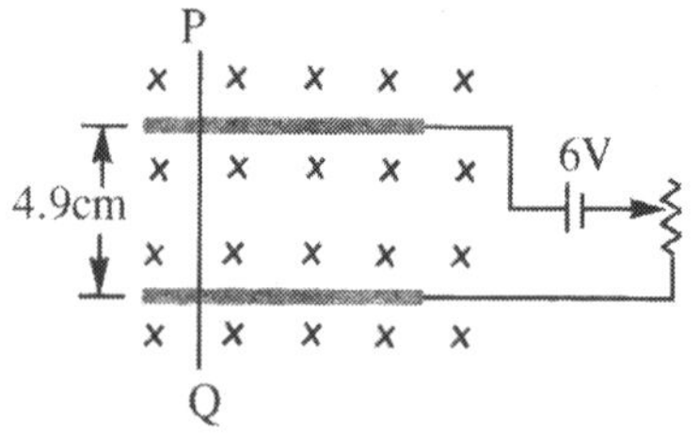

Sol. Wire PQ begins to slide when magnetic force is just equal to the force of friction, i.e.

$$\mu mg = il\,B\sin\theta \quad (\theta = 90^0)$$

Here, $i = \frac{E}{R} = \frac{6}{20} = 0.3A \qquad \mu = \frac{ilB}{mg}$

$$= \frac{(0.3)(4.9\times10^{-2})(0.8)}{(10\times10^{-3})(9.8)} = 0.12$$

WE -38:

A current carrying conductor of mass m, length l carrying a current i hangs by two identical springs each of stiffness k. For an outward magnetic field B find the deformation of the springs. Put $m = 50gm$.

$g = 10m/s^2, l = \frac{1}{2}m, i = 1A$ and $B = 1T$ and $k = 50N/m$

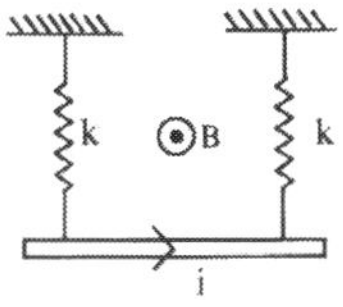

Sol. The forces acting on the rod are 'mg' downwards, $F_{mag} = ilB$ downwards

and $F_{spring} = 2kx$ upwards

Under the action of these forces the rod is in equilibrium. Then, $F_{net} = 0$

or $mg + ilB = 2kx$

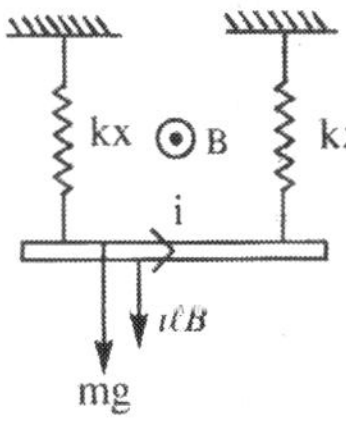

or $x = \frac{mg + ilB}{2k}$

$= \frac{\left(\frac{1}{20}\right)(10) + (1)\left(\frac{1}{2}\right)(1)}{2 \times 50}$

$= \frac{1}{200}m = 0.5cm$

WE -39:

A rod CD of length b carrying a current I_2 is placed in a magnetic field due to a thin long wire AB carrying current I_1 as shown in fig. Then find the net force experienced by the wire

Sol.

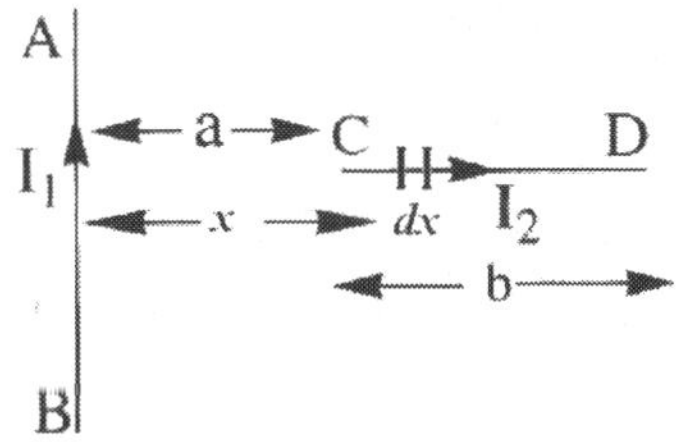

Magnetic induction due to a straight wire at a position of small element dx at a distance x from the conductor AB is $\frac{\mu_0}{2\pi}\frac{I_1}{x}$

Force on the current element is $\overrightarrow{dF} = dBI_2 dx \sin 90^0$

$$dF = \frac{\mu_0}{2\pi}\frac{I_1 I_2}{x} dx$$

Net force on conductor is $F = \int_a^{a+b} \frac{\mu_0}{2\pi}\frac{I_1 I_2}{x} dx$

$$= \frac{\mu_0}{2\pi} I_1 I_2 [\log x]_a^{a+b} = \frac{\mu_0}{2\pi} I_1 I_2 \log\left(1 + \frac{b}{a}\right)$$

13.1 FORCE BETWEEN TWO PARALLEL CURRENT CARRYING LONG STRAIGHT CONDUCTORS

i) Force per unit length on each wire is given by $\frac{F}{l} = \frac{\mu_0}{2\pi}\frac{i_1 i_2}{r}$

If $i_1 = i_2 = 1amp, r = 1m,$ then force per unit length of te conducter is $2\times10^{-7} N/m$

ii) If currents in the two wires are in **same direction,** then the force of **attraction** takes place between them.

iii) If currents in the two wires are in **opposite direction,** then the force of **repulsion** takes place between them

iv) A stright and very long wire carries current i_1 and rectangular loop of wire carrying current i_2 is placed nearby it. The force on the loop is

$$F = \frac{\mu_0 i_1 i_2 l}{2\pi}\left[\frac{1}{a} - \frac{1}{b}\right]$$

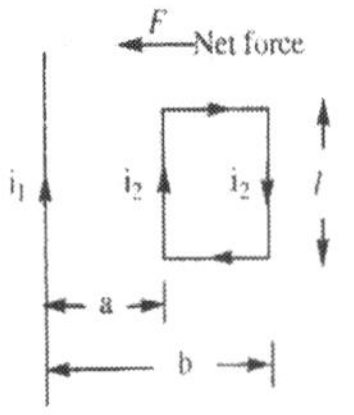

v) A very long horizontal wire carries a current i_1 is rigidly fixed. Another wire is placed directly above and parallel to it carries a current i_2. r is the perpendicular distance of seperation between the wires and currents are in opposite directions for the second wire remains stationary, the condition is

$$F = mg \Rightarrow \frac{\mu_0 i_1 i_2 \ell}{2\pi r} = mg$$

$$\Rightarrow \boxed{\frac{m}{\ell} = \frac{\mu_0 i_1 i_2}{2\pi r g}}$$

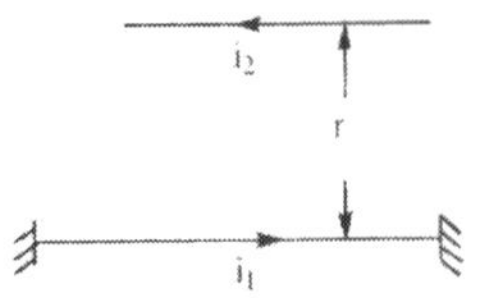

vi) Three long parallel conductors carry currents as shown

a) Resultant force per unit length on the wire **'C'** is

$$F = \frac{\mu_0}{2\pi}\left[\frac{i_1 i_2}{a} \sim \frac{i_2 i_3}{b}\right]$$

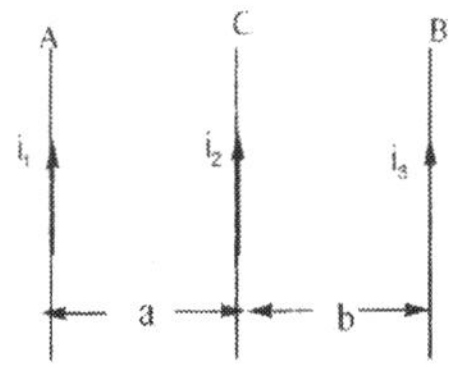

b) If the resultant force on the wire 'C' is zero, the condition is

$$\frac{i_1 i_2}{a} = \frac{i_2 i_3}{b} \Rightarrow \frac{i_1}{a} = \frac{i_3}{b}$$

Note: Here the resultant force per unit length on the A and B wire can be also determined in the similar way. The currents can be along different directions.

13.2 NULL POINTS DUE TO TWO CURRENT CARRYING PARALLEL WIRES.

i) Two straight parallel conductors are carrying currents $i_1, i_2\left(i_1 < i_2\right)$ in the **same direction,** and are seperated by a distance r, the null point is formed in between them. The distance of the null point from the conductor carrying smaller current is

$$x = \frac{r}{\frac{i_2}{i_1} + 1}$$

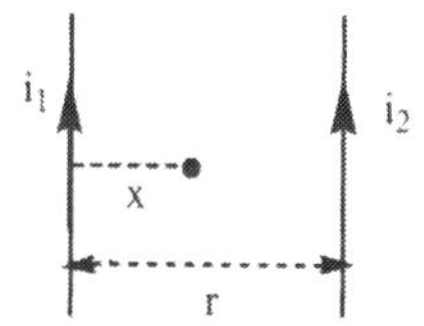

ii) Two straight parallel conductors are carrying currents $i_1, i_2 (i_1 < i_2)$ in **opposite directions,** and are seperated by a distance r, then the null point is formed **out side** the conductors, the distance of the null point from the conductor carrying smaller current is given by

$$x = \frac{r}{\frac{i_2}{i_1} - 1}$$

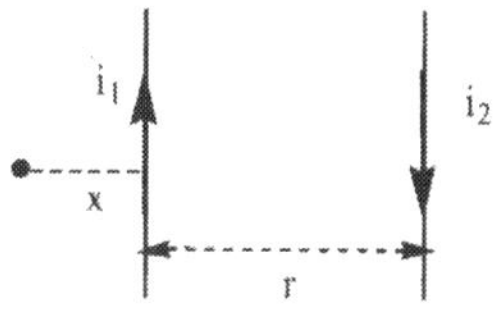

WE -40 :

A long straight conductor carrying a current of 2A is in parallel to another conductor of length 5cm. and carrying a current 3A. They are separated by a distance of 10cm. Calculate (a) B due to first conductor at second conductor (b) the force on the short conductor.

Sol. Given $i_1 = 2A; i_2 = 3A$

$r = 10cm = 10\times10^{-2} m; l_2 = 5cm$

a) $B = \frac{\mu_0 i_1}{2\pi r} = 2\times10^{-7}\times\frac{2}{10\times10^{2}} = 4\times10^{-6}$ Tesla

b) $F = \frac{\mu_0 i_1 i_2}{2\pi r}\times\ell_2 = 2\times10^{-7}\times\frac{2\times3}{10\times10^{-2}}\times5\times10^{-2} = 6\times10^{-7} N$

WE -41 :

Two long stright parallel current carrying conductors each of length l and current i are placed at a distance r_0. Show that the total work done by an external agent in slowly reducing their distance of seperation to $\frac{r_0}{2}$ is $\frac{\mu_0}{2\pi} i^2 l\, n(2)$

Sol.

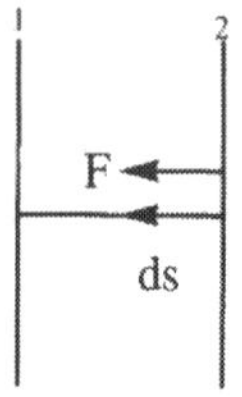

The force acting on the conductor 2 is F = ilB

$$= il\frac{\mu_0 i}{2\pi r} = \frac{\mu_0 i^2 l}{2\pi r}$$

This force does a work dW in displacing the conductor 2 by a distance dr

$$dW = \vec{F}.\vec{dr}$$

$$= \frac{\mu_0 i^2 l}{2\pi r}(-dr) \quad (\therefore \theta = 180^0)$$

Then, the total work done is

$$W = \int dW$$

$$= -\frac{\mu_0 i^2 l}{2\pi}\int_{r_0}^{\frac{r_0}{2}} \frac{dr}{r} = \frac{\mu_0 i^2 l}{2\pi}\ln 2$$

WE -42 :

Two parallel horizontal conductors are suspended by light vertical threads 75.0 cm long. Each conductor has a mass of 40.0gm per metre, and when there is no current they are 0.5 cm apart. Equal magnitude current in the two wires result in a separation of 1.5cm. Find the values and directions of currents

Sol.

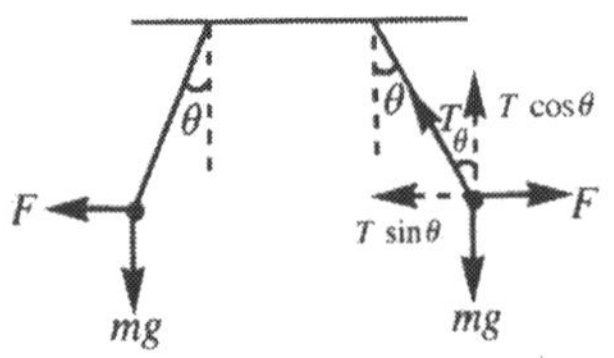

The situation is shown in figure

Here, we have $T\cos\theta = mg$

$$T\sin\theta = F = \frac{\mu_0}{4\pi}.\ell.\frac{2i_1 i_2}{d} \quad \text{or}$$

$$T\sin\theta = \frac{\mu_0}{4\pi}.\ell.\frac{2i^2}{d}$$

from the above equations

$$\tan\theta = \frac{\mu_0}{4\pi}.l.\frac{2i^2}{d}.\frac{1}{mg}$$

where θ is small, $\tan\theta \approx \sin\theta$

From figure $\sin\theta = \dfrac{0.5\times10^{-2}}{75\times10^{-2}}$

$m = 40.0\times10^{-3}1kg$

Where $l =$ length of conductor in meter

Substituting We get $\dfrac{0.5\times10^{-2}}{75\times10^{-2}} =$

$$10^{-7}.I.\frac{2i^2}{\left(1.5\times10^{-2}\right)}\times\frac{1}{\left(40\times10^{-3}\right)1\times9.8}$$

Solving, we get $i = 14amp.$

As conductgors are repelled, the currents in them are in opposite directions.

WE -43 :

A conductor AB of length 10cm at a distance of 10cm from an infinitely long parallel conductor carrying a current 10A. What work must be done to move AB to a distance of 20cm if it carries 5A?

Sol. Force on a conductor at a distance X is $F = \dfrac{\mu_0 i_1 i_2 l}{2\pi x}$

Wone doen to displace it through a small distance

$$dx = dW = \vec{F}.\vec{dx}$$

$$dW = \frac{\mu_0 i_1 i_2 l}{2\pi x}dx$$

$$W = \int_{0.1}^{0.2}\frac{\mu_0 i_1 i_2 l}{2\pi x}dx$$

$$W = \frac{\mu_0 i_1 i_2 l}{2\pi}\left[\log_e x\right]_{0.1}^{0.2}$$

$$W = \frac{4\pi\times10^{-7}\times10\times5\times10\times10^{-2}}{2\pi}\log e^2$$

$$W = 0.693\times10^{-6}J$$

14. FORCE BETWEEN TWO STREAMS OF ELECTRIC CHARGES:

i) If two streams of electrons or protons are moving with velocity 'v' in parallel and same directions, there will be both electric repulsive force and magnetic attractive force. Since electric force predominates the magnetic force, there will be repulsion.

ii) If they move parallel and opposite directions, there will be electric repulsive force and magnetic repulsive force and hence there will be repulsion again.

15.1 TORQUE ACTING ON A CURRENT LOOP KEPT IN UNIFORM MAGNETIC FIED :

i) When a coil carrying current is placed in unifrorm magnetic field, the net force on it is zero but it experiences a torque or couple.

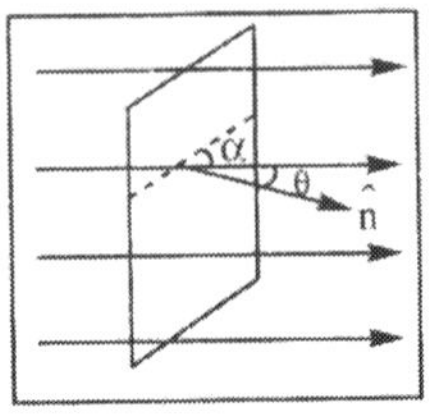

ii) Torque acting on a current carrying coil placed in uniform magnetic field is $\vec{\tau} = \vec{M} \times \vec{B}$

iii) Torque acting on the coil is $\tau = BiNA\sin\theta$

$= BiNA\cos\alpha$ Here A = are of coil carrying current i

N = number of turns of the coil

B = Magnetic induction of the field

α = Angle made by the plane of the coil with $\vec{B}$

θ = Angle made by the normal to the plane of the coil with $\vec{B}$

iv) If the plane of coil is parallel to the direction of magnetic field $\tau = \tau_{max} = BiNA$

v) If the plane of coil is perpendicular to the direction of magnetic field, $\tau = 0$

vi) If current carrying coil is placed in a non-uniform magnetic field it experiences both force and torque.

vii) For a given area, torque is independent of shape of the coil

viii) Torque is directly proportional to area of the coil.

15.2 SPECIAL CASES

i) When a current carrying coil is placed in uniform magnetic field, net force on it F = 0. But net torque may acts.

ii) When a current carrying coil is placed in non-uniform magnetic field, net force, net torque both acts.

$\tau_{net} \neq 0$ $F_{net} \neq 0$.

iii) If the angle made by $\vec{M}$ of the coil with $\vec{B}$ in uniform magnetic field is 'θ', then its potential energy

$P.E = -\vec{M}.\vec{B}$

$\boxed{P.E = -MB\cos\theta}$

iv) If a current carrying coil is rotated in a uniform field such that the angle made by $\vec{M}$ with $\vec{B}$ is changes from θ_1 to θ_2. $W = MB(\cos\theta_1 - \cos\theta_2)$

v) If ext field is along, the direction of $\overline{M}$, then $\theta = 0°$.

$\underline{\tau = 0}$ $\underline{P.E = -MB}$ (min)

This position corresponds to stable equilibrium.

vi) If external magnetic field is opposite to $\overrightarrow{M}$ then, $\theta = 180°$

$\underline{\tau = 0}$. $\underline{P.E. = +MB}$ (max)

vii) This corresponds to unstable equilibrium.

WE - 44 :

A circular loop of area $1cm^2$ carrying a current of 10A is placed in a magnetic field of $2T\hat{i}$. The loop is in xy plane with current in clock wise direction. Find the torque on the loop.

Sol. $\vec{\tau} = \vec{M} \times \vec{B} \quad = \left(ni\vec{A} \times \vec{B}\right)$

$$= \left[10 \times 10^{-4}\left(-\hat{k}\right)\right] \times \left(2\hat{j}\right) = 2 \times 10^{-3}\, Nm\left(\hat{i}\right)$$

WE -45 :

A metallic wire is folded to form a square loop a side 'a'. It carries a current 'i' and is kept perpendicular to a uniform magnetic field. If the shape of the loop is changed from square to a circle without changing the length of the wire and current, the amount of work done in doing so is

Sol. $W = Find\ P.E - \text{initial P.E}$

$$W = -M_f B - (M_i B)$$

$$W = iB\left(A_i - A_f\right)$$

$$W = iBa^2\left[1 - \frac{4}{\pi}\right]$$

WE -46 :

A flat insulating disc of radius 'a' carries an excess charge on its surface is of surface charge density $\sigma C/m^2$. Consider disc to rotate around the axis passing through its centre and perpendicualr to its plane with angular speed ω rad/s. If magnetic field $\vec{B}$ is directed perpendicualr to the rotation axis, then find the torque acting on the disc.

Sol. Suppose the disc is placed in xy-plane and is rotated about the z-axis. Consider an annular ring of radius r and of thickness dr, the charge on this ring.

$$dq = \sigma\left(2\pi r dr\right)$$

As the ring rotates with angular velocity ω, so the current

$$i = \frac{dq}{dt} = \frac{\sigma\left(2\pi r dr\right)}{\frac{2\pi}{\omega}} = \sigma\omega r dr$$

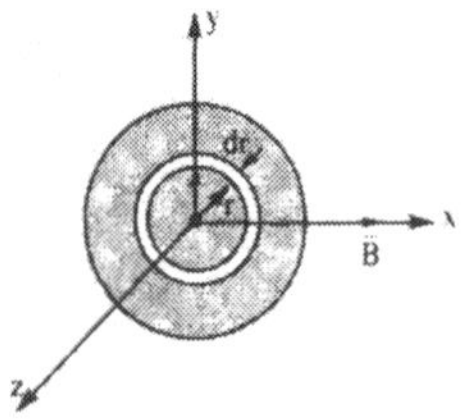

The torque on the current loop

$\vec{\tau} = i\vec{A} \times \vec{B}$

Hence the torque on this annular ring

$d\tilde{\tau} = i\left(d\vec{A} \times \vec{B}\right)$

$= \sigma\omega r dr\left(\pi r^2 B \sin 90^0\right)$

$= \pi\sigma\omega r^3 B dr$

and $\tau = \pi\sigma\omega B\int_0^a r^3 dr = \dfrac{\pi\sigma\omega B a^4}{4}$

WE - 47 :

A loop carrying current 'i' is lying inthe plane of the paper. It is the field of a long straight wire with constant current i_0 (inward) as shown in fig. Find the torque acting on the loop.

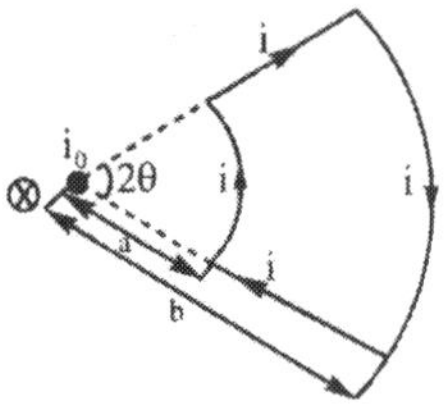

Sol. The field due to current carrying wire is tangential to every point on the circular portion of the loop and hence the forces acting on these segments are zero.

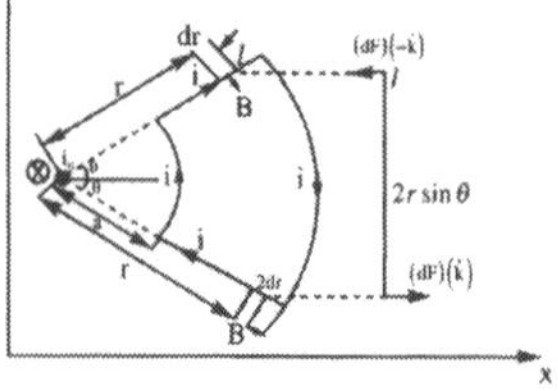

Now consider two small elements of length dr at a distance r from the axis symmetrically as shown in fig.

The magnitude of the force experienced by each element is $dF = B\ i\ dr = \left(\dfrac{\mu_0}{2\pi}\dfrac{i_0}{r}\right)idr$

On element 1 it is into the page and on 2 it out of the page, $d\tau = dF \times 2r\sin\theta$

$= \left(\dfrac{\mu_0 i_0 i}{2\pi r}dr\right) \times 2r\sin\theta$

Now total torque

$$\tau = \frac{\mu_0 i_0 i \sin\theta}{\pi}\int_a^b dr = \frac{\mu_0 i_0 i}{\pi}\sin\theta(b-a)$$

16. MOVING COIL GALVANOMETER

i) **Principle of moving coil galvanometer:** When a current carrying coil suspended in a unifrom magnetic field, it experiences a torque and hence it rotates.

ii) Poles of magnet are concave is shape, to make the magnetic field radial so that at all orientations the plane of the coil is parallel to the field, and hence torque acting on it is maximum.
This makes the relation between current and deflection linear.

iii) Soft iron cylider is kept at the center of magnetic field to increase the flux.

iv) Phosphor Bronze has
a) high Young's modulus so that the wire will not be stretched easily.
b) low rigidity modulus so that the wire can be twisted easily.
c) small elastic after effect so that it comes back quickly to orginal position after withdrawing current.

v) Small mirror is attached on the phosphor Bronze wire, to measure the deflection using lamp and scale arrangement.

vi) If 'θ' is the deflection for passage of current 'i', then $C\theta = BiAN \Rightarrow i = \left(\frac{C\theta}{BAN}\right)$

where $k = \left(\frac{C}{BNA}\right)$ = Galvanometer constant or figure of merit. It is independent of B_H . Where 'C' is couple per unit twist.

vii) a) **Current sensitivity** of a galvanometer is defined as the deflection produced in the galvanometer per unit current flowing through it.

$$S_I = \frac{d\theta}{di} = \frac{BAN}{C}$$

b) **Voltage sensitivity** of a galvanometer is defined as the deflection produced in the galvanometer per unit voltage applied to it.

$$S_V = \frac{\theta}{V} = \frac{\theta}{iG} \Rightarrow \frac{\theta}{V} = \frac{BAN}{CG}$$

Where G is resistance of galvanometer

i) Increasing B   ii) Increasing A
iii) Increasing N   iv) Decreasing C

viii) It is used to measure current upto a minimum of 10^{-9} Amp.
a) Plane of coil need not be along the magnetic meridian
b) Galvanometer constant is independent of B_H. So it can be used to measure currents even at poles.
c) External magnetic fields have no effect on deflection. So, it can be used to measure current even in the environment of stray magnetic fields.

WE -48 :

The area of the coil in a moving coil galvanometer is 16 cm^2 and has 20 turns. The magnetic induction is 0.2T and the couple per unit twist of the suspended wire is 10^{-6} Nm per degree. If the deflection is

45^0 calculate the current passign through it

Sol. Given, $A = 16cm^2 = 16 \times 10^{-4} m^2$

$B = 0.2T; N = 20, C = 10^{-6} Nm/$ degree; $\theta = 45^0$

From, $C\theta = BiAN$ $\quad i = \frac{C\theta}{BAN} = \frac{10^{-6} \times 45}{0.2 \times 16 \times 10^{-4} \times 20} = 9.94 \times 10^{-1} A.$

WE -49 :

A coil area $100cm^2$ having 500 turns carries a current of 1mA. It is suspended in a uniform magnetic field of induction 10^{-3} Wb/m². Its plane makes an angle of 60^0 with the lines of induction. Find the torque acting on the coil.

Sol. Given

$i = 1mA = 10^3 = 10^{-3} A; N = 500; B = 10^{-3} Wb/m^2$

$\theta = 60^0, \tau = ?\ A = 100cm^2 = 100 \times 10^{-4} m^2$

Couple acting on the coil is given by

$\tau = BiAN \sin\phi$

Where ϕ is angle made by normal to the plane of coil with B.

$\phi = 90 - 60 = 30^0$

$\therefore C = 10^{-3} \times 10^{-3} \times 100 \times 10^{-4} \times 500 \times \sin 30$

$= 250 \times 10^{-8} Nm$

WE -50 :

A galvanometer of resistance 95Ω, shunted by a resistance of 5 ohm gives a deflection of 50 divisions when joined in series with a resistance of $20k\Omega$ and a 2 volt accumulator. What is the current sensitivity of the galvanometer (in div/ μA)

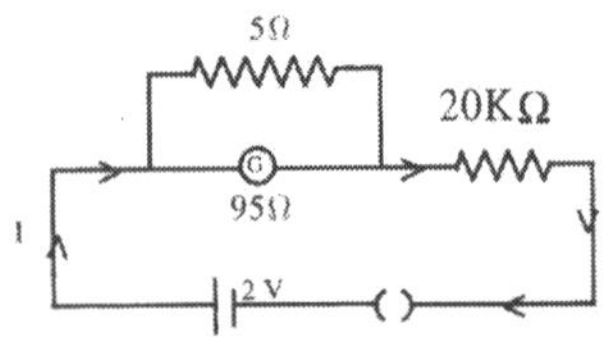

In accordance with given problem, the situation is depicted by the circuit diagram in fig. As here $20k\Omega$ is much greater than the resistance of shunted galvanometer $(< 5\Omega)$, the current in the circuit will be

$$I = \frac{2}{20 \times 10^3} = 10^{-4} A = 100\mu A$$

and as this current produces deflection of 50 divisions in the galvanometer

$$CS = \frac{\theta}{I} = \frac{50div}{100\mu A} = \frac{1div}{2\mu A}$$

17. SHUNT

i) A low resistance connected in parallel to galvanometer to protect it from large current is known as shunt.

ii) When shunt is connected range increase but sensitivity decreases.

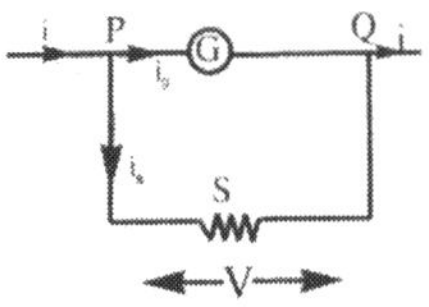

iii) $R_{equivalent} = \frac{GS}{G+S}$

iv) $V = iR_{eq} = i\frac{GS}{G+S}$

v) $V_{PQ} = i_g G = i_s S$

18. AMMETER

i) Galvanometer can be converted in to Ammeter by connecting low resistance parallel to it.

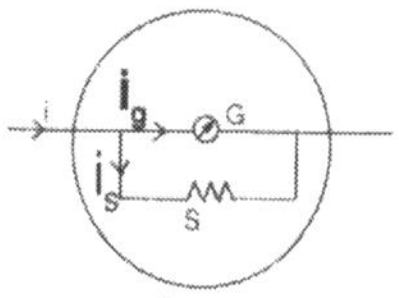

ii) To increase the range by 'n' times or to decrease the sensitivity by 'n' times, shunt to be connected across Galvanometer is

$$S = \frac{G}{\left(\frac{i}{i_g} - 1\right)} \Rightarrow S = \frac{G}{n-1}$$

Here $n = \frac{i}{i_g} = \frac{new\ range}{old\ range} = \frac{old\ \ division/amp}{new\ \ division/amp}$

iii) Equivalent resistance of ammeter $= \frac{GS}{G+S}$

iv) The relation between currents is

a) $i = i_g + i_s$

b) $i_g = \frac{iS}{G+S}$

c) $i_s = \frac{iG}{G+S}$

d) $\frac{i_g}{i_s} = \frac{S}{G}$; $\frac{i_g}{i} = \frac{S}{G+S}$; $\frac{I_s}{I} = \frac{G}{G+S}$

v) It is a device used to measure current in electrical circuits.

vi) Resistance of an ammeter is very small and it is zero for an ideal ammeter. Potential drop across ideal ammeter is zero.

vii) Ammeter must always be connected in series to the circuit

viii) Among low range and high range ammeters, low range ammeter has more resistance.

WE -51 :

A galvanometer of resitance 20Ω is shunted by a 2Ω resistor. What part of the main current flows through the galvanometer?

Sol. $\frac{i_g}{i}=\frac{G}{G+S}$. Given $G=20\Omega; S=2\Omega$

$\therefore \frac{i_g}{i}=\frac{2}{22}=\frac{1}{11};\frac{1}{11}th$ part of current is passing through galvanometer.

WE -52 :

A galvanometer has resistance 500 ohm. It is shunted so that its sensitivity decreases by 100 times. Find the shunt resistance.

Sol. Sensitivity $\propto \frac{1}{range}$ $\therefore n=100$

$$S=\frac{G}{(n-1)}=\frac{500}{(100-1)}=\frac{500}{99}\Omega \Rightarrow S=5.05\Omega$$

WE -53 :

The resistance of galvanometer is 999Ω. A shunt of 1Ω is connected to it. If the main current is $10^{-2} A$, what is the current flowing through the galvanometer.

Sol. $G=999\Omega, S=1\Omega\ i=10^{-2} A; i_g=?$

$$i_g=i\left(\frac{S}{G+S}\right)=10^{-2}\times\left(\frac{1}{999+1}\right)=10^{-5} A$$

WE -54 :

A galvanometer has a resistance of 98Ω. If 2% of the main current is to be passed through the meter, what should be the value of the shunt?

Sol. $G=98\Omega; \frac{i_g}{i}\times 100=2\%$ $\quad s=\frac{G}{\left(\frac{i}{i_g}-1\right)}; \therefore \frac{i}{i_g}=\frac{100}{2}=50 \quad \therefore S=\frac{98}{(50-1)}=2\Omega$

19. VOLTMETER

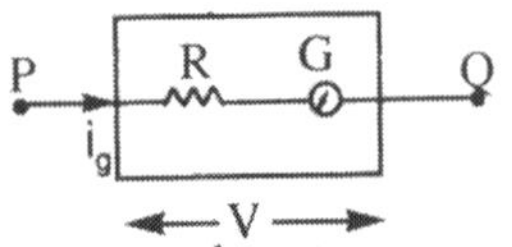

volt meter

i) Galvanometer is converted into voltmeter by connecting high resistance in series to it.

ii) Voltmeter is alwasy connected in parallel to the conductor [P.D. across which is to be measured) in the circuit.

iii) P.D. across the ends of voltmeter is, $V=i_g(G+R)$

iv) Voltmeter is used to measure P.D. across the conductor in electric circuits.

v) Resistance of a voltmeter is very high and that of an ideal volmeter is infinity. Current drawn by an ideal voltmeter is zero.

vi) Among low range and high range voltmeters, high range voltmeter has more resistance.

vii) Equivalent resistance of voltmeter = G+R

viii) Resistance to be connected in series to galvanometer to convert into voltmeter of range $0-V$ volt is

$$R=\frac{V}{i_g}-G$$

ix) To increase the range by n times,

$$n=\frac{new\ range\ V_2}{old\ range\ V_1}=\frac{i_g(G+R)}{i_g(G)}=1+\frac{R}{G}$$

Hence resistance to be conncted in series to galvanometer is $R=G(n-1)$

WE -55 :

A maximum current of 0.5mA can be passed through a galvanometer of resistance 20Ω, Calculate the resistance to be connected in series to convert it into a voltmeter of range $(0-5)V$.

Sol. $R=G(n-1), where\ n=\frac{V}{V_g}$

$V=5V; V_g=i_gG=0.5\times10^{-3}\times20=10^{-2}V$

$\therefore n=500$ and $R=20(500-1)=9980\Omega$

WE -56 :

A galvanometer has a resistance of 100Ω. A current of 10^{-3} A pass through the galvanometer How can it be converted into (A) ammeter of range 10A and (b) voltmeter of range 10v

Sol. $G=100\Omega; i_g=10^{-3}A$

a) $i=10A;\ n=\frac{i}{i_g}=10^4$

$$S=\frac{G}{(n-1)}=\frac{100}{(10^4-1)}=\frac{100}{999}\Omega$$

b) $V_g=i_gG=10^{-3}\times100=10^{-1}V$

$$V=10V\Rightarrow n=\frac{V}{V_g}=\frac{10}{10^{-1}}=100$$

$\therefore R=G(n-1)=100(100-1)=9900\Omega$

WE -57 :

A galvanometer having 30 divisions has current sensitivity of 20 $\mu A / division.$ It has a resistance of 25 ohm. How will you convert it into an ammeter measuring voltmeter reading upto 1V?

Sol. The full scale deflection current

$$i_g = 30 \times \left(20 \times 10^{-6}\right) = 6 \times 10^{-4} A.$$

If S is the required value of the shunt connected in parallel with galvanometer, then

$$i_g = \frac{S}{S+G} i \Rightarrow 6 \times 10^{-4} = \frac{S}{S+25} \times 1$$

After solving, we get $S = \frac{150}{9994} \Omega = 0.0150\Omega$

The resistance of the ammeter

$$R_A = \frac{SG}{S+G} = \frac{0.0150 \times 25}{0.0150 + 25} = 0.0150\Omega$$

To convert this ammeter into the voltmeter, we can use

$V = i_g \left(R_A + R_0\right) \qquad Here V = 1V, i_g = 1A$

$\therefore 1 = 1\left(0.0150 + R_0\right)$ or $R_0 = 0.985\Omega$

WE -58 :

What is the value ot shunt which passes 10% of the main current through a galvanometer of 99 ohm?

Sol. As shunt is a small resistance S in parallel with a galvanometer (of resistance G) as shown in fig.

$$\left(I - I_G\right) S = I_G G$$

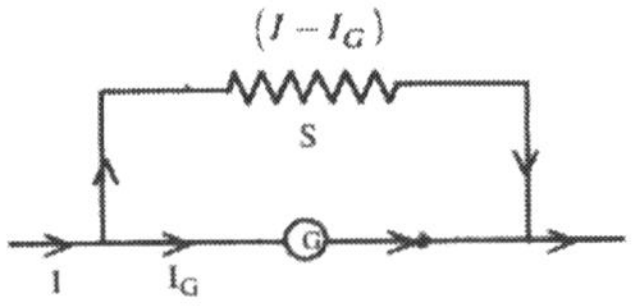

i.e., $S = \frac{I_G G}{\left(I - I_G\right)}$

And as here, $G = 99\Omega$ and

$$I_G = \left(\frac{10}{100}\right) I = 0.1I$$

$$S = \frac{0.1I \times 99}{\left(I - 0.1I\right)} = \frac{0.1}{0.9} \times 99 = 11\Omega$$

CONCEPTUAL QUESTIONS

AMPERE'S CIRCUITAL LAW MAGNETIC FIELD DUE TO STRAIGHT CONDUCTOR

1. **The work done in moving a unit n-pole round a conductor carrying current in a circle of radius 10cm is w. The work done in moving it in a circle of radius 20cm is.**

   1) w 2) 2w 3) w/2 4) 4w

2. **A current I flows along an infinitely long straight thin walled tube. The magnetic induction at a point inside the tube at a distance r from its wall is**

   1) Infinite 2) Zero 3) $\frac{\mu_o}{4\pi}.\frac{2I}{r}$ 4) $\frac{2I}{r}$

3. **A long straight thin conductor has a current of 'i' ampere. The magnetic induction B away from the conductor at a distance 'r' from its axis varies as shown in**

   1) 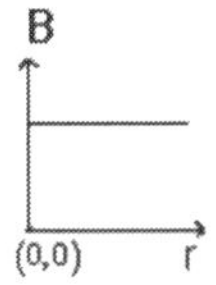

   2) 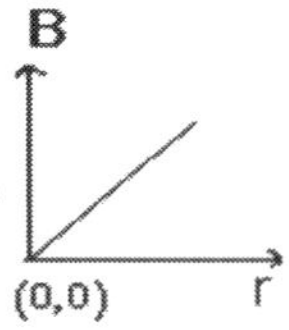

   3) 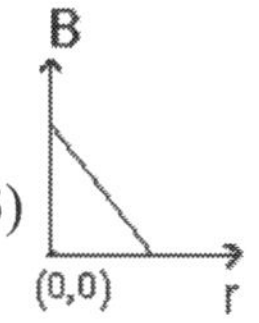

   4) 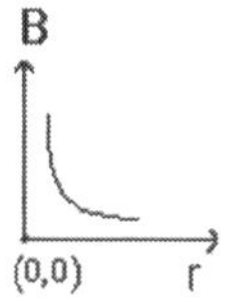

4. **A current carrying wire produces in the neighbourhood**

   1) Electric and magnetic fields 2) Electric field only
   3) Magnetic field only 4) No field

5. **A vertical straight conductor carries a current vertically upwards. A point P lies to the east of it at a small distance and another point Q lies to the west at the same distance. The magnetic field at P is**

   1) Greater than at Q 2) Same as at Q 3) Less than at Q
   4) Greater or less than at Q depending upon the magnetic field of the current.

6. **A current carrying power line carries current from west to east .The direction of magnetic field 1m above is**

   1) north to south 2) south to north 3) east to west 4) west to east

7. **Magnetic lines of force due to a straight conductor carrying current are**

   1) Straight lines 2) Elliptical 3) Circular 4) Parabolic

8. **A current 'i' flows along an infinitely long straight conductor. If r is the perpendicular distance of a point, very far from the ends of the conductor then the magnetic induction B is given by**

   1) $B=\frac{\mu_0}{4\pi}\frac{2i}{r}$ 2) $B=\frac{\mu_0}{4\pi}\frac{i}{r}$ 3) $B=\frac{\mu_0}{4\pi}\frac{\pi i}{r}$ 4) $B=\frac{\mu_0}{4\pi}\frac{2\pi i}{r}$

9. **A current 'i' flows along an infinitely long straight conductor. If 'r' is the perpendicular distance of a point from the lower end of the conductor, then the magnetic induction B is given by**

   1) $B=\frac{\mu_0}{4\pi}\frac{2i}{r}$ 2) $B=\frac{\mu_0}{4\pi}\frac{i}{r}$ 3) $B=\frac{\mu_0}{4\pi}\frac{\pi i}{r}$ 4) $B=\frac{\mu_0}{4\pi}\frac{2\pi i}{r}$

10. **A long straight wire of circular cross-section carries a current along its length. On the axis inside the wire, it follows that**

    1) strength of electric and magnetic fields are zero
    2) strength of electric field is zero but magnetic field is non-zero
    3) strength of electric and magnetic fields is non-zero
    4) strength of electric field is non-zero but magnetic field is zero

BIOT-SAVART LAW AND ITS APPLICATIONS

11. Magnetic field at a point on the line of current carrying conductor is

1) maximum 2) infinity 3) zero 4) finite value

12. If a long hollow copper wire carries a current, then the magnetic field produced will be

1) Inside the pipe only
2) Outside the pipe only
3) Neither inside nor outside the pipe.
4) Both inside and outside the pipe.

13. The magnetic field $\overline{dB}$ due to a small current element $\overline{dl}$ at a distance $\vec{r}$ and carrying current 'i' is

1) $\overline{dB} = \frac{\mu_0}{4\pi} i\left(\frac{\overline{dl}\times\overline{r}}{r}\right)$
2) $\overline{dB} = \frac{\mu_0}{4\pi} i^2\left(\frac{\overline{dl}\times\overline{r}}{r^2}\right)$
3) $\overline{dB} = \frac{\mu_0}{4\pi} i^2\left(\frac{\overline{dl}\times\overline{r}}{r}\right)$
4) $\overline{dB} = \frac{\mu_0}{4\pi} i\left(\frac{\overline{dl}\times\overline{r}}{r^3}\right)$

14. For a given distance from a current element, the magnetic induction is maximum at an angle measured with respect to axis of the current carrying conductor.

1) $3\pi/4$ 2) $\pi/4$ 3) $\pi/2$ 4) 2π

15. If we double the radius of a current carrying coil keeping the current unchanged ,the magnetic field at its centre

1) becomes four times 2) doubled 3) remains unchanged 4) halved

16. A current carrying coil is placed with its plane in the magnetic meridian of the earth. When seen from the east side a clockwise current is set up in the coil.The magnetic field at its centre will be directed towards

1) north 2) south 3) west 4) east

17. A unit N-pole is placed on the axis of a circular coil carrying current in anti-clockwise direction. It experiences a force

1) towards the coil
2) perpendicular to the coil
3) inclined to axis
4) parallel to the coil.

18. If we double the radius of the coil keeping the current through it unchanged,the magnetic field on its axis at very very far away points

1) becomes four times 2) is doubled 3) remains unchanged 4) halved

19. Two concentric circular loops of radii r_1 and r_2 carry clockwise and anticlockwise currents i_1 and i_2. If the centre is a null point, i_1/i_2 must be equal to

1) r_2/r_1 2) r_2^2/r_1^2 3) r_1^2/r_2^2 4) r_1/r_2

20. A length of wire carries a steady current. It is first bent to form a circular coil of one turn. The same length is now bent more sharply to give a loop of two turns of smaller radius. The magnetic field at the centre caused by the same current now will be

1) a quarter of its first value
2) same as that of the first value
3) four times the first value
4) double of its first value

21. A charge 'e' moves round a circle of radius 'r' with a uniform speed 'v'. The magnitude of the magnetic induction at the centre of the circle is

1) $\mu_0 ev/4\pi r$ 2) $\mu_0 ev/4\pi r^2$ 3) $\mu_0 ev/4\pi r^3$ 4) $\mu_0 er/4\pi v$

22. **An electron of charge e moves in a circular orbit of radius r round a nucleus the magnetic field due to orbit motion of the electron at the site of the nucleus is B. The angular velocity ω of the electron is**

1) $\omega = \frac{\mu_0 eB}{4\pi r}$ 2) $\omega = \frac{\mu_0 eB}{\pi r}$ 3) $\omega = \frac{4\pi rB}{\mu_0 e}$ 4) $\omega = \frac{2\pi rB}{\mu_0 e}$

23. **A particle of charge q and mass m moves in a circular orbit of radius r with angular speed ω. The ratio of the magnitude of its magnetic moment to that of its angular momentum depends on**

1) ω : q 2) ω q : m 3) q : 2m 4) ω : m

24. **A loosely wound helix made of stiff wire is mounted vertically with the lower end just touching a dish of mercury when a current from the battery is started in the coil through the mercury**

1) the wire oscillates 2) the wire continues making contact

3)the wire breaks contact just when the current is passed

4) the mercury will expand by heating due to passage of current

25. **Two very thin metallic wires placed along X and Y axes carry equal currents as shown AB and CD are lines at 45^0 with the axes having origin at O the magnetic field will be zero on the line**

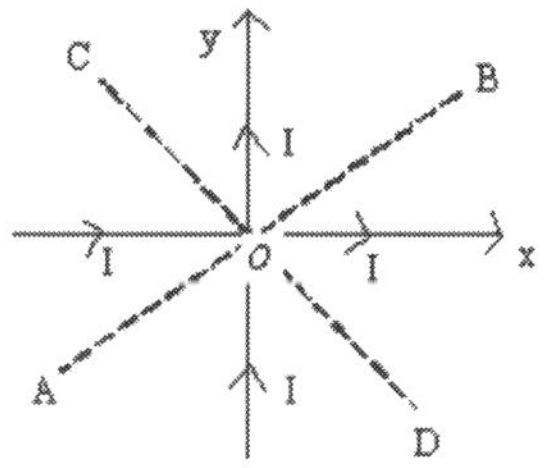

1) AB 2) CD

3) straight segment OB only of line AB 4) straight segment OC only of line CD

26. **A positively charged particle enters at the middle as shown in Fig. with speed 10^5 m/s will bend**

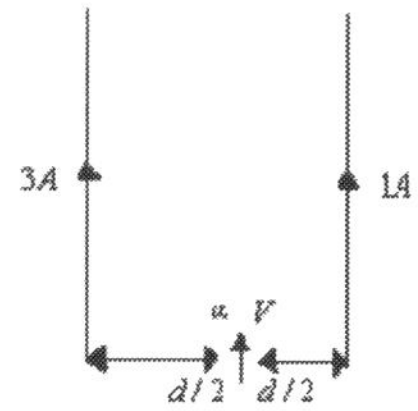

1) towards 1 A wire 2) upwards the plane of wires

3) towards 3 A wire 4) down wards the plane of wires

27. **Net magnetic field at the center of the circle o due to a current through loop ABC as shown in figure is ($\theta < 180^0$)**

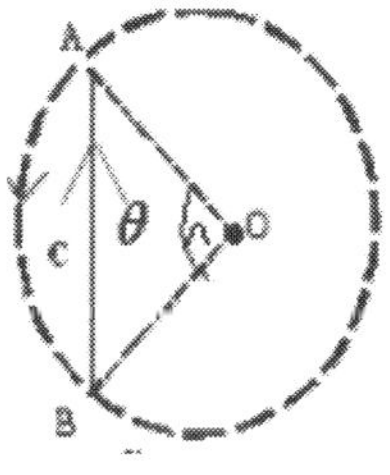

1) zero
2) perpendicular to paper inwards
3) perpendicular to paper outwards
4) perpendicular to the paper inwards if $\theta \le 90^0$ and perpendicular to paper outwards if $90^0 \le \theta \le 180^0$

28*. Match the following and find the correct pairs.

List-I	List-II
a) Fleming's left hand rule	**e) Direction of induced current**
b) Right hand thumb rule	**f) Magnitude and direction of magnetic induction**
c) Biot-Savart law	**g) Direction of force due to magnetic induction**
d) Fleming's right hand rule	**h) Direction of magnetic lines due to current**

1) a-g, b-e, c-f, d-h 2) a-g, b-h, c-f, d-e 3) a-f, b-h, c-g, d-e 4) a-h, b-g, c-e, d-f

TANGENT GALVANOMETER

29. The reduction factor of a tangent galvanometer may be defined as the current passing through it to produce a deflection of

1) 90° 2) 45° 3) 30° 4) 60°

30. A tangent galvanometer is taken from equator to the north pole. During this the sensitivity of the tangent galvanometer

1) decreases because its reduction factor decreases
2) increases because its reduction factor decreases
3) decreases because its reduction factor increases
4) increases because its reduction factor increases

31. In a tangent galvanometer, the circular coil is unwound and rewound to have twice the previous radius. As a result of this the reduction factor (K) of the tangent galvanometer is

1) unaffected 2) doubled 3) quadrupled 4) halved

32. The sensitivity of a tangent galvanometer increases if

1) number of turns decreases
2) number of turns increases
3) field increases
4) number of turns remains same.

33. The plane of the coil of tangent galvanometer is parallel to the magnetic meridian

1) to avoid the influence of earth's magnetic field.
2) to increase the magnetic field due to the current in the coil.
3) to make earth's magnetic field perpendicular to that due to the current in the coil.
4) for some other reason.

34. The galvanometer constant of a tangent galvanometer depends upon

1) earth's magnetic field
2) current in the coil
3) magnetic field of the coil
4) deflection of the magnetic needle

35. The sensitiveness of tangent galvanometer will be maximum if deflection in it is tending to .....and reading is accurate for $\theta =$ ......

1) $0^0, 45^0$ 2) $30^0, 0^0$ 3) $45^0, 0^0$ 4) $60^0, 45^0$

36. A tangent galvanometer of reduction factor 1A is placed with plane of its coil perpendicular to the magnetic meridian when a current of 1A is passed through it. The deflection produced is

1) 45^0 2) Zero 3) 30^0 4) 60^0

Chapter - 7

(Part - 2) MOVING CHARGES AND MAGNETISM

FORCE ON A MOVING CHARGE IN MAGNETIC FIELD

37. An electric charge in uniform motion produces

1) an electric field only
2) a magnetic field only
3) both electric and magnetic fields.
4) no such field at all

38. The force acting on a charge 'q' moving with a velocity V in a magnetic field of induction B is given by

1) $\frac{q}{\vec{V}x\vec{B}}$
2) $\frac{\vec{V}x\vec{B}}{q}$
3) $q(\vec{V}x\vec{B})$
4) $(\vec{V}.\vec{B})q$

39. A magnetic field exerts no force on

1) a stream of electrons
2) a stream of protons
3) unmagnetised piece of iron
4) stationary charge.

40. If the direction of the initial velocity of a charged particle is neither along nor perpendicular to that of the magnetic field, then the orbit will be

1) a straight line
2) an ellipse
3) a circle
4) a helix.

41. A charge moving with velocity V in X direction is subjected to a field of magnetic induction in the negative X direction.As a result the charge will

1) remain unaffected
2) start moving in a circular path in y-z plane
3) retard along X - axis
4) move along a helical path around X - axis

42. The mono energetic beams of electrons moving along +y direction enter a region of uniform electric and magnetic fields. If the beam goes straight through these simultaneously then field B and E are directed respectively along

1) -y axis and -z axis
2)+z axis and -x axis
3) + x axis and - x axis
4)-x axis and -y axis

43. An α - particle moves from E to W in a magnetic field perpendicular to the plane of the paper and into the paper. The particle is deflected towards

1) East
2) West
3) South
4) North

44. A positively charged particle falls vertically downwards. The horizontal component of earth's magnetic field will deflect it towards

1) West
2) East
3) South
4) North

45. An electron and a proton enter a magnetic field with equal velocities. The particle that experiences more force is

1) electron
2) proton
3) both experience same force
4) it cannot be predicted.

46. An electron and a proton enter a magnetic field at right angle to the field with same kinetic energy. The correct statement is

1) trajectory of electron is less curved
2) trajectory of proton is less curved
3) both are equally curved
4) both move along straight line paths

47. A charged particle moving in a magnetic field experiences a resultant force

1) in the direction opposite to that of the field.
2) in the direction opposite to that of its velocity
3) in the direction perpendicular to both field & its velocity
4) in the direction parallel to the field

48. **An electron of mass 'm' is accelerated through a potential difference of V and then it enters a magnetic field of induction B. normal to the lines of force. Then the radius of the circular path is**

1) $\sqrt{\frac{2eV}{m}}$ 2) $\sqrt{\frac{2Vm}{eB^2}}$ 3) $\sqrt{\frac{2Vm}{eB}}$ 4) $\sqrt{\frac{2Vm}{e^2B}}$

49. **Among the following, one characteristic is independent of the angle between $\vec{V}$ and $\vec{B}$**

1) Momentum 2) Radius of helical path
3) Angular speed 4) Both 1 and 2.

50. **A charged particle enters into a uniform magnetic field the parameter that remains constant is**

1) velocity 2) momentum
3) kinetic energy 4) angular velocity

51. **A free charged particle moves through a magnetic field. The particle may undergo a change in**

1) speed 2) energy
3) direction of motion 4) magnitude of the velocity

52. **An electron is projected parallel to electric and uniform magnetic fields acting simultaneously in the same direction .The electron**

1) gains kinetic energy 2) loses kinetic energy
3) moves along circular path 4) moves along a parabolic path

53. **A charged particle is moving with velocity 'V' in a magnetic field of induction B. The force on the particle will be maximum when**

1) V and B are in the same direction 2) V and B are in Opposite direction
3) V and B are perpendicular 4) V and B are at an angle of 45^0

54. **If electron velocity is 2i+4j and it is subjected to magnetic field of 4k, then its**

1) speed will change 2) path will change
3)velocity is Constant 4) momentum is Constant

55. **A proton moving with a velocity V is acted upon by electric field E and magnetic field B. The proton will move undeflected if**

1) E is perpendicular to B and E parallell to V
2) E is parallel toV and perpendicular to B
3) E and B both are parallel to V
4) E, V and B are mutually perpendicular and V = E/B

56. **A proton and an alpha particle enter into a uniform magnetic field with the same velocity.The period of rotation of the alpha particle will be**

1) four times that of proton 2) two times that of proton
3) three times that of proton 4) same as that of proton

57. **A uniform electric field and a uniform magnetic field exist in a region in the same direction An electron is projected with velocity pointed in the same direction the electron will**

1) turn to its right 2) turn to its left
3) keep moving in the same direction but its speed will increase
4) keep moving in the same direction but its speed will decrease

58. **Imagine that you are seated in a room and there is a uniform magnetic field pointing vertically down wards in it at the centre of the room an electron is projected horizontally from left to right with a certain speed. Discuss the speed and the path of the electron in this field**

1) electron moves in anticlockwise path
2) electron moves in clockwise path
3) electron moves left wards
4) electron moves right wards

59. **A charged particle with charge q enters a region of constant uniform and mutually orthogonal fields $\vec{E}$ and $\vec{B}$ with a velocity $\vec{v}$ perpendicular to both $\vec{E}$ and $\vec{B}$ and comes out without any change in magnitude and direction of $\vec{v}$. Then**

1) $\vec{v} = \left(\vec{E} \times \vec{B}\right) / B^2$
2) $\vec{v} = \left(\vec{B} \times \vec{E}\right) / B^2$
3) $\vec{v} = \left(\vec{E} \times \vec{B}\right) / E^2$
4) $\vec{v} = \left(\vec{B} \times \vec{E}\right) / E^2$

60. **An electron of charge e, revolves round in an orbit of radius r with a uniform angular velocity ω. The magnetic dipole moment of the electron in the orbit is**

1) $e\omega r / 2$
2) $e\omega r^2 / 2$
3) $e\omega^2 r / 2$
4) $e\omega^2 r^2 / 2$

61. **A proton and a deuteron are projected with same speeds normal to a uniform magnetic field. Which of the following statements is/are true**

a) The ratio of their respective time periods is 1 : 2

b) The ratio of their respective angular momenta about the centres of their circular path is 1 : 4

c) The ratio of their respective radii of their circular paths is 1 : 2

1) only a
2) only c
3) only a, b
4) All are true

62. **If a charged particle is projected perpendicular to a uniform magnetic field, then**

a) it revolves in circular path
b) its K.E. remains constant
c) its momentum remains constant
d) its path is spiral

1) only a, b are correct
2) only a, c are correct
3) only b, d are correct
4) only a, d are correct

63. **When a positively charged particle enters a uniform magnetic field with uniform velocity, its trajectory can be**

a) a straight line
b) a circle
c) a helix

1) a only
2) a or b
3) a or c
4) any one of a, b and c

FORCE ON A CURRENT CARRYING CONDUCTOR IN A MAGNETIC FIELD

64. **A circular coil of wire carries a current. PQ is a part of very long wire carrying a current and passing close to the circular coil. if the directions of currents are those shown in figure, then the direction of the force acting on PQ is**

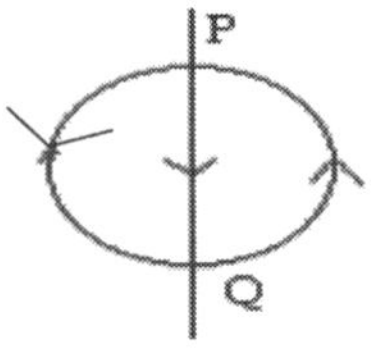

1) parallel to PQ, towards p
2) parallel to PQ, towards Q
3) at right angles to PQ, to the right
4) at right angles to PQ, to the left

65. **A conductor AB of length l carrying a current i is placed perpendicular to a long straight conductor XY carrying a current I, as shown. the force on AB will act**

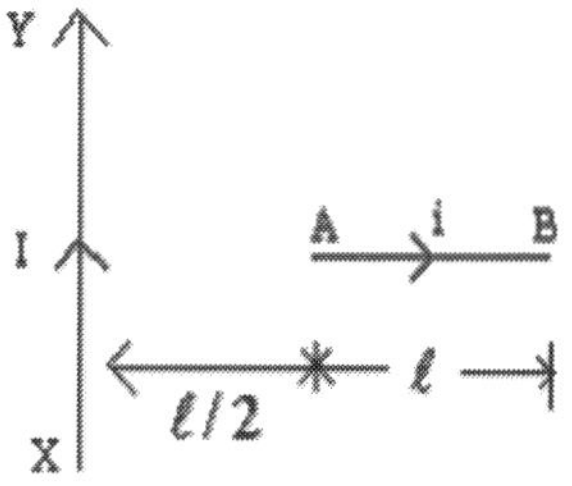

1) along x to y 2) along y to x 3) to the right 4) to the left

66. **A conducting circular loop of radius r carries a constant current I it is placed in a uniform magnetic field B such that B is perpendicular to the plane of the loop the magnetic force acting on the loop is**

1) B I R 2) 2π (B I R) 3) zero 4) π (B I R)

FORCE BETWEEN TWO PARALLEL CURRENT CARRYING CONDUCTORS

67. **Two coplanar circular coils of equal radius carrying currents i_1, i_2 in opposite directions are at a large distance 'd'. The distance from the first coil where the resultant magnetic induction is zero is**

1) $\frac{d}{1+\frac{i_2}{i_1}}$ 2) $\frac{d}{1+\sqrt{\frac{i_2}{i_1}}}$ 3) $\frac{d}{1+\left(\frac{i_2}{i_1}\right)^{\frac{1}{3}}}$ 4) $\frac{d}{1+\left(\frac{i_2}{i_1}\right)^{2}}$

68. **A rectangular loop carrying current I is located near an infinite long straight conductor carrying current I as shown in the figure. The loop,**

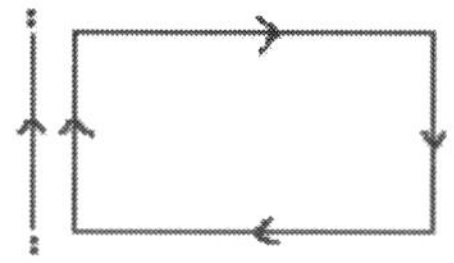

1) remain stationary 2) is attracted towards the wire
3) is repelled away from the wire 4) will rotate about an axis parallel to the wire

69. **Two streams of protons move parallel to each other in the same direction. They will**

1) attract each other 2) repel each other 3) neither attract nor repel 4) rotate

70. **Two streams of electrons are moving parallel to each other in the same direction. They**

1) attract each other 2) repel each other
3) cancel the electric field of each other 4) cancel the magnetic field of each other

71. **A light body is hanging at the lower end of a vertical spring. On passing current in the spring, the body**

1) rises up 2) goes down 3) no change 4) oscillates up & down

72. **A current carrying wire is placed along east and west in a magnetic field directed north wards. If the current in the wire is directed east wards, the direction of force on the wire is**

1) due west 2) due south
3) vertically upwards 4) vertically downwards

73. **Two parallel, long wires carry currents i_1 & i_2 ($i_1 > i_2$) when the currents are in the same direction, the magnetic induction at a point midway between the two wires is X. If the direction of i_2 is reversed, the magnetic induction becomes 2x, then i_1 / i_2 is**

1) 1 2) 2 3) 3 4) 4

74. **Two straight long conductors AOB and COD are perpendicular to each other and carry currents i_1 and i_2. The magnitude of magnetic induction at a point P at a distance a from the point O in the direction perpendicular to the plane ABCD is**

1) $(\mu_0 / 2\pi a)(i_1 + i_2)$ 2) $(\mu_0 / 2\pi a)(i_1 - i_2)$

3) $(\mu_0 / 2\pi a)(i_1^2 + i_2^2)^{1/2}$ 4) $(\mu_0 / 2\pi a)[i_1 i_2/(i_1+i_2)]$

75. **Two parallel wires carrying current I and 2I in same direction have magnetic field B at the midpoint between them. If the current 2I is switched off, the magnetic field at that point will be**

1) B/2 2) B 3) 2B 4) 3B

76. **Two long straight horizontal parallel wires one above the other are separated by a distance '2a'. If the wires carry equal currents in opposite directions, the magnitude of the magnetic induction in the plane of the wires at a distance 'a' above the upper wire is**

1) $\frac{\mu_o i}{2\pi a}$ 2) $\frac{\mu_o i}{2\pi a} + \frac{\mu_o i}{4\pi a}$ 3) $\frac{\mu_o i}{2\pi a} - \frac{\mu_o i}{4\pi a}$ 4) $\frac{\mu_o i}{3\pi a}$

77. **Choose the correct statement. There will be no force experienced if**

1) two parallel wires carry currents in the same direction

2) two parallel wires carry currents in the opposite direction

3) a positive charge is projected between the pole pieces of bar magnet

4) a positive charge is projected along the axis of a solenoid carrying current

CURRENT LOOP AND MAGNETIC DIPOLE

78. **If the angular momentum of an electron revolving in a cirular orbit is L,then its magnetic moment is**

1) eLm 2) eL/m 3) eL/2m 4) zero

79. **The dipole moment of a current loop is independent of**

1) current in the loop 2) number of turns

3) area of the loop 4) magnetic field in which it is situated

80. **Tangent law is applicable to a dipole placed in two magnetic fields $\vec{B}$ and $\vec{Bo}$ when**

1) $\vec{B} = \vec{B}_o$

2) $\vec{B}$ and $\vec{B}_o$ are perpendicular to each other 3) $\vec{B}$ makes any angle with $\vec{B}_o$

4) $\vec{B}$ is directed opposite to $\vec{B}_o$

81. **A magnetic dipole placed in two perpendicular magnetic fields $\vec{B}$ and $\overline{B}_0$ is in equilibrium making an angle θ with $\vec{B}$ then.**

1) $B = B_0$ 2) $B \cos\theta = B_o \sin\theta$

3) $B \sin\theta = B_o \cos\theta$ 4) $B = B_o \tan\theta$

82. A current loop placed in a magnetic field behaves like a

1) magnetic dipole
2) magnetic substance
3) magnetic pole
4) non magnetic substance

83. Singly ionized helium (x), ionized deuteron(y), alpha(z) particles are projected into a uniform magnetic field 3×10^{-4} tesla with velocities 10^5 ms^{-1}, 0.4×10^4 ms^{-1} and 2×10^3 ms^{-1} respectively. The correct relation between the ratio of the angular momentum to the magnetic moment of the particles is

1) $x > y = z$
2) $x < y < z$
3) $y < x < z$
4) $z > x > y$

84. A small coil of N turns has an area A and a current 'i' flows through it. The magnetic dipole moment of the coil will be

1) iNA
2) i^2NA
3) iN^2A
4) iN/A

TORQUE ON CURRENT LOOP

85. A straight horizontal conductor of length L meter and mass m kg carries a current 'i' ampere. The minimum magnetic induction which must exist in the region to balance its weight

1) mg/iL
2) iL/mg
3) mgL/i
4) mL/ig

86. A current carrying loop in a uniform magnetic field will experience

1) force only
2) torque only
3) both torque and force
4) neither torque nor force

87. The torque acting on a magnetic dipole of moment P_m when placed in a magnetic field is

1) P_mB
2) $\overline{P}_m \times \overline{B}$
3) $\overline{P}_m . \overline{B}$
4) P_m/B

88. A coil of area A, turns N and carrying current i is placed with its face parallel to the lines of magnetic induction B. The work done in rotating the coil through an angle of 180^0 is

1) iNAB
2) 2iNAB
3) iNAB/2
4) zero

89. A conducting circular loop of radius 'r' carries a constant current 'i'. It is placed in a uniform magnetic field $\overline{B}_0$ such that $\overline{B}_0$ is perpendicular to the plane of the loop. The magnetic force acting on the loop is

1) irB_0
2) $2\pi riB_0$
3) zero
4) πriB_0

90. A current carrying loop is free to turn in a uniform magnetic field. The loop will then come into equilibrium when its plane is inclined at

1) 0^0 to the direction of the field.
2) 45^0 to the direction of the field.
3) 90^0 to the direction of the field.
4) 60^0 to the direction of the field.

91. When a current carrying coil is placed in a uniform magnetic field of induction B, then a torque τ acts on it. If I is the current, n is the number of turns and A is the face area of the coil and the normal to the coil makes an angle θ with B, Then

1) $\tau = BInA$
2) $\tau = BInA\sin\theta$
3) $\tau = BInA\cos\theta$
4) $\tau = BInA\tan\theta$

92. A moving coil type of galvanometer is based upon the principle that

1) a coil carrying current experiences a torque in magnetic field.
2) a coil carrying current produces a magnetic field.
3) a coil carrying current experiences impulse in a magnetic field.
4) a coil carrying current experiences a force in magnetic field.

93. **Four wires each of length 2m are bent into four loops P,Q,R, and S and then suspended into a uniform magnetic field same current is passed in each loop.The correct statement is**

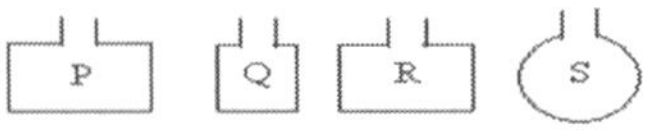

1) couple on loop P will be highest
2) couple on loop Q will be highest
3) couple on loop R will be highest
4) couple on loop S will be highest

94. **Two circular coils carrying currents are of nearly same radius have common centre and released from rest with their planes perpenducular. Assuming that they can freely rotate about their diameters, select the wrong alternative**

1) Each will exert a torque on the other
2) Through out their rotation, angular momentum of the system is conserved
3) Angular momentum of system initially increases and then decreases
4) Potential energy of system first decreases

95. **A current carrying circular coil, suspended freely in a uniform external magnetic field orients to a position of stable equilibrium.In this state:**

1) the plane of the coil is normal to the external magnetic field
2) the plane of the coil is parallel to the external magnetic field
3) flux through the coil is minimum
4) torque on the coil is maximum

96. **A conducting wire of length l is turned in the form of a circular coil and a current i is passed through it.For torque due to external uniform magnetic field to be maximum, the number of turns in the coil will be**

1) 1
2) 2
3) infinity
4)0

97. **When a current loop is placed in a uniform magnetic field**

i) $\overrightarrow{F_R} = 0\, and\, \overline{\tau}$
ii) $\overrightarrow{F_R} = 0\, but\, \overline{\tau} \neq 0$
iii) $\overrightarrow{F_R} \neq 0\, but\, \tau = 0$
iv) $\overrightarrow{F_R} \neq 0 \;\; and \;\; \tau \neq 0$

1) only i & ii are ture
2) only ii & iii are true
3) only iii & iv are true
4) only i & iv are true

98. **When a current carrying coil is situated in a uniform magnetic field with its magnetic moment antiparallel to the field**

i) Torque on it is maximum
ii) Torque on it is minimum
iii) PE of loop is maximum
iv) PE of loop is minimum

1) only i and ii are true
2) only ii and iii are true
3) only iii and iv are true
4) only i, ii and iii are true

MOVING COIL GALVANOMETER

99. **In a moving coil galvanometer a radial magnetic field is applied with concave magnetic poles, to have**

A) uniform magnetic field
B) the plane of the coil parallel to field

1) A, B true
2) A,B false
3) A true , B false
4) A false B true

100. **A current carrying coil tends to set itself**

1) parallel to an external magnetic field.
2) parallel to its own magnetic field
3) perpendicular to the external magnetic field.
4) perpendicular to the geographic meridian

101. **The restoring couple in the moving coil galvanometer is due to**

1) current in the coil
2) magnetic field of the magnet.
3) material of the coil.
4) twist produced in the suspension wire.

102. **A wire of length 'L' is made in the form of a coil in a moving coil galvanometer. To have maximum sensitiveness the shape of the coil is**

1) circular
2) elliptical
3) rectangular
4) square

103. The relation between voltage sensitivity σ_V and the current sensitivity σ_i of a moving coil galvammeter is (Given that G is the resistance of the galvanometer)

1) $\sigma_V = G\sigma_i$ 2) $\sigma_V = \sigma_i / G$ 3) $\sigma_V \sigma_i = G$ 4) $\sigma_V \sigma_i = 1/G$

104. The resistance of an ideal voltmeter is

1) zero 2) infinity 3) finite, very small 4) finite and large

105. The sensitivity of a moving coil galvanometer increases with the decrease in

1) number of turns 2) area of coil 3) magnetic field 4) couple per unit twist

106. If a galvanometer is shunted then among the following which statement is not true

1) effective range increases. 2) equivalent resistance decreases.
3) galvanometer becomes more sensitive 4) galvanometer becomes more protective.

107. The purpose of soft iron cylinder between the pole pieces of the horse-shoe magnet in a moving coil galvanometer is

1) to increase the magnetic induction in the polar gap
2) to evenly distribute the magnetic lines of force
3) to provide a radial magnetic field
4) to reduce the magnetic flux leakage in the polar gap

108. The radial magnetic field is used in a suspended coil galvanometer to provide

1) a uniform torque on the coil 2) maximum torque on the coil in all positions
3) a uniform and maximum torque in all positions of the coil
4) a non uniform torque on the coil

109. Assertion (A): In M.C.G., the deflection 'θ' is directly proportional to the strength of the current

Reason (R): In M.C.G., the the torque experienced by the loop is BiANcosθ

1) Both A and R are correct , R is correct reason of A
2) Both are wrong
3) Both A and R are correct and R is not the correct reason of A
4) A is correct, R is wrong

CONVERSION OF MCG INTO AMMETER

110. To measure the resistance of a device using Ohm's law the mode of connection used is

1) ammeter in series, voltmeter in parallel 2) voltmeter in series, ammeter in parallel
3) both ammeter and voltmeter in series 4) both ammeter and voltmeter in parallel

111. To increase the range of an ammeter, we need to connect a suitable

1) low resistance in parallel 2) low resistance in series
3) high resistance in parallel 4) high resistance in series.

112. An ammeter has a resistance of G ohm and a range of 'i' ampere. The value of resistance used in parallel, to convert into an ammeter of range 'ni' ampere is

1) nG 2) (n-1)G 3) G/n 4) G/n-1

CONVERSION OF MCG INTO VOLTMETER

113. Among the following the false statement is

1) ammeter is connected in series and maximum current flows through it
2) voltmeter is connected in parallel and potential is maximum
3) ammeter is connected in series and current through it is negligible
4) voltmeter is connected in parallel and current through it is negligible.

114. A voltmeter has a resistance of g ohm and range of V volt. The value of resistance used in series to convert it into voltmeter of range nV volt is

1) ng 2) $g(n-1)$ 3) $\frac{g}{n}$ 4) $\frac{g}{(n-1)}$

115. In an electrical circuit containing a source of emf and a load resistance, the voltmeter is connected by mistake in series with the load across the source and ammeter is connected parallel to the load. Then which meter burns out

1) ammeter 2) voltmeter 3) both ammeter and voltmeter 4) neither ammeter nor voltmeter

116. If a voltmeter, in advertently mistaken for an ammeter, were inserted into the circuit, the current

1) increases 2) remains same 3) decreases 4) becomes zero

117. An ammeter and a voltmeter of resistance R are connected in series to an electric cell of negligible internal resistance. Their readings are A and V respectively. If another resistance R is connected in parallel with the voltmeter, then

1) Both A and V increases 2) Both A and V decreases
3) A decreases but V increases 4) A increases but V decreases.

118. A moving coil voltmeter is generally used to measure the potential difference across a conductor of resistance 'r' carrying a current i. The resistance of voltmeter is R. For more correct measurement of potential difference

1) $R = r$ 2) $R >> r$ 3) $R << r$ 4) $R = 0$

119. The resistance of an ideal voltmeter is

1) Zero 2) infinity 3) 1000 Ω 4) 10000 Ω

120. If G, r_A and r_V denote the internal resistances of a galvanometer, an ammeter and a voltmeter among the following the correct relationship is

1) $G < r_A < r_V$ 2) $r_A < r_V < G$ 3) $r_A < G < r_V$ 4) $r_V < r_A < G$

121. Among the following the true statement is,

1) ammeter is a high resistance galvanometer and voltmeter is a low resistance galvanometer
2) ammeter is a low resistance galvanometer and voltmeter is a high resistance galvanometer.
3) ammeter and voltmeter cannot be distinguished on the basis of their resistance.
4) ammeter and voltmeter have same resistance

KEY CONCEPTUAL

1) 1	2) 2	3) 4	4) 3	5) 1	6) 1	7) 3	8) 1	9) 2	10) 1
11) 3	12) 2	13) 4	14) 3	15) 4	16) 3	17) 3	18) 1	19) 4	20) 3
21) 2	22) 3	23)3	24) 1	25)1	26)3	27) 3	28) 2	29) 2	30) 2
31) 3	32) 2	33) 3	34) 1	35) 1	36) 2	37) 3	38) 3	39) 4	40) 4
41) 1	42) 2	43) 3	44) 2	45) 3	46) 2	47) 3	48) 2	49) 3	50) 3
51) 3	52) 2	53) 3	54) 2	55) 4	56) 2	57) 4	58) 2	59) 1	60) 2
61) 4	62) 1	63) 4	64) 4	65) 1	66) 3	67) 3	68) 2	69) 2	70) 2
71) 1	72) 3	73) 3	74) 3	75) 2	76) 4	77) 4	78) 3	79) 4	80) 2
81) 3	82) 1	83)1	84) 1	85) 1	86) 2	87) 2	88) 2	89) 1	90) 3
91) 2	92) 1	93) 4	94) 3	95) 1	96) 1	97) 2	98) 2	99) 1	100) 3
101) 4	102) 1	103)2	104)2	105) 4	106) 3	107) 1	108)1	109) 4	110) 1
111) 1	112) 4	113) 3	114) 2	115) 4	116) 3	117) 4	118) 2	119) 2	120) 3
121) 2									

LEVEL - I (CW)

AMPERE'S CIRCUITAL LAW BIOT-SAVART LAW AND ITS APPLICATIONS

1. A north pole of strengt π A m, is moved around a circle of radius 10 cm which lies around a long straight conductor carrying a current of 10 A. The work done is nearly

   1) 4 μ J 2) 40 μ J 3) 400 μ J 4) 0.4 μ J

2. There is 10 units of charge at the centre of a circle of radius 10m. The workdone in moving one unit of charge around the circle once is

   1) Zero 2) 10 units 3) 100 units 4) 1unit

3. A current of $\frac{1}{4\pi}$ A is flowing in a long straight conductor. The line integral of magnetic induction around a closed path enclosing the current carrying conductor is

   1) $4\pi\times10^{-7}$ Wb/m 2) 10^{-7} Wb/m 3) $16\pi^2\times10^{-7}$ Wb/m 4) zero

4. A wire in the form of a square of side '2m' carries a current 2A. Then the magnetic induction at the centre of the square wire is (magnetic permeability of free space = μ_0)

   1) $\frac{\mu_0}{2\pi}$ 2) $\frac{\mu_0\sqrt{2}}{\pi}$ 3) $\frac{2\sqrt{2}\mu_0}{\pi}$ 4) $\frac{\mu_0}{\sqrt{2}\pi}$

5. An electric current passes through a long straight wire. At a distance 5 cm from the wire, the magnetic field is B. The field at 20 cm from the wire would be.

   1) 2B 2) B/4 3) B/2 4) B

6. A current of one ampere is passed through a straight wire of length 2 metre. The magnetic field at a point in air at a distance of 3 m from one end of the wire but lying on the axis of the wire will be

   1) $\mu_0/2\pi$ 2) $\mu_0/4\pi$ 3) $\mu_0/8\pi$ 4) Zero

7. A straight vertical conductor carries a current. At a point 5 cm due north of it, the magnetic induction is found to be 20 μT due east. The magnetic induction at a point 10 cm east of it will be

   1) 5μT north 2) 10 μT north 3) 5μT south 4) 10μT south

8. A circular coil of radius 25 cm, carries a current of 50 ampere. If it has 35 turns, the flux density at the centre of the coil is (in Wb/m^2)

   1) $\pi\times10^{-3}$ 2) $1.4\pi\times10^{-3}$ 3) $14\pi\times10^{-3}$ 4) $2\pi\times10^{-3}$

9. A circular coil of radius R carries a current i. The magnetic field at its centre is B. The distance from the centre on the axis of the coil where the magnetic field will be B/8 is

   1) $\sqrt{2}$R 2) $\sqrt{3}$R 3) 2R 4) 3R

10. Two circular coils are made of two identical wires of same length and carry same current. If the number of turns of the two coils are 4 and 2, then the ratio of magnetic induction at the centres will be

    1) 2:1 2) 1:2 3) 1:1 4) 4:1

11. Two circular coils have diameters 10cm and 20cm with same number of turns. The ratio of the magnetic field induction produced at the centre of the coils when connected in series is

    1) 1:2 2) 2:1 3) 4:1 4) 1:4

12. A wire carrying a current of 4 A is in the form of a circle. It is necessary to have a magnetic field of induction $\pi\times10^{-5}$ T at the centre. The radius of the circle must be

    1) 0.08 cm 2) 0.8 cm 3) 8 cm 4) 80 cm

13. Two concentric circular coils A and B have radii 25 cm and 15 cm and carry currents 10A and 15A respectively. A has 24 turns and B has 18 turns. The direction of currents are in opposite order. The magnetic induction at the common centre of the coil is

1) 120 μ_0 T 2) 480 μ_0 T 3) 420 μ_0 T 4) μ_0

14. A wire carrying a current of 140 ampere is bent into the form of a circle of radius 6 cm. The flux density at a distance of 8 cm on the axis passing through the centre of the coil and perpendicular to its plane is (in Wb/m^2 (approximately))

1) $\pi\times10^{-4}$ 2) $2\pi\times10^{-4}$ 3) $\frac{\pi}{2}\times10^{-4}$ 4) $\frac{1}{\pi}\times10^{-4}$

15. The magnetic induction at a point at a large distance d on the axial line of circular coil of small radius carrying current is 120 μ T. At a distance 2d the magnetic induction would be

1) 60 μ T 2) 30 μ T 3) 15 μ T 4) 240 μ T

16. A particle having charge 100 times that of an electron is revolving in a circular path of radius 0.8m with one rotation per second. The magnetic field produced at the centre is

1) $10^{-7}/\mu_0$ 2) $10^{-17}\ \mu_0$ 3) $10^{-6}\ \mu_0$ 4) $10^{-15}/\mu_0$

17. A battery is connected between two points A and B on the circumference of a uniform conducting ring of radius r and resitance R. One of the arcs AB subtends an angle θ at the centre. The value of the magnetic induction at the centre due to the current in the ring

1) proportional to 2 $(180^0-\theta)$ 2) Inversely proportional to r

3) zero only if $\theta=180^0$ 4) zero for all values of θ

TANGENT GALVANOMETER

18. A TG has 500 turns, each of radius $2\pi cm$. If $B_H=3.6\times10^{-5}Wb/m^2$, The deflection due to 7.2 mA current is

1) 60^0 2) 30^0 3) 45^0 4) Zero

19. In a properly adjusted tangent galvanometer, the deflection for 1 A current is found to be 30°. Now the coil is turned through 90° about the vertical axis, the deflection for the same current will be

1) 60° 2) 30° 3) 90° 4) 0°

20. Two tangent galvanometers are connected in series across a battery. The deflections in them are found to be 30° and 60° respectively. The ratio of their reduction factors is

1) $\sqrt{3}:1$ 2) $1:\sqrt{3}$ 3) 3 : 1 4) 1 : 3

21. In a tangent galvanometer, the magnetic induction produced by the coil of wire situated in the magnetic meridian is found to be equal to the horizontal component of the earth's magnetic field. The deflection produced in it will be

1) 30° 2) 60° 3) 45° 4) 90°

CURRENT LOOP AND MAGNETIC DIPOLE

22. If an electron is revolving in a circular orbit of radius 0.5 A^0 with a velocity of 2.2×10^6 m/s. The magnetic dipole moment of the revolving electron is

1) 8.8×10^{-24} Am 2) 8.8×10^{-23} Am 3) 8.8×10^{-22} Am 4) 8.8×10^{-21} Am

23. Magnetic induction at the centre of a circular loop of area π square meter is 0.1 tesla. The magnetic moment of the loop is (μ_0 is permeability of air)

1) $\frac{0.1\pi}{\mu_0}$ 2) $\frac{0.2\pi}{\mu_0}$ 3) $\frac{0.3\pi}{\mu_0}$ 4) $\frac{0.4\pi}{\mu_0}$

SOLENOID & TOROID

24. The length of a solenoid is 0.1 m and its diameter is very small. A wire is wound over in two layers. The number of turns in the inner layer is 50 and that on the outer layer is 40. The strength of current flowing in two layers in the same direction is 3 ampere. The magnetic induction in the middle of the solenoid will be

1) $3.4\times10^{-3}T$ 2) $3.4\times10^{-3}gauss$ 3) $3.4\times10^{3}T$ 4) $3.4\times10^{3}gauss$

25. The magnetic induction at the centre of a solenoid is B. If the length of the solenoid is reduced to half and the same wire is wound in two layers the new magnetic induction is

1) B 2) 2B 3) B/2 4) 4B

FORCE ACTING ON A MOVING CHARGE IN MAGNETIC FIELD

26. A proton is fired with a speed of $2\times10^{6}m/s$ at an angle of 60° to the X- axis. If a uniform magnetic field of 0.1 tesla is applied along the Y- axis, the force acting on the proton is

1) 1.603×10^{-14} N 2) 1.6×10^{-14} N 3) 3.203×10^{-14} N 4) 3.2×10^{-14} N

27. A closed circuit is in the form of a regular hexagon of side a. If the circuit carries a current I, what is magnetic induction at the centre of the hexagon?

1) $\frac{\sqrt{3}\mu_0 I}{4\pi a}$ 2) $\frac{\sqrt{3}\mu_0 I}{2\pi a}$ 3) $\frac{\sqrt{3}\mu_0 I}{3\pi a}$ 4) $\frac{\sqrt{3}\mu_0 I}{\pi a}$

28. A proton enters a magnetic field with a velocity of $2.5X10^7ms^{-1}$ making an angle 30^0 with the magnetic field. The force on the proton is (B=25T)

1) $1.25X10^{-11}$N 2) $2.5X10^{-11}$N 3) $5.0X10^{-11}$N 4) $7.5X10^{-11}$N

29. A doubly ionised He^{+2} atom travels at right angles to a magnetic field of induction 0.4T with a velocity of 10^5ms^{-1} describing a circle of radius r. A proton travelling with same speed in same direction in the same field will describe a circle of radius

1) r/4 2) r/2 3) r 4) 2r

30. A proton is projected with a velocity $10^7 ms^{-1}$, at right angles to a uniform magnetic field of induction 100 mT. The time (in seconds) taken by the proton to traverse 90^0 arc is:(Mass of proton $=1.65\times10^{-27}$ kg and charge of proton $=1.6\times10^{-19}C$)

1) 0.81×10^{-7} 2) 1.62×10^{-7} 3) 2.43×10^{-7} 4) 3.24×10^{-7}

31. A 2MeV proton is moving perpendicular to a uniform magnetic field of 2.5 tesla. The force on the proton is

1) $2.5 X 10^{-10}$newton 2) $8 X 10^{-11}$newton
3) $2.5 X 10^{-11}$newton 4) $8 X 10^{-12}$newton

CYCLOTRON

32. A cyclotron in which protons are accelerated has a flux density of 1.57T. The variation of frequency of electric field is (in Hz)

1) 4.8×10^{8} 2) 8.4×10^{8} 3) 2.5×10^{7} 4) 4.8×10^{6}

33. Cyclotron is adjusted to give proton beam, magnetic induction is 0.15 wbm^{-2} and the extreme radius is 1.5 m. The energy of emergent proton in MeV will be

1) 34.2 2) 3.42 3) 2.43 4) 24.3

34 A cyclotron has an oscillator frequency 12MHz and a dee of radius 50 cm. Calculate the magnetic induction needed to accelerate deuterons of mass $3.3\times10^{-27}kg$ and charge $1.6\times10^{-19}C$

1) $1.55wb/m^2$ 2) $2.55wb/m^2$ 3) $0.55wb/m^2$ 4) $3.55wb/m^2$

FORCE ON A CURRENT CARRYING CONDUCTOR IN A MAGNETIC FIELD

35. A straight wire (conductor) of length 10 cm is kept in a uniform magnetic field of induction 0.02 T. The angle between the conductor and the field direction is 30^0. A current of 5A is passed through the conductor. The force on the conductor is (in N)

1) 4×10^{-3} 2) 5×10^{-3} 3) 6×10^{-3} 4) 7×10^{-3}

36 A circular coil of 20 turns and radius 10 cm is placed in a uniform magnetic field of 0.1 T normal to the plane of the coil .If the current in the coil is 5.0 A what is the average force on each electron in the coil due ot the magnetic field (The coil is made of copper wire of cross-sectional area $10^{-5}m^2$ and the free electron density in copper is given to be about $10^{29}m^{-3}$.)

1) $2.5\times10^{-25}N$ 2) $7.5\times10^{-25}N$ 3) $5\times10^{-25}N$ 4) $10^{-25}N$

37. A thin 50 cm long metal bar with mass 750 g rests on,but is not attached to , two metallic supports in a uniform 0.45T magnetic field as shown in Fig .A battery and a 25Ω resistor in series are onnceted to the supports. The largest voltage the battery can have without breaking the circuit at the supports (units are in"V") is

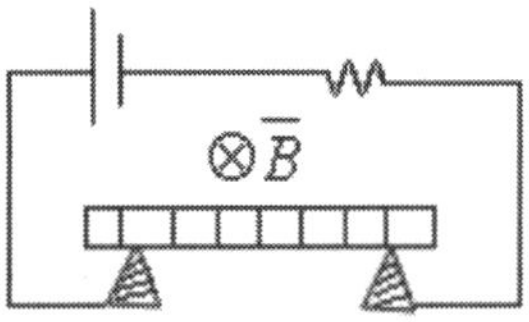

1) 817 2) 718 3) 827 4)837

FORCE BETWEEN TWO PARALLEL CURRENT CARRYING CONDUCTORS

38. The magnitude of the force between a pair of conductors, each of length 110 cm, carrying a current of 10A and seperated by a distance of 10 cm is

1) $55\times10^{-5}N$ 2) $44\times10^{-5}N$ 3) $33\times10^{-5}N$ 4) $22\times10^{-5}N$

39. Two parallel conductors A and B separated by 5 cm carry electric current of 6 A and 2 A in the same direction. The point between A and B where the field is zero at

1) 0.25 cm from B 2) 1 cm from B 3) 1.25 cm from B 4) 3.75 cm from B

40. The distance between the wires of electric mains is 12cm. These wires exprience 4 mgwt per unit length. The value of current flowing in each wire will be if they carry current in same direction

1) $4.85A$ 2) $zero$ 3) $4.85X10^{-2}A$ 4) $85X10^{-4}A$

41. Two long straight conductors are held parallel to each other 7cm apart. The conductors carry currents of 9A and 16A in opposite directions. The distance of neutral point from the conductor carrying 16A current is

1) 9cm 2) 16cm 3) 25cm 4) 63/25cm

TORQUE ON CURRENT LOOP

42. A rectangular coil of wire of 500 turns of area $10\times5\ cm^2$ carries a current of 2 A in a magnetic field of induction 2×10^{-3} T. If the plane of the coil is parallel to the field. The torque on the coil is (in) Nm.

1) 0.1 2) 0.01 3) 0.001 4) 1

43. A coil of area 100 cm^2 having 500 turns carries a current of 1 m A. It is suspended in a uniform magnetic field of induction $10^{-3}wb/m^2$. Its plane makes an angle of 60^0 with the lines of induction. The torque acting on the coil is

1) $250\times10^{-8}Nm$ 2) $25\times10^{-8}Nm$ 3) $2.5\times10^{-8}Nm$ 4) $0.2\times10^{-8}Nm$

44. A circular coil of 1 turn and area 0.01 m^2 carries a current of 10 A. It is placed in a uniform magnetic field of induction 0.1 tesla such that the plane of the circle is perpendicular to the direction of the field, the torque acting on the coil is

1) 0.1N m 2) 0.001Nm 3) 0.01N m 4) Zero

MOVING COIL GALVANOMETER

45. A current of 10^{-5} A produces a deflection of 10^0 in a moving coil galvanometer. A current of 10^{-6} amp in the same galvanometer produces a deflection of

1) 1^0 2) 0.1^0 3) 10^0 4) $(1/100)^0$

46. The coils made of same material in two moving coil galvanometers have their areas in the ratio of 2:3 and number of turns in the ratio 4:5. These two coils carry the same current and are situated in the same field. The deflections produced by these two coils will be in the ratio of

1) 8:15 2) 15:8 3) 8:1 4) 1:4

CONVERSION OF MCG INTO
AMMETER & VOLTMETER

47. A galvanometer has a resistance of 400 Ω. The value of shunt so that its sensitivity is to be reduced by 1/50 times

1) 6.16 Ω 2) 7.16 Ω 3) 8.16 Ω 4) 9.16 Ω

48. A galvanometer of resistance 20 Ω is to be shunted so that only 1% of the current passes through it. Shunt connected is

1) 99/20 Ω 2) 9/20 Ω 3) 20/99 Ω 4) 2/99 Ω

49. The resistance of a moving coil galvanometer is 5 ohm. The maximum current it can measure is 0.015 A. To convert it into an ammeter to meausre 1.5 A.

1) connected 5/99 ohm in series 2) connected 99/50 ohm in parallel
3) connected 5/99 ohm in parallel 4) connected 99/50 ohm in series

50. A galvanometer of coil resistance 100 Ω is connected to a shunt of resistance 10 Ω. The current through the galvanometer is i_1, the current through the shunt is i_2 and the total current into the combination is i_3, then the ratio $i_1 : i_2 : i_3$ is

1) 1 : 10 : 11 2) 10 : 1 : 11 3) 11 : 10 : 1 4) 10 : 11 : 1

51. The resistance of a galvanometer is 100 Ω. A shunt of 5 Ω is connected to it to convert it into an ammeter. The internal resistance of the ammeter is

1) 5.2 Ω 2) 4.8 Ω 3) 4.6 Ω 4) 4.2 Ω

52. A galvanometer of resistance 100 ohms is shunted so that only 1/11 of the main current flows through the galvanometer. The resistance of the shunt is

1) 1 ohm 2) 11 ohms 3) 10 ohms 4) 9 ohms

53. If a shunt is to be applied to a galvanometer of resistance 50 Ω so that only 5% of total current passes through the galvanometer. The resistance of shunt should be

1) 1.63 Ω 2) 4.2 Ω 3) 3.5 Ω 4) 2.63 Ω

54. If only 2% of the main current is to be passed through a galvanometer of resistance G, then the resistance of the shunt will be

1) G/50 2) G/49 3) 50G 4) 49 G

55. A maximum current point 0.5 mA can be passed through a galvanometer of resistance 20 Ω. The resistance to be connected in series to convert it in to voltmeter of range 0-5V. is

1) 980 Ω 2) 9980 Ω 3) 990 Ω 4) 9990 Ω

56. The maximum potential that can be measured with a voltmeter of resistance 1000 Ω is 6V. Resistance that must be connected to measure a potential of 30V with it is

1) 4000 Ω in Series 2) 6000 Ω in Series 3) 12000 Ω in Series 4) 2000 Ω in Series

57. A voltmeter has an internal resistance of 1000Ω and gives full scale deflection when 2 V is applied across the terminals. Now a resistance of 4000 Ω is connected in series with it. Then it gives full scale deflection with

1) 8 V 2) 10 V 3) 6 V 4) 4 V

58. To convert a voltmeter measuring 15 V into a voltmeter measuring 150V, if the resistance of the voltmeter is 1000 Ω, the resistance to be connected is

1) 10,000 Ω in Series 2) 9,000 Ω in Series

3) 11,000 Ω in Series 4) 8,000 Ω in Series

KEY LEVEL-I (CW)

1) 2	2) 1	3) 2	4) 3	5) 2	6) 4	7) 4	8) 2	9) 2	10) 4
11) 2	12) 3	13) 3	14) 1	15) 3	16) 2	17) 4	18) 3	19) 4	20) 3
21) 3	22) 1	23) 2	24) 1	25) 2	26) 2	27) 4	28) 3	29) 2	30) 2
31) 4	32) 3	33) 3	34) 1	35) 2	36) 3	37) 1	38) 4	39) 3	40) 1
41) 2	42) 2	43) 1	44) 4	45) 1	46) 1	47) 3	48) 3	49) 3	50) 1
51) 2	52) 3	53) 4	54) 2	55) 2	56) 1	57) 2	58) 2		

HINTS LEVEL-I (CW)

1. $w = \mu_0 mni$

2. $w = \mu_0 mni$

3. $W = \mu_0 i^0$

4. $B = 8\sqrt{2}\frac{\mu_0 i}{4\pi a}$

5. $B\alpha\frac{1}{r}$

6. $B = \frac{\mu_0 i}{2\pi r}$

7. $B = \frac{\mu_0 i}{2\pi r}$

8. $B = \frac{\mu_0 ni}{2r}$

9. $B = \frac{\mu_0 nir^2}{2\left(r^2+x^2\right)^{3/2}} = \frac{1}{8}\times\frac{\mu_0 ni}{2r}$

10. $B = \frac{\mu_0 \pi n^2 i}{l}$

11. $B\alpha\frac{1}{r}$

12. $r = \frac{\mu_0 ni}{2B}$

13. $B = B_1 \sim B_2$

14. $B = \frac{\mu_0 nir^2}{2\left(r^2+x^2\right)^{3/2}}$

15. $B\alpha\frac{1}{x^3}$

16. $B = \frac{\mu_0 ne}{2r}$ where n is the frequency

17. $B = \frac{\mu_0 i\theta}{4\pi r}$

18. $I = \frac{2rB_H}{\mu_0 n}\tan\theta$

19. $B = B_H Tan\theta$

20. $\frac{K_1}{K_2} = \frac{\tan\theta_2}{\tan\theta_1}$

21. $B = B_H Tan\theta$

22. $M = inA\, and\ i = \frac{q}{T}$

23. $M = inA\, and\ i = \frac{q}{T}$

24. $B = \mu_o ni$

25. $B = \mu_o ni$

26. $F = BqV\sin\theta$

27. Magnetic field at O due to AB,i.e $B = \frac{\mu_0 I}{4\pi d}\left(\sin\phi_1 + \sin\phi_2\right)$

28. $F = BqV\sin\theta$

29. $r = \frac{mv}{Bq}$

30. $T = \frac{2\pi m}{Bq}$

31. $F = Bqv, v = \sqrt{\frac{2KE}{m}}$

32. $T = \frac{2\pi m}{Bq}$

33. $r = \frac{mv}{Bq}$

$KE = \frac{1}{2}mv^2$

34. $n = \frac{Bq}{2\pi m}$

35. $F = BIl \sin\theta$

36. Force acting on each electron, i.e..,

Lorentz force, $F_m = ev_d B$ or $F_m = e\left(\frac{1}{Ane}\right)B = \frac{IB}{An}$ $(as I = neAv_d)$

or $F_m = \frac{5 \times 0.10}{10^{-5} \times 10^{29}} N = 5 \times 10^{-25} N$

(as A = cross-sectional area of the wire $= 10^{-5} m^2, n =$ free electron density $= 10^{-29} m^{-3}$)

37 As F- mg when the bar is just ready to levitate,

$IlB = mg$ or $I = \frac{mg}{lB} = \frac{0.750 \times 9.8}{0.5 \times 0.45} A = 32.67 A$

$\varepsilon = IR = (32.67)(25)V = 817V$

38. $F = \frac{\mu_0}{2\pi} \frac{i_1 i_2}{r} \times l$

39. $x = \frac{d}{\frac{i_2}{i_1} + 1}$

40. $\frac{F}{l} = \frac{\mu_0 i^2}{2\pi d}$

41. $x = \frac{d}{\frac{i_2}{i_1} - 1}$

42. $\tau = BAIN \cos\theta$

43. $\tau = BAIN \cos\theta$

44. $\tau = BAIN \cos\theta$

45. $\theta \alpha I$

46. $\tau\theta = BAIN$

47. $S = \frac{G}{n-1}$

48. $S = \frac{G}{\frac{i}{i_g} - 1}$

49. $S = \frac{G}{\frac{i}{i_g} - 1}$

50. $Ig = \frac{S}{G+S}(I)$

51. $R_{eff} = \frac{GS}{G+S}$

52. $S = \frac{G}{\frac{i}{i_g} - 1}$

53. $S = \frac{G}{n-1}$

54. $S = \frac{G}{\frac{i}{i_g} - 1}$

55. $R = \left(\frac{V}{I_g} - G\right)$

56. $R = \left(\frac{V}{I_g} - G\right)$

57. $R = \left(\frac{V}{I_g} - G\right)$

58. $R = G(n-1)$

LEVEL - I

AMPERE'S CIRCUITAL LAW BIOT-SAVART LAW AND ITS APPLICATIONS

1. A north pole of strength π Am, is moved around a circle or radius 10 cm which lies around a long straight conductor carrying a current of 10 A. The work done in moving it by 2 revolutions is

1) 4 μ J 2) 80 μ J 3) 400 μ J 4) 0.4 μ J

2. Two long straight conductors carry currents 4 A and 2 A into the plane of paper. A circular path is imagined to be enclosing these currents. The value of $\oint \overline{B}.\overline{dl}$ is

1) $6\mu_0$ 2) $7\mu_0$ 3) $5\mu_0$ 4) $2\mu_0$

3. Three long straight conductors are kept perpendicular to the plane of paper. Currents 2 A, 3 A are passing through the two conductors into the plane of paper in first two conductors and 5 A current passes through third conductor, directed out of the paper. A closed loop encloses the conductors, then the value of $\oint \overline{B}.\overline{dl}$ over the closed loop is (assume current into the paper as negative and out of the paper as positive)

1) $2\mu_0$ 2) zero 3) $-\mu_0$ 4) $+\mu_0$

4. A long straight wire carries an electric current of 2A. The magnetic induction at a perpendicular distance of 5 m from the wire is ($\mu_0 = 4\pi \times 10^{-7}$ Hm^{-1})

1) 4×10^{-8} T 2) 8×10^{-8} T 3) 12×10^{-8} T 4) 16×10^{-8}T

5. A current of 2 amperes is flowing through a circular coil of radius 10 cm containing 100 turns. The magneitc flux density at the centre of the coil is (in wb/m^2)

1) 0.126×10^{-2} 2) 1.26×10^{-2} 3) 1.26×10^{-4} 4) 1.26×10^{-5}

6. A coil of radius π meters, 100 turns carries a current of 3A. The magnetic induction at a point on its axis at a distance equal to $\sqrt{3}$ times its radius from its centre is

1) $7.2\text{X}10^{-6}$wbm^{-2} 2) $7.4\text{X}10^{-6}$wbm^{-2} 3) $7.5\text{X}10^{-6}$wbm^{-2} 4) $7.83\text{X}10^{-6}$wbm^{-2}

7. The coils A and B having the radii in the ratio 1:2 carrying currents in the ratio 5:1 and have the number of turns in the ratio 1:5. The ratio of magnetic inductions at their centres is

1) 1:2 2) 2:1 3) 1:5 4) 5:1

8. The intensity of magnetic induction at the centre of a current-carrying circular coil is B_1 and at a point on its axis at a distance equal to its radius from the centre is B_2, then B_1/B_2 is

1) $2\sqrt{2}$ 2) $\frac{1}{2\sqrt{2}}$ 3) $\frac{1}{\sqrt{2}}$ 4) $\sqrt{2}$

9. The magnetic induction at a point distance 'X' from the centre, on the axis of a circular current carrying coil is inversely proportional to (if X>> radius of coil)

1) X 2) X^2 3) X^3 d. $X^{3/2}$

10. A circular coil of radius 5cm has 169 turns carries a current of 2.6A. The magnetic induction at a point on the axis at a distance of 12cm from the centre of the coil is

1) 2T 2) 4.2T 3) $3.14\text{X}10^{-4}$T 4) $3.14\text{X}10^{-2}$T

TANGENT GALVANOMETER

11. A TG has 50 turns each of diameter 15 cms. When a certain current is passed through its coil, the defleciton of the needle is 45^0. If $B_H = 4\pi\ \mu T$ the value of current is

1) 77.44 mA 2) 44.77 mA 3) 74.74 mA 4) 30mA

12. A tangent galvanometer properly adjusted gives a deflection of 30° when a certain current is passed through it. When the current is changed, then it gives a deflection of 45°. The ratio of the currents in the two cases is

1) 2 : 3 2) $1:\sqrt{2}$ 3) $1:\sqrt{3}$ 4) $\sqrt{3}:1$

13. In a tangent galvanometer a deflection of 30° is obtained with a certain current flowing through the coil. If the current is tripled, the deflection obtained will be

1) 45° 2) 60° 3) 90° 4) 30°

CURRENT LOOP AND MAGNETIC DIPOLE

14. An electron of charge "e" has a time period of revolution of "T" in a Bohr orbit of radius "r". The dipole moment of the electron is

1) $\pi r^2 eT$ 2) $\pi r^2 e/T$ 3) $\pi r^2 T/e$ 4) $T/\pi r^2 e$

15. A circular coil of wire of "n" turns has a radius "r" and carries a current "i". Its magnetic dipole moment is "M". Now the coil is unwound and again rewound in to a circular coil of half the initial radius and the same current is passed through it, then the dipole moment of this new coil is

1) M / 2 2) M / 4 3) M 4) 2 M

SOLENOID & TOROID

16. A long solenoid has 200 turns per cm and carries a current i. the magnetic field at its centre is $6.28\times10^{-2}Wb/m^2$. another long solenoid has 100 turns per cm and it carries a current i/3. The value of the magnetic field at its centre is

1) $1.05\times10^{-4}Wb/m^2$ 2) $1.05\times10^{-2}Wb/m^2$

3) $1.05\times10^{-5}Wb/m^2$ 4) $1.05\times10^{-6}Wb/m^2$

FORCE ACTING ON A MOVING CHARGE IN MAGNETIC FIELD

17. A proton and an α-particle enter a uniform magnetic field at right angles to the field with same speed. The ratio of the periods of α-particle and proton is

1) 1:1 2) 1:4 3) 1:2 4) 2:1

18. A proton takes 10^{-12}s to complete one revolution in uniform magnetic field. The time taken in another orbit of double the radius in the same field is

1) $0.5\text{X}10^{-12}$ sec 2) $2\text{X}10^{-12}$sec 3) $4\text{X}10^{-12}$sec 4) 10^{-12}sec

19. A charged particle, having charge q accelerated through a potential difference V enters a perpendicular magnetic field in which it experiences a force F. If V is increased to 5 V, the particle will experience a force

1) F 2) 5 F 3) F/ 5 4) $\sqrt{5}F$

20. A proton moving with a velocity of $2\times10^6 ms^{-1}$ describes a circle of radius R in a magnetic field. The speed of an α - particle to describe a circle of same radius in the same magnetic field is

1) $1\text{X}10^6$m/s 2) $2\text{X}10^6$m/s 3) $4\text{X}10^6$m/s 4) $8\text{X}10^6$m/s

21. A particle of charge 16×10^{-18} coulomb moving with velocity 10m/s along the x-axis enters a region where a magnetic field of induction B is along the y-axis, and an electric field of magnitude $10Vm^{-1}$ is along the negative Z-axis. If the charged particle continues moving along the X-axis, the magnitude of B is

1) 1 Wb/m^2 2) $10^5 Wb/m^2$ 3) $10^6 Wb/m^2$ 4) $10^{-3} Wb/m^2$

22. A proton is rotating along a circular path with kinetic energy K in a uniform magnetic field B.If the magnetic field is made four times, the kinetic energy of rotation of proton is

1) 16K 2) 8K 3) 4K 4) K

CYCLOTRON

23. In a cyclotron, if the frequency of proton is 5MHz, the magnetic field necessary for resonance is

1) 0.528T 2) 2.32T 3) 0.389T 4) 0.327T

24. A cyclotron's oscillator frequency is 10 MHz. The operating magnetic field for accelerating protons is

1) 0.66 T 2) 0.12 T 3) 1.67 T 4) 0.36 T

FORCE ON A CURRENT CARRYING CONDUCTOR IN A MAGNETIC FIELD

25. A straight wire of length 0.5 metre and carrying a current of 1.2 ampere is placed in a uniform magnetic field of induction 2 tesla. If the magnetic field is perpendicular to the length of the wire, the force acting on the wire is

1) 2.4 N 2) 1.2 N 3) 3.0 N 4) 2.0 N

FORCE BETWEEN TWO PARALLEL CURRENT CARRYING CONDUCTORS

26. Two parallel conductors carrying 5A each, repel with a force of $0.25Nm^{-1}$. The distance between them is

1) $4\times10^{-5}m$ 2) $3\times10^{-5}m$ 3) $2\times10^{-5}m$ 4) $1\times10^{-5}m$

27. Two straight parallel wires carry currents of 200 mA and 1A in opposite directions. If the wires are 20cm apart, the distance of the neutral point from the 1A wire is(in cm)

1) 5 2) 15 3) 20 4) 25

28. Two long parallel copper wires carry currents of 5A each in the opposite direction. If the wires are seperated by a distance of 0.5 m, then the force between the two wires is

1) 10^{-5} N/m attractive 2) 10^{-5} N/m repulsive
3) 2×10^{-6} N/m attractive 4) 2×10^{-5} N/m repulsive.

TORQUE ON CURRENT LOOP

29. A rectangular coil of wire of area 400 cm^2 contains 500 turns. It is placed in a magnetic field of induction 4×10^{-3} T and it makes an angle 60° with the field. A current of 0.2 A is passed through it. The torque on the coil is

1) $8\sqrt{3}\times10^{-3}$ N m 2) 8×10^{-3} N m
3) $8\sqrt{3}\times10^{-4}$ N m 4) 8×10^{-4} N m

30. A rectangular coil of wire carrying a current is suspended in a uniform magnetic field. The plane of the coil is making an angle of 30° with the direction of the field and the torque experienced by it is τ_1 and when the plane of the coil is making an angle of 60° with the direction of the field the torque experienced by it is τ_2. Then the ratio $\tau_1:\tau_2$ is

1) $1:\sqrt{3}$ 2) $\sqrt{3}:1$ 3) 1 : 3 4) 3 : 1

31. A vertical rectangular coil of sides 5cm x 2cm has 10turns and carries a current of 2A.The torque(couple) on the coil when it is placed in a uniform horizontal magnetic field of 0.1T with its plane perpendicular to the field is

1) $4\text{X}10^{-3}$N-m 2) Zero 3) $2\text{X}10^{-3}$N-m 4) 10^{-3}N-m

MOVING COIL GALVANO METER

32. The coil in a MCG has an area of 4 cm^2 and 500 turns. The intensity of magnetic induction is 2T. When a current of $10^{-4}A$ is passed through it, the deflection is 20^0. The couple per unit twist is (N-m)

1) 3×10^{-6} 2) 2×10^{-6} 3) 4×10^{-6} 4) 5×10^{-6}

33. The area of the coil in a moving coil galvanometer is 15 cm^2 and has 20 turns. The magnetic induction is $0.2T$ and the couple per unit twist of the suspended wire is $10^{-6}Nm$ per degree. If the deflection is 45^0, the current passing through it is

1) $75\times10^{-4}A$ 2) $7.5\times10^{-4}A$ 3) $0.75\times10^{-4}A$ 4) $750\times10^{-4}A$

CONVERSION OF MCG INTO AMMETER & VOLTMETER

34. The sensitivity of a galvanometer of resistance 990 Ω is increased by 10 times. The shunt used is

1) 100Ω 2) 120Ω 3) 110Ω 4) 50 Ω

35. A galvanometer of resistance 50Ω gives full scale defleciton when a current 10^{-3} A is passed through it converted into an ammeter to measure 0.5 A current.

1) 50/499 Ω in parallel 2) 9/20Ω in parallel
3) 2/99 Ω in parallel 4) 20/99 Ω in parallel

36. A galvanometer of resistance 20 Ω is shunted by 2 Ω resister. The part of the main current flows through the meter is

1) 1/10 part 2) 1/11 part 3) 1/12 part 4) 1/13 part

37. A galvanometer has a resistance 50 Ω and is shunted by a 0.5 Ω resistor. The fraction of the main current that flows through the galvanometer is

1) 1/100 2) 1/101 3) 1/10 4) 1/11

38. A galvanometer with a coil resistance of 100Ω gives a full scale deflection when a current of 1 mA is passed through it. The resistance of the shunt needed in ohm to convert this galvanometer into an ammeter of range 10 A is nearly

1) 0.01 2) 0.001 3) 0.1 4) 0.099

39. A galvanometer has a resistance of 49 Ω. If 2% of the main current is to be passed through the meter, The value of the shunt will be

1) 2 Ω 2) 1Ω 3) 1/2Ω 4) 1/4 Ω

40. In a galvanometer 5% of the total current in the circuit passes through it. If the resistance of the galvanometer is G, the shunt resistance 'S' connected to the galvanometer is

1) 19G 2) G/19 3) 20G 4) G/20

41. A galvanometer of internal resistance 100Ω has a full scale deflection current of 1mA. To convert it into a voltmeter of range 0-10V, the resistance to be connected is

1) 9000 Ω in Series 2) 10,000 Ω in Series
3) 9,900 Ω in Series 4) 9,800 Ω in Series

42. When a high resistance 'R' is connected in series with a voltmeter of resistance G, the range of the voltmeter increases 5 times. Then G:R will be

1) 4:1 2) 1:2 3) 8:1 4) 1:4

KEY LEVEL-I

1) 2	2) 1	3) 2	4) 2	5) 1	6) 3	7) 2	8) 1	9) 3	10) 3
11) 4	12) 3	13) 2	14) 2	15) 1	16) 2	17) 4	18) 4	19) 4	20) 1
21) 1	22) 4	23) 4	24) 1	25) 2	26) 3	27) 4	28) 2	29) 2	30) 2
31) 2	32) 2	33) 1	34) 3	35) 1	36) 2	37) 2	38) 1	39) 2	40) 2
41) 3	42) 4								

HINTS LEVEL-I

1. $W = \mu_0 mni$

2. $\mu_0 \sum i$

3. $\oint B.dl = \mu_o i$

4. $B = \frac{\mu_0 i}{2\pi r}$

5. $B = \frac{\mu_0 ni}{2r}$

6. $B = \frac{\mu_0 nir^2}{2(r^2 + x^2)^{3/2}}$

7. $B \propto \frac{ni}{R}$

8. $B = \frac{\mu_0 nir^2}{2(x^2 + r^2)^{\frac{3}{2}}}$

9. $B\alpha \frac{1}{x^3}$

10. $B = \frac{\mu_0 nir^2}{2(r^2 + x^2)^{3/2}}$

11. $I = \frac{(2r)B_H}{\mu_0 n}.\tan\theta$

12. $\frac{I_1}{I_2} = \frac{\tan\theta_1}{\tan\theta_2}$

13. $I = K\tan\theta$

14. $M = iA = \frac{e}{T}A = \frac{e}{T}\pi r^2$

15. M = niA

16. $B = \mu_o ni$

17. $\frac{T_1}{T_2} = \frac{m_1}{m_2} \times \frac{q_2}{q_1}$

18. $\frac{T_1}{T_2} = \frac{m_1}{m_2} \times \frac{q_2}{q_1}$

19. $F = Bqv\sin\theta, \ \frac{1}{2}mv^2 = Vq$

20. $\frac{V_1}{V_2} = \frac{q_1}{q_2} \times \frac{m_2}{m_1}$

21. $V = \frac{E}{B}$

22. $E = \frac{Q^2B^2R^2}{2m}$

23. $f = \frac{Bq}{2\pi m}$

24. $f = \frac{Bq}{2\pi m}$

25. $F = Bqv$

26. $F = \frac{\mu_0}{2\pi}\frac{i_1 i_2}{r} \times l$

27. $x = \frac{d}{\frac{i_2}{i_1} - 1}$

28. $F = \frac{\mu_0 I_1 I_2}{2\pi r} l$

29. $\tau = BAIn\cos\theta$

30. $\tau = BAIn\cos\theta$

31. $\tau = BAIn\cos\theta$

32. $\tau\theta = BAIn$

33. $\tau\theta = BAIn$

34. $S = \frac{G}{(n-1)}$

35. $S = \frac{G}{(n-1)}$

36. $i_g = \frac{iS}{G+S}$

37. $i_g = \frac{iS}{G+S}$

38. $i_g = \frac{iS}{G+S}$

39. $i_g = \frac{iS}{G+S}$

40. $i_g = \frac{iS}{G+S}$

41. $R = \frac{V}{i_g} - G$

42. $S = \frac{G}{n-1}$

LEVEL - II

AMPERE'S CIRCUITAL LAW, BIOT-SAVART LAW AND ITS APPLICATIONS

1. **A thin straight vertical conductor has 10amp current flows vertically upwards. It is present at a place where $B_H = 4\times10^{-6}T$. Arrange the net magnetic induction at the following points in ascending order**

   a) at 0.5m on south of conductor **b) at 0.5m on west of conductor**

   c) at 0.5m on east of conductor **d) at 0.5m on north-east of conductor**

   1) a,b,c,d 2) a,b,d,c 3) a,c,b,d 4) b,a,d,c

2. **A long straight wire carrying a current of 30A is placed in an external uniform magnetic field of induction $4\times10^{-4}T$. The magnetic field is acting parallel. to the direction of current. The magnitude of the resultant magnetic induction in tesla at a point 2 cm away from the wire is** $[\mu_0 = 4\pi\times10^{-7} Hm^{-1}]$

   1) 10^{-4} 2) 3×10^{-4} 3) 5×10^{-4} 4) 6×10^{-4}

3. **A straight section PQ of a circuit lies along the X axis from x = -a/ 2 to x = a/2 and carries a steady current I .Then the magnetic field due to the PQ section at a distance x = +a will be**

   1) proportional to a 2) proportional to I/a 3) proportional to a^2 4) zero

4. **ABCD is a square of side L. A very long straight conductor carrying a current i passes through the vertex A of the square and is perpendicular to its plane. The minimum magnetic induction at a vertex of the square is**

   1) $\frac{\mu_0}{4\pi}\frac{2\sqrt{2}i}{L}$ 2) $\frac{\mu_0}{4\pi}\frac{\sqrt{2}i}{L}$ 3) $\frac{\mu_0}{4\pi}\frac{4\sqrt{2}i}{L}$ 4) $\frac{\mu_0}{4\pi}\frac{2i}{L}$

5. **The magnetic field at the centre of circular loop in the circuit shown below is**

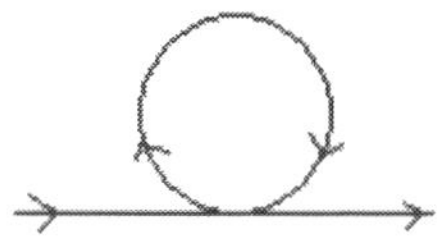

   1) $\frac{\mu_0}{4\pi}\frac{2I}{r}(1+\pi)$ 2) $\frac{\mu_0}{4\pi}\frac{2I}{r}(\pi-1)$ 3) $\frac{\mu_0}{4\pi}\frac{2I}{r}$ 4) $\frac{\mu_0}{4\pi}\frac{I}{r}(\pi+1)$

6. **A current IA is flowing in the sides of equilateral triangle of side $4.5\times10^{-2}m$ .The magnetic induction at centroid of the traingle is**

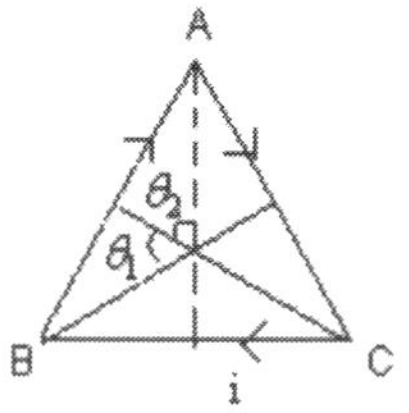

   1) $4\times10^{-5}T$ 2) 40T 3) $0.4\times10^{-3}T$ 4) $4\times10^{-2}T$

7. In the given figure the magnetic induction at the point O is

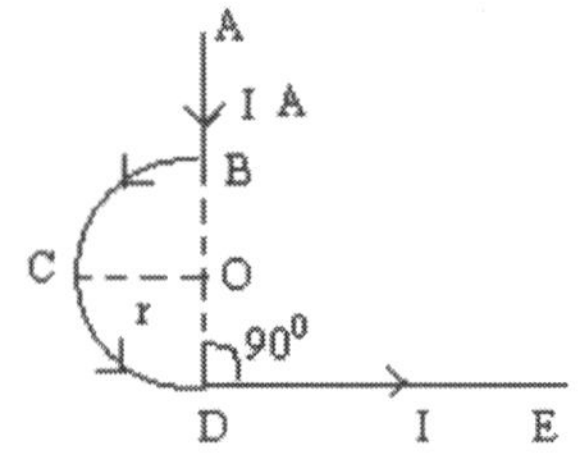

1) $\frac{\mu_0 I}{4\pi r}$ 2) $\frac{\mu_0 I}{4r}+\frac{\mu_0 I}{2\pi r}$ 3) $\frac{\mu_0 I}{4r}+\frac{\mu_0 I}{4\pi r}$ 4) $\frac{\mu_0 I}{4r}-\frac{\mu_0 I}{4\pi r}$

8. Figure shows a coil of radius 2 cm concentric with a coil of radius 7 cm Each coil has 1000 turns with a current of 5 A. In larger coil, then the current needed in the smaller coil to give the total magnetic field at the centre equal to 2 mT is

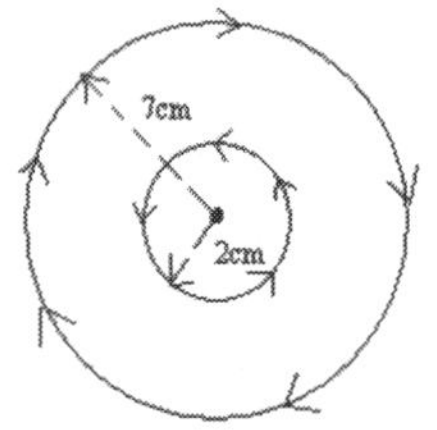

1) 1.49A 2) 1.84A 3) 2.88A 4) 3.4A

9. Two identical coils carry equal currents have a common centre and their planes are at right angles to each other. The ratio of the magnitude of the resultant magnetic field at the centre and the field due to one coil is

1) 2 : 1 2) 1 : 2 3) $\sqrt{2}$:1 4) 1:$\sqrt{2}$

10. A uniform wire of resistance 12 Ω is bent in the form of a square. A cell of emf 6V having negligible internal resistance connected across the diagonal of the square. The magnetic induction at its centre is (in tesla).

1) 0 2) 10^{-7} 3) 5 x 10^{-7} 4) $\frac{\mu_0}{4\pi}\times 5\times 10^{-7}$

11. A wire of length 10 cm is bent into an arc of a circle such that it subtends an angle of 1 radian at the centre. If a current of 1 A is passed through the wire, the magnetic induction at the centre of the circle will be

1) 2 $\times$ 10^{-4} tesla 2) 1 $\times 10^{-6}$ tesla 3) 1 $\times 10^{-4}$ tesla 4) 2 $\times 10^{-6}$ tesla

12. A circular coil of radius 'r' having 'n' turns carries a current 'i'. The magnetic induction at the center of the coil is 'B'. Now the coil is unwound and rewound with half the original radius. If the magnetic induction at the center of the coil is to be the same, the current that should be passed through the coil is

1) 2i 2) i 3) i/2 4) i/4

13. Two wires A and B are of lengths 40cm and 30cm. A is bent into a circle of radius r and B into an arc of radius r. A current i_1 is passed through A and i_2 through B. To have the same magnetic inductions at the centre, the ratio of $i_1 : i_2$ is

1) 3 : 4 2) 3 : 5 3) 2 : 3 4) 4 : 3

14. Two long straight conductors with currents I_1 and I_2 are along X and Y axes. The equation of locus of points of zero magentic induction is

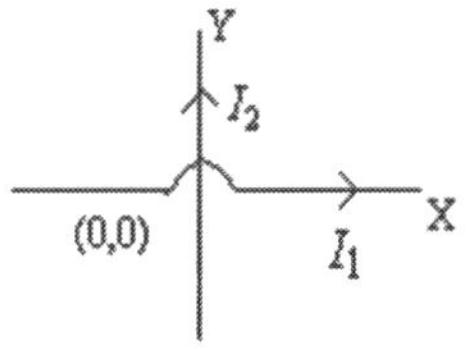

1) Y=X
2) $Y = \frac{I_2 X}{I_1}$
3) $Y = \frac{I_1}{I_2} X$
4) $Y = \frac{X}{I_1 I_2}$

15. Magnetic field induction at the center of a circular coil of radius 5cm and carrying a current 0.9 A is (in S.I. units) ($\in_0$ = absolute permittivity of air in S.I. units: velocity of light $= 3 \times 10^8 ms^{-1}$)

1) $\frac{1}{\in_0 10^{16}}$
2) $\frac{10^{16}}{\in_0}$
3) $\frac{\in_0}{10^{16}}$
4) $10^{16} \in_0$

16. The magnetic induction at the centre of a current carrying circular coil of radius 10 cm is $5\sqrt{5}$ times the magnetic induction at a point on its axis. The distance of the point from the centre of the coil in cm is

1) 5
2) 10
3) 20
4) 25

17. Same current 'i' is flowing in three infinitely long wire along positive x, y and z directions. The magnetic field at a point (0,0-a) would be

1) $\frac{\mu_0 i}{2\pi a}(\hat{j}-\hat{i})$
2) $\frac{\mu_0 i}{2\pi a}(\hat{i}+\hat{j})$
3) $\frac{\mu_0 i}{2\pi a}(\hat{i}-\hat{j})$
4) $\frac{\mu_0 i}{2\pi a}(\hat{i}+\hat{j}+\hat{k})$

18. Two long straight wires are connected by a circular section which has a radius R .All the three segments lie in the same plane and carry a current I. The magnetic induction at the centre O of the circular segments is

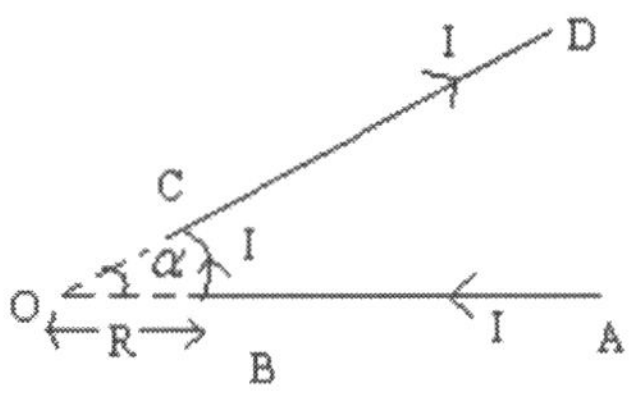

1) $\frac{\mu_0 I}{4\pi R}$
2) $\frac{\alpha \mu_0 I}{4\pi R}$
3) $\frac{\alpha \mu_0 I}{R}$
4) $\frac{\alpha \mu_0 I}{2\pi R}$

19. If B is the magnetic induction, at the centre of a circular coil of rdius 'r' carrying a current is 1 T, then its value at a distance of $\sqrt{3}r$ on the axis from the centre of the coil is :

1) $\frac{1}{8}T$
2) $\frac{1}{16}T$
3) $8T$
4) $\frac{1}{4}T$

TANGENT GALVANOMETER

20. A cell of negligible internal resistance is connected to a tangent galvanometer and the deflection produced is 30°. If three such cells are connected in series and the combination is connected to the same galvanometer, the deflection will be

1) 30° 2) 60° 3) 90° 4) 45°

21. The resistance of the coil of a tangent galvanometer is $60\,\Omega$. It is connected to a battery of negligible internal resistance. The deflection is found to be 60°. Now a shunt resistanceof $30\,\Omega$ is connected across the coil of the tangent galvanometer. The deflection produced will be

1) 30° 2) 45° 3) 60° 4) 37°

CURRENT LOOP AND MAGNETIC DIPOLE

22. Magnetic induction at the center of a circular loop carrying a current is 'B'. If 'A' is the area of the coil, the magnetic dipole moment of the loop is

1) $\frac{BA^2}{\mu_o \pi}$ 2) $\frac{BA\sqrt{A}}{\mu_o}$ 3) $\frac{BA\sqrt{A}}{\mu_o \pi}$ 4) $\frac{2BA}{\mu_o}\sqrt{\frac{A}{\pi}}$

23. A circular current loop of magnetic moment M is in an arbitrary orientation in an external magnetic field $\vec{B}$. The work done to rotate the loop by 30^0 about an axis perpendicular to its plane is :

1) MB 2) $\sqrt{3}\frac{MB}{2}$ 3) $\frac{MB}{2}$ 4) zero

SOLENOID & TOROID

24. A solenoid of length 20 cm and radius 2 cm is closely wound with 200 turns. The magnetic field intensity at either end of the solenoid when the current in the winding is 5 amp. is

1) 2500 Amp/m 2) 2000 Amp/m 3) 1750 Amp/m 4) 2940 Amp/m

25. A solenoid of length 0.5 m has a radius of 1 cm and is made up of 500 turns. It carries a current of 5 A. The magnetic field inside the solenoid is

1) 3.14×10^{-3}T 2) 6.28×10^{-3}T 3) 9.14×10^{-3}T 4) 1.68×10^{-3}T

26. The length of a solenoid is 0.1m and its diameter is very small. A wire is wound over in two layers. The number of turns in the inner layer is 50 and that on the outer layer is 40. The strength of current flowing in two layers in the same direction is 3 ampere. The magnetic induction in the middle of the solenoid will be

1) $3.4\times10^{-3}T$ 2) $3.4\times10^{-3}\,gauss$ 3) $3.4\times10^{3}T$ 4) $3.4\times10^{3}\,gauss$

27. A long solenoid has 200 turns per cm and carries a current i. The magnetic field at its centre is $6.28\times10^{-2}Wb/m^2$. Another long solenoid has 100 turns per cm and it carries a current i/3. The value of the magnetic field at its centre is

1) $1.05\times10^{-4}\,Wb/m^2$ 2) $1.05\times10^{-2}\,Wb/m^2$

3) $1.05\times10^{-5}\,Wb/m^2$ 4) $1.05\times10^{-4}\,Wb/m^2$

28. A toroidal solenoid has 3000 turns and a mean radius of 10cm . It has a soft iron core of relative permeability 2000. Find the magnetic field in the core when a current of 1.0A is passed through the solenoid .

1) 20T 2) 12T 3) 6T 4) 3T

FORCE ACTING ON A MOVING CHARGE IN MAGNETIC FIELD

29. **A particle of mass** 1×10^{-26} **kg and charge** 1.6×10^{-19} **C travelling with a velocity** $1.28\times10^{6}ms^{-1}$ **along the positive X-axis enters a region in which a uniform electric field** $\vec{E}=-102.4\times10^{3}k\ NC^{-1}$ **and magnetic field** $B=8\times10^{-2}jWbm^{-2}$**, the direction of motion of the particles is:**

1) x-axis 2) y-axis 3) z-axis 4) -x-axis

30. **Two particles X and Y having equal charges, after being accelerated through the same potential differences enter a region of uniform magnetic field and describe circular paths of radii R_1 and R_2 respectively. The ratio of the mass of X to that of Y is**

1) $(R_1/R_2)^{1/2}$ 2) (R_2/R_1) 3) $(R_1/R_2)^2$ 4) (R_1/R_2)

31. **A charged particle moving at right angles to a uniform magnetic field and starts moving along a circular arc of radius of curvature 'r'. In the field it now penetrates a layer of lead and loses 3/4th of its initial kinetic energy. The radius of curvature of its path now will be**

1) 4r 2) 2r 3) r/4 4) r/2

32. **Two particles having same charge and KE enter at right angles into the same magnetic field and travel in circular paths of radii 2 cm and 3 cm respectively. The ratio of their masses is.**

1) 2 : 3 2) 3 : 2 3) 4 : 9 4) 9 : 4

33. **Two electrons move parallel to each other with equal speed 'V' the ratio of magnetic & electric force between them is**

1) V/C 2) C/V 3) V^2 / C^2 4) C^2 / V^2

34. **A proton, a deuteron and an α particle are accelerated through same potential difference and then they enter a normal uniform magnetic field, the ratio of their kinetic energies will be**

1) 2:1:3 2)1:1:2 3) 1:1:1 4)1:2:4

35. **A proton of energy 8eV is moving in a circular path in a uniform magnetic field. The energy of an alpha particle moving in the same magnetic field and along the same path will be**

1) 4eV 2) 2eV 3) 8eV 4) 6eV

36. **A charged particle of charge 10mC enters a uniform magnetic field of induction** $\overline{B}=4\hat{i}+y\hat{j}+z\hat{k}$ **tesla with a velocity** $\overline{V}=2\hat{i}+3\hat{j}-6\hat{k}\ m\sec^{-1}$. **If the particle continues to move undeviated then the strength of the magnetic field induction in tesla**

1) 4 2) 8 3) 14 4) 30

37. **Magnetic induction field is existing along +Z axis in a region** $0\le x\le a$ **(a is positive). A point charge q is projected with a velocity 'v' at origin along positive x-axis, choose the correct alternative regarding its deviation**

1) maximum deviation is π rad independent of value of a

2) maximum deviation is $\frac{\pi}{2}$ rad independent of value of a

3) maximum deviation is $\frac{\pi}{2}$ rad if a is greater than its radius of curvature

4) Maximum deviation is $\frac{\pi}{2}$ rad only if a is equal to its radius of curvature

38. A proton moving with a velocity of (6i + 8j) x 10^5 **ms^{-1} enters uniform magnetic field of induction 5 x 10^{-3} $\ddot{k}$ tesla. The magnitude of the force acting on the proton is (i, j and k are unit vectors along X, Y, Z directions respectively)**

1) zero 2) 8×10^{-16} N 3) 3×10^{-16} N 4) 4×10^{-16} N

39. A proton of energy 2 MeV is moving perpendicular to uniform magnetic field of 2.5 T. The force on the proton is ($M_p = 1.6 \times 10^{-27}$ Kg and $q_p = e = 1.6 \times 10^{-19}$C)

1) 2.5×10^{-16}N 2) 8×10^{-11}N 3) 2.5×10^{-11} N 4) 8×10^{-12} N

40. Acceleration experienced by a particle with specific charge 1×10^7 C/kg when fired perpendicular to a magnetic field of induction 100 μT with a velocity 10^5 ms^{-1} is

1) 10^8 ms^{-2} 2) 10^{-6} ms^{-2} 3) 10^{14} ms^{-2} 4) 10^{-8} ms^{-2}

41. When two electrons enter into a magnetic field with different velocities, they deflect in different circular paths, in such a way that the radius of one path is double that of the other. $1X10^7 ms^{-1}$ is the velocity of the electron in smaller circle of radius $2X10^3 m$. The velocity of electron in the other circular path is :

1) $4X10^7 ms^{-1}$ 2) $4X10^6 ms^{-1}$ 3) $2X10^7 ms^{-1}$ 4) $2X10^6 ms^{-1}$

42. A beam of charged particle, having kinetic energy $10^3 eV$, contains masses $8X10^{-27}$ kg and $1.6X10^{-26}$ kg emerge from the end of an accelerator tube. There is a plate at distance 10^{-2} m from the end of the tube and placed perpendicular to the beam. The magnitude of the smallest magnetic field which can prevent the beam from striking the plate is.

1) 1.414 T 2) 2.414 T 3) 3.414 T 4) 4.414 T

43. A beam of mixture of α particles and protons are acceleratged through same potential difference before entering into the magnetic field of strength B. If $r_1 = 5$ cm, then r_2 is

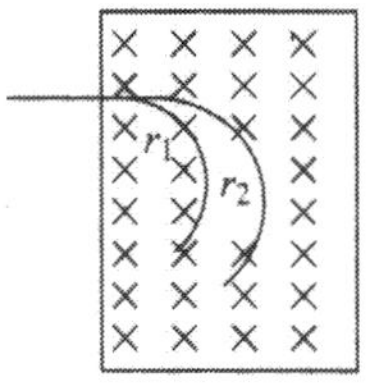

1) 5 cm 2) $5\sqrt{2}cm$ 3) $10\sqrt{2}cm$ 4) 20 cm

FORCE BETWEEN TWO PARALLEL CURRENT CARRYING CONDUCTORS

44. A horizontal wire of length 0.05m carries a current of 5A. If the mass of the wire is 10mg, the minimum magnetic field required to support the weight of the wire is (g=10m/s^2)

1) 4×10^{-4} T 2) 25×10^{-4}T 3) 4×10^{-1} T 4) 25×10^{-1}T

45. Currents of 10 A, 2 A are passed through two parallel wires A and B respectively in opposite directions. If the wire A is infinetely long and the length of the wire B is 2 metre, the force on the conductor B, which is situated at 10cm distance from A will be

1) 8×10^{-5} newton 2) 5×10^{-5} newton

3) $8\pi \times 10^{-7}$ newton 4) $4\pi \times 10^{-7}$ newton

46. Two long parallel conductors carry currents I and 2I in the same direction. The magnetic induction at a point exactly mid way between them is B. If the current in the first conductor is reversed in direction, the magnetic induction at the same point will be

1) $B/3$ 2) $2B$ 3) $3B$ 4) $B/2$

47. A horizontal wire carries 200 amp current below which another wire of linear density 20×10^{-3} kgm^{-1} carrying a current is kept at 2 cm distance. If the wire kept below hangs in air. The current in this wire is

1) 100A 2) 9.8 A 3) 98A 4) 48A

48. **Two long parallel conductors are placed at right angles to a metre scale at the 2cm and 6 cm marks, as shown in the figure**

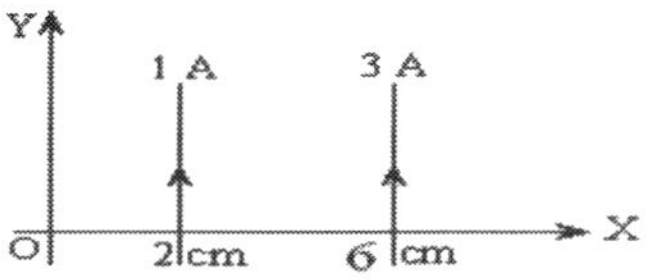

They carry currents of 1 A and 3 A respectively. They will produce zero magnetic field at the (ignore the earth's magnetic field)

1) 5 cm mark 2) 3 cm mark 3) 1 cm mark 4) 8 cm mark

49. **A rectangular loop of wire of size** $4cm \times 10cm$ **carries a steady current of 2A. A straight long wire carrying 5A current is kept near the loop (as shown in fig). If the loop and the wire are coplanar, find the net force on the loop**

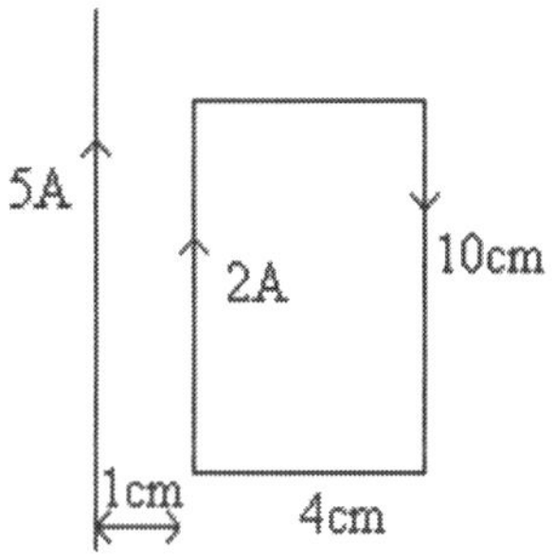

1) $3.2\times10^{-5}N$ 2) $1.6\times10^{-5}N$ 3) $0.4\times10^{-5}N$ 4) $4\times10^{-5}N$

50. **Wire-1 in fig is oriented along the Y-axis and carries a steady current** I_1 **A rectangular circuit loacated to the right of the wire carries a current** I_2 **Find the force experienced by the top of the horizontal wire (wire-2) of the rectangular circuit.**

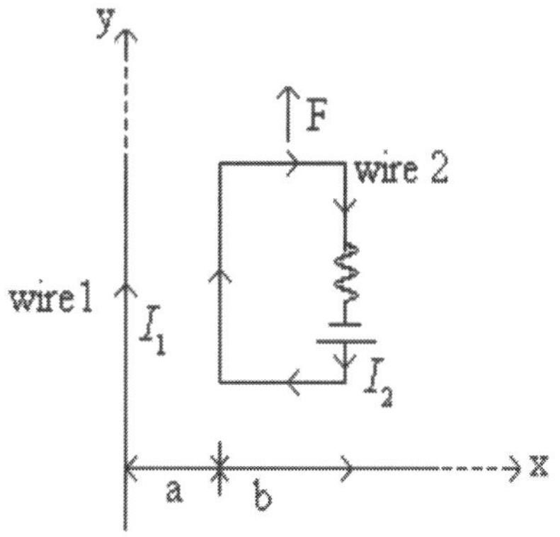

1) $\left[\frac{\mu_0 I_1 I_2}{2\pi} ln\left(1+\frac{b}{a}\right)\hat{j}\right]$ 2) $\left[\frac{\mu_0 I_1 I_2}{2\pi} ln\left(1+\frac{b}{a}\right)\left(-\hat{j}\right)\right]$

3) $\left[\frac{\mu_0 I_1 I_2}{2\pi} ln\left(1+\frac{b}{a}\right)\left(\ \hat{i}\right)\right]$ 4) $\left[\frac{\mu_0 I_1 I_2}{2\pi} ln\left(1+\frac{b}{a}\right)\left(\hat{i}\right)\right]$

TORQUE ON CURRENT LOOP

51. A square loop of side L carries a current I. Another smaller square loop of side $l(l << L)$ carrying a current i is placed inside the bigger loop such that they are coplanar with their centres coinciding. If the currents in the loops are in the same direction, the magnitude of the torque on the smaller loop is

1) $\frac{\mu_0 Iil^2}{\sqrt{2}\pi L}$ 2) $\frac{\mu_0 Iil^2}{2\pi L}$ 3) $\frac{\mu_0 Iil^2}{\sqrt{3}\pi L}$ 4) Zero

52. A coil in the shape of an equilateral triangle of side 0.02 m is suspended from a vertex such that it is hanging in a vertical plane between the pole pieces of a permanant magent producing a horizontal fields of $5\times10^{-2}T$. Find the couple acting on the coil when a current of 0.1A is passed through it and the magnetic field parallel its plane

1) $0.866\times10^{-6}N-M$ 2) $1.732\times10^{-4}N-M$
3) $0.422\times10^{-6}N-M$ 4) $0.866\times10^{-2}N-M$

MOVING COIL GALVANOMETER

53. A moving coil galvanometer A has 200 turns and resistance 100 Ω. Another meter B has 100 turns and resistance 40 Ω. All the other quantities are same in both the cases. The current sensistivity of

1) B is double as that of A 2) A is 2 times of B
3) A is 5 times of B 4) B is 5 times of A.

54. A rectangular coil of size 3x3 cm consisting of 100 turns caries 0.1 A. If it produces a deflection 10^0, in a field of induction 0.1T, the couple per unit twist is

1) $9\text{x}10^{-2}$ N-m/Degree 2) $9\text{x}10^{-5}$ N-m/Degree
3) $9\text{x}10^{-5}$ N-m/rad 4) 0.9 N-m/Degree

55. To increase the current sensitivity of a moving coil galvanometer by 50%, its resistance is increased so that the new resistance becomes twice its initial resistance. By what factor does the voltage sensitivity change?

1) decreased by 75% 2) Increased by 75% 3) decreased by 25% 4) Increased by 25%

CONVERSION OF MCG INTO AMMETER & VOLTMETER

56. A galvanometer of resistance 25 ohm is connected to a battery of 2 V along with a resistance in series. When the value of this resistance is 3000 Ω a full scale deflection of 30 units is obtained in the galvanometer. In order to reduce this deflection to 20 units, the resistance in series will be

1) 4513 ohm 2) 5413 ohm 3) 2000 ohm 4) 6000 ohm.

57. A voltmeter has range 0 $\rightarrow$V with a series resistance R. With a series resistance 2R, the range is 0$\longrightarrow$V'. The correct relation between v and v' is

1) $v^1=2v$ 2) $v^1>2v$ 3) $v^1<2v$ 4) $v^1>>2v$

58. A 100 V voltmeter having an internal resistance of 20 KΩ when connected in series with a large resistance R across a 110 V line reads 5 V. The magnitude of R is

1) 210 KΩ 2) 315 KΩ 3) 420 KΩ 4) 440 KΩ

59. **A galvanometer has resistance G and Current I_g produces full scale deflection. S_1 is the value of the shunt which converts it into an ammeter of range 0 - I and S_2 is the value of shurnt for the range 0 - 2I. The ratio of S_1 and S_2 is**

1) $\frac{1}{2}\left(\frac{I-I_g}{2I-i_g}\right)$ 2) $\frac{2I-I_g}{I-I_g}$ 3) 1/2 4) 2

60. **The scale of a galvanometer is divided into 150 equal division. The galvanometer has the current sensitivity of 10 divisions per mA and the voltage sensitivity of 2 divisions per mV. How the galvanometer can be designed to read (a) 6A / division and (b) $1V$ / division ?**

1) $S = 8.3\times10^{-5}\Omega, R = 9995\Omega$ 2) $S = 8.3\times10^{-2}\Omega, R = 995\Omega$

3) $S = 4.3\times10^{-5}\Omega, R = 9950\Omega$ 4) $R = 8.3\times10^{-5}\Omega, S = 995\Omega$

61. **In a galvanometer 5% of the total current in the circuit passes through it. If the resistance of the galvanometer is G, the shunt resistance 'S' connected to the galvanometer is**

1) 19G 2) G/19 3) 20G 4) G/20

62. **The sensitivity of a galvanometer is 60 divisions/Amp. When a shunt is used, its sensitivity becomes 10 divisions Amp. If the galvanometer is of resistance 20Ω, the value of shunt used is:**

1) 4Ω 2) 5Ω 3) 20Ω 4) 2Ω

63. **The sensitivity of a galvanometer that measures current is decreased by 1/40 times by using shunt resistance of 10Ω. Then, value of resistance of the galvanometer is :**

1) 400Ω 2) 410Ω 3) 30Ω 4) 390Ω

KEY LEVEL - II (CW)

1) 4	2) 3	3) 4	4) 2	5) 2	6) 1	7) 3	8) 1	9) 3	10) 1
11) 2	12) 4	13) 1	14) 3	15) 1	16) 3	17) 1	18) 2	19) 1	20) 2
21) 3	22) 4	23) 4	24) 1	25) 2	26) 1	27) 2	28) 2	29) 1	30) 3
31) 4	32) 3	33) 3	34) 2	35) 3	36) 3	37) 3	38) 2	39) 4	40) 1
41) 3	42) 1	43) 2	44) 1	45) 1	46) 3	47) 3	48) 2	49) 2	50) 1
51) 4	52) 1	53) 2	54) 2	55) 3	56) 1	57) 3	58) 3	59) 2	60) 1
61) 2	62) 2	63) 4							

HINTS LEVEL - II (CW)

1. $B = \frac{\mu_0 i}{2\pi r}$ 2. $B = \sqrt{B_1^2 + B_2^2}$ 3. $d\bar{l}\times\bar{r} = 0$

4. $B = \frac{\mu_0 i}{2\pi r}$. The maximum distance of a vertex from the conductor is $\sqrt{2}L$

5. $B_1 = \frac{\mu_0 i}{2\pi r}$ $B_2 = \frac{\mu_0 i}{2r}$ $B = B_1 - B_2$ 6. $B = \frac{\mu_0 i}{4\pi r}(\sin\theta_1 + \sin\theta_2)\times 3$

7. $B = B_1 + B_2$

$$B = \frac{\mu_0 I}{4r} + \frac{\mu_0 I}{4\pi r}$$

8. $B = B_1 - B_2 = \frac{\mu_0 i}{2r_1} - \frac{\mu_0 i}{2r_2}$

$$B = \frac{\mu_0 ni}{2r}$$

9. $\frac{B_C}{B_a} = \frac{(x^2 + R^2)^{\frac{3}{2}}}{R^3}$

10. $B_1 = \frac{\mu_0 i}{4\pi r}(\sin\theta_1 + \sin\theta_2)$

11. $B = \frac{\mu_0 i\theta}{4\pi r}$

12. $n_1 r_1 = n_2 r_2$; $B = \frac{\mu_0 ni}{2r}$

13. $B_1 = B_2$

14. $B = 0$

$$\frac{\mu_0 i_1}{2\pi y} = \frac{\mu_0 i_2}{2\pi x}$$

15. $B = \frac{\mu_0 ni}{2r}$; $c = \frac{1}{\sqrt{\mu_0 E_0}}$

16. $B = \frac{\mu_0 nir^2}{2(r^2 + x^2)^{3/2}}$

17. Magnetic field due to current along the z-axis is 0.

and due to rest two wires is $\frac{\mu_0}{2\pi}\frac{i}{a}$

$$B_x = \frac{\mu_0 i}{2\pi a}\hat{j}$$

$$B_y = \frac{\mu_0 i}{2\pi a}(-\hat{i})$$

$$B_z = 0$$

18. $B = \frac{\mu_0 I}{R} \times \frac{\theta}{4\Pi}$

19. $\frac{B_a}{B_c} = \frac{R^3}{(x^2 + R^2)^{3/2}}$

20. $\frac{I_1}{3I_1} = \frac{\tan 30^0}{\tan\theta_2}$

21. $i = k\tan\theta$

22. $B = \frac{\mu_0 i}{2r}$

23. No work is done to rotate the loop about an axis perpendicular to its plane as $\overrightarrow{M}$ is directed along the axis. Work is done only when the planeof the loop rotates.

24. $H = \frac{B}{\mu_0} = \frac{ni}{2}$

25. $B = \mu_0 in$

26. $B = \frac{\mu_0 i}{L}(N_1 + N_2)$

27. $\frac{B_1}{B_2} = \frac{n_1}{n_2}\left(\frac{i_1}{i_2}\right)$

28. $B = \frac{\mu Ni}{2\pi r}$

29. $V = \frac{E}{B}$

30. $r = \frac{mV}{Bq}$

31. $\frac{KE_1}{KE_2} = \frac{r_1^2}{r_2^2}$

32. $r = \frac{mv}{Bq} = \frac{\sqrt{2mk}}{Bq}$

33. $\frac{F_m}{F_e} = \frac{V^2}{C^2}$

34. $E_k = qV$

$E_K \propto q \; QV = \text{constant}$

35. $Bqv = \frac{mv^2}{r}$

$Bq = \frac{mv}{r} = \frac{p}{r}$

$E_k = \frac{q^2 r^2 B^2}{2m}$

$E_k \propto \frac{q^2}{m}$

$\frac{E_{k_\alpha}}{E_{K_P}} \propto \frac{q_\alpha^2}{q_P^2} \times \frac{m_P}{m_d}$

36. $\vec{v} // \vec{B} \; y = 6 \; z = -12 \; |\vec{B}| = 14$

37. $r = \frac{mv}{qB}$

38. $F = q(\bar{V} \times \bar{B})$

39. $F = Bqv = Bq\sqrt{\frac{2KE}{m}} \quad (\because \frac{1}{2}mv^2 = KE)$

40. $F = qvB$

41. $v = \frac{qBr}{m}$

42. $r = \frac{mv}{qB} = \frac{\sqrt{2mE}}{qB}$

43. $r = \frac{mv}{qB} = \frac{\sqrt{2mqv}}{qB}$

44. $B = \frac{mg}{il}$

45. $F = \frac{\mu_0}{2\pi} \frac{i_1 i_2}{r} \times l$

46. In the first case the two fields are in opposite directions and take the difference as the resultant field. In the next case, the two fields are in the same direction and add up

47. $\frac{\mu_0 i_1 i_2}{2\pi r} = \frac{mg}{l}$

48. $x = \frac{d}{\frac{i_2}{i_1} + 1}$

49. As $\vec{F}_{AB} = -\vec{F}_{DC}, \vec{F}_{AB} + \vec{F}_{DC} = \vec{0}$

$F_{AD} = k_m\left(\frac{2I_1 I_2}{a}\right)(AB) = \frac{(10^{-7} N/A)(2 \times 5A \times 2A)(10cm)}{(1cm)} = 2 \times 10^{-5} N$

Similarly, $F_{BC} = 0.4\times10^{-5} N$

Thus . $F_{net} = F_{AD} - F_{BC} = 1.6\times10^{-5} N$ (towards right)

50. $\left[\frac{\mu_0 I_1 I_2}{2\pi} In\left(1+\frac{b}{a}\right)\hat{j}\right]$

Magnetic field $(\vec{B})$ created by wire - I at a distancex from it , i.e ..

$$\vec{B} = \frac{\mu_0 I_1}{2\pi x}(-k)$$

Force acting on a small element dx of wire - 2, i.e

$$\overrightarrow{dF} = I_2 \overrightarrow{dx}\times\vec{B} \text{ or } \overrightarrow{dF} = I_2\overrightarrow{dx}\times\left(\frac{\mu_0 I_1}{2\pi x}(-k)\right)$$

$$= \frac{\mu_0 I_1 I_2}{2\pi x}dx\left[\hat{i}\times\left(-\hat{k}\right)\right] \quad \left(as\vec{dx} = dx\hat{i}\right)$$

$$= \left(\frac{\mu_0 I_1 I_2}{2\pi x}dx\right)\hat{j}$$

$$\vec{F} = \int \vec{dF} = \left[\frac{\mu_0 I_1 I_2}{2\pi}\int_a^{a+b}\frac{1}{x}dx\right]$$

$$\hat{j} = \left[\frac{\mu_0 I_1 I_2}{2\pi} In\left(1+\frac{b}{a}\right)\right]\hat{j}$$

$\vec{F}$ point upward as indicated by $\hat{j}$

51. $\tau = BiAn\cos 90^0$

52. Here $l = 0.02m,\ B = 5\times10^{-2} T,\ I = 0.1A$

Area of the triangular coil,

$$A = \frac{1}{2}\left|\vec{l}\times\vec{l}\right| = \frac{1}{2}\left(l\times l\sin 60^0\right)$$

$$= \frac{1}{2}l^2\sin 60^0 = \frac{1}{2}(0.02m)^2(0.865) = 1.732\times10^{-4} m^2$$

(as angle between the sides of an equilateral triangle is 60^0)

Couple acting on the coil,

$$\tau = IAB\cos\alpha$$

$$= (0.1A)\left(1.732\times10^{-4} m^2\right)\left(5\times10^{-2} T\right)\cos 0^0$$

$$= 0.866\times10^{-6} N\,m$$

($\alpha = 0^0$ as magnetic field is parallel to the plane of the coil)

53. $\frac{\theta}{i} = \frac{NAB}{C}$

54. $c\theta = NiAB$

55. Let CS and VS be the original current sensitivity and voltage sensitivity of MCG

Changed current sensitivity

$$CS^1 = CS + \frac{50}{100}CS = \frac{3}{2}CS$$

Since $VS = \frac{CS}{R}$

changed voltage sensitivity, i.e.,

$$VS^1 = \frac{CS^1}{2R} = \frac{(3/2)CS}{2R}$$

$$= \frac{3}{4}\left(\frac{CS}{R}\right) = 0.75VS = 75\% = 75\%VS$$

Thus, voltage sensitivity decreases by 25%

56. $R = \frac{V}{i_g} - G$

57. $R = (n-1)G$

58. $V = iR$

59. $S = \frac{G}{n-1}$

60. Here $I_g = \left(\frac{1mA}{10div}\right)(150div) = 15mA$

$$V_g = \left(\frac{1mA}{2div}\right)(150div) = 75mV$$

$$G = \frac{V_g}{I_g} = \frac{75mV}{15mA} = 5\Omega$$

(a) I (current to be measured)

$$= \left(\frac{6A}{div}\right)(150div) = 900A$$

As $n = \frac{I}{I_g} = \frac{900A}{15mA} = \frac{900A}{15\times10^{-3}A} = 6\times10^4$

$$S = \frac{G}{(n-1)} = \frac{5\Omega}{(6\times10^4 - 1)} \approx \frac{5\Omega}{6\times10^4} = 8.3\times10^{-5}\Omega$$

(b) V (voltage to be measured)

$$= \left(\frac{1V}{div}\right)(150div) = 150V$$

As $n=\frac{V}{V_g}=\frac{150V}{75mV}=\frac{150V}{75\times10^{-3}V}=2\times10^3$

$R=G(n-1)=5\Omega(2\times10^3-1)=5\Omega(2000-1)=9995\Omega$

61. $S=\frac{G}{n-1}$ 62. $S=\frac{G}{n-1}$ 63. $S=\frac{G}{n-1}$

LEVEL - II

AMPERE'S CIRCUITAL LAW BIOT-SAVART LAW AND ITS APPLICATIONS

1. **A current of 30A is flowing in a vertical straight wire. If the horizontal component of earths magnetic field is 2×10^{-5} T, then the position of null point will be**

   1) 0.9m 2) 0.3mm 3) 0.3cm 4) 0.3m

2. **The magnetic field at the centre of the coil in the figure shown below is (the wires crossing at P are insulated from each other)**

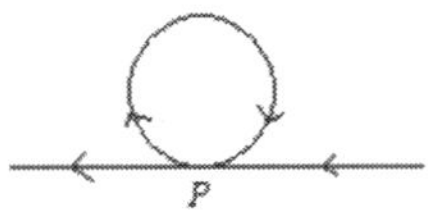

   1) $\frac{\mu_0}{4\pi}\frac{2I}{r}(1+\pi)$ 2) $\frac{\mu_0}{4\pi}\frac{2I}{r}(\pi-1)$ 3) $\frac{\mu_0}{4\pi}\frac{2I}{r}(\pi^2+1)$ 4) $\frac{\mu_0}{4\pi}\frac{2\pi I}{r}$

3. **Due to a straight vertical current carrying conductor, a null point occurred at P on east of the conductor. The net magnetic induction at a point 'Q' which is at same distance on north of the conductor is**

   1)0 2) $\sqrt{3}B_H$ 3) B_H 4) $\sqrt{2}B_H$

4. **The wire shown in figure carries a current of 40A. If r=3.14cm the magnetic field at point p will be**

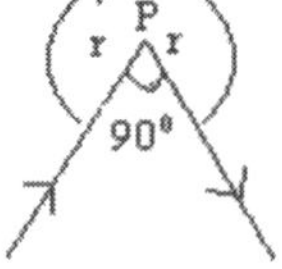

   1) $1.6\times10^{-3}T$ 2) $3.2\times10^{-3}T$ 3) $6\times10^{-4}T$ 4) $4.8\times10^{-3}T$

5. **The magnetic field at the centre of semicircle o in the figure is**

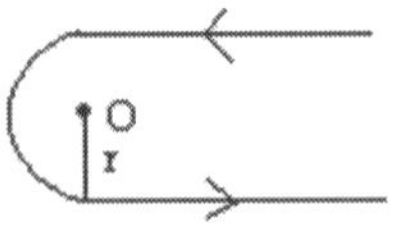

   1) $B=\frac{\mu_0 i}{4r}$ 2) $\frac{\mu_0 i}{4r}(1+2\pi)$ 3) $\frac{\mu_0 i}{4\pi r}(\pi-2)$ 4) $\frac{\mu_0 i}{4\pi r}(\pi+2)$

6. A cube made of wire of equal length is connected to a battery as shown in the figure. The magnetic field at the centre of the cube is

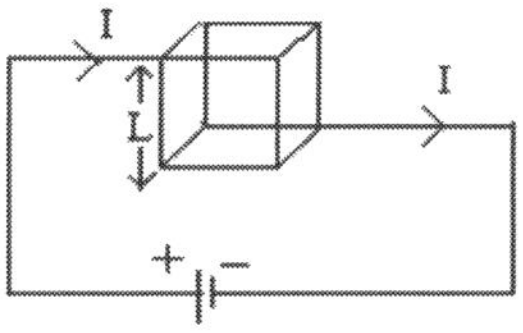

1) $\frac{12}{\sqrt{2}}\frac{\mu_0 I}{\pi L}$ 2) $\frac{6\mu_0 I}{\sqrt{2}\pi L}$ 3) $\frac{6\mu_0 I}{\pi L}$ 4) zero

7. In Bohr's model of hydrogen atom, the electron circulates round the nucleus in a path of radius 5×10^{-11}m at a frequency of 6.8×10^{15} revolutions per second. The value of magnetic induction at the centre of the orbit is

1) 12.27 T 2) 10.8 T 3) 13.2 T 4) 13.6T

8. In the given loop the magneticfield at the centre O is

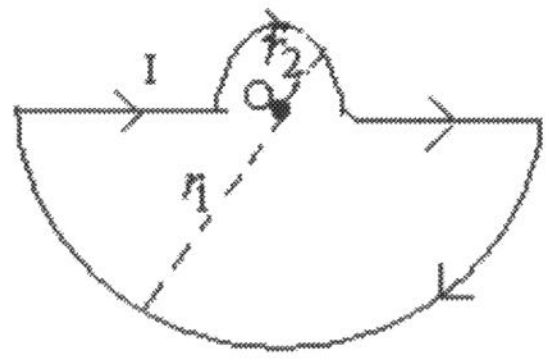

1) $\frac{\mu_0 I}{4}\left(\frac{r_1+r_2}{r_1 r_2}\right)$ out of the page 2) $\frac{\mu_0 I}{4}\left(\frac{r_1+r_2}{r_1 r_2}\right)$ into the page

3) $\frac{\mu_0 I}{4}\left(\frac{r_1-r_2}{r_1 r_2}\right)$ out of the page 4) $\frac{\mu_0 I}{4}\left(\frac{r_1-r_2}{r_1 r_2}\right)$ into the page

9. The field normal to the plane of a coil of n turns and radius r which carries a curent i is measured on the axis of the coil at a small distance h from the centre of the coil. This is smaller than the field at the centre by the fraction

1) $\frac{3}{2}\frac{h^2}{r^2}$ 2) $\frac{2}{3}\frac{h^2}{r^2}$ 3) $\frac{3}{2}\frac{r^2}{h^2}$ 4) $\frac{2}{3}\frac{r^2}{h^2}$

10. Figure shows the cross-sectional view of the hollow cylindrical conductor with inner redius 'R' and outer radius '2R'. Cylinder is carrying uniformly distributed current along it's axis. The magnetic induction at point 'P' at a distance $\frac{3R}{2}$ from the axis of the cylinder will be

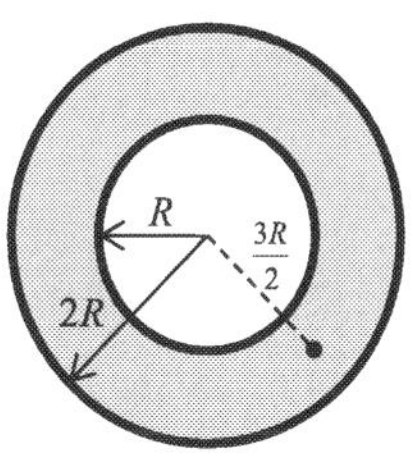

1) Zero 2) $\frac{5\mu_0 i}{72\pi R}$ 3) $\frac{7\mu_0 i}{18\pi R}$ 4) $\frac{5\mu_0 i}{36\pi R}$

11. In the following figure a wire bent in the form of a regular polygon of n sides is inscribed in a circle of radius a. Net magnetic field at centre will be

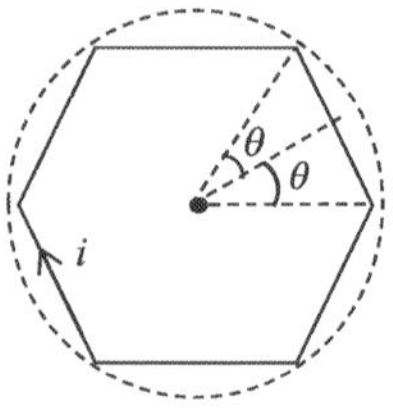

1) $B = \frac{\mu_0 i}{2\pi a} \tan\frac{\pi}{n}$
2) $B = \frac{\mu_0 ni}{2\pi a} \tan\frac{\pi}{n}$
3) $B = \frac{2\mu_0 ni}{\pi a} \tan\frac{\pi}{n}$
4) $B = \frac{\mu_0 ni}{2a} \tan\frac{\pi}{n}$

12. Two concentric coils of 10 turns each are placed in the same plane. Their radii are 20cm and 40cm and carry 0.2 and 0.3 amp. current respectively in opposite directions. The magnetic induction (in Tesla) at the centre is

1) $\frac{3}{4}\mu_0$
2) $\frac{5}{4}\mu_0$
3) $\frac{7}{4}\mu_0$
4) $\frac{9}{4}\mu_0$

TANGENT GALVANOMETER

13. Two tangent galvanometers having coils of the same radius are connected in series. A current flowing in them produces deflections of 60^0 and 45^0 respectively. The ratio of the number of turns in the coils are respectively

1) $\frac{\sqrt{3}}{1}$
2) $\frac{1}{\sqrt{3}}$
3) $\frac{\sqrt{3}+1}{1}$
4) $\frac{1}{\sqrt{3}+1}$

14. A tangent galvanometer carrying a certain current gave a deflection of 30^0. If the glavanometer is taken to another location where the earth's horizontal component of magnetic induction is one third of the previous value, the deflection for the same current will be

1) 60^0
2) 45^0
3) 90^0
4) 30^0

CURRENT LOOP AND MAGNETIC DIPOLE

15. A wire of length 6.28 m is bent into a circular coil of 2 turns. If a current of 0.5A exists in the coil, the magnetic moment of the coil is, in Am^2:

1) $\frac{\pi}{4}$
2) $\frac{1}{4}$
3) π
4) 4π

16. A wire of length 6.28 m is bent into a circular coil of 2 turns. If a current of 0.5 A exists in the coil, the magnetic moment of the coil is, in $A-m^2$:

1) $\frac{\pi}{4}$
2) $\frac{1}{4}$
3) π
4) 4π

SOLENOID & TOROID

17. A solenoid of 1000 turns is wound uniformly on a glass tube 50 cm long and 10 cm diameter. The strength of magnetic field at the centre of solenoid when a current of 0.1 A. flows through it is

1) 100 A/m
2) 200 A/m
3) 400 A/m
4) 50 A/m

18. **A long solenoid has 200 turns per cm and carries a current i.The magnetic field at its centre is $6.28\times10^{-2}\ wb/m^2$ .Another long solenoid has 100 turns per cm and it carries a current i/3.The value of magnetic field at its centre is**

1) $1.05\times10^{-2}\ wb/m^2$ 2) $1.05\times10^{-5}\ wb/m^2$

3) $1.05\times10^{-3}\ wb/m^2$ 4) $1.05\times10^{-4}\ wb/m^2$

FORCE ACTING ON A MOVING CHARGE IN MAGNETIC FIELD

19. **A proton moving in a perpendicular magnetic field possesses kinetic energy E. The magnetic field is increased 8 times. But the proton is constrained to move in the path of same radius. The kinetic energy will increase**

1) 1/8 times 2) 8 times 3) 16 times 4) 64 times

20. **Electrons accelerated by a potential difference V enter a uniform magnetic field of flux density B at right angles to the field. They describe a circular path of radius 'r'. If now V is doubled and B is also doubled, the radius of the new circular path is**

1) 4r 2) 2r 3) $2\sqrt{2r}$ 4) $r/\sqrt{2}$

21. **An electron is shot in steady electric and magnetic fields such that its velocity is V. Electric field E and magnetic field B are mutually perpendicular. The magnitude of E is 1 volt/cm and that of B is 2 tesla. Now it happens that the Lorentz (Magnetic) force cancels with the electro static force on the electron , then the velocity of the electron is**

1) 50 ms^{-1} 2) 2 cms^{-1} 3) 0.5 cms^{-1} 4) 200ms^{-1}

22. **A beam of protons enters a uniform magnetic field of 0.3 T with a velocity of $4\times10^5\ m/sec$ in a direction making an angle of 60^0 with the direction of magnetic field, the pitch of the helix will be**

1) 4.7m 2) 0.47m 3)0.047m 4) 0.0047 m

23. **A uniform magnetic field B is acting from south to north and is of magnitude 1.5 Wb / m². If a proton hiving mass = $1.7\times10^{-27}\ kg$ and charge = $1.6\times10^{-19}\ C$ moves in this field vertically doqnwards with energy 5 MeV, then the force acting onit will be**

1) $7.4\times10^{12}\ N$ 2) $7.4\times10^{-12}\ N$ 3) $7.4\times10^{19}\ N$ 4) $7.4\times10^{-19}\ N$

24. **An electron travelling with a velocity $\overline{V}=10^7$ im/s enter a magnetic field of induction $\overline{B}=\overline{2j}$. The force on electron is**

1) 1.6x$10^{-12}\ \overline{j}$N 2) 3.2x$10^{-12}\ \overline{k}$ N 3) 6.4x$10^{-12}\overline{k}$ N 4) -3.2x$10^{-12}\overline{k}$ N

25. **A magnetic field $4X10^{-3}\hat{K}T$ exerts a force $(4\hat{i}+3\hat{j})\times10^{-10}N$ on a particle having a charge $10^{-9}C$ and going in the X-Y plane. The velocity of the particle is**

1) $-75\hat{i}+100\hat{j}$ 2) $100\hat{i}+75\hat{j}$ 3) $75\hat{i}+100\hat{j}$ 4) $100\hat{i}-75\hat{j}$

FORCE BETWEEN TWO PARALLEL CURRENT CARRYING CONDUCTORS

26. A uniform conducting wire ABC has a mass of 10g. A current of 2A flows through it. the wire is kept in a uniform magnetic field B = 2T. The acceleration of the wire will be

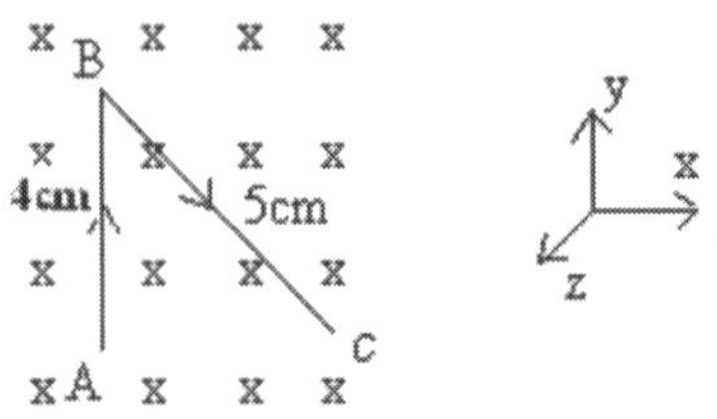

1) Zero
2) 12 ms^{-2} along y-axis
3) $1.2\times10^{-3}ms^{-2}$ along y-axis
4) $0.6\times10^{-3}ms^{-2}$ along y-axis

27. A straight conductor carrying a current is kept in a uniform magnetic field so as to experience maximum force. If now the conductor is turned in its own plane such that the force acting on it is half of the maximum force, then the angle made by the conductor in the final position with respect to the field is

1) 60°
2) 45°
3) 30°
4) 90°

28. Two long parallel wires are separated by a distance of 2 m. They carry a current of 1A each in opposite direction. The magnetic induction at the midpoint of a straight line connecting these two wires is

1) zero
2) $2\text{x}10^{-7}$ T
3) $4\text{x}10^{-5}$T
4) $4\text{x}10^{-7}$T

29. Three very long straight thin wires are connected parallel to each other through a battery of negligible internal resistance. The resistance of the wires are $2\Omega, 3\Omega$ and 4Ω. The ratio of distances of middle wire from the first and third wires if resultant magnetic force on the middle wire is zero is

1) 2 : 1
2) 3 : 4
3) 2 : 3
4) 3 : 5

30. Three long straight conductors are arranged parallel to each other in the same plane and carry currents of 1 A, 2 A and 3 A all in the same direction. The distance between the first two conductors is "x" and the distance between the second and third conductors is "y". If the middle conductor is in equilibrium, the ratio x : y is

1) 1 : 3
2) 3 : 1
3) 1 : $\sqrt{3}$
4) $\sqrt{3}$: 1

TORQUE ON CURRENT LOOP

31. A wire of length L is bent in the form of a circular coil of some turns. A current I flows through the coil. The coil is placed in a uniform magnetic field of induction B. The maximum torque on the coil can be

1) $\frac{IBL^2}{4\pi}$
2) $\frac{IBL^2}{\pi}$
3) $\frac{IBL^2}{2\pi}$
4) $\frac{2BIL^2}{\pi}$

MOVING COIL GALVANOMETER

32. A moving coil galvanometer A has 100 turns and resistance 10 Ω. Another galvanometer B has 50 turns and 5 Ω. The other quantities are same in both the cases. Then the voltage sensitivity of

1) A is greater than that of B
2) B is greater than that of A
3) A and B is Same
4) cannot be compared

33. The coil of a galvanometer consists of 100 turns and effective area $1cm^2$. The restoring couple is $10^{-8} N-m/rad$. The magnetic field between the pole pieces is 5 tesla. The current sensitivity of the galvanometer will be

1) 5×10^4 radian per micro ampere
2) 5×10^6 radian per micro ampere
3) 2×10^{-7} radian per micro ampere
4) 5 radian per micro ampere

CONVERSION OF MCG INTO AMMETER & VOLTMETER

34. The deflection in a moving coil galvanometer falls from 100 divisions to 20 divisions when a shunt of 12Ω is used. The resistance of the galvanometer coil is

1) 3Ω
2) 12Ω
3) 48Ω
4) $\frac{48}{5}\Omega$

35. A galvanometer requires $10\,\mu A$ for one division of its scale. It is to be used to measure a current of 1 amp to the full scale deflection. The scale has 100 divisions. The value of shunt if the resistance of the galvanometer is 999 Ω

1) 2 Ω
2) 3 Ω
3) 1 Ω
4) 4 Ω

36. The scale of a galvanometer of resistance 100 ohms contains 25 divisions. It gives a deflection of one division on passing a current of 4×10^{-4} amperes. The resistance in ohms to be added to it, so that it may become a voltmeter of range 2.5 volts is

1) 100
2) 150
3) 250
4) 300

37. A micro ammeter has a resistance of 100 Ω and a full scale range of 50 μ A. It can be used as a voltmeter or as a higher range ammeter provided a resistance is added to it. Pick the current range and resistance combinations.

1) 50 Volt range with 10 KΩ resitance in series
2) 5 Volt range with 200KΩ resistance in series
3) 5 mA range with 1 Ω resistance in parallel
4) 10 mA range with 1 Ω resistance in parallel

LEVEL - II KEY

1) 4	2) 1	3) 4	4) 3	5) 4	6) 4	7) 4	8) 2	9) 1	10) 4
11) 2	12) 2	13) 1	14) 1	15) 1	16) 1	17) 2	18) 1	19) 4	20) 4
21) 1	22) 3	23) 2	24) 4	25) 1	26) 2	27) 3	28) 4	29) 1	30) 1
31) 1	32) 3	33) 4	34) 3	35) 3	36) 2	37) 3			

LEVEL - II (HINTS)

1. $H=\frac{\mu_0 i}{2\pi r}$

2. $B=B_1+B_2$; $B_1=\frac{\mu_0 ni}{2r}$; $B_2=\frac{\mu_0 i}{2\pi r}$

3. $R=\sqrt{B^2+B_H^2}=\sqrt{B_H^2+B_H^2}=\sqrt{2}B_H$

4. $B=\frac{\mu_0 i}{4\pi r}(\theta)$

$B=\frac{3}{4}\left(\frac{\mu_0 I}{2r}\right)$

5. $B = B_1 + B_2 + B_3 = \frac{\mu_0 i}{4\pi r} + \frac{\mu_0 i}{4r} + \frac{\mu_0 i}{4\pi r}$

6. Magnetic field is zero at the centre due to symmetrical current distribution

7. $B = \frac{\mu_0 ne}{2r}$ where n is the frequency

8. $B = B_1 + B_2 = \frac{\mu_0 i}{4r_1} + \frac{\mu_0 i}{4r_2}$

$B_{total} = \frac{\mu_0 I}{4}\left(\frac{r_1 + r_2}{r_1 r_2}\right)$

9. Field at the centre $B_1 = \frac{\mu_0 NI}{2r}$

Field at distance h from the centre $B = \frac{\mu_0 nir^2}{2(r^2 + h^2)^{3/2}} = B_1\left(1 + \frac{h^2}{r^2}\right)^{-3/2}$

$B_1 = \left(1 - \frac{3}{2}\frac{h^2}{r^2}\right)$

10. $B = \frac{\mu_0 i}{2\pi R}\left(\frac{r^2 - a^2}{b^2 - a^2}\right)$ and $r = \frac{3R}{2}$, $a = R, b = 2R$

11. Magnetic field at the centre due to one side $B_1 = \frac{\mu_0 i}{2\pi r}\sin\theta$, where $r = a\cos\theta$

$\therefore B_1 = \frac{\mu_0 i}{2\pi a\cos\theta}\sin\theta = \frac{\mu_0 i}{2\pi a}\tan\theta$

$\therefore B_{net} = nB_1$

12. $B = \frac{n\mu_0 i}{2r}$

13. $I = K\tan\theta, K = \frac{2B_H r}{\mu_0 n}$; $\tan\theta \propto n$

14. If K is the reduction factor of the galvanometer, then $K = \frac{2rB_H}{\mu_0 n}$. As B_H is reduced to $1/3^{rd}$, K becomes K/3. Use $i = K\tan\theta$

15. $M = NI\,\pi r^2$

$l = (2\pi r)N$

16. $M = niA$

17. $H = ni$

18. $\frac{B_2}{B_1} = \frac{i_2 n_2}{i_1 n_1}$

19. $r = \frac{\sqrt{2m(K.E)}}{Bq}$

20. $\frac{r_1}{r_2} = \sqrt{\frac{V_1}{V_2} \times \frac{B_2^2}{B_1^2}}$

21. $r = \frac{mv}{Bq}$

22. $p = v\cos\theta \times T$

$T = \frac{2\pi r}{v\sin\theta}$

23. $F = Bqv$ and $KE = \frac{1}{2}mv^2 \Rightarrow F = qB\sqrt{\frac{2KE}{m}}$

24. $F = q(\vec{v} \times \vec{B})$ 25. $F = q(v \times B)$

26. Force on the wire $F = Bil = 2\times 2\times 3\times 10^{-2}$

27. $F = \frac{F_{max}}{2}$

$\therefore \sin\theta = \frac{1}{2}$

28. $B = B_1 + B_2$

29. As the wires are parallel, $I_1 : I_2 : I_3 = \frac{1}{2}:\frac{1}{3}:\frac{1}{4} = 6:4:3$

Force between first second & second, third wires are equal $\frac{F}{l} = \frac{\mu_0 I_1 I_2}{2\pi r}$

30. $F = \frac{\mu_o i_1 i_2}{2\pi r}$ & $f_{12} + f_{23} = 0$ 31. $\tau_{max} = BINA$

32. $\frac{\theta}{V} = \frac{NAB}{CG}$ 33. $\frac{\theta}{i} = \frac{NAB}{C}$ 34. $S = \frac{G}{n-1}$

35. $S = \frac{G}{\frac{i}{ig} - 1}$ 36. $R = \frac{V}{i_g} - G$ 37. $S = \frac{G}{\frac{i}{ig} - 1}$

LEVEL-III

AMPERE'S CIRCUITAL LAW

1. Two wires AO and OC carry currents i as shown in figure. One end of both the wires extends to infinity $\angle AOC = \alpha$, the magnitude of magnetic field at a point P on the bisector of the two wires at a distance r from O is

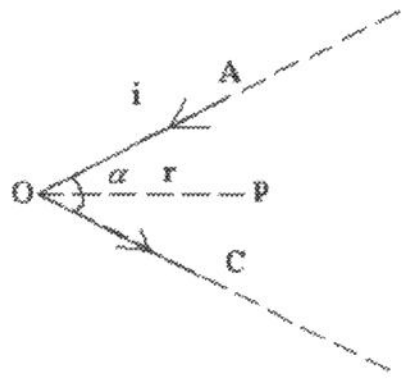

1) $\frac{\mu_o i}{2\pi r}\cot\left(\frac{\alpha}{2}\right)$ 2) $\frac{\mu_o i}{4\pi r}\cot\left(\frac{\alpha}{2}\right)$

3) $\frac{\mu_o i}{2\pi r}\frac{\left(1+\cos\frac{\alpha}{2}\right)}{\sin\left(\frac{\alpha}{2}\right)}$ 4) $\frac{\mu_o i}{4\pi r}\sin\left(\frac{\alpha}{2}\right)$

2. A current of 1A is flowing through a straight conductor of length 16cm. The magnetic induction (in tesla) at a point 10cm from the either end of the wire is :

1) $\frac{8}{3}\times10^{-6}$ 2) $\frac{1}{6\sqrt{2}}\times10^{-5}$ 3) $\frac{1}{6\sqrt{3}}\times10^{-6}$ 4) $\frac{\sqrt{3}}{6}\times10^{-6}$

3. Each of two long parallel wires carries a constant current i along the same direction. The wires are separated by a distance 2l. The magnitude of resultant magnetic induction in the symmetric plane of this system located between the wires at a distance R from each wire will be

1) $\frac{\mu_0 i}{\pi R}$ 2) zero 3) $\frac{\mu_0 i}{\pi\sqrt{R^2-l^2}}$ 4) $\frac{\mu_0 i}{\pi R}\sqrt{1-\frac{l^2}{R^2}}$

4. A long straight wire of radius 'a' carries a steady current i. The current is uniformly distributed across section. The ratio of the magnetic field at $\frac{a}{2}$ and 2a is

1) 1/4 2) 4 3) 1 4) 1/2

5. PQ is a uniform rod of length ℓ and mass m carrying current i and is suspended in uniform magnetic field of induction $\vec{B}$ acting inward as shown in figure. The tension in each string is

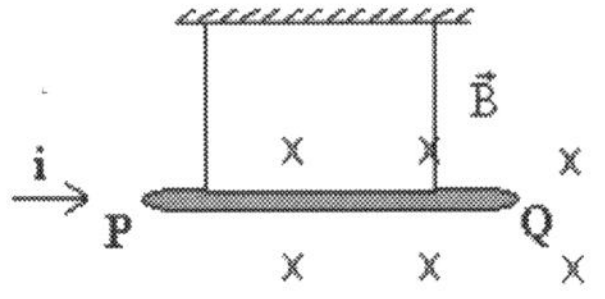

1) $mg - Bi\ell$ 2) $mg + Bi\ell$ 3) $\frac{mg - Bi\ell}{2}$ 4) $\frac{mg + Bi\ell}{2}$

6. Wires 1 and 2 carrying currents i_1 and i_2 respectively are inclined at an angle θ to each other. What is the force on a small element *dl* of wire 2 at a distance of r from wire 1 (as shown in figure) due to the magnetic field of wire 1

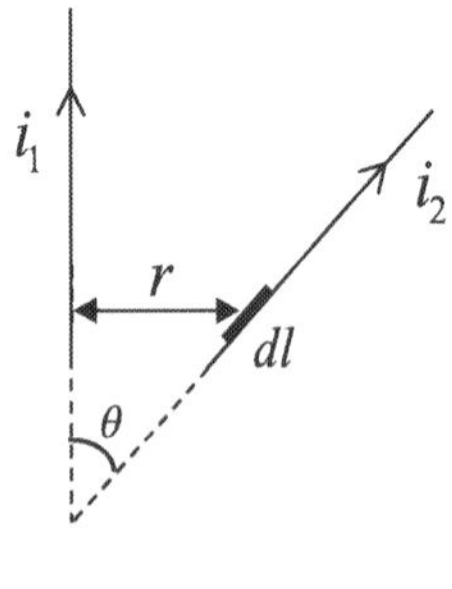

1) $\frac{\mu_0}{2\pi r}i_1i_2dl\tan\theta$ 2) $\frac{\mu_0}{2\pi r}i_1i_2dl\sin\theta$

3) $\frac{\mu_0}{2\pi r}i_1i_2dl\cos\theta$ 4) $\frac{\mu_0}{4\pi r}i_1i_2dl\sin\theta$

7. **A current $i = 2A$ be flowing in each part of a wire frame as shown in fig. The frame is a combination of two equilateral triangles ACD and CDE of side 1m. It is placed in uniform magnetic field B=4T acting perpendicular to the plane of frame. The magnitude of magnetic force acting on the frame is**

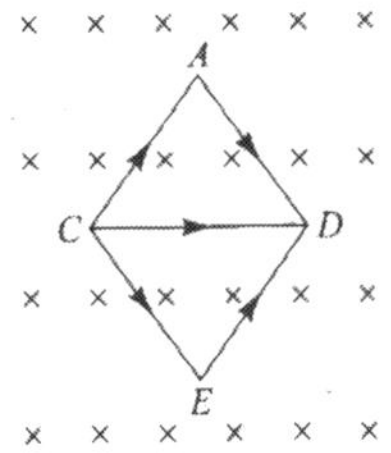

1) 24 N 2) Zero 3) 16N 4) 8N

8. **A wire bent in the form a right angled triangle PQR, earries a current 1 A. It is placed in a region of a uniform magnetic field B= 0.2T. If PR = 1m, the net force on the wire is**

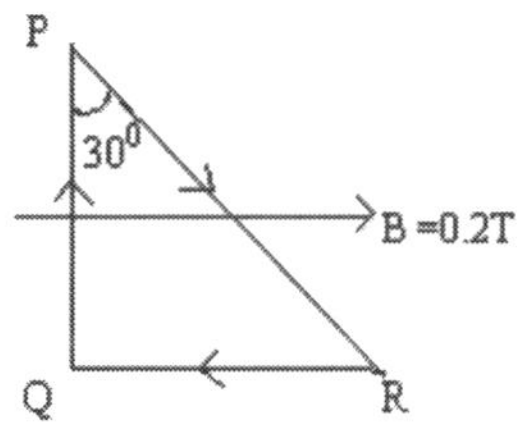

1)1.73N 2)3.46N 3).2.732N 4) Zero

9. **Two long parallel conductors carry currents $i_1 = 3A$ and $i_2 = 3A$ both are directed into the plane of paper. The magntidue of resultant magnetic field at point 'P; is**

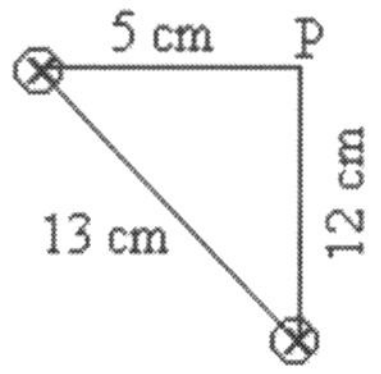

1) $12\mu T$ 2) $5\mu T$ 3) $13\mu T$ 4) $7\ \mu T$

10. **The figure shows a unirom conducting structure which carries current i with each small square has side a. The structure is kept in a uniform magnetic field B. Then the magnetic force on the structure will be**

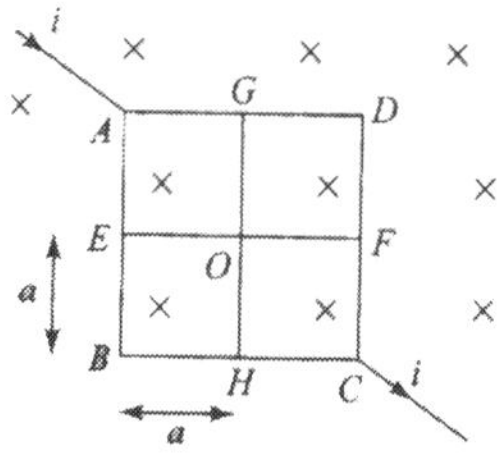

1) $2\sqrt{2}$ **iBa** 2) $\sqrt{2}$ **iBa** 3) **2 iBa** 4) **iBa**

11. A current-carrying circular loop of radius R is placed in the XY-plane with centre at the origin. Half of the loop with x>0 is now bent so that it now lies in the YZ-plane.

1) The magnitude of magnetic moment now diminishes
2) The magnetic moment does not change
3) The magnitude of $\vec{B}$ at (0,0,z) ,z>>R increases
4) The magnitude of $\vec{B}$ at (0,0,z),z>> R is unchanged

MAGNETIC INDUCTION FIELD DUE TO CURRENT CARRYING CIRCULAR COIL

12. The magnetic induction at O due to a current in conductor shaped as shown in fig. is

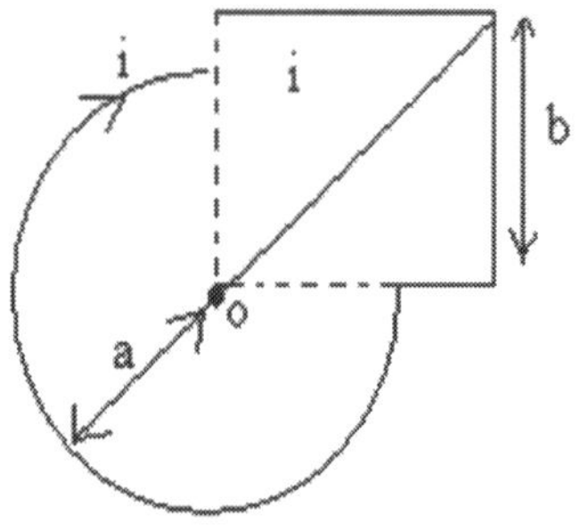

1) $\frac{\mu_o i}{4\pi}\left[\frac{3\pi}{2a}+\frac{\sqrt{2}}{b}\right]$ 2) $\frac{\mu_o i}{4\pi}\left[\frac{3\pi}{4a}-\frac{\sqrt{2}}{b}\right]$ 3) $\frac{\mu_o i}{2\pi}\left[\frac{3\pi}{4a}+\frac{1}{\sqrt{2}b}\right]$ 4) $\frac{\mu_o i}{4\pi}\left[\frac{1}{a}+\frac{1}{b}\right]$

13. A wire is bent in the form of a circular arc with a straight portion AB. If the current in the wire is i, then the magnetic induction at O is

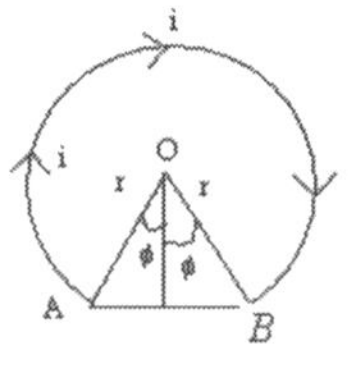

1) $\frac{\mu_0 i}{2\pi r}(\tan\phi)$ 2) $\frac{\mu_0 i}{2\pi r}(\pi-\phi)$ 3) $\frac{\mu_0 i}{2\pi r}(\pi-\phi+\tan\phi)$ 4) $\frac{\mu_0 i}{2\pi r}(\pi-\tan\phi)$

14. A wire of length 'L' is bent in to an arc of a circle and found to subtend and angle of 'θ' radians at the center. If a current of 'i' is passed through it, the magnetic induction at the center of the circle is

1) $\frac{\mu_o i\theta^2}{2\pi L}$ 2) $\frac{\mu_o i\theta^2}{4\pi L}$ 3) $\frac{\mu_o i\theta}{4\pi L}$ 4) $\frac{\mu_o i\theta}{2\pi L}$

15. A wire of length L is shaped into a circle and then bent in such a way that the two semi-circles are perpendicular. The magnetic moment of the system when current I flows through the system is

1) $\frac{\sqrt{2}iL^2}{8\pi}$ 2) $\frac{\sqrt{3}iL^2}{4\pi}$ 3) $\frac{iL^2}{4\pi}$ 4) $\frac{iL^2}{2\pi}$

16. A square frame of side ℓ carries a current produces a field B at its centre. The same current is passed through a circular loop having same perimeter as the square. The field at its centre is B', the ratio of $\frac{B}{B'}$ is

1) $\frac{8\sqrt{2}}{\pi^2}$ 2) $\frac{8}{\pi^2}$ 3) $\frac{\sqrt{2}}{\pi^2}$ 4) $\frac{\pi^2}{8\sqrt{2}}$

17. A circular coil of 100 turns and effective diameter 20cm carries a current of 0.5 A. It is to be turned in a magnetic field of B = 2.0 T from a position in which the normal to the plane of the coil makes an angle θ equals to zero to one in which θ equals to 180°. The work required in this process is

1) π J 2) 2 π J 3) 4 π J 4) 8 π J

18. A long solenoid has 200 turns per centimetre and carries a current i. The magnetic field at its centre is $6.28X10^{-2}Wb/m^2$. Another long solenoid has 100 turns per centimetre and it carries a current i/3. The value of the magnetic field at its centre is :

1) $1.05X10^{-3}Wb/m^2$ 2) $1.05X10^{-4}Wb/m^2$

3) $1.05X10^{-2}Wb/m^2$ 4) $1.05X10^{-5}Wb/m^2$

19. The magnetic induction at the point O, if the wire carrying current I is

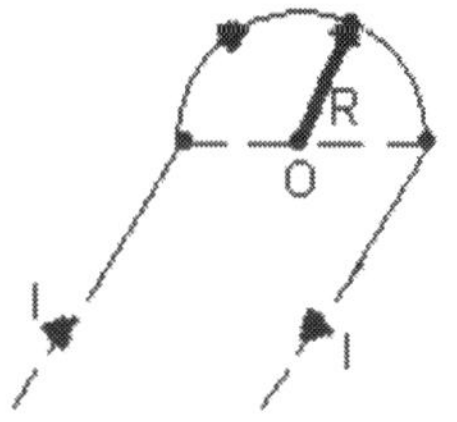

1) $\frac{\mu_0 I}{2R}$ 2) $\frac{\mu_0 I}{2\pi R}$ 3) $\frac{\mu_0 I(\pi^2+4)^{1/2}}{4\pi R}$ 4) $\frac{\mu_0 I(\pi^2+4)}{4\pi R}$

20. Two circular coils P and Q are made from similar wire but radius of Q is twice that of P. Relation between the values of potential difference across them so that the magnetic induction at their centers may be the same is

1) $V_q = 2V_p$ 2) $V_q = 3V_p$ 3) $V_q = 4V_p$ 4) $V_q = \frac{1}{4}V_p$

MOTION OF A CHARGED PARTICLE IN A MAGNETIC FIELD

21. A charged particle of charge 5mc and mass 5gm is moving with a constant speed 5 m/s. in a uniform magnetic field on a curve $x^2 + y^2 = 25$. Where x and y are in meter. The value of magnetic filed required will be

1) 1 Tesla 2) 1 T along z-axis

3) 5 KT along the x-axis 4) 1 KT along any line in the x-y plane

22. Velocity and acceleration vectors of a charged particle moving in a magnetic field at some instant are $\vec{v} = 3\hat{i} + 4\hat{j}$ and $\vec{a} = 2\hat{i} + x\hat{j}$. Select the wrong alternative

1) x = –1.5 2) x = 3

3) magnetic field is along z-direction 4) Kinetic energy of the particle is constant

23. **A charged particle enters a magnetic field at right angles to the field. The field exists for a length equal to 1.5 times the radius of circlar path of particle. The particle will be deviated from its path by**

1) 90° 2) $\sin^{-1}\left(\frac{2}{3}\right)$ 3) 30° 4) 180°

24. **A particle of mass m and charge q, moving with velocity V enters Region II normal to the boudary as shown in fig. Region II has a uniform magnetic field B perpendicular to the plane of the paper. The length of the Region II is l. Then :**

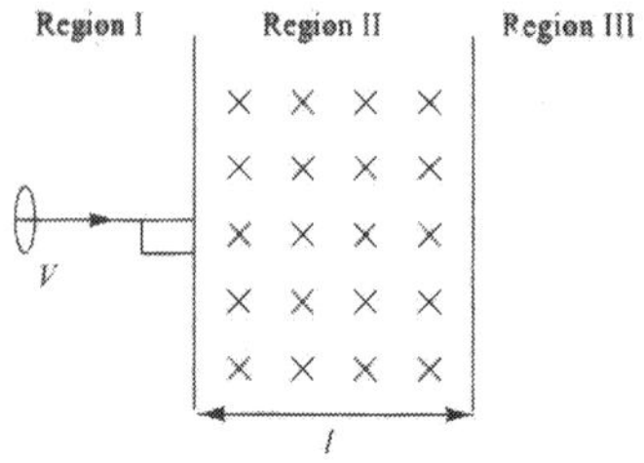

1) The particle enters Region III only if its velocity $V > \frac{qlB}{m}$

2) Path length of the particle in region II is maximum when velocity $V = \frac{qlB}{m}$

3) Time spent in Region II is same for any velocity V as long as the particle returns to Region I.

4) All the above are correct

25. **A proton accelerated by a pd V=500 KV moves through a transverse magnetic field B=0.51 T as shown in figure. Then the angle θ through which the proton deviates from the initial direction of its motion is (approximately)**

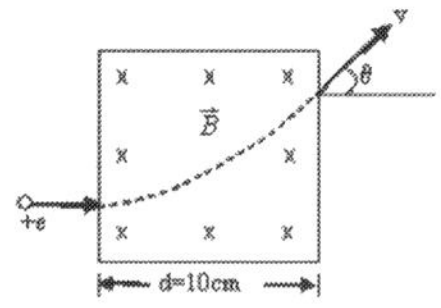

1) 15° 2) 30° 3) 45° 4) 60°

26. **Two identical charged particles enter a uniform magnetic field with same speed but at angles 30^0 and 60^0 with field. Let a, b and c be the ratio of their time periods, radii and pitches of the helical paths then:**

1) abc =1 2) abc>1 3) abc<1 4) ac = b

27. **An electron whose e/m is 1.76X10^{11} c/kg enter a region where there is a uniform magnetic field of induction $2X10^{-3}$ tesla with a velocity of $3X10^6$ m/sec in a direction making an angle of 45^0 with the field. The pitch of its helical path in the region is**

1) 1.5cm 2) 3.8cm 3) 5.36cm 4) 8.4cm

28. **Two charged particles traverse identical helical paths in a completely opposite sense in a uniform magnetic field $\vec{B} = B_0\hat{K}$**

1) They have equal z-components of momenta. 2) They must have equal charges

3) They necessarily represent a partical-antipar ticle pair

4) The charge to mass ratio satisfy: $(e/m)_1 + (e/m)_2 = 0$

FORCE BETWEEN PARALLEL CURRENT CARRYING CONDUCTORS

29. Three very long straight conductors are arranged parallel to each other in a plane and have resistances in the ratio 1:2:3. They are connected in parallel to a battery of negligible internal resistance such that the currents in all three wires are in the same direction. The distance between the first two conductors is x and the distance between the second and third conductors is y. If the middle conductor is in equilibrium, the ratio $x:y$ is

1) 1 : 3   2) $3:1$ 3) $1:\sqrt{3}$ 4) $\sqrt{3}:1$

30. Two long wires are placed parallel to each other 10cm apart as shown in fig. The magnetic field at piont P is

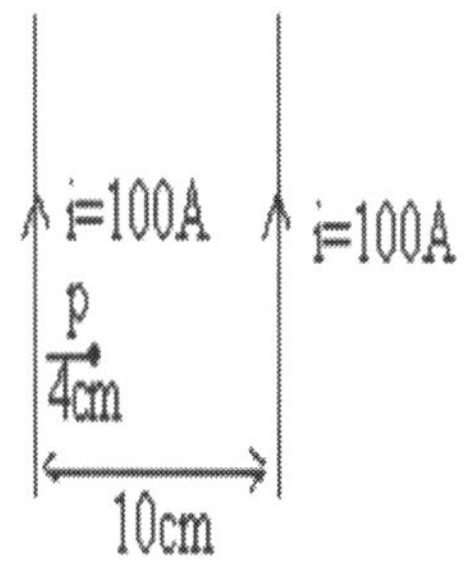

1) 5/6 x 10^{-3} T directed perpendicular into the paper
2) 1/3 x 10^{-3} T directed perpendicular out of the paper
3) 5/6 x 10^{-3} T directed perpendicular out of the paper
4) 1/6 x 10^{-3} T directed perpendicular into the paper

31. Two long parallel wires are at a distance 2d apart. They carry steady equal currents flowing out of the plane of the paper as shown. The variation of the magnetic field B along the line XX^1 is given by

1) 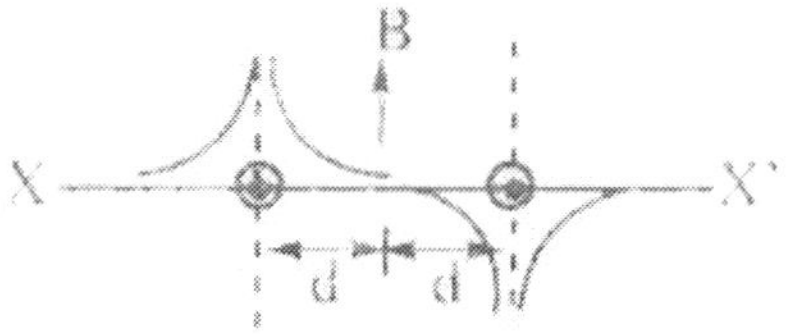

2) 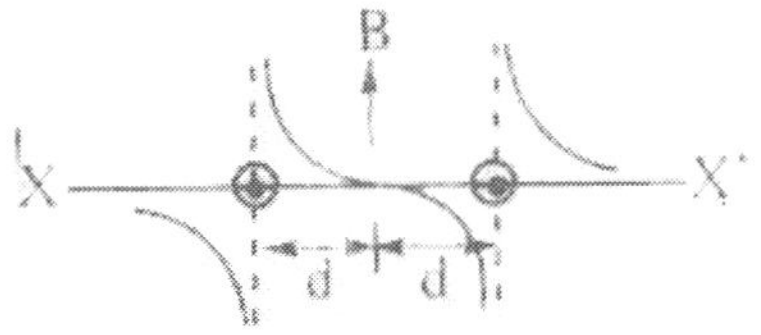

3) 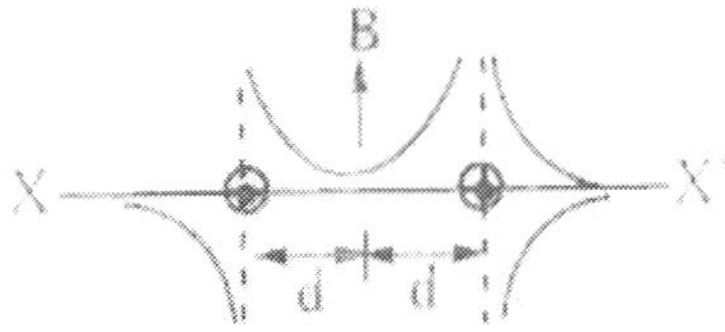

4) 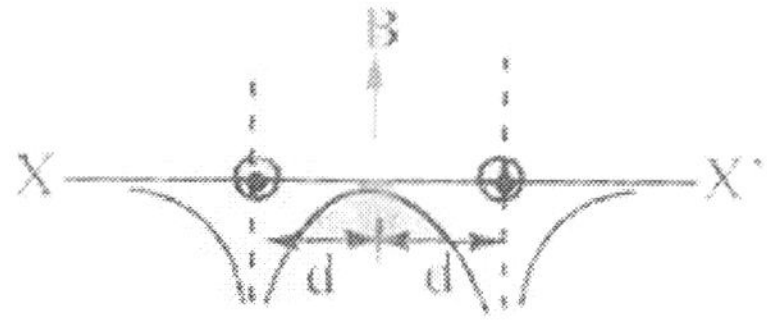

TORQUE ON A CURRENT CARRYING LOOP IN A MAGNETIC FIELD AND MCG

32. A circular loop of radius 20cm is placed in a uniform magnetic field $\vec{B} = 2T$ in X-Y plane. the loop carries a current 1A in the direction shown in figure. The magnitude of torque acting on the loop is nearly

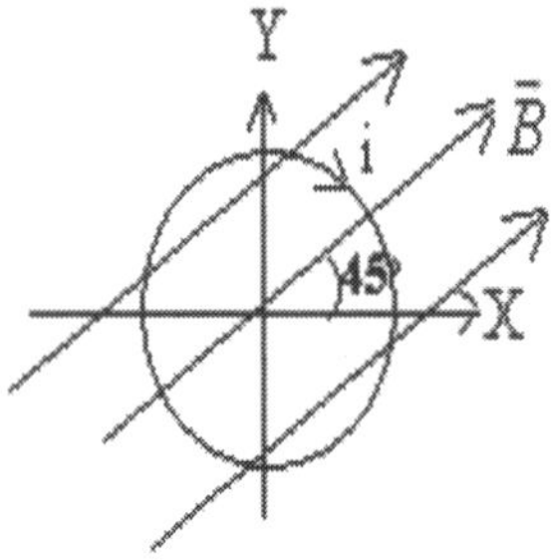

1) 0.25 N-m　2) $\frac{0.25}{\sqrt{2}}$N-m　3)0.75N-m　4) $\frac{0.75}{\sqrt{2}}$N-m

33. A non conducting disc of radius R, charge q is rotating about an axis passing through its centre and perpendicular to its plane with an angular velocity ω charge q is uniformly distributed over its surface. The magnetic moment of the disc is

1) $\frac{1}{4}q\omega R^2$　2) $\frac{1}{2}q\omega R$　3) $q\omega R$　4) $\frac{1}{2}q\omega R^2$

34. A galvanometer of resistance 5Ω is connected in series with a resistance of 0.2Ω to a battery of negligible internal resistance. The deflection is noted. If the 0.2Ω resistance is replaced by 2Ω resistance, the value of shunt resistance to be connected to the galvanometer to maintain the same deflection is

1) 1Ω　2) 8.9Ω　3) 0.01Ω　4) 10Ω

35. A thin wire of length L is made of an insulating material. The wire is bent to form a circular loop, and a positive charge q is distributed uniformly around the circumference of the loop. The loop is then set into rotation with angular speed ω around an axis through its centre. If the loop is in the region where there is a uniform magnetic field $\vec{B}$ directed parallel to the plane of the loop, calculate the magnitude of the magnetic torque on the loop.

1) $\frac{q\omega L^2 B}{8\pi^2}$　2) $\frac{q\omega L^2 B}{4\pi^2}$　3) $\frac{q\omega L^2 B}{2\pi^2}$　4) $\frac{q\omega L^2 B}{\pi^2}$

36. A rigid circular loop of radius r and mass m lies in the XY- plane of a flat table and has a current I flowing in it. At this particular place. the Earth's magnetic field $\vec{B} = B_x i + B_z k$. . The value of I so that the loop starts tilting is :

1) $\frac{mg}{\pi r\sqrt{B_x^2 + B_z^2}}$　2) $\frac{mg}{\pi r B_x}$　3) $\frac{mg}{\pi r B_z}$　4) $\frac{mg}{\pi r\sqrt{B_x B_z}}$

37. A hollow sphere has charge 'q' which is uniformly distributed over its surface and rotationg about its diameter with angular velocity 'ω'. The magnetic moment of the sphere is

1) $\frac{qr^2\omega}{2}$　2) $\frac{qr^2\omega}{3}$　3) $qr^2\omega$　4) $\frac{qr^2\omega}{4}$

38. A loop carrying current I lies in the x-y plane as shown in fig. The unit vector $\hat{k}$ is coming out of the plane of the paper. The magnetic moment of the current loop is

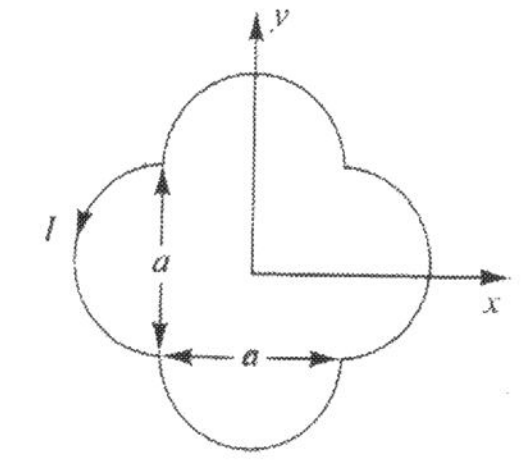

1) $a^2 I\hat{k}$ 2) $\left(\frac{\pi}{2}+1\right)a^2 I\hat{k}$ 3) $-\left(\frac{\pi}{2}+1\right)a^2 I\hat{k}$ 4) $(2\pi+1)a^2 I\hat{k}$

39. Electric field strength $\overline{E} = E_0\hat{i}$ and magnetic induction $\overline{B} = B_0\hat{i}$ exists in a region. A charged particle 'q' is released from rest at origin. Work done by both the fields is after certain time is

1) qEx 2) qEy 3) qEz 4) $qE\sqrt{x^2+y^2}$

40. A long current carrying wire, carrying current such that it is flowing out from the plane of paper, is placed at O. A steady state current is flowing in the loop ABCD. Then

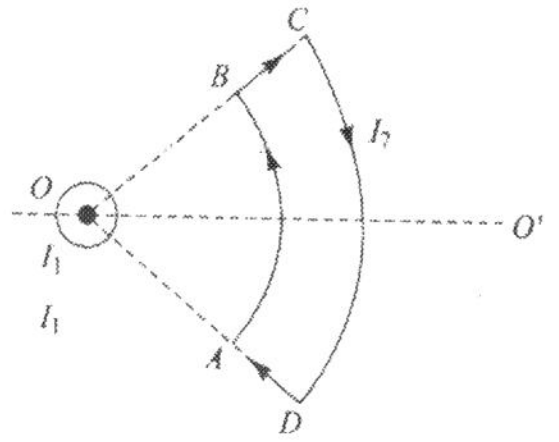

1) the net force is zero
2) as seen from O, the loop will rotate in clockwise direction along axis OO'.
3) both 1 & 2 are correct.
4) as seen from O,the loop will rotate in anticlock wise direction along axis OO'.

KEY LEVEL-III

1) 3	2) 1	3) 2	4) 3	5) 3	6) 3	7) 1	8) 3	9) 3	10) 1
11) 1	12) 1	13) 3	14) 2	15) 1	16) 1	17) 2	18) 3	19) 3	20) 3
21) 2	22) 2	23) 4	24) 4	25) 2	26) 1	27) 2	28) 4	29) 2	30) 4
31) 2	32) 1	33) 1	34) 2	35) 1	36) 2	37) 2	38) 2	39) 1	40) 3

HINTS LEVEL-III

1. $B = \frac{\mu_0}{4\pi}\frac{I}{a}\left[\sin(90-\theta/2)+\sin 90^0\right]\times 2$

2. $B = \frac{\mu_o I}{4\pi r}(\sin\phi_1 + \sin\phi_2)$

3. $B_1 = \frac{\mu_0 i}{2\pi R}$; $B_2 = \frac{\mu_0 i}{2\pi r}$

4. $B_1 = \frac{\mu_0}{2\pi}\frac{ir_1}{R}$ $B_2 = \frac{\mu_0}{2\pi}\frac{i}{r_2}$

5. $2T = mg - Bil$

6. Length of the component dl which is parallel to wire (1) is $dl\cos\theta$, so force on it.

$$F = \frac{\mu_0}{2\pi}\frac{i_1 i_2}{r}(dl\cos\theta)$$

7. $F = il_{CD}B$

8. $F = Bil\sin\theta$ and net force $= F_1 - F_2$

9. $B = \frac{\mu_0 I}{2\pi r}$

$$B_1 = \sqrt{B_1^2 + B_2^2 + 2B_1B_2\cos\theta}$$

10. F=ilB Where l is effective length AC.

11. The magnetic moment due to two halves of the bent circular loop are at right angle to each other. If m is the magnetic moment of the unbent loop, then that of the bent loop is $\sqrt{(m/2)^2 + (m/2)^2} = m/\sqrt{2} < m$

12. $B_1 = \frac{\mu_o i\phi}{4\pi a}, B_2 = \frac{\mu_o i}{4\pi b}(\sin 0° + \sin 45°)$

$\therefore B = B_1 + B_2$

13. $B = B_1 + B_2 = \frac{\mu_0 i}{4\pi r}(2\pi - 2\theta) + \frac{\mu_0 i}{4\pi d}(2\sin\theta)$

where $\cos\theta = d/r$ $\quad d = r\cos\theta$

14. $B = \frac{\mu_o i\theta}{4\pi r}$; $L = r\theta$

15. $B = \frac{\mu_0 i\phi}{2r}$

16. $B = 4\left[\frac{\mu_o}{4\pi r}(\sin\theta_1 + \sin\theta_2)\right], B = \frac{\mu_o ni}{2r}$

17. $w = MB(1 - \cos\theta)$

18. $B = \mu_0 ni$

19. $B_1 = \frac{\mu_0 i}{4\pi R} + \frac{\mu_0 i}{4\pi R}$

$B_2 = \frac{\mu_0 i}{4R}$

$B = \sqrt{B_1^2 + B_2^2}$

20. $B = \frac{\mu_0 i}{2r}$ where $i = \frac{V}{R}$

21. $r = \frac{mV}{Bq}$

22. $\vec{F}.\vec{v} = 0$

23. If the length of the field is more than the radius then the particle deviates by 180^0

24. $\bar{v} \perp \bar{B}$ in region II. Therefore, path of particle is circle in region II. Particle enters in region III, if radius of circular path, r > l or

$$\frac{mv}{Bq} > l \text{ or } v > \frac{Bql}{m}$$

If $v=\frac{Bql}{m}, r=\frac{mv}{Bq}=l$, particle will turn back and path length will be maximum. If particle returns to region 1, time spent in region II will be

$t=\frac{T}{2}=\frac{\pi m}{Bq}$, which is independent of v

25. $Bqv=\frac{mv^2}{r}$

$\frac{1}{2}mv^2=Vq$

and $\sin\theta=\frac{d}{r}$

26. $T=\frac{2\pi m}{qB}; R=\frac{mv_{\perp}}{qB}\ pitch=T\times\cos 45^0$

27. $T=\frac{2\pi m}{Bq}$

28. Since the charged particles traverse identical helical paths in a completely opposite sense in a uniform magnetic field B, their radii $(r=mv\sin\theta/eB)$ and frequencies $(v=eB/2\pi m)$ must be equal For this

$(e/m)_1=-(e/m)_2\ or\ (e/m)_1+(e/m)_2=0$

29. $F=\frac{\mu_0 i_1 i_2 l}{2\pi r}$

30. $B=\frac{\mu_0 I}{2\pi}\left[\frac{1}{r_1}-\frac{1}{r_2}\right]$

31. $B_{net}=B_1\sim B_2$

32. $\vec{\tau}=\vec{M}\times\vec{B}$

33. $M=(I\omega)\frac{q}{2m}$

34. To keep i same, $R_1=R_2$

35. $\tau=mB\sin\theta=IAB\sin\theta=IAB\ (as\theta=90^0)$

Further ,as $I=\frac{q}{T}=\frac{q}{2\pi/\omega}=\frac{q\omega}{2\pi}$,

$A=\pi r^2=\pi\left(\frac{L}{2\pi}\right)^2=\frac{L^2}{4\pi}$

$$\tau = \left(\frac{q\omega}{2\pi}\right)\left(\frac{L^2}{4\pi}\right)B = \frac{q\omega L^2 B}{8\pi^2}$$

36. For the point O $B_1 = \left(\frac{\mu_0 I_c}{2R}\right)_{\odot}$

(due to circular loop)

$B_2 = \left(\frac{\mu_0 I_c}{2\pi H}\right)_{\otimes}$ (due to long straight wire)

Since $\overrightarrow{B_1}$ and $\overrightarrow{B_2}$ act in opposite direction for resulatant magnetic field to be zero $B_1 = B_2$

whence, $H = \frac{I_1 R}{I_c \pi}$

37. Select an annular ring of radius $r\sin\theta$ and thickness $rd\theta$. $d\mu$ is its magnetic moment.

$$d\mu = \frac{(r\sin\theta)^2\,\omega dq}{2};$$

$$dq = \frac{q}{4\pi r^2}\left(2\pi r^2 \sin\theta\, d\theta\right)$$

$$\mu = \int_0^{\pi} d\mu$$

38. Area $= a^2 + 4\times\frac{\pi\left(\frac{a}{2}\right)^2}{2} = a^2 + \frac{\pi a^2}{2}$

$A = \left(1+\frac{\pi}{2}\right)a^2\hat{k}$; M = niA

39. $\overline{B}$ cannot do any work. q will move only along x-axis. Hence $W_E = qEx$.

40. Net force on AB and CD will be zero because magnetic field due to I_1 is along these parts. Force on BC due to I_1 will be outwards and on AB will be inwards. This will provide clockwise torque as seen from O.

LEVEL-IV

1. **Assertion (A): The plane of the coil of tangent galvanometer should be parallel to the magnetic meridian**

Reason (R): It makes the magnetic field of the coil perpendicular to the horizontal component of earth's magnetic field so that tangent law can be applicable

1) A and R are true and R is the correct explanation of A

2) A is true and R is true

3) both A and R are false

4) A and R are true and R is not a correct explanation of A

2. **Assertion (A) : For a charged particle describing circular path in a uniform transverse magnetic field, the kinetic energy remain constant**

Reason (R) : The work done by the force due to magnetic field on a moving charge is always zero

1) A and R are true and R is the correct explanation of A

2) A and R are true and R is not correct explanation of A

3) A is true, R is false

4) A is false , R is true

3. **Assertion (A): If an electron is not deflected while passing through a certain region of space, then only possibility is that there is no magnetic field in this region.**

Reason (R): Force on an electron moving in a magnetic field is directly proportional to the magnetic field applied.

1) Both A and R are correct , R is correct reason of A

2) Both are wrong

3) Both A and R are correct and R is not the correct reason of A

4) A is correct, R is wrong

4. **Assertion (A): If a proton and an α -particle enter a uniform magnetic field perpendicularly, with the same speed, the time period of revolution of α -particle is double that of proton.**

Reason (R): In a magnetic field, the time period of revolution of a charged particle is inversely proportional its specific charge.

1) Both A and R are correct , R is correct reason of A

2) Both are wrong

3) Both A and R are correct and R is not the correct reason of A

4) A is wrong, R is correct

5. **Assertion (A): If an electron while coming vertically from outerspace enter the earth's magnetic field, it is deflected towards west.**

Reason (R): Direction of force experienced by a charged particle in a magnetic field is determined by flemings left hand rule.

1) Both A and R are correct , R is correct reason of A

2) Both are wrong

3) Both A and R are correct and R is not the correct reason of A

4) A is correct, R is wrong

6. **Assertion (A): Torque on the coil is always maximum, when coil is suspended in a radial magnetic field.**

   Reason (R): Torque depends upon the magnitude of the applied magnetic field.

   1) Both A and R are correct , R is correct reason of A

   2) Both are wrong

   3) Both A and R are correct and R is not the correct reason of A

   4) A is correct, R is wrong

7. **Assertion (A): Magnetic field interacts with a moving charge and not with a stationary charge.**

   Reason (R): Magnetic field will be produced by a moving charge but not by a stationary charge.

   1) Both A and R are correct , R is correct reason of A

   2) Both are wrong

   3) Both A and R are correct and R is not the correct reason of A

   4) A is correct, R is wrong

8. **Assertion (A): No net force acts on a rectangular coil carrying a steady current when suspended freely in a uniform magnetic field.**

   Reason (R): Forces acting on each pair of the opposite sides of the coil are equal and opposite.

   1) Both A and R are correct , R is correct reason of A

   2) Both are wrong

   3) Both A and R are correct and R is not the correct reason of A

   4) A is correct, R is wrong

9. **Assertion (A): Out of galvanometer, ammeter and voltmeter, resistance of ammeter is lowest and resistance of voltmeter is highest.**

   Reason (R): An ammeter is connected in series and a voltmeter is connected in parallel, in a circuit.

   1) Both A and R are correct , R is correct reason of A

   2) Both are wrong

   3) Both A and R are correct and R is not the correct reason of A

   4) A is correct, R is wrong

10. **Assertion (A): In ammeter, current in shunt is always greater than current in galvanometer.**

    Reason (R): Value of shunt resistance is smaller than galvanometer resistance and is parallel to it.

    1) Both A and R are correct , R is correct reason of A

    2) Both are wrong

    3) Both A and R are correct and R is not the correct reason of A

    4) A is correct, R is wrong

11. **Assertion (A): The range of given voltmeter can be both increased and decreased.**

    Reason (R): By adjusting the value of resistance in series with galvanometer the range of voltmeter can be adjusted.

    1) Both A and R are correct , R is correct reason of A

    2) Both are wrong

    3) Both A and R are correct and R is not the correct reason of A

    4) A is correct, R is wrong

12. **Consider the following statements:**

A) Time period of a charged particle in uniform magnetic field is independent of K.E of the particle

B) Time period of a charged particle in uniform magnetic field depends on the angle between velocity and magnetic field

C) Time period of a charged particle in uniform magnetic field is inversely propor tional to specific charge

1) A, B are correct and C is wrong
2) A is correct, B and C are wrong
3) A, C are correct and B is wrong
4) A,B& C are correct

13. **A proton, an alpha particle and an electron are projected perpendicularly into uniform transverse electric and magnetic fields. It is observed that proton travels undeflected. Then**

a) deflection for alpha particle > deflection for proton

b) alpha particle travels in clockwise, and electron in anti-clockwise directions.

c) both alpha particle and electron travel without deflection

d) alpha particle gets deflected but not electron.

1) Only a is correct
2) Only b, c are wrong
3) Only d is correct
4) a, b, d are wrong

14. **A flat circular coil carrying current has a magnetic moment μ.**

a. μ has only magnitude, it does not have direction

b. The direction of μ is along the normal to the plane of the coil

c. The direction of μ depends on the direction of current flow

d. The direction of μ does not change if the current in the coil is reversed

1) only a is correct 2) b and c are correct 3) only d is correct 4) b and d are correct

15. **If a current carrying loop is placed in non uniform magnetic field, then the loop**

a) experiences a force
b) experiences a torque
c) will develop induced current
d) oscillates

1) a, c are correct 2) a, b, c are correct 3) b,c,d are correct 4) a,b,d are correct

16. **Statement A : Ammeter reads more current than galvanometer**

Statement B : Voltmeter reads less pd than galvano meter

1) A, B are true
2) A, B are false
3) A is true, B is false
4) A is false, B is true

17. **An electron beam produces**

i) Electrical field around the beam
ii) Magnetic field around the beam
iii) Electric field is more stronger than the magnetic field
iv) Electric and magnetic fields are not produced

1) only i and ii are true
2) only ii and iii are true
3) only i, ii and iii are true
4) only i and iii are true

18. **A charged particle is projected into a transfers uniform magnetic field. The area enclosed by its path**

i) Directly proportional to its KE

ii) Inversely proportional to square of charge

iii) Directly proportional to velocity **iv) inversely proportional to time**

1) only i and ii are true 2) only ii and iii are true

3) all are correct 4) i and iv are correct

19. **A charged particle goes undeflected in a region containing electric and magnetic fields. It is possible that**

i) $\bar{E} \parallel \bar{B}; \bar{V} \parallel \bar{E}$ **ii)** $\bar{E}$ **is perpendicular to** $\bar{B}$ **and both are perpendicular to** $\bar{V}$

iii) $\bar{V} \parallel \bar{B}$ **but** $\bar{E}$ **is not parallel to** $\bar{B}$ **iv)** $\bar{E} \parallel \bar{B}$ **but** $\bar{V}$ **is not parallel to** $\bar{E}$

1) only i and ii are true 2) only ii and iii are true

3) only iii and iv are true 4) only i, ii & iii are true

20. **A charged particle describes a circle under the influence of the magnetic field. The quantities that remain constant are**

i) KE **ii) Velocity** **iii) Time period** **iv) Momentum**

1) only i and iii are true 2) only i and iv are true

3) only i, ii and iii are true 4) all are true

21. **When a current loop is placed in a uniform magnetic field**

(i) $\bar{F}_R = 0$ **and** $\bar{\tau} = 0$ **(ii)** $\bar{F}_R = 0$ *but* $\bar{\tau} \neq 0$

(iii) $\bar{F}_R \neq 0$ *but* $\bar{\tau} = 0$ **(iv)** $\bar{F}_R \neq 0$ *and* $\bar{\tau} \neq 0$

1) only (i) & (ii) are true 2) only (ii) & (iii) are true

3) only (iii) & (iv) are true 4) only (i) & (iv) are true

22. **Which of the following statement (s) are true**

i) Moving coil galvanometer can be used in mines

ii) The coil of tangent galvanometer need not be arranged in magnetic meridian

iii) The reduction factor of a tangent galvanometer varies from place to place

iv) In tangent galvanometer, current is proportional to tangent of deflection

1) i, ii and iii are correct 2) i, ii and iv are correct

3) i, ii and iv are correct 4) i, iii and iv are correct

23. **The correct option**

a) Electric field and magnetic field are basically independent

b) Electric field and magnetic field are to aspects of the electro magnetic field

c) Electric field and magnetic field may be produced by charge at rest

d) A moving charge produces both electric and magnetic fields

1) a and b are correct 2) b and d are correct

3) b, c and d are correct 4) a, c and d are correct

24. If a charged particle is projected perpendicular to uniform magnetic field, then

a) force experienced will be perpendicular to the magnetic field and initial velocity.

b) force experienced will be perpendicular to the magnetic field and instantaneous velocity.

c) the work done by the magnetic field is zero.

d) the particle experiences both radial and tangential accelerations.

1) a, b, c are correct d is wrong
2) all are correct
3) a, b are correct, c, d are wrong
4) a, b, c are wrong, d is correct

25. A proton is travelling due north. If a uniform magnetic field is applied vertically down then the particle

a) deflected towards west
b) revolves in anti clock wise direction
c) acquires acceleration
d) comes to rest

1) a, b, c are correct
2) a, d are wrong
3) b, d are wrong
4) b, c, d are correct

26. A charged particle is projected perpendicular to a uniform magnetic field. Arrange the following in correct order

A) The direction of momentum of the particle changes

B) The particle experiences a force

C) The particle revolves in a circular path.

1) A,B,C
2) B,A,C
3) C,B,A
4) C,A,B

27. Resistance of galvanometer and shunt are given below for 3 ammeters A,B,C

A) 200 Ω and 50 Ω B) 100 Ω and 20 Ω C) 300 Ω and 50 Ω

The increasing order of sensitivity of ammeter constructed is

1) A, B, C
2) A, C, B
3) C, B, A
4) C, A, B

28. A thin straight vertical conductor has 10 amp current, vertically upwards. It is present at a place where $B_H = 4\times10^{-6}T$. Arrange the net magnetic inductions at the following points in ascending order

a) at 0.5m on south of conductor
b) at 0.5m on west of conductor
c) at 0.5m on east of conductor
d) at 0.5m on north-east of conductor

1) a, b, c, d
2) b, a, c, d
3) a, d, c, b
4) b, a, d, c

29. The following particles projected into a uniform transverse magnetic field with velocities v, 2v, 3v respectively. Then arrange their time periods of revolutions in decreasing order

a) electron
b) proton
c) alpha particle

1) a, b, c
2) c, b, a
3) a, c, b
4) c, a, b

30. Arrange the following in ascending order of their resistances

a) Ammeter of range 10A
b) Ideal voltmeter
c) Voltmeter of range 100V
d) Ammeter of range 1mA

1) a, b, c, d
2) a, d, b, c
3) a, d, c, b
4) b, c, a, d

31. In magnetic field $\vec{B} = (2\hat{i} + 3\hat{j})T$, velocity of a charged particle is given in table -1. Corresponding magnetic force is given in table-2. Match the two tables:

Table - 1	Table - 2
A) $(4\hat{i} - 3\hat{j})m/s$	P) zero
B) $(4\hat{i} + 6\hat{j})$ m/s	Q) maximum
C) $(6\hat{k})$ m/s	R) may be along positive z-direction
	S) must be along positive z-direction

1. A-R, B-P, C-Q 2. A-P, B-R, C-Q 3. A-Q, B-R, C-P 4. A-R, B-Q, C-P

32.

List-I	List-II
a) Tangent galvanometer	e) Uniform calibration of scale
b) Radial magnetic field	f) High resistance
c) Ammeter	g) Reduction factor
d) Voltmeter	h) Low resistance

The correct match is

1) a-h, b-g, c-f, d-e 2) a-g, b-e, c-h, d-f 3) a-h, b-e, c-f, d-g 4) a-h, b-g, c-e, d-f

33. Charged particle projected into a transeverse uniform magnetic field. Its

List-I	List-II
a) radius	e) Directly proportional to specific charge
b) Time period	f) Component of velocity perpendicular to magnetic field
c) Centripetal force	g) Independent of velocity
d) Angular frequency	h) Directly proportional to square root of it KE

The correct match is

1) a-h, b-g, c-f, d-e 2) a-h, b-f, c-g, d-e 3) a-f, b-g, c-h, d-e 4) a-h, b-f, c-e, d-g

34. A particle of charge +q and mass m moving under the influnce of a uniform electric field $E\hat{i}$ and a uniform magnetic field $B\hat{k}$ follows trajectory from P to Q as shown in figure. The velocities at P and Q $v\hat{i}$ and $-2v\hat{j}$ respectively. Which of the following statement(s) is/are correct

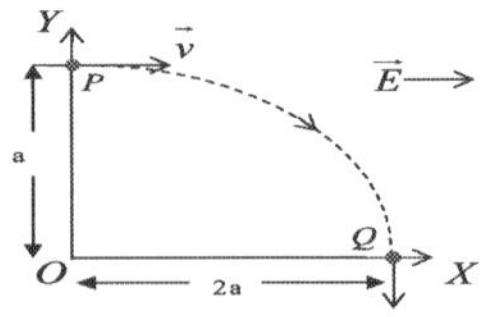

1) $E = \frac{3}{4}\frac{mv^2}{qa}$ 2) Rate of work done by electric field at P is $\frac{3}{4}\frac{mv^3}{a}$

3) Rate of work done by electric field at P is zero

4) Rate of work done by both the fields at Q is zero

35. **Two coaxil solenoids 1 and 2 of the same length are set so that one is inside the other. The number of turns per unit length are n_1 and n_2. The currents i_1 and i_2 are flowing in opposite directions. The magnetic field inside the inner coil is zero. This is possible when**

1) $i_1 \neq i_2$ *and* $n_1 = n_2$ 2) $i_1 = i_2$ *and* $n_1 \neq n_2$ 3) $i_1 = i_2$ *and* $n_1 = n_2$ 4) $i_1 n_1 = i_2 n_2$

36. **Paragraph: A current loop ABCD is fix on the plane of the paper as shown in the figure. The arcs BC (radius = b) and DA (radius = a) of the loop are joined by two straight wires AB and CD. A steady current *I* is flowing in**

the loop. Angle made by AB and CD at the

origin O is 30°. Another straight thin wire with steady current I_1, flowing out of the plane of the paper is kept at the origin.

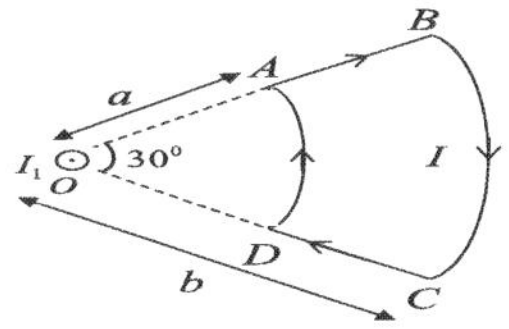

a) **The magnitude of the magnetic field (B) due to loop ABCD at the origin (O) is**

1) zero 2) $\dfrac{\mu_0(b-a)i}{24ab}$

3) $\dfrac{\mu_0 I}{4\pi}\left[\dfrac{b-a}{ab}\right]$ 4) $\dfrac{\mu_0 I}{4\pi}\left[2(b-a)+\dfrac{\pi}{3}(a+b)\right]$

b) **Due to the presence of the current I_1 at the origin**

1) The forces on AB and DC are zero 2) The forces on AD and BC are zero

3) The magnitude of the net force on the loop is given by $\dfrac{\mu_0 I I_1}{4\pi}\left[2(b-a)+\dfrac{\pi}{3}(a+b)\right]$

4) The magnitude of the net force on the loop is given by $\dfrac{\mu_0 I I_1}{4\pi}\left[\dfrac{b-a}{ab}\right]$

37. **A current carrying loop is placed in the non-uniform magnetic field whose variation in space is shown in figure. Direction of magnetic field is into the plane of paper. the magnetic force experienced by the loop is**

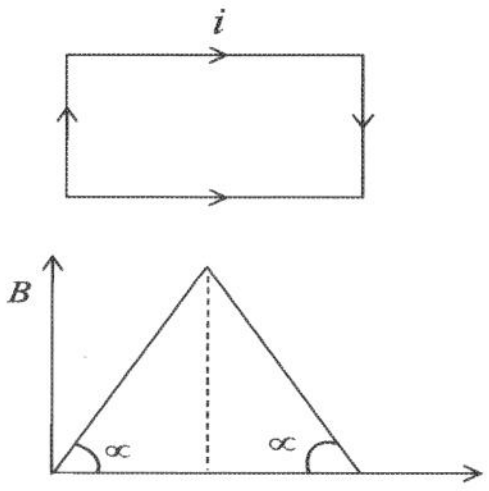

1) non-zero 2) zero 3) cannot say anything 4) none of the above

38. **Electric field strength $\overline{E} = E_0\hat{i}$ and $\overline{B} = B_0\hat{i}$ exists in a region. A charge is projected with a velocity $\overline{v} = v_0\hat{j}$ at origin, then**

1) It moves along helix with constant pitch 2) It moves along circular path in YZ plane

3) It moves along helix with increasing pitch 4) It moves along helix with decreasing pithch

KEY LEVEL - IV

1) 1	2) 1	3) 4	4) 1	5) 1	6) 3	7) 1	8) 1	9) 3	10) 1
11) 1	12) 4	13) 4	14) 2	15) 2	16) 3	17) 3	18) 1	19) 1	20) 1
21) 1	22) 4	23) 2	24) 1	25) 1	26) 2	27) 3	28) 4	29) 2	30) 3
31) 1	32) 2	33) 1	34) 1,2,3		35) 3,4	36a) 3	36. b) 2	37) 1	38) 3

HINTS LEVEL - IV

33. $\frac{mv^2}{r} = qvB$

34. Kinetic energy of the particle at point $P = \frac{1}{2}mv^2$

K.E. of the particle at point $Q = \frac{1}{2}m(2v)^2$

Increase in K.E. $= \frac{3}{2}mv^2$

It comes from the work done by the electric force qE on the particle as it covers a distance 2a along the x-axis.

Thus $\frac{3}{2}mv^2 = qE \times 2a \Rightarrow E = \frac{3}{4}\frac{mv^2}{qa}$.

The rate of work done by the electric field at $P = F \times v = qE \times v = \frac{3}{4}\frac{mv^2}{a}$

At Q, $\vec{F}_e = q\vec{E}$ is along x-axis while velocity is along negative y-axis. Hence reate of work done by electric field $= \vec{F}_e \cdot \vec{v} = 0$ $(\because \theta = 90^0)$. Similarly, according to equation $\vec{F}_m = q(\vec{v} \times \vec{B})$ force $\vec{F}_m$ is also perpendicular to velocity vector $\vec{v}$.

Hence the rate of work done by the magnetic field = 0.

35. $B_{net} = B_1 - B_2 \Rightarrow B_1 - B_2 = 0 \Rightarrow B_1 = B_2$

$\Rightarrow B \propto ni$. So $n_1i_1 = n_2i_2$ or $n_1=n_2$ and $i_1=i_2$

36 a. Net magnetic field due to loop ABCD at O is $B = B_{AB} + B_{BC} + B_{CD} + B_{DA}$

$$B = 0 + \frac{\mu_0 I}{4\pi a} \times \frac{\pi}{6} + 0 - \frac{\mu_0 I}{4\pi b} \times \frac{\pi}{6}$$

36b. The forces on AD and BC are zero because magnetic field due to a straight wire on AD and BC is parallel to elementary length of the loop.

37. Force due to non-uniform magnetic field

38. It moves along a circular path in YZ plane due to $\overline{B}$ and along a straight line path due to $\overline{E}$. This combination is helix but with increasing pitch.

Chapter - 8

MAGNETISM AND MATTER

MAGNETISM AND MATTER

SYNOPSIS

1. INTRODUCTION:

Magnet: A piece of matter, which when suspended freely rests itself in a particular direction (north-south) and which possesses a net magnetic moment and which attracts pieces of fernomagnetic materials like Iron, Cobalt, Nickel etc towards it, is called a magnet

2. TYPES OF MAGNETS:

a) Natural magnets b) Artificial magnets

i) **Natural Magnets :**

a) The magnet which is found in nature is called a natural magnet

e.g. magnetite. (Fe_2O_3).

b) Generally they are weak magnets

ii) **Artificial magnets:**

The magnets which are artificially prepared are known as artificial magnets. These are generally made of iron, steel and nickel.

3. MAGNETIC POLES:

i) If a piece of magnet is dipped in iron filings, they cling to it, especially at certain points. These points are called the poles of the magnet

ii) If a bar-shaped permanent magnet (or) bar magnet is free to rotate, one end points north. This end is called north-seeking end north pole (or) N-pole. The other end is a south pole (or) S-pole.

iii) Unlike poles attract each other and like poles repel each other

iv) The two poles of a magnet can not be isolated.

v) Earth is also a big magnet. Its geographical north pole is close to magnetic south pole. Its geographical south pole is close to its magnetic north pole.

vi) The two poles of the magnet generally are of equal strength and near the ends of the magnet.

vii) The straight line joining the two poles of a bar magnet is called axial line.

viii) The line passing through the midpoint normal to the axial line is called equatorial line.

ix) The vertical plane passing through the magnetic axis of a freely suspended bar magnet is called magnetic meridian

4. COULOMB'S INVERSE SQUARE LAW :

i) The force between two isolated magnetic poles is expressed by

$F = \frac{\mu_0}{4\pi}\frac{m_1 m_2}{d^2}$ when the poles are in air or vacuum. Where μ_0 is the permeability of free space and m_1 and m_2 are the pole strengths of the pole. S.I unit of pole strength is A-m. Pole strength is scalar.

ii) Dimensional formula for pole strength is IL

$\mu_o = 4\pi \times 10^{-7}$ henry/meter or $\frac{Weber}{amp - m}$ or $\frac{N}{A^2}$ Dimensional formula of μ_0 is $MLT^{-2} I^{-2}$

$F = \frac{\mu}{4\pi}\frac{m_1 m_2}{d^2}$ (S I system)

iii) Where μ is the absolute permeability of the medium between the poles. $\mu = \mu_0 \mu_r$

μ_r is relative permeability of the medium

iv) Relative permeability of the medium

$$(\mu_r) = \frac{\text{Permeability of the medium}}{\text{Permeability of air or vacuum}} = \frac{\mu}{\mu_0}$$

μ_r has no units and no dimensions

5. **RESULTANT FORCE IN DIFFERENT CASES :**

i) Force on a magnetic pole due to other poles surrounding it can be found by the principle of super position. i.e. by the vector sum of the individual forces on it due to surrounding poles.

ii) Two magnetic poles of same strength (m) are placed at two vertices of an equilateral triangle of side 'a' the force on a similar magnetic pole on third corner is given by $\boxed{F^1 = \sqrt{3}\frac{\mu_0}{4\pi}\frac{m^2}{a^2}}$

iii) Two unlike magnetic poles of same strength (m) are placed at two vertices of an equilateral triangle of side 'a'. If a north pole of same strength is placed at the thirdvertex, it experiences a force ofmagnitude which is given by $\boxed{F^1 = \frac{\mu_0}{4\pi}\frac{m^2}{a^2}}$

iv) Four like poles each of pole strength 'm' are kept at the four cornes of a square of side 'a'. The net magnetic force on the pole at any one corner is given by

$$\boxed{F^1 = \frac{\mu_0}{4\pi}\frac{m^2}{a^2}\left[\sqrt{2}+\frac{1}{2}\right]}$$

6. **MAGNETIC LENGTH** (2L) **:**

i) The distance between the two poles of a magnet is called magnetic length of a magnet. It is generally represented as '*2l*'

ii) It is directed along the south pole to north pole and it is a vector

iii) Magnetic length = 5/6 x geometric length.

7. **MAGNETIC MOMENT (M) :**

i) It is measured as the product of magnetic length and pole strength $\boxed{M = 2\ell \times m}$.

ii) Its S.I unit is ampere-metre^2 or Joule/Tesla or N-m^3/weber, dimensional formula of M = [IL^2]

iii) It is a vector with its direction from south pole to north pole along its axial line (S $\rightarrow$ N)

A) Magnet divided into parts:

i) When a bar magnet is cut into 'n' equal parts parallel to its axis

a) Pole strength of each part = m^1 = m/n

b) Length of each part = *(2l)*^{1} = *2l*

c) Magnetic moment of each part $\boxed{M^1 = M/n}$

ii) When a bar magnet is cut into 'n' equal parts normal to its axis

a) Pole strength of each part m^1 = m

b) Length of each part *(2l)*^{1} = *2l*/n

c) Magnetic moment of each part $\boxed{M^1 = M/n}$

iii) When a bar magnet is cut into 'xy' equal parts x parts parallel to its axis and y parts normal to the axis.

a) Pole strength of each part m^1 = $\frac{m}{x}$

b) Length of each part $(2\ell)^1 = 2\ell/y$

c) Magnetic moment of each part $\boxed{M^1 = M / xy}$

B) Resultant magnetic moment:

i) When two magnets of magnetic moments M_1 and M_2 are kept at an angle 'θ' with like poles touching each other, then the resultant magnetic moment $\boxed{M^1 = \sqrt{M_1^2 + M_2^2 + 2M_1M_2 \, Cos\theta}}$

ii) When two magnets of magnetic moments M_1 and M_2 are kept at an angle 'θ' with unlike poles touching each other, then the resultant magnetic moment

$\boxed{M^1 = \sqrt{M_1^2 + M_2^2 - 2M_1M_2 \, Cos\theta}}$

iii) When two magnets of magnetic moments M_1 and M_2 are placed one above the other with like poles touching each other (or) placed side by side with unlike poles touching each other then $M^1 = M_1 + M_2 \left(\theta = 0^0\right)$

iv) When two magnets of magnetic moments M_1 and M_2 are placed one above the other with unlike poles touching each other (or) placed side by side with like poles touching each other then $M^1 = M_1 \sim M_2$ $\left(\theta = 180^0\right)$

v) i) When 'n' identical bar magnets form a closed polygon with unlike poles nearer, the resultant magnetic moment is zero.

ii) If one magnet is removed from the polygon the resultant magnetic moment becomes equal to the magnetic moment of one magnet (M).

vi) If one magnet in the polygon is reversed and kept at the same position, the resultant magnetic moment becomes '2M'.

C) Bending of magnets:

On bending a magnet its pole strength remains unchanged where as its magnetic moment changes.

i) When a thin bar magnet of magnetic moment 'M' and length '2ℓ' is bent at its mid point with an angle 'θ' between the two parts (a) its effective length becomes $2\ell \sin\frac{\theta}{2}$ (b) its new magnetic moment $\boxed{M^1 = M\sin(\theta/2)}$

ii) When a Magnetised wire of length '2ℓ' and magnetic moment 'M' is bent in to an arc of a circle, that makes an angle 'θ' at the centre of the circle

Its magnetic moment decreases and becomes

$\boxed{M^1 = \left[\frac{2MSin(\theta/2)}{\theta}\right]}$ where θ is in radian

NOTE :

i) Force between two co-axial magnetic dipoles $\boxed{F = \frac{\mu_0}{4\pi}\frac{6M_1M_2}{d^4}}$

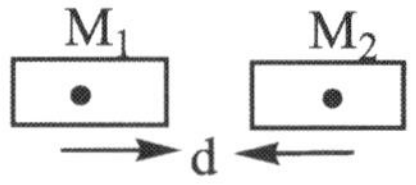

ii) The force between two parallel magnets

$\boxed{F = \frac{\mu_0}{4\pi}\frac{3M_1M_2}{d^4}}$

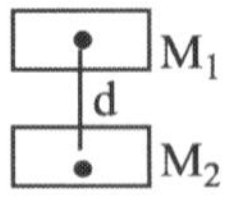

8. MAGNETIC FIELD :

The space (or) region in which a magnetic pole will experience force.

9. MAGNETIC INDUCTION OR INDUCTION FIELD STRENGTH (B)

i) Magentic induction field strength at a point in a magnetic field is defined as force experienced by a unit north pole placed at that point in magnitude as well as direction $\boxed{\vec{B} = \frac{\vec{F}}{m}}$

ii) S.1 unit is Tesla or weber/m^2
Dimentional formula of B is $[M T^{-2} I^{-1}]$

iii) B at a distance 'd' in air due to a pole of strength 'm' is $\boxed{B = \frac{\mu_0}{4\pi}\frac{m}{d^2}}$

Direction : If it is a north pole $\vec{B}$ is away from the pole and if it is a south pole $\vec{B}$ is towards the pole .

A) Formulae of magnetic induction in diffrent cases:

i) Magnetic induction field strength at a point due to a number of poles can be found by the principle of superposition i.e. at a point field is vector sum of magnetic induction field strengths due to individual poles.

ii) Three identical magnetic poles each of strength 'm' placed at the three vertices of an equilateral triangle. The resultant magnetic induction at the centre is equal to zero

iii) The magnetic induction at the centre of the line joining the two poles of a horse shoe magnet of pole strength 'm' and separated by a distance 'd' is

$\boxed{B = 8\frac{\mu_0}{4\pi}\frac{m}{d^2}}$ directed from N to S pole

iv) When a pole of pole strength m is placed in a magnetic field of induction $\vec{B}$,the force $\vec{F}$ experienced by it 1is given by $\boxed{\vec{F} = m\vec{B}}$

v) North pole experiences the force in the direction of field.

vi) South pole experiences the force in a direction opposite to that of the field.

10. MAGNETIC LINES OF FORCE :

i) A magnetic line of force is the path followed by a free unit north pole in a magnetic field.

ii) Magnetic field can be represented by magnetic lines of force. They are imaginary.

iii) They diverge from north pole and converge at south pole.

iv) Two magnetic lines of force never intersect; If they intersect it means that field has two directions at the point of intersection which is impossible.

v) Inside the magnet they are from south pole to north pole. Outside the magnet they are from north pole to south pole. They are closed curves.

vi) They contract longitudinally and expand laterally.

vii) The tangent at any point to the line of force gives the direction of magnetic induction field strength at that point.

viii) If lines of force are crowded it implies a stronger field and weak if they are far apart

11. DIFFERENCE BETWEEN MAGNETIC LINES AND ELECTRIC LINES:

i) Magnetic lines of force pass through the magnetic substance of the magnet. Electric lines do not pass through a conductor.

ii) Magnetic lines of force are closed curves where as electric lines of force are open curves

12. TYPES OF MAGNETIC FIELD:

A) Uniform magnetic field:

i) The magnetic field, in which the magnetic induction field strength is same both in magnitude and direction at all points, is known as uniform magnetic field.

ii) In such a magnetic field the magnetic lines of force are equidistant and parallel straight lines.

Ex: Horizontal component of earth's magnetic field in a limited region

B) Non uniform magnetic field:

i) The magnetic field, in which the magnetic induction or field strength differs either in magnitude, in direction or both is known as non uniform magnetic field

ii) It is represented by non-parallel lines of force

Ex: The magnetic field near the pole of any magnet

13. MAGNETIC FLUX :

i) Magnetic flux $)\phi($linked through a cross section is defined as dot product of $\vec{B}$ and $\vec{A}$

$$\boxed{\phi = \vec{B}.\vec{A} = BA\cos\theta}$$

where 'θ' is the angle made by magnetic field $(\vec{B})$ with the normal to the area $(\hat{n})$

$\vec{A} = A\hat{n}$ A = area of the coil

ii) i) It is a scalar (ii)S.I unit is weber

iii)Dimensional Formula is $[ML^2T^{-2}I^{-1}]$

14. MAGNETIC INDUCTION DUE TO A BAR MAGNET:

A) Field on axial line.

i) B at a point 'p' on the axial line of a bar magnet at a distance 'd' from the centre of the magnet is

$$\boxed{B_a = \frac{\mu_0}{4\pi}\frac{2Md}{(d^2 - l^2)^2}}$$

ii) In case of short bar magnet $(\because l << d)$ $\boxed{B_a = \frac{\mu_0}{4\pi}\frac{2M}{d^3}}$

iii) Direction of $\vec{B}$ is from S to N along the axial line. i.e. along $\overrightarrow{SN}$

B) Field on equatorial line.

i) B at a point 'p' on the equatorial line of a bar magnet at a distance 'd' from centre of the magnet is

$$\boxed{B_e = \frac{\mu_0}{4\pi}\frac{M}{(d^2 + l^2)^{\frac{3}{2}}}}$$

ii) In case of short magnet $\boxed{B_e = \frac{\mu_0}{4\pi}\frac{M}{d^3}}$ $(\because l << d)$

iii) Direction of $\vec{B}$ is parallel to the axis and is directed from N to S pole i.e $\overrightarrow{NS}$.

iv) The ratio of fields at a distance 'd' on the axial and equatorial lines of a short bar magnet is

$$\frac{B_{axial}}{B_{equitorial}} = 2$$

NOTE: Magnetic induction at a point 'p' and whose position vector from the mid point of the magnet makes an angle θ with magnetic ax

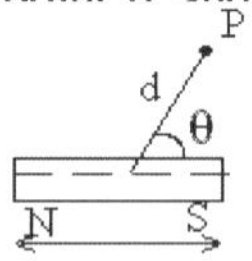

$$\boxed{B = \frac{\mu_0}{4\pi}\frac{M}{d^3}\sqrt{1+3\cos^2\theta}}$$

Case -I: If $\theta = 0^0$ $B = \frac{\mu_0}{4\pi}\frac{2M}{d^3}$ (Axial line)

Case -II: If $\theta = 90^0$ $B = \frac{\mu_0}{4\pi}\frac{M}{d^3}$ (equatorial line)

15. COUPLE ACTING ON A BAR MAGNET IN A UNIFORM MAGNETIC FIELD :

i) When a magnet with magnetic moment M is suspended in a uniform field of induction B at an angle θ with the field direction then the couple (C) acting on the magnet is given by $\boxed{C = MB\sin\theta}$

Vectorially $\boxed{\vec{C} = \vec{M} \times \vec{B}}$

Case – I : When $\theta = 90^0$, C is maximum

$|\vec{C}_{max}| = MB.$

Case -II: When $\theta = 0°$ (or) 180^0 $|\vec{C}_{min}| = \vec{0}$

ii) In an uniform magnetic field a bar magnet experiences only a couple but no net force. Therefore it undergoes only rotatory motion.

iii) In a non-uniform magnetic field a bar magnet experiences a couple and also a net force. So it undergoes both rotational and translational motion.

iv) Two magnets of magnetic moments M_1 and M_2 are joined in the form of a cross and this arrangement is pivoted so that it is free to rotate in a horizontal plane under the influence of earth's magnetic field. If θ is the angle made by the magnetic meridian with M_1 then

$$\tan\theta = \frac{M_2}{M_1}$$

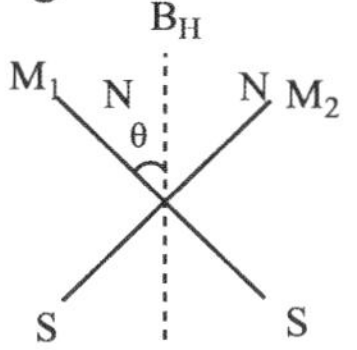

A) Potential energy of bar magnet : The P.E. of a bar magnet of magnetic moment $\vec{M}$ placed in a uniform magnetic field $\vec{B}$ is given by

$U = -\vec{M}.\vec{B}$ $\quad U = -MB\cos\theta$

Where θ is the angle between $\vec{M}(\vec{SN})$ and $\vec{B}$.

i) If $\vec{M}$ is parallel to $\vec{B}$ then P. E. is minimum, then U = – MB. Then magnet is said to be in **stable equilibrium.**

ii) If $\vec{M}$ is anti–parallel to $\vec{B}$ then P. E. is maximum, then U = + MB. Then magnet is said to be in **unstable equilibrium.**

B) Work done by external agent in deflecting a magnet:

i) The work done in deflecting a magnet from angular positions θ_1 to an angular position θ_2 with the field is given by W = ΔU. (or) $W = MB(\cos\theta_1 - \cos\theta_2)$

ii) The work done in deflecting a bar magnet through an angle θ from its stable equilibrium position in a uniform magnetic field is given by $W = MB(1-\cos\theta)$

16. NEUTRAL POINTS OR NULL POINTS :

i) A point where magnetic induction field strength is zero is called null point.

ii) A compass needle placed at null point comes to rest in any direction.

iii) If two poles of pole strengths m_1 and m_2 ($m_1<m_2$) are seperated by distance d, then the distance of the neutral point from the first pole m_1 is

$$x = \frac{d}{\sqrt{\frac{m_2}{m_1}} \pm 1} \quad \begin{pmatrix} + \text{ for like poles} \\ - \text{ for unlike poles} \end{pmatrix}$$

a) For like poles the neutral point is situated in between the poles

b) For unlike poles the neutral point is situated on line joining the poles. But not in between them.

c) In either case null point is always closer to the weaker pole.

iv) If two short bar magnets of magnetic moments M_1 and M_2 ($M_1 < M_2$) are placed along the same line with like poles facing eachother and 'd' is the distance between their centres, the distance of null point from M_1 is

$$x = \frac{d}{\left(\frac{M_2}{M_1}\right)^{1/3} \pm 1}$$

a) + for null point formed between the magnets

b) – for null point formed outside the magnets.

c) When unlike poles face each other, null point is not possible.

v) When a short bar magnet is placed in the earth's magnetic field with its north pole towards geographic north, then two null points are formed on the equatorial line, one on either side at equal distances from mid point of the magnet.

At the null point $\vec{B} = -\vec{B}_H$ $\boxed{B_H = \frac{\mu_0}{4\pi}\frac{M}{d^3}}$

B_H is the horizontal component of earth's magnetic field.

vi) When a short bar magnet is placed in earth's field with south pole pointing geographic north, then two null points are formed on axial line one on either side at equal distances from mid point of the magnet..

At the null point $\vec{B} = -\vec{B}_H$ $\boxed{B_H = \frac{\mu_0}{4\pi}\frac{2M}{d^3}}$

vii) a) If a very long magnet is placed vertically with one pole on the table then a single neutral point will be formed

b) If 'm' is the polestrength and 'd' is the distance from the pole of the magnet where the neutral point is formed then

$$\boxed{B_H = \frac{\mu_0}{4\pi}\frac{m}{d^2}}$$

c) If the north pole is on the table, then the neutral point lies towards geographic south.

d) If the south pole is on the table, then the

neutral point lies towards geographic north.

17. TANGENT LAW :

If a magnet is placed in a region where two perpendicular fields $\vec{B_1}$ and $\vec{B_2}$ are present, then magnet $\left(\vec{SN}\right)$ will align itself in the direction of resultant of $\vec{B_1}$ and $\vec{B_2}$, Such that $\boxed{B_2 = B_1 \tan\theta}$. This is called Tangent Law.

$$\boxed{\tan\theta = \frac{B_2}{B_1}}$$

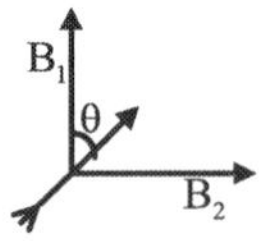

18. DEFLECTION MAGNETOMETER (DMM):

DMM works on the principle of tangent law.

i) A long aluminium pointer in the compass box provides stability to the needle and enables us to have a large circular scale.

ii) The plane mirror is provided below the aliminium pointer to note the deflections without parallax error.

iii) Both in Tan A as well as in Tan B position the bar magnet is in east west direction.

A) Tan - A position or Gauss - A position or End on position

i) Arms of the magnetometer are perpendicular to the magnetic meridian i.e. arms lie in East-West direciton.

ii) Aluminium pointer is parallel to the arm

iii) The bar magnet is placed in such a way that its axial line passes through the centre of the needle. Bar magnet is parallel to the arms then

For long bar magnets

$$\boxed{\frac{\mu_0}{4\pi}\frac{2Md}{\left(d^2 - l^2\right)^2} = B_H \tan\theta}$$

Where θ is the average deflection of alluminium pointer (or) needle

iv) For short bar magnets

$$\frac{\mu_0}{4\pi}\frac{2M}{d^3} = B_H \tan\theta$$

B) Tan - B position, or Gauss B position or broad side on position

i) Arms of the magnetometer are parallel to the magnetic meridian i.e. arms lie in north-south direciton.

ii) Aluminium pointer is perpendicular to the arms.

iii) The bar magnet is placed in such a way that its equatorial line passes through the centre of the needle. The bar magnet is perpendicular to the arms. then

For long bar magnet $\frac{\mu_0}{4\pi}\frac{M}{(d^2+l^2)^{\frac{3}{2}}} = B_H \tan\theta$

iv) For short bar magnet $\frac{\mu_0}{4\pi}\frac{M}{d^3} = B_H \tan\theta$

v) **Deflection magnetometer is used to**

a) Compare the magnetic moments of two bar magnets

b) Compare the magnetic field strengths

c) Verify the inverse square law

C. Comparing the magnetic moments:

i) **By equal distance method :** Two magnets with magnetic moments M_1 & M_2 are placed on the arms of DMM at equal distances separately then, either in Tan A or Tan B position we have,

$$\frac{M_1}{M_2} = \frac{\tan\theta_1}{\tan\theta_2}$$

Where θ_1 & θ_2 are the average deflectons when M_1 is placed on the arm and M_2 is placed on the arm respectively

ii) **By null deflection method**: A magnet with magnetic moment M_1 is placed at a distance d_1 on one arm of DMM. If 'd_2' is the distance on the other arm of an another magnet with magnetic moment M_2 such that there is no deflection of the needle, then either in TanA or TanB positions we have

$$\frac{M_1}{M_2} = \frac{d_1^3}{d_2^3}$$

NOTE:

i) Tan A position is preferred to Tan - B position because, for the same distance between the centre of compass and magnet deflection is more in tan-A compared to tan-B ($\because$ the field is more on axial line than equatorial line)

ii) Null deflection method is preferred to equal distance method, because % error is less and hence accuracy is more in the null deflection method.

iii) In tan - A position, null deflection method the bar magnets are placed with their like poles facing each other.

iv) In tan B position, null deflection method the bar magnets are placed such that their unlike poles lie on the same side of the deflection magnetometer.

D. Verification of inverse square law:

i) When a short bar magnet is placed at a distance 'd' in tanA and then in tanB positions and θ_A and θ_B are the average deflection respectively then

$$\frac{\tan\theta_A}{\tan\theta_B} = 2$$

ii) Only one bar magnet is used while verifying inverse square law.

NOTE:

i) Deflection magnometer does not work at magnetic poles of the earth because $B_H = 0$

ii) Deflection magnometer is more accurate when the deflection is 45°. Therefore the reading should lie between 30° to 60° for accurate results.

iii) The relative error in measuring magnetic moment is minimum when $\theta = 45°$

iv) The point of pivot of the magnetic needle may not be exactly at the centre of the circular scale. This error is eliminated by taking the reading of both ends of the pointer.

v) The magnet may not be uniformly magnetised. To eliminate error due to it, magnet is reversed at a given position so that the poles are interchanged and reading of pointer are taken

vi) There may be error due to the centre of the arms not coinciding with the point of pivot of the magnetic needle. This error is eliminated by placing the bar magnet at an equal distance from the centre of the DMM on both arms separatly..

19. VIBRATION MAGENETOMETER(VMM):

i) **Principle:**

When a bar magnet is suspended freely in a uniform magnetic field and is displaced from its equilibrium, it starts vibrating and executes angular SHM about the equilibrium position.

ii) The plane mirror with the index line helps to count the vibrations of the magnet and avoids the error due to parallax.

iii) The time period of vibrating magnet is

$$\boxed{T = 2\pi\sqrt{\frac{I}{MB_H}}}$$

and The frequency of vibrating magnet $\boxed{n = \frac{1}{2\pi}\sqrt{\frac{MB_H}{I}}}$

Where $I = m\left(\frac{\ell^2 + b^2}{12}\right)$ is the moment of inertia of the bar magnet about the axis of oscillation and for thin bar magnet $I = \frac{m\ell^2}{12}$

m = mass of the magnet

ℓ = length of the magnet

b = breadth of the magnet

b_h = horizontal component of earth's magnetic field

M = Magnetic moment of the magnet.

iv) A magnet is oscillating in a magnetic field and its time period is T. If another identical magnet is placed over that magnet with like poles together, then time period remains unchanged.

v) A magnet is oscillating in a magnetic field B and its time period is T. If another identical magnet is placed over that magnet with unlike poles together, then time period becomes infinite since resultant magnetic moment is equal to zero

vi) A thin bar magnet is oscillating in a magnetic field, its time period is T.

a) If magnet is cut into 'n' equal parts, parallel to length then time period of each part remains same $T^1 = T$

b) If magnet is cut into 'n' equal parts perpendicula to length then time period of each part is $\boxed{T^1 = \frac{T}{n}}$

vii) Two magnets of magnetic moments M_1 and M_2 $(M_1 > M_2)$ are placed one over the other. If T_1 is time period when like poles touch each other and T_2 is the time period when unlike poles touch each other, then $\boxed{\frac{M_1}{M_2} = \frac{T_2^2 + T_1^2}{T_2^2 - T_1^2}}$

In terms of frequencies of oscillating magnet

$$\boxed{\frac{M_1}{M_2}=\frac{n_1^2+n_2^2}{n_1^2-n_2^2}}$$

viii) **Uses of vibration magnetometer.**

a) To compare the magnetic moments of two bar magnets.

b) To compare the horizontal components of earth's magnetic field at two different places.

ix) Together with deflection magnetometer vibration magnetometer can be used to find the magnetic moment of a magnet and the horizontal component of earth's magnetic field at a given place.

a) Using the same bar magnet in deflection magnetometer and vibration magnetometer

From VMM $MB_H=\left[\frac{4\pi^2 I}{T^2}\right]=X\rightarrow \boxed{1}$

From DMM $\frac{M}{B_H}=\left[\frac{4\pi\left(d^2-\ell^2\right)^2 \operatorname{Tan}\theta}{\mu_0 2d}\right]=Y\rightarrow\boxed{2}$

x) From 1 & 2 The horizontal component of earth's field at a given place is given by

$$B_H=\frac{1}{T\left(d^2-\ell^2\right)}\sqrt{\frac{\mu_0 2\pi Id}{\tan\theta}}=\sqrt{\frac{X}{Y}}$$

xi) The magnetic moment of the given magnet is given by

$$M=\frac{2\pi\left(d^2-l^2\right)}{T}\sqrt{\frac{4\pi I\tan\theta}{\mu_0 2d}}=\sqrt{XY}$$

θ =deflection in the deflection magnetometer

I = Moment of inertia of the bar magnet

$2l$ = magnetic length of the magnet

T = time period of vibration

d = distance between magnet and centre of needle in deflection magnetometer.

20. PROPERTIES OF MAGNETIC MATERIALS :

A) Intensity of Magnetising field or magnetising force (H)

i) 'H' is an auxillary field which is measured as the ratio of magnetic induction to the permeability of the medium at the given point.

ii) $\boxed{B=\mu_0 H}$ in vacuum

$B=\mu H=\mu_0\mu_r H$ in any medium

iii) S.I unit is $\frac{A}{m}$, CGS unit is oersted

$1\ Am^{-1}=4\pi \times 10^{-3}$ oersted

Dimensional formula of H is $[I L^{-1}]$

iv) H is a vector in the direction of magnetic field (B). H is independent of the medium

B) Intensity of magnetisation (I) :

i) It is the magnetic moment per unit volume or polestrength per unit area

$\boxed{I=\frac{M}{V}=\frac{2\ell.m}{2\ell.a}=\frac{m}{a}}$ a = area of cross section

ii) I is a vector, it is in the direction of H (ferro, paramagnetics) or opposite to H (diamagnetics)

iii) I and H have same units and dimensions.

C. Magnetic susceptibility (χ) :

i) The ratio of magnitude of intensity of magnetisation (I) in a material to that of magnetising field(H) is called magnetic susceptibility of that material(χ). $\boxed{\chi = I/H}$

ii) ▢ has no units and no dimensions

D. Relation between μ and χ :

i) When a magnetic material is placed in a magnetising field for its magnetisation the field inside the magnetic material is the resultant of the magnetising field $\overrightarrow{B_0}$ and the induced field $\overrightarrow{B_i}$ i.e $\vec{B} = \overrightarrow{B_0} + \overrightarrow{B_i}$

$$= \mu_0 H + \mu_0 I = \mu_0 (H + I)$$

$$\therefore B = \mu_0 \mu_r H = \mu_0 (H + I)$$

$$\Rightarrow \mu_0 \mu_r H = \mu_0 (H + \chi H) \Rightarrow \boxed{\mu_r = 1 + \chi}$$

22. ELECTRON THEORY OF MAGNETISM :

i) Molecular theory of magnetism was first given by Weber and was later developed by Ewing.

ii) Electron theory of magnetism was proposed by Langevin.

iii) The main reason for the magnetic property of a magnet is spin motion of electron. Most of the magnetic moment is produced due to electron spin. The contribution of the orbital revolution is very small.

A) Explanation of diamagnetism :

i) Since diamagnetic substance have paired electrons, magnetic moments cancel each other and there is no net magnetic moment.

ii) When a diamagnetic substance is placed in an external magnetic field each electron experiences radial force F = Bev either inwards or outwards. Due to this the angular velocity, current, and magnetic moment of one electron increases and of the other decreases. This results in a non-zero magnetic moment in the substances in a direction opposite to the field.

iii) Since the orbital motion of electrons in atoms is an universal phenomenon, diamagnetism is present in all materials. Hence diamagnetism is a universal property.

B) Explanation of Paramagnetism :

i) Paramagnetic materials have a permanent magnetic moment in them. The moments arise from both orbital motion of electrons and the spinning of electrons in certain axis.

ii) In atoms whose inner shells are not completely filled, there is a net moment in them since more number of electrons spin in the same direction. This permanent magnet behaves like a tiny bar magnet called atomic magnet.

iii) In absence of external magnetic field atomic magnets are randomly oriented due to the thermal agitation and the net magnetic moment of the substance is zero.

iv) When it is placed in an external magnetic field the atomic magnets align in the direction of the field and thermal agitation oppose them to do so.

v) At low fields the total magnetic moment would be directly proportional to the magnetic field B and inversely proportional to temperature T.

vi) At high field and low temperature the total moment reaches the saturation when all the atomic magnets align in the direction of the field. $M = C\frac{B}{T}$ C is a constant

$I \alpha \frac{B}{T}$

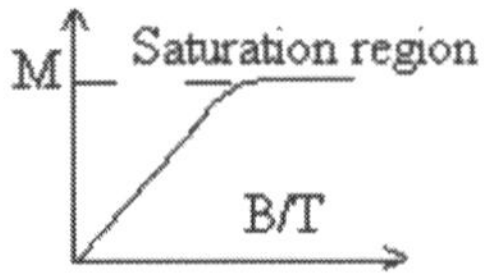

I = Intensity of magnetisation

23. EXPLANATION OF FERRO MAGNETISM (DOMAIN THEORY) :

i) Every ferromagnetic material is made of a number of very small region which are known as domains

ii) Domain theory proposed by Weiss explains ferro magnetism.

iii) **Exchange interaction:-** It is the interaction through which the magnetic moments of individual atoms are coupled to neighbouring moments very strongly.

iv) Domain : A small volume of the order 10^{-6} to 10^{-2} cm^3 containing 10^{17} to 10^{21} atoms whose magnetic moments are aligned in the same direction.

v) Due to exchange interaction domains are formed in ferro magnetic substance. Each domain possesses a resultant magnetic moment.

vi) In the absence of external magnetic field the magnetic moments of domains are randomly oriented giving zero resultant magnetic moment.

vii) When the ferro magnetic substance is placed in an external magnetic field the domains having the magnetic moment in the direction of external magnetic field increase their volume at the cost of those which are in opposite direction till saturation state of magnetization.

24. CURIE'S LAW:

i) According to this law, the magnetic susceptibility of paramagnetic substances is inversely proportional to absolute temperature, i,e.,

$$\chi \propto \frac{1}{T} \Rightarrow \chi = \frac{C}{T}$$

Where C = curie constant, 'T' absolute temperature.

ii) On increasing temperature, magnetic susceptibility of paramagnetic substances decreases or vice-versa.

iii) The magnetic susceptibility of ferro magnetic substances does not change according to curie law.

A) Curie Temperature or Cuire Point (Tc)

i) The magnetic susceptibility of these substances decreases on increasing the temperature and above a particluar tempeture, a ferromagnetic substance behaves like a paramagnetic substance. This particular temperature is called the **Curie** temperature of the substance.

ii) For various ferromagnetic materials its values are different,

Ex: For Ni, $T_{C_{Ni}} = 358^0C$ for Fe, $T_{C_{Fe}} = 770^0C$

for CO, $T_{C_{CO}} = 1120^0C$

iii) At this temperature the ferromagnetism of the substances suddenly vanishes.

B) Curie - weiss law:

i) At temperatures above Curie temperature the magnetic susceptibility of ferromagnetic materials is inversely proportional to $(T - T_C)$

i.e. $\chi \propto \frac{1}{T - T_C} \Rightarrow \chi = \frac{C}{(T - T_C)}$

Here T_C - Curie temperature χ -T curve is shown (for Curie - Weiss Law)

ii) Susceptibility – absolute temperature curve

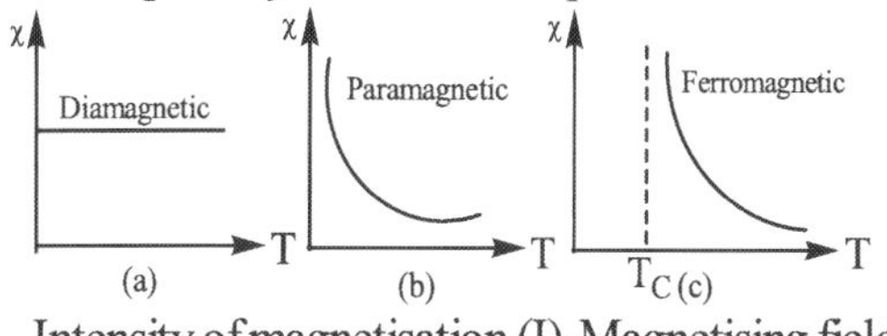

iii) Intensity of magnetisation (I)-Magnetising field (H) curve

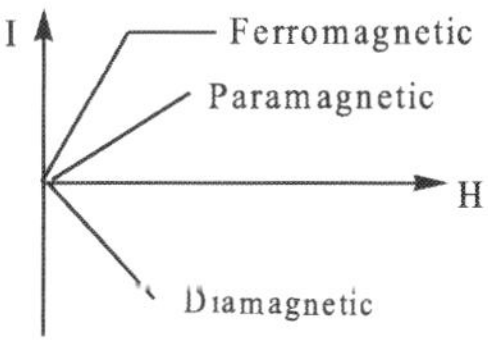

LEVEL - I

1. **A magnet is placed in iron powder and then taken out, then maximum iron powder is at**
   1) North pole only 2) South pole only
   3) The middle of the magnet 4) The ends of the magnet
2. **Which of the following, the most suitable material for making permanent magnet is**
   1) Steel 2) Soft iron 3) Copper 4) Nickel
3. **A permanent magnet**
   1) Attracts all substances 2) Attracts only ferromagnetic substances
   3) Attracts magnetic substances and repels all others 4) Attracts some substances and repels others
4. **The best method for demagnetisation is**
   1) dropping the magnet from certain height 2) Hammering the magnet
   3) Heating the magnet 4) Sending alternating currents through the magnet
5. **Examples for artificial magnets are**
   1) bar magnets 2) horse shoe magnets
   3) potshaped & cylindrical magnets 4) all the above
6. **Statement A: Horse shoe magnets are used in cycle dynamos.**
   Statement B : Pot shaped magnets are used in loud speakers
   1) A true, B true 2) A false , B true 3) A false , B false 4) A true , B false
7. **A sensitive magnetic instrument can be shielded very effectively from outside magnetic fields by placing it inside a box of:**
   1) Teak wood 2) Plastic material
   3) Soft iron of high permeability 4) A metal of high conductivity
8. **A bar magnet having pole strength 'm' is bent in the form of an arc, then its pole strength**
   1) increases 2) decreases 3) may increase or decrease 4) remains same
9. **If a piece of metal was thought to be magnet, which one of the following observations would offer conclusive evidence**
   1) It attracts a known magnet 2) It repels a known magnet
   3) Neither 1 nor 2 4) It attracts a steel screw driver
10. **A bar magnet is cut into two equal halves by a plane parallel to the magnetic axis. Of the following physical quantities the one which remains unchanged is**
    1) Pole strength 2) Magnetic moment
    3) Intensity of magnetisation 4) Moment of interia
11. **Two similar bar magnets P and Q each of magnetic moment M are taken. If P is cut along it axial line and Q is cut along its equatorial line, all the four pieces obtained have each of**
    1) Equal pole strength 2) Magnetic moment M/4
    3) Magnetic moment M/2 4) Magnetic moment M
12. **If two bar magnets of different magnetic lengths have equal moments, then the pole strength is**
    1) Equal for both the magnets 2) Lesser for shorter magnet
    3) More for longer magnet 4) More for shorter magnet
13. **A magnetised wire of magnetic length '2l', pole strength 'm' and magnetic moment 'M' is bent at angle is 'θ' radian at the centre of the circle, then**
    1) Its pole strength remains same 2) Its length decreases and becomes $\left[\frac{4l\,\text{Sin}\left(\frac{\theta}{2}\right)}{\theta}\right]$
    3) Its new magnetic moment becomes $\left[\frac{2M\,\text{Sin}\left(\frac{\theta}{2}\right)}{\theta}\right]$ 4) All the above are correct

14. **If a bar magnet of moment M is bent as arc its magnetic moment**
1) decreases 2) increases 3) doesnot change 4) may change

15. **When two identical bar magnets each of magnetic moment 'M' are inclined to each other. The resultant magnetic moment is equal to 'M'. If**
1) The angle between them is 60° with like poles nearer.
2) The angle between them is 120° with unlike poles near
3) The angle between them is '60' with unlike polesnear
4) The angle between them is 90° with like or unlike poles near

16. **Two magnets have the same length and the same pole strength. But of the magnets has a small hole at its centre. Then**
1) Both have equal moment 2) One with hole has large moment
3) One with hole has smaller moment 4) One with hole loses magnetism through the hole

17. **What happens to the force between magnetic poles when their pole strength and the distance between them are both doubled**
1) Force increases to two times the previous value 2) No change
3) Force decreases to half the previous value 4) Force increases to four times the previous value

18. **The earth's magnetic field**
1) varies in direction but in magnitude 2) varies in magnitude but not in direction
3) varies in both magnitude and direction 4) is centered exactly about the center of earth

19. **Earth's magnetic field always has a horizontal component except at**
1) Equator 2) Magnetic pole 3) A latitude of 60^0 4) None of the above

20. **In the case of bar magnet, lines of magnetic induction**
1) Start from the north pole and end at the south pole
2) Run continuously through the bar and outside
3) Emerge in circular paths from the middle of the bar
4) Are produced only at the north pole like rays of light from a bulb

21. **The lines of force due to the earths horizontal magnetic field are**
1) parallel and straight 2) Concentric circles 3) elliptical 4) curved lines

22. **Magnetic lines of force**
1) Always intersect 2) Are always closed
3) Tend to drowd far away from the poles of magnet 4) Do not pass through vacuum

23. **The direction of the line of magnetic field of bar magnet is**
1) From south pole to north pole 2) From north pole to south pole
3) Across the bar magnet
4) From south pole to north pole inside the magnet and from north pole to south outside the magnet

24. **The magnetic lines of force inside a bar magnet**
1) do not exist
2) depends on area of cross section of the bar magnet
3) are from N-pole to S-pole of the magnet 4) are from S-pole to N-pole of the magnet

25. **Find out the wrong statement**
1) The lines of force may intersect 2) The lines of force never intersect
3) The tangent drawn to the line of induction at any point gives the direction of magnetic inductionat that point.
4) None

26. **The incorrect statement regarding the lines of force of the magnetic field B is**
1) Magnetic intensity is a measure of lines of force passing through unit area held normal to it
2) Magnetic lines of force are closed curves
3) Inside a magnet, its magnetic lines of force are from north pole of the magnet to its south pole
4) Due to a magnet, magnetic lines of force never cut each other

27. **Find the wrong statement among the following.**
Two unlike isolated magnetic poles are at some distance apart in air.
1) The resultant induction at a point between the poles is $B_1 + B_2$ on the line joining them
2) The resultant induction is $B_1 \sim B_2$ at any point out side the poles on the line joining them.
3) No neutral point is formed on the line joining them if the pole strengths are equal.
4) A neutral point is formed in between the poles and nearer to weak pole on the line joining them.

28. **When a bar magnet is suspended freely in a uniform magnetic field it has**
1) Only translatory motion
2) Only rotatory motion
3) Both 1 & 2
4) None

29. **When a bar magnet is suspended freely in a non uniform magnetic field it has**
1) Only translatory motion
2) Only rotatory motion
3) Both 1 & 2
4) None

30. **When a bar magnet is suspended freely in a uniform magnetic field, the correct statement is**
1) The magnet experiences only couple and undergoes only rotatory motion.
2) The direction of Torque is along the suspension wire.
3) The magnitude of Torque is maximum when the magnet is normal to the field direction.
4) All the above.

31. **A magnetic field is produced and directed along y-axis. A magnet is placed along x-axis. The direction of torque on the magnet is**
1) in the x-y plane
2) along y-axis
3) along z-axis
4) Torque will be zero.

32. **The effect due to uniform magnetic field on a freely suspended magnetic needle is as follows**
1) both torque and net force are present
2) torque is present but no net force
3) both torque and net force are absent
4) net force is present but no torque

33. **A magnetic dipole is free to rotate in a uniform magnetic field. For what orientation of the magnet with repect to the magnetic field**
a) Torque is maximum
b) Potential energy is maximum
1) 45^0 and 45^0
2) 0^0 and 180^0
3) 90^0 and 180^0
4) 0^0 and 90^0

34. **The rate of change of torque with deflection $\left(\frac{d\tau}{d\theta}\right)$ is maximum for a magnet suspended freely in a uniform magnetic field when 'θ' is**
1) 0^0
2) 45^0
3) 60^0
4) 90^0

35. **A magnetic needle is kept in a non- uniform magnetic field. It experiences**
1) a torque but not a force
2) neither a force nor a torque
3) a force and a torque
4) a force but not a torque

36. **A magnetic needle, suspended horizontally by an unspun silk fibre, oscillates in the horizontal plane, because of a restoring force originating mainly from**
1) the torsion of the silk fibre
2) the force of gravity
3) the horizontal component of earths' magnetic field
4) all of the above

37. **There is no couple acting when two bar magnets are placed coaxially separated by a distance because:** [

1) There are no forces on the poles 2) The forces are parallel and their lines of action do not coincide

3) The forces are perpendicular to each other 4) The forces act along the same line

38. **A bar magnet of magnetic moment $\vec{M}$ is placed in a magnetic field of induction $\vec{B}$. The torque exerted on it is**

1) $\vec{M}.\vec{B}$ 2) $-\vec{M}.\vec{B}$ 3) $\vec{M}\times\vec{B}$ 4) $\vec{B}\times\vec{M}$

39. **A bar magnet of magnetic moment M is freely suspended in a uniform magnetic field of strength B. The work done in rotating the magnet through an angle θ is**

1) $MB(1-\sin\theta)$ 2) $MB\sin\theta$ 3) $MB\cos\theta$ 4) $MB(1-\cos\theta)$

40. **Two short magnets are placed on a piece of cork which floats on water. The magnets are so placed that the axis of one produced bisects the axis of other at right angles. Then the cork,**

1) Rotates only 2) Moves along a straight line only

3) Has rotational as well as translational motion 4) Has neither translational nor rotational motion

41. **Two small bar magnets are placed in a line with like poles facing each other at a certain distance 'd' apart. If the length of each magnet is negligible as compared to 'd', the force between them will be inversely proportional to**

1) d 2) d^2 3) $1/d^2$ 4) d^4

42. **The magnetic field due to a small dipole of a given magnetic moment varies as inverse of the**

1) Square of the distance 2) Cube of the distance

3) Fourth power of the distance 4) None of the above

43. **Points A and B are situated along the extended axis of a 2cm long bar magnet at distances 'x' and 'dx' cm respectively from the pole nearer to the point. The ratio of the magnetic fields at A and B will be**

1) d : 1 exactly 2) d^4 : 1 exactly

3) d^2 :1 approximately 4) d^3 : 1 approximately

44. **A point at which two or more magnetic fields cancel one another is called**

1) inversion point 2) curie point 3) faraday 4) neutral point

45. **When a magnet is placed vertically on a horizontal board, number of neutral points obtained on the board is**

1) four 2) three 3) two 4) one

46. **A very long magnet is held vertically with its south pole on a table. A single neutral point is located on the table on**

1) The east of the magnet 2) The north of the magnet

3) The west of the magnet 4) The south of the magnet

47. **A magnet is placed with its north pole facing north in north south direction. In its map of magnetic field the null points will be**

1) on its equatorial line 2) on its axial line

3) on a line making 45°with axial line

4) At a distance equal to twice the length of magnet, both on axial and equatorial line

48. **The null points are on the axial line of a bar magnet, when it is placed such that its south pole points**

1) South 2) East 3) North 4) West

49. **At a neutral point in the neighbourhood of a bar magnet placed in earth's magnetic field**

1) The magnetic field of the magnet is zero

2) The fields of the earth and the magnet are equal in magnitude and oppositely directed

3) A magnetic compass needle will point perpendicular to the bar magnet

4) A magnet compass needle will align, itself parallel to the bar magnet

50. Deflection magnetometer works on

1) Coulomb's law 2) Curie law 3) tangent law 4) The law of vibration of a magnet

51. Deflection magnetometer is held in tan B position. A magnet placed on one of its arms produces no deflection.This implies that the axis of the magnet is

1) in the east-west direction
2) in the north-south direction
3) perpendicular to the wooden bench
4) None

52. In working with deflection magnetometer, the relative error will be minimum when the deflection is

1) 30^0 2) 45^0 3) 60^0 4) 90^0

53. In the following, the most sensitive method to compare the magnetic moment of two bar magnets with a deflection magnetometer is

1) tan A, null method 2) tan A, equidistant 3) tan B, equidistant 4) tan B, null method

54. The purpose of the plane mirror in the compass box of a DMM is

1) to eliminate pivotal error
2) To eliminate zero error
3) to eliminate parallax error
4) to eliminate all types of errors

55. The error due to imperfect centering of the pivot of the magnetic needle in the compass box in deflection magnetometer experiment is eliminated by taking the readings

1) at both ends of aluminium pointer
2) by reversing the magnet
3) by inter changing the arms of the deflection magnetometer
4) with out parallax

56. Deflection magnetometer cannot be used at

1) The magentic north pole of earth
2) The magnetic south pole of earth
3) The magnetic equator of earth
4) at magnetic north and south poles of earth

57. The direction of magnet in tanB position is along

1) North-South 2) East-West 3) North-West 4) South-West

58. Two magnets have same magnetic moment but have different lengths. If the field at the same distance from the centre of the magnets in Tan A positions are measured, which of the following in correct.

1) Both fields are equal
2) shorter magnet produces stronger field
3) Longer magnet produces stronger field
4) None

59. StatementA: Magnet should be placed at the same level as that of the needle of campass in DMM.
StatementB: Other magnets and magnetic materials should be removed from the vicinity of D.M.M

1) A True, B true 2) A true, B false 3) A false, B false 4) A false, B true

60. The arms of a delfection magnetometer in the tanB position are placed along

1) The east-west
2) The north-south direction
3) The north east-south west direction
4) The north west south direction

61. A small bar magnet produces a deflection of 45^0 in tan-A position. What will be the deflection when the arms are rotated through 90^0 in anti-clock wise direction.

1) 45° 2) 60° 3) Zero 4) 90°

62. The restoring torque for a magnet oscillating in the vibration magnetometer is provided by

1) torsion in the suspension thread
2) gravity
3) horizontal component of the earth's magnetic field
4) by all these

63. VMM is used to

1) Compare the magnetic moments of two bar magnets.
2) Compare the horizontal components of earth's magnetic induction at different places
3) Determine the magnetic moment of a magnet & B_H at a given place when used to along with DMM
4) All the above

64. Vibration magnetometer works on the priciple of

1) Torque acting on the bar magnet suspended in magnetic field
2) Force acting on the bar magnet
3) Both the force and the torque acting on the bar magnet
4) None of the above

65. The time period of a freely suspended magnet does not depend upon

1) length of the magnet
2) the polestrength of the magnet
3)the horizontal component of earths' magnetic field
4) the length of the suspension

66. Using a bar magnet P, a vibration magnetometer has time period 2 seconds. When a bar Q (identical to P in mass and size) is placed on top of P, the time period is unchanged. Which of the following statements is true

1) Q is of non magnetic substance
2) Q is a bar magnet identical to P and its north pole is placed on the top of P's north pole
3) Q is an unmagnetised ferromagnetic material
4) Nothing can be said about Q's property

67. One pole of a magnet is placed near the needle of a small compass box. The needle is allowed to vibrate and the magnet is moved further and the magnet is moved further away. The new period of vibrations is

1) Unrelated to the original period
2) The same as the original period
3) Longer than the original period
4) Smaller than the original period

68. The magnetism of magnet is due to

1) The spin motion of electron
2) Earth
3) Pressure of big magnet inside the earth
4) Cosmic rays

69. Magnetic field produced by electrons in atoms and molecules is due to their

1) spin motion only
2) orbital motion only
3) spin and orbital motion
4) neither spin nor orbital motion

70. When all the molecules in a magnet arrange themselves in the direction of the magnetic field the condition is called

1) permeability
2) saturation
3) retentivity
4) reluctance

71. Magnetic lines of force prefer to pass through ferro magnetic materials because their

1) μ and X_m are low
2) μ is high, X_m is low
3) μ and X_m both are high
4) none of these

72. The process in which the length of a ferro magnetic material increases when magnetised is known as

1) magnetisation
2) demagnetisation
3) magnetostriction
4) hysterisis

73. Metals getting magnetised by orientation of atomic magnetic moments in external magnetic field are called:

1) Diamagnetics
2) Paramagnetics
3) Ferromagnetics
4) Antimagnetic

74. When a material is used in a magnetic field B, a magnetic moment proportional to B but opposite in direction is induced. The metal is:

1) Diamagnetics
2) Paramagnetics
3) Ferromagnetics
4) Antimagnetic

75. Susceptibility is positive and large for a

1) paramagnetic substance
2) ferromagnetic substance
3) diamagnetic substance
4) non magnetic substance

76. Liquid and gases never show

1) diamagnetic properties
2) paramagneitc properties
3) ferro magnetic properties
4) none of the above

77. For a paramagnetic material, the dependence of the magnetic susceptibility χ on the absolute temperature T is given by

1) $\chi \alpha T$
2) χ=constant $\sqrt{T}$
3) $\chi \, \alpha \frac{1}{T}$
4) χ=constant

78. All magnetic materials lose their magnetic properties when

1) Dipped in water
2) Dipped in oil
3) Brought near a piece of iron
4) Strongly heated

79. For a diamagnetic material

1) $\mu_r > 1, x_m > 1$
2) $\mu_r > 1, x_m < 1$
3) $\mu_r < 1, x_m < 0$
4) $\mu_r < 1, x_m > 0$

80. The permanent magnetic moment of atoms or molecules of a diamagnetic materials is

1) small
2) large
3) negative
4) zero

81. Of dia, para and ferromagnetism, the universal property of all substance is

1) diamagnetism
2) Paramagnetism
3) Ferromagnetism
4) All of these

82. The substances have been classifield as ferromagnetic, paramagnetic and diamagnetic according to

1) electronic theory
2) molecular theory
3) ohm's law
4) formation of electronic pairs

83. Correct increasing order of susceptibility of magnetic material is

1) Dia, Ferro, Para
2) Dia, Para, Ferro
3) Ferro, Para, Dia
4) Para, Dia, Ferro

84. The property of allowing magnetic lines of force through material is

1) relative permeability
2) permeability
3) intensity of magnetisation
4) susceptibility

85. Which of the following is best for permanent magnets

1) alnico
2) iron
3) steel
4) brass

86. To shield an instrument from external magnetic field, it is placed inside a cabin made from

1) wood
2) ebonite
3) iron
4) diamagnetic substance

87. Screens to prevent the magnetic effects are made of

1) Glass
2) Wood
3) Iron
4) Brass

88. Consider the following two statements A and B and identify the correct choice in the given answers:

(A) Paramagnetism is explained by domain theory.

(B) Sussceptibility of a diamagnetic substance is independent of temperature

1) A and B correct
2) A and B wrong
3) A is correct but B is wrong
4) A is wrong and B is correct

89. Which of the following materials is repelled by an external magnetic field

1) Iron
2) Cobalt
3) Steel
4) Copper

90. Dia magnetic substances are

1) Feebly attracted by magnets
2) Strongly attracted by magnets
3) Feebly repelled by magnets
4) Strongly repelled by magents

91. Curie temperature is the temperature above which

1) A ferromagnetic material becomes para magnetic
2) A paramagnetic material becomes diamagnetic
3) A ferromagnetic material becomes diamagnetic
4) A paramagnetic material becomes ferromagnetic

92. **The material suitable for making electro-magnets should have**

1)high retentivity and high coercivity   2)low retentivity and low coercivity

3)high retentivity and low coercivity   4) low retentivity and high coercivity

93. **Needles N_1,N_2,N_3 are made of a ferromagnetic ,pamagnetic and a diamagnetic substance respectively. A magnet when brought close to them will**

1) attract all three of them   2) attract N_1 and N_2 strongly but repel N_3

3) attract N_1strongly, N_2weakly and repelN_3 weakly   4)attract N_1strongly,but repel N_2 N_3 weakly

94. **A magnet of moment M is cut along its length and breadth into four equal parts.**

i) magnetic moment becomes M/ 4   ii) magnetic length becomes half

iii)pole strength becomes m/2   iv) magnetic moment becomes M/2

1) i only   2) i and ii only   3) i, ii and iii   4) i, ii and iv only

95. **Identify the correct choice**

i)lines of force start from the north pole and end at the south pole outside the magnet

ii) no two lines of force intersect

iii) lines of force are real curves which are invisible

iv) the strength of magnetic field is the largest where the lines of force are crowded together

1) i only   2) i and ii only   3) i, ii and iii   4) i , ii and iv only

ASSERTION AND REASON

In each of the following questions, a statement of Assertion (A) is given followed by a corresponding statement of reason (R) just below it. Of the statement mark the correct answer.

1) Both 'A' and 'R' are true and 'R' is the correct explanation of 'A'

2) Both 'A' and 'R' are true and 'R' is not the correct explanation of 'A'

3) 'A' is true and 'R' is false   4) 'A' is false and 'R' is true

96. **consider the following statements :**

(A) : The poles of magnets can not be separated by breaking into two pieces

(R) : The magnetic moment will be reduced to half when a magnet is broken into two equal pieces

97. **(A) :** A magnet suspended freely in a uniform magnetic field experiences no net force, but a torque that tends to align the magnet along the field

(R) : For a freely suspended magnet in a uniform magnetic field the force on N-Pole and S-pole being equal unlike and parallel make up a couple that tends to align the magnet along the field.

98. **(A) :** A freely suspended magnet comes to rest in the direction of magnetic meridaian

(R) : Earth acts a huge magnet

99. **(A) :** There is no couple acting when two bar magnets are placed with their magnetic axes along the same line separated by a distance

(R) : The lines of action of forces are along the same line when the magnets are placed coaxially.

100. **(A) :** The pointer in the compass box is made of aluminium

(R) : Aluminium is a paramagnetic substance

101. **(A) :** In the experiment with DMM magnet is reversed at a given position pole to pole

(R) : The magnet may not be uniformly magnetised

102. **(A) :** In the experiment with DMM the magnet is kept at the equal distances on both arms

(R) : The centre of circular scale may not coicide with the zero of the linear scale

103. **(A) :** In the experiment with DMM unwanted magnetic substances are kept away

(R) : The stray magnetic fields will influence the readings in DMM experiment

104. **(A) :** The magnet in a vibration magnetometer executes periodic but not S.H.M. for larger amplitudes.

(R) : For larger amplitude angular acceleration of the suspended magnet is not proporational to deflection in VMM.

MATCH THE FOLLOWING

105. Match the following :

List - 1	**List - 2**
a) permeability	e) [$M^0 L^2 T^0 I$]
b) magnetic induction	f) [$MLT^{-2} I^{-2}$]
c) magnetic moment	g) [$M^0 L^{-1} T^0 I$]
d) intensity of magnetization	h) [$ML^0 T^{-2} I^{-1}$]

1) a - f, b - h , c - g , d - e
2) a - f, b - h, c - e, d - g
3) a - h, b - f , c - g , d - e
4) a - h, b - f, c - e, d - g

106. Match the following

List - 1	**List - 2**
a) magnetic moment	e) H/m
b) magnetic induction	f) A/m
c) permeability	g) N - m^3 / Wb
d) ntensity of magnetization	h) N / (A - m)

1) a - f, b - h, c - g, d - e
2) a - f, b - h, c - e, d - g
3) a - g, b - h, c - e, d -f
4) a - h, b - f , c - e, d - g

107. Some physical quantities are given in the list I the related units are given in the list II. Match the correct pairs in the lists.

List - 1	**List - 2**
a) Magnetic field intensity	e) Wbm^{-1}
b) Magnetic flux	f) Wbm^{-2}
c) Magnetic potential	g) Wb
d) Magnetic induction	h) Am^{-1}

1) a - e, b - f , c - g , d - h
2) a - h , b - g, c - e , d - f
3) a - h , b - g , c - e , d - f
4) a - f, b - g, c - e, d - h

ANSWERS

1) 4	2) 1	3) 4	4) 4	5) 4	6) 1	7) 3	8) 4	9) 2	10) 3
11) 3	12) 4	13) 4	14) 1	15) 3	16) 2	17) 2	18) 3	19) 2	20) 2
21) 1	22) 2	23) 4	24) 4	25) 1	26) 3	27) 4	28) 2	29) 3	30) 4
31) 3	32) 2	33) 3	34) 1	35) 3	36) 3	37) 4	38) 3	39) 4	40) 4
41) 4	42) 2	43) 4	44) 4	45) 4	46) 2	47) 1	48) 3	49) 2	50) 3
51) 2	52) 2	53) 1	54) 3	55) 1	56) 4	57) 2	58) 3	59) 1	60) 2
61) 3	62) 3	63) 4	64) 1	65) 4	66) 2	67) 3	68) 1	69) 3	70) 2
71) 3	72) 3	73) 2	74) 1	75) 2	76) 3	77) 3	78) 4	79) 3	80) 4
81) 1	82) 1	83) 2	84) 2	85) 1	86) 3	87) 3	88) 4	89) 4	90) 3
91) 1	92) 2	93) 3	94) 3	95) 4	96) 2	97) 1	98) 1	99) 1	100) 1
101) 1	102) 1	103) 1	104) 1	105) 2	106) 3	107) 2			

LEVEL-II(A)

MAGNETIC MOMENT

1. **The magnetic length of a bar magnet is 10cm then its geometric length is**

1) 10cm 2) 12cm 3) 5cm 4) 6cm

2. **Two magnets have their lengths in the ratio 2 : 3 and their pole strengths in the ratio 3 : 4. The ratio of their magnetic moment is**

1) 2 :1 2) 4 :1 3) 1 : 2 4) 1 : 4

3. **A bar magnet of magnetic moment $5A\text{–}m^2$ has poles 20cm apart its pole strength is**

1) 50A–m 2) 15A–m 3) 25A–m 4) 45A–m

4. **If a bar magnet of pole strength 'm' and magnetic moment 'M' is cut equally 5 times parallel to its axis and 4 times perpendicular to its axis then the pole strength and magnetic moment of each piece are respectively**

1) $\frac{m}{20},\frac{M}{20}$ 2) $\frac{m}{4},\frac{M}{20}$ 3) $\frac{m}{5},\frac{M}{20}$ 4) $\frac{m}{5},\frac{M}{4}$

COULOMB FORCE

5. **A magnet of moment $80Am^2$ is placed in a uniform magnetic field of induction $1.8\times10^{-5}T$. If each pole of the magnet experiences a force of $25\text{x}10^{-3}$N, the length of the magnet is**

1) 0.292cm 2) 5.76cm 3) 0.362cm 4) 2.262cm

6. **An isolated unit north pole experiences a force of $2\times10^{-7}N$ when kept at a distance of 1m from a magnetic pole in air. The pole strength of the pole is**

1) $2\times10^{7}A-m$ 2) 2Am 3) $2\times10^{-7}A-m$ 4) 4Am

7. **Two magnetic poles one of which is 4 times as strong as the other exert a force of 64 mg.wt. When placed 10cm apart in air. Strength of the poles are (g= $10ms^{-2}$)**

1) 3A–m and 12Am 2) 4A–m and 16Am 3) 5Am and 20Am 4) 6Am and 10 Am

8. **Two magnetic poles of strengths 40 Am and 10 Am are separated by a distance of 20 cm in air. Find the force between them. If the distance is reduced to 10 cm, find the force.**

1) 10^{-3} N, 4 x 10^{-3} N 2) 2 x 10^{-3} N, 2 x10^{-3} N

3) 4 x 10^{-3} N, 10^{-3} N 4) 10^{-3} N, 10^{-3} N

9. **Two poles of same strength attract each other with a force of magnitude F when placed at the two corners of an equilateral triangle. If a north pole of same strength is placed at the third vertex, it experiences a force of magnitude equal to**

1) $F\sqrt{3}$ 2) F 3) $F\sqrt{2}$ 4) 2F

RESULTANT MAGNETIC MOMENT

10. **A magnetised straight wire has a magnetic moment M. It is bent into a semi circular arc, its magnetic moment becomes**

1) $\frac{M}{\pi}$ 2) $2\pi\frac{M}{3}$ 3) $\frac{2M}{\pi}$ 4) $\frac{2\pi}{M}$

11. **A thin bar magnet of length 'ℓ' and magnetic moment 'M' is bent at the mid point so that the two parts are at right angles. The new magnetic length and magnetic moment are respectively**

1) $\sqrt{2}\ell,\sqrt{2}M$ 2) $\frac{\ell}{\sqrt{2}},\frac{M}{\sqrt{2}}$ 3) $\sqrt{2}\ell,\frac{M}{\sqrt{2}}$ 4) $\frac{\ell}{\sqrt{2}},\sqrt{2}M$

12. **A magnetised wire of moment 'M' is bent into an arc of a circle subtending an angle 60^0 at the centre. The new magnetic moment will be (2001M)**

1) $\frac{3M}{\pi}Am^2$ 2) $\frac{2M}{\pi}Am^2$ 3) $\frac{M}{\pi}$ Am^2 4) $\frac{4M}{\pi}$ Am^2

13. If three identical bar magnets each of magnetic moment 'M' are arranged in the form of an equilateral triangle such that unlike poles are in contact. The resultant magnetic moment is

1) $\sqrt{3}M$ 2) M/3 3) 3M 4) Zero

14. The resultant magnetic moment for the following arrangement is (the vectors are non coplanar)

1) M 2) 2M

3) 3M 4) 4M

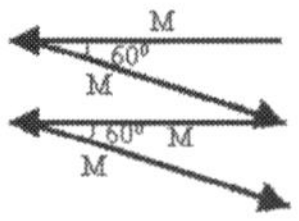

15. A bar magnet of magnetic moment M is bent in the 'U' shape such that all the parts are of equal lengths. Then the new magnetic moment is

1) $\frac{M}{3}$ 2) $\frac{M}{\sqrt{2}}$ 3) $\sqrt{3}M$ 4) $3\sqrt{3}M$

16. A magnet of moment 'M' is cut into two equal parts, the two parts are placed perpendicular to each other. The resultant magnetic moment is

1) $\sqrt{2}M$ 2) $\frac{M}{\sqrt{2}}$ 3) $\sqrt{3}M$ 4) $3\sqrt{3}M$

17. Three magnets of same length but moments M,2M and 3M are arranged in the form of an equilateral triangle with opposite poles nearer, the resultant magnetic moment of the arrangement is

1) 6M 2) zero 3) $\sqrt{3}M$ 4) $\frac{\sqrt{3}}{2}M$

18. A bar magnet of moment M is cut into two identical pieces along the length. One piece is bent in the form of a semi circle. If two pieces are perpendicular to each other, then resultant magnetic moment is

1) $\left(\frac{M}{\pi}\right)^2+\left(\frac{M}{2}\right)^2$ 2) $\sqrt{\left(\frac{M}{\pi}\right)^2+\left(\frac{M}{2}\right)^2}$

3) $\sqrt{\left(\frac{M}{\pi}\right)^2-\left(\frac{M}{2}\right)^2}$ 4) $\frac{M}{\pi}+\frac{M}{2}$

MAGNETIC INDUCTION

19. The magnetic induction field due to a magnetic north pole of strength 10Am and 5cm away from the pole in weber/m^2 is.

1) $4\text{x}10^{-4}$ radially inward 2) $4\text{x}10^{-4}$ radially outward

3) $2\text{x}10^{-4}$ radially inward 4) $2\text{x}10^{-4}$ radially outward

20. Two points A and B are located at distances 20cm and 30 cm from the centre of a short bar magnet on its axial line.The ratio of magnetic induction at A and B is

1) 2 : 3 2) 3 : 2 3) 9 : 4 4) 27 : 8

21. Magnetic moment of a dipole is 8Am^2. The magnetic induction on its equitorial line of short bar magnet at a distance of 20 cm is

1) $0.1\times10^{-4}T$ 2) $0.01\times10^{-4}T$ 3) $10\times10^{-4}T$ 4) $10^{-4}T$

22. Two points A and B are located at equal distance from the centre of the short magnet on the axial line and equitorial line. The ratio of inductions at A and B is

1) 1 : 1 2) 2 : 1 3) 1 : 2 4) 4 : 1

23. The force experienced by a magnetic pole of strength m kept at a distance of d from the centre of a short magnet of magnetic moment M on its equatorial line is

1) $m\frac{\mu_0}{4\pi}\frac{M}{d^3}$ 2) $m\frac{\mu_0}{4\pi}\frac{2M}{d^3}$ 3) $\frac{1}{m}\frac{\mu_0}{4\pi}\frac{2M}{d^3}$ 4) $\frac{1}{m}\frac{\mu_0}{4\pi}\frac{2}{Md^3}$

COUPLE

24. **At what angle a bar magnet should be inclined to the magnetic field so that the couple acting on it would be half of the maximum couple.**
1) 30^0 2) 60^0 3) 40^0 4) 45^0

25. **The maximum torque acting on the magnet of length 2 x 10^{-1} m in a uniform magnetic field having induction 2x10^{-1} tesla is 10Nm. Calculate its pole strength.**
1) 250 A–m 2) 125 A–m 3) 500 A–m 4) 50 A–m

26. **A magnet of length 10cm and pole strength 0.5Am is placed normal to a uniform magnetic field of strength 5×10^{-5} weberm^{-2}. The moment of the couple acting on it is**
1) $2.5\times10^{-6}N-m$ 2) $5\times10^{-5}Nm$ 3) zero 4) $2.5\times10^{-7}N-m$

27. **A bar magnet of length 10cm and pole strength 20A-m is deflected through 30^0 from the magnetic meridian. If earth's horizontal component field is $\frac{320}{4\pi}$ A/m. The moment of the deflecting couple is**
1) $160\times10^{-7}Nm$ 2) $320\pi\times10^{7}Nm$ 3) $320\times10^{-7}Nm$ 4) $160\pi\times10^{-7}Nm$

28. **A bar magnet when placed at an angle of 30^0 to the direction of magnetic field induction of $5\times10^{-5}T$, experiences a moment of couple $2.5\times10^{-6}N-m$. If the length of the magnet is 5cm its pole strength is**
1) $2\times10^{-2}Am$ 2) $5\times10^{-2}Am$ 3) 2A–m 4) 5A–m

29. **Calculate the moment of couple required to keep a bar magnet of magnetic moment 2 x 10^2 Am^2 in a uniform field of induction 0.36 T at an angle of 30^0 with the direction of uniform field.**
1) 36 Nm 2) 3.6 Nm 3) 72 Nm 4) 7.2 Nm

30. **Two magnets of magnetic moments M and $\sqrt{3}$ M are joined to form a cross +. The combination is suspended freely in a uniform magnetic field. In the equlibrium position, the angle between the magnetic moment M and the field is**
1) 30^0 2) 45^0 3) 60^0 4) 90^0

31. **A magnet of magnetic moment $50\hat{i}$ Am^2 is placed along the x-axis in a magnetic field $\vec{B}=(0.5\hat{i}+3.0\hat{j})$ T. The torque acting on the magnet is**
1) 175 $\hat{K}$ N- m 2) 150 $\hat{K}$ N- m 3) 75 $\hat{K}$ N - m 4) 25 $\sqrt{37}\hat{K}$ N -m

32. **The torque required to keep a magnet of length 20 cm at 30° to a uniform field is 2x10^{-5} N-m. The magnetic force on each pole is**
1) 2 x 10^{-4} N 2) 4 x 10^{-4} N 3) 2 x 10^{-3} N 4) 2 x 10^{-6} N

33. **A magnetic needle of pole strength 20 $\sqrt{3}$ Am is pivoted at its centre. Its N-pole is pulled eastward by a string. The horizontal force required to produce a deflection of 30° from magnetic meridian (take $B_H=10^{-4}T$) is**
1) $4\times10^{-3}N$ 2) $2\times10^{-3}N$ 3) $\frac{2}{\sqrt{3}}\times10^{-3}N$ 4) $4\sqrt{3}\times10^{-3}N$

WORK DONE

34. **A bar magnet of magnetic moment1.5J/Tis along the direction of the uniform magnetic field of 0.22T. The work done in turning the magnet opposite to the field direction and the torque requried to keep in that position are**
1) 0.33J and 0.33N-m 2) 0.66J and 0.66N-m
3) 0.33J and 0 N-m 4) 0.66J and 0N-m.

35. **A bar magnet of magnetic moment 5 Am^2 is suspended in a uniform magnetic field of strength 2T making an angle 60^0 with the direction of the field. The potential energy of the magnet in that position is**
1) 5 J 2) -5 J 3) 10 J 4) -10 J

36. **Amagnet is parallel to a uniform magnetic field. The work done in rotating the magnet through $60^0$is 24 x10^{-5} J. The work done in rotating through another 60^0 is**

1) 12 x10^{-5} J 2) 6 x10^{-5} J 3) 24 x 10^{-5} J 4) 48 x 10^{-5} J

37. **A bar magnet of moment 2A-m² is free to rotate about a Vertical axis passing through its centre. The magnet is released form rest from east west direction. The K.E of the magnet as it takes north-south direction is (B_H=25 x10^{-6}T)**

1) 25 x10^{-6}J 2) 50 x10^{-6}J 3) 75 x10^{-6J} 4) 100x10^{-6}J

38. **A bar magnet of magnetic moment 100 Am² is divided into 5 equal parts cutting perpendicular to length. Then one part is suspended in a uniform magnetic field of strength 12 gauss and held making an angle 120^0 with the direction of the field when the magnet is released. Then the kinetic energy acquired by the magnet in the equilibrium position is**

1) 0.012 J 2) 0.024 J 3) 0.036 J 4) 0.072 J

NEUTRAL POINT

39. **Two isolated north poles of pole strengths 16 A-m and 4 A-m are at 30 cm apart in air. The distance of neutral point from the weaker pole is**

1) 5 cm 2) 10 cm 3) 25 cm 4) 15 cm

40. **A short magnet of moment 6.4 A-m² lies in the magnetic meridian with north pole facing south (B_H = 1.6 gauss). The distance between the two neutral points is**

1) 20 cm along the axial line 2) 20 cm along the equatorial line

3) 40 cm along the axial line 4) 40 cm along the equatorial line

41. **A short bar magnet with its north pole facing north forms a neutral point at P in a horizontal plane. If the magnet is rotated by 90^0 in the horizontal plane, the net magnetic induction at P is (Horizontal component of earth's magnetic field = B_H)**

1) $\sqrt{5}B_H$ 2) 2 B_H 3) $\frac{\sqrt{5}}{2}B_H$ 4) 0

42. **A very long magnet of pole strength 16 A-m is placed vertically with its one pole on the table. At what distance from the pole, there will be a neutral point on the table. (B_H = 4 x 10^{-5} Wbm^{-2})**

1) 0.4 m 2) 0.2 m 3) 0.4 m 4) 0.8 m

TANGENT LAW AND D M M

43. **A small magnet produces a deflection of 45^0 in TanB position when used at 10 cm from the magnetic needle. If B_H is $0.36x10^{-4}T$,the magnetic moment of magnet is**

1) $0.72Am^2$ 2) $0.36Am^2$ 3) $0.18Am^2$ 4) $0.9Am^2$

44. **A short magnet produces a deflection of 45^0 in a deflection magnetometer in TanB position, when placed at a distance of 10cm from the needle. Where should the same bar magnet be placed in Tan A position of the magnetometer to produce same deflection 45^0**

1) $10(2^{1/3})cm$ 2) 10cm 3) 30cm 4) $10x2^3 cm$

45 **A short bar magnet placed end on to a compass needle with its cemtre at 0.15 m from the centre of the needle deflects the needle through 45^0. By how far the magnet should be moved away so that the deflection may fall to 30^0 ?**

1) 0.18m 2) 0.33 m 3) 0.15 m 4) 0.36 m

46. **Two short magnets are kept on opposite arms of a deflection magnetometer at 20cm and 40cm from the needle. If the needle lies in the magnetic meridian and moment of first magnet is 100Am², the moment of other magnet is**

1) 200Am² 2) 800Am² 3) 2700Am² 4) 900Am²

47. In a deflection magnetometer experiment the deflections produced separately by two short bar magnets kept at the same distance are 45° and 30°. Then the ratio of the magnetic moments of the two magnets is

1) $\sqrt{3}:2$ 2) $\sqrt{3}:1$ 3) $\sqrt{2}:1$ 4) $1:\sqrt{3}$

48. A magnet shows a deflection of 30° when it is arranged properly in Tan A position . If magnet is rotated by 90°, the new deflection will be

1) 0^0 2) 180^0 3) 60^0 4) 30^0

49. The needle of a deflection magnetometer shows a deflection of 30^0 due to a short bar magnet at a certain distance in TanB position. If distance is halved, the deflection is

1) 45° 2) $Tan^{-1}\frac{1}{8\sqrt{3}}$ 3) $Tan^{-1}8/\sqrt{3}$ 4) $Tan^{-1}\sqrt{3}$

50. A small magnet of moment M in tan A position at a distance 'd' produces a deflection 60^0. Another short magnet with moment 8M when kept at a distance 2d in tanA position produces a deflection of

1) 60° 2) 30° 3) 45° 4) 27°

51. A short bar magnet of moment M produces a deflection of 45^0 in tan A position.The deflection produced by another short bar magnet of moment 2M when placed at the same distance in tan B position is

1) 30^0 2) 60^0 3) 90^0 4) 45^0

52. Two bar magnets of same length kept one over the other gives a deflection of 60^0 in Tan A position, where 'd' is the distance from the centre of compass box, then the poles of one of them is reversed the deflection is 30^0 . the ratio of magnetic moments is

1) $\sqrt{3}:1$ 2) $\sqrt{2}:1$ 3)2:3 4) 2:1

53. A short magnet when placed at a distance of 20 cm from the magnetic needle of a deflection magnetometer in Tan A position produces a deflection of 60^0. If the magnet is cut into four equal parts and one of them is placed at same distance, the deflection will be

1) 30° 2) $Tan^{-1}\frac{\sqrt{3}}{4}$ 3) $Tan^{-1}\frac{3}{4}$ 4) $Tan^{-1}\frac{4}{3}$

VIBRATION MAGNETO METER (VMM)

54. A long bar magnet of time period T is cut into four equal parts by cutting it perpendicular to both length and breadth.The time period of each part is

1) $T/4$ 2) $T/16$ 3) $T/32$ 4) $T/2$

55. Two magnets are placed one above the other with like poles together the time period of this combination is 10seconds. Now they are placed one above the other with unlike poles together the time period of this combination in the same field is 20sec. The ratio of magnetic moments is

1) 3:1 2) 5:1 3) 5:3 4) 3:5

56. A bar magnet suspended in a field of induction $1.5x10^{-6}T$ has a time period $\frac{\pi}{5}$sec. The moment of inertia about the axis of suspension is $1.5x10^{-4}kgm^2$. The magnetic moment of the magnet is

1) $10000Am^2$ 2) $\frac{1}{16}Am^2$ 3) $10Am^2$ 4) $1Am^2$

57. With a standard rectangular bar magnet of length(ℓ) breadth (b << ℓ) and magnetic moment M, the time period of magnet in a vibration magnetometer is 4 s. If the magnet is cut normal to its length into 4 equal pieces, the time period (in seconds) with one of the pieces is (2005E)

1) 16 2) 2 3) 1 4) 1/4

58. A bar magnet is oscillating in a magnetic field and its frequency is 10Hz. If another identical piece of brass be placed over that magnet, the frequency of that combination in that field is

1)20 Hz 2)10 Hz 3) $\frac{10}{\sqrt{2}}$ Hz 4) $10\sqrt{2}$ Hz

59. **The moments of inertia as well as the time periods of two magnets are in the ratio of 1:2. The ratio of their magnetic moments is**

1)2:1 2)1:2 3)4:3 4)3:4

60. **A bar magnet freely suspended in a vibration magnetometer has time period T at a place A and T/2 at a place B. If horizantal Component of earth's magnetic field at A is $36x10^{-6}T$, then its value at 'B' is**

1) $100x10^{-6}T$ 2) $72x10^{-6}T$ 3) $144x10^{-6}T$ 4)None

61. **A magnet has time period 'T' in a certain magnetic field. If an identical magnet be kept over it with their moments perpendicular to each other now the time period is**

1) $\sqrt{2}T$ 2) $2^{1/4}T$ 3) $2^{1/3}T$ 4)2T

62. **A bar magnet suspended freely in uniform magnetic field is vibrating with a time period of 3 seconds.If the field strength is increased to four times of the earlier field strength,the time period will be (in seconds)**

1) 12 2) 6 3) 1.5 4) 0.75

63. **A bar magnet of magnetic moment M and moment of inertia I is freely suspended such that the magnetic axis is in the direction of magnetic meridian. If the magnet is displaced by a very small angle (θ), the angular acceleration is (Magnetic induction of earth's horizontal field $=B_H$). (2007E)**

1) $\frac{MB_H\theta}{I}$ 2) $\frac{IB_H\theta}{M}$ 3) $\frac{M\theta}{IB_H}$ 4) $\frac{I\theta}{MB_H}$

64. **The period of oscillation of a magnet at a place is 4 seconds . When it is remagnetised, so that the pole strength becomes four times the initial value, the period of oscillation in seconds is**

1) 1/2 2) 1 3) 2 4) 4

65. **The magnet of a vibration magnetometer is heated so as to reduce its magnetic moment by 19%. By doing this, the time period of the magnet will**

1) increases by 11% 2) decreases by 19%

3) increases by 19% 4) decreases by 4%

PROPERTIES OF MAGNETIC MATERIALS

66. **The magnetic susceptibility of a material of a rod is 499. Permeability of vaccum is $4\pi\times10^{-7}$ H/m. The absolute permeability of the material of the rod in henry / meter is**

1) $\pi\times10^{-4}$ 2) $2\pi\times10^{-4}$ 3) $3\pi\times10^{-4}$ 4) $4\pi\times10^{-4}$

67. **The magnetic induction and the intensity of magnetic field inside an iron core of an electromagnet are 1 Wbm^{-2} and 150 Am^{-1} respectively. The relative permeability of iron is : $(\mu_0=4\pi\times10^{-7}$ Henry/m$)$**

1) $\frac{10^6}{4\pi}$ 2) $\frac{10^6}{6\pi}$ 3) $\frac{10^5}{4\pi}$ 4) $\frac{10^5}{6\pi}$

68. **The mass of an iron rod is 80 gm and its magnetic moment is 10 Am^2. If the density of iron is $8gm/c.c.$ Then the value of intensity of magnetisation will be**

1) $10^6\ A/m$ 2) $10^4\ A/m$ 3) $10^2\ A/m$ 4) $10\ A/m$

69. **A rod of cross sectional area $10cm^2$ is placed with its length parallel to a magnetic field of intensity 1000 A/M the flux through the rod is 10^4 webers. Then the permeability of material of rod is**

1) $10^4\ wb/Am$ 2) $10^3\ wb/Am$ 3) $10^2\ wb/Am$ 4) $10\ wb/Am$

70. **A bar magnet of magnetic moment $10Am^2$ has a cross sectional area of $2.5x10^{-4}m^2$ .If the intensity of magnetisation of the magnet is $10^6\ A/m$, then the length of magnet is**

1) 0.4m 2) 0.04cm 3) 0.04m 4) 40 cm

ANSWERS

1) 2	2) 3	3) 3	4) 3	5) 2	6) 2	7) 2	8) 1	9) 2	10) 3
11) 2	12) 1	13) 4	14) 2	15) 1	16) 2	17) 3	18) 2	19) 2	20) 4
21) 4	22) 2	23) 1	24) 1	25)1	26) 1	27) 3	28) 3	29) 1	30) 3
31) 2	32) 1	33) 1	34) 4	35) 2	36) 4	37) 2	38) 3	39) 2	40) 3
41) 1	42) 2	43) 2	44) 1	45) 1	46) 2	47) 2	48) 1	49) 3	50) 1
51) 4	52) 4	53) 2	54) 4	55) 3	56) 1	57) 3	58) 3	59) 1	60) 3
61) 2	62) 3	63) 1	64) 3	65) 1	66) 2	67) 4	68) 1	69) 1	70) 3

EXERCISE-II(B)

MAGNETIC MOMENT

1. **The mgnetic length of a bar magnet is 5cm. The distance of each pole from the nearest ends is**

   1) 1cm 2) 1.5cm 3) 0.5cm 4) 2cm

2. **Two magnets have moments are in the ratio X: Y and their lengths in the ratio X : Z. The ratio of pole strength is**

   1) Y : Z 2) ZY : 1 3) 1 : YZ 4) Z : Y

3. **A bar magnet of pole strength is 25Am having length 5cm, then magnetic moment is**

   1) 125 Am^2 2) 1.25Am^2 3) 25Am^2 4) 50Am^2

4. **If a bar magnet is cut into 4 times parallel to its axial line and 4 times perpendicular to its axial line then the pole strength and length of each piece are**

   1) $\frac{m}{5},\frac{2\ell}{4}$ 2) $\frac{m}{4},\frac{2\ell}{5}$ 3) $\frac{m}{5},\frac{2\ell}{5}$ 4) $\frac{m}{4},\frac{2\ell}{4}$

COULOMB FORCE

5. **The force experienced by each pole of a magnet kept in the magnetising field of strength 30N/weber is $6\times10^{-3}N$. The magnetic moment of the magnet of length 10 cm is**

   1) $25\pi\,Am^2$ 2) $\frac{25}{\pi}J/T$ 3) $50\pi Am^2$ 4) $\frac{50}{\pi}J/T$

6. **Two magnetic poles are separated by a distance 20cm in air experience a force $4\times10^{-7}N$. The product of pole strengths of two poles are (in SI units)**

   1) 16 2) 1.6 3) 0.16 4) 0.016

7. **Two magnetic poles one of which is three times as strong as the other exert on each other, a force equal to 3 x 10^{-3} N when separated by a distance of 10 cm. Find the strength of each pole.**

   1) 10 Am. 30 Am 2) 20 Am, 60 Am 3) 5 Am, 15 Am 4) 10 Am, 10 Am

8. **Two similar poles repel each other with a force of 40N, when separated by 10 cm. If the distance is changed to 30 cm, then the ratio of forces are**

   1) 1 : 8 2) 8 : 1 3) 9 : 1 4) 1 : 9

9. **Two magnetic isolated north poles each of stength 'm' are placed at the two vertices of an equilateral triangle of side 'a'. If a third north pole is at the third vertex, then the resultant force at the third vertex is (μ_0 is permeability)**

   1) $\frac{\mu_0}{4\pi}\frac{m^2}{a^2}$ 2) $\frac{\mu_0}{4\pi}\frac{\sqrt{2}m^2}{a^2}$ 3) $\frac{\mu_0}{4\pi}\frac{\sqrt{3}m^2}{a^2}$ 4) $\frac{\mu_0}{2\pi}\frac{m^2}{a^2}$

RESULTANT MAGNETIC MOMENT

10. A magnetised wire of moment $1.57 Am^2$ is bent in the form of a semi circle. The new magnetic moment will be

1) $0.1 Am^2$ 2) $10 Am^2$ 3) $1 Am^2$ 4) $0.01 Am^2$

11. A bar magnet of magnetic moment $100 Am^2$ is bent at its mid point such that, the two parts make an angle of 60^0. The new magnetic moment is

1) $100\sqrt{2}Am^2$ 2) $\frac{100}{\sqrt{2}}Am^2$ 3) $50 Am^2$ 4) $50\sqrt{3}Am^2$

12. A magnet of magnetic moment 'M' is in the form of a quadrant of a circle. If it is straightend its new magnetic moment will be

1) $\frac{M\pi}{\sqrt{2}}$ 2) $\frac{2\sqrt{2}M}{\pi}$ 3) $\frac{\sqrt{2}M}{\pi}$ 4) $\frac{M\pi}{2\sqrt{2}}$

13. Three identical bar magnets each of magnetic moment M are arranged in the form of an equilateral triangle such that at two vertices like poles are in contact. The resultant magnetic moment will be

1) zero 2) 2M 3) $\sqrt{2}M$ 4) $\sqrt{3}M$

14. Find the resultant magnetic moment for the following arrangement (non coplanar vectors)

1) M
2) 2M
3) 3M
4) 4M

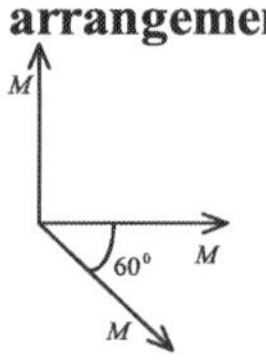

15. A bar magnet of moment 'M' is bent into a shape'**'. If the length of the each part is same, its new magnetic moment will be**

1) $\frac{M}{\sqrt{3}}$ 2) $\frac{M}{\sqrt{5}}$ 3) $\frac{M}{\sqrt{2}}$ 4) $\frac{2}{3}M$

16. If two identical bar magnets each of length 'ℓ' and poles strength 'm' and magnetic moment M are placed perpendicular to each other with their unlike poles in contact, the magnetic moment of combination is (2005E)

1) $\frac{M}{\sqrt{2}}$ 2) $\ell m\sqrt{2}$ 3) $2\ell m\sqrt{2}$ 4) 2M

17. Four magnets of equal magnetic moments M, 2M, 3M and 4M are arranged in the form of a square such that unlike poles are in contact. Then the resultant magnetic moment is

1) $2\sqrt{2}M$ 2) $\sqrt{2}M$ 3) 10M 4) 2M

18. If a bar magnet has magnetic moment M is divided into two equal parts. Half of the bar magnet is bent in the form of a semi circle (**) its new magnetic moment becomes**

1) $\frac{M}{\pi}$ 2) $\frac{M}{2\pi}(2+\pi)$ 3) $\frac{2M}{\pi}$ 4) $\frac{M}{\pi}(2+\pi)$

MAGNETIC INDUCTION

19. The magnetic induction field due to a magnetic south pole of strength 50Am at a point in air 10cm away from the pole in weber m^{-2} is.

1) 5×10^{-4} radially outward 2) 5×10^{-4} radially inward
3) 2.5×10^{-4} radially outward 4) 2.5×10^{-4} radially inward

20. Points A and B are situated along the extended axis of a 2cm long bar magnet at a distance 'x' and '2x' cm respectively from the pole nearer to the points. The ratio of magnetic fields at A and B will be

1) 4 : 1 exactly 2) 8 : 1 exactly 3) 4:1 approximately 4) 8 : 1approximately

21. **The magnetic moment of a short bar magnet is 4Am². The magnetic induction on its axial line at a distance of 20 cm is**
1) 10^{-4}T 2) 10^{-2}T 3) 10^{-3}T 4) 10^{-1}T

22. **The magnetic induction due to a short bar magnet at the point A on the axial line and at the point B on the equitorial line are equal. The ratio of distance of A and B from the centre of the magnet is**
1) 1 : 2 2) 2 :1 3) $2^{1/3}$: 1 4) 1: $2^{1/3}$

23. **The force experienced by a magnetic pole of strength 10A-m kept at a distance of 20cm from the centre of a short magnet of magnetic moment 10Am² on its axial line is**
1) 0.0025N 2) 0.25N 3) 0.005N 4) 0.05N

COUPLE

24. **A bar magnet is suspended in a uniform magnetic field in a position such that it experiences maximum torque. The angle through which it must be rotated from this position such that it experiences half of the maximum troque is**
1) 60° 2) 30° 3) 45° 4) 37°

25. **If the maximum couple acting on a magnet in a field of induction 0.2 tesla is 10 Nm, what is its magnetic moment ?**
1) 50 Am² 2) 2 Am² 3) 5 Am² 4) 20 Am²

26. **A torque of 2 x 10^{-4} Nm is required to hold a magnet at right angle to the magnetic meridian. The torque required to hold it at 30^0 to the magnetic meridian in N-m is**
1) 0.5 x 10^{-4} 2) 1 x 10^{-4} 3) 4 x 10^{-4} 4) 8 x 10^{-4}

27. **A bar magnet of length 10 cm experiences a torque of $\frac{\sqrt{2}}{10}$ N-m in a uniform magnetic field of induction 0.4 wb/m², when it is suspended making an angle 45^0 with the field, the pole strength of the magnet is**
1) 5 A-m 2) 2.5 A-m 3) 10 A-m 4) 15 Am

28. **When a bar magnet is placed at 90^0 to a uniform magnetic field, it is acted upon by a couple which is maximum. For the couple to be half of the maximum value, at what angle should the magnet be inclined to the magnetic field (B) ?**
1) 30^0 2) 60^0 3) 45^0 4) 0^0

29. **A bar magnet of pole strength 2 A-m is kept in a magnetic field of induction 4×10^{-5} wb m^{-2} such that the axis of the magnet makes an angle 30^0 with the direction of the field. The couple acting on the magnet is found 80×10^{-7}N-m. Then the distance between the poles of the magnet is**
1) 20 m 2) 2 m 3) 3 cm 4) 20 cm

30. **Two identical bar magnets are joined to form a cross. If this combination is suspended freely in a uniform field the angles made by the magnets with field direction are respectively**
1) 60°, 30° 2) 37°, 53° 3) 45°, 45° 4) 20°, 70°

31. **A magnet of magnetic moment $20\hat{k}$ Am² is placed along the z - axis in a magnetic field $\vec{B} = (0.4\hat{j} + 0.5\hat{k})$ T. The torque acting on the magnet is**
1) $8\hat{i}$ N-m 2) $6\hat{j}$ N-m 3) $-8\hat{i}$ N-m 4) $-6\hat{j}$ N-m

32. **The torque required to keep a magnet of length 10cm at 45^0 to a uniform magnetic field is $\sqrt{2}\times10^{-5}$Nm. The magnetic force on each pole is**
1) 0.2mN 2) 20 μN 3) 0.02N 4) 2N

33. **A magnetic needle of pole strength 'm' is pivoted at its centre. Its N-pole is pulled eastward by a string. Then the horizontal force required to produce a deflection of θ from magnetic meridian (B_H horizontal componet of earths magnetic field)**
1) $mB\cos\theta$ 2) $mB\sin\theta$ 3) $2mB\tan\theta$ 4) $mB\cot\theta$

WORK DONE

34. The work done in rotating a magnet of pole strength 1 A-m and length 1 cm through an angle of 60^0 from the magnetic meridian is (H=30 A/m)

1) 9.42×10^{-8} J 2) 3.14×10^{-8} J 3) 18.84×10^{-8} J 4) 10×10^{-8} J

35. The work done in turning a magnet normal to field direction from the direction of the field is 40×10^{-6} J. The kinetic energy attained by it when it reaches the field direction when released is

1) Zero 2) 30×10^{-6}J 3) 10×10^{-6} J 4) 40×10^{-6} J

36. A magnet is parallel to a uniform magnetic field. The work done in rotating the magnetic through $60^0$is 8×10^{-5} J. The work done in rotating through another 30^0 is

1) 4×10^{-5}J 2) 6×10^{-5}J 3) 8×10^{-5}J 4) 2×10^{-5}J

37. A bar magnet of moment 40 A-m^2 is free to rotate about a vertical axis passing through its centre. The magnet is released from rest from east west direction. The kinetic energy of the magnet as it takes north-south direction is ($B_H = 30\mu T$)

1) 0.6 mJ 2) 1.2 mJ 3) 2.4 mJ 4) 0.3 mJ

38. A bar magnet of magnetic moment M is divided into 'n' equal parts cutting parallel to length. Then one part is suspended in a uniform magnetic field of strength 2T and held making an angle 60^0 with the direction of the field. When the magnet is released the K.E of the magnet in the equilibrium position is

1) $\frac{M}{n}$J 2) Mn J 3) $\frac{M}{n^2}$J 4) Mn^2 J

NEUTRAL POINT

39. Two south poles of pole strengths 0.4 A-m and 6.4 A-m are separated by a distance of 10 cm. Find the distance of the neutral point between them from pole of greater pole strength.

1) 4 cm 2) 6 cm 3) 8 cm 4) 10cm

40. A short bar magnet of moment 3.2 A-m^2 lies in the magnetic meridian with north pole pointing north (B_H = 0.4 gauss). The distance between the two neutral points is

1) 20 cm along the axial line 2) 20 cm along the equatorial line
3) 40 cm along the axial line 4) 40 cm along the equatorial line

41. A bar magnet of magnetic moment 0.4 Am^2 is placed in the magnetic meridian with its north pole pointing north. A neutral point is obtained at a distance of 10 cm from the centre of the magnet. If the length of the magnet is also 10 cm what is the value of horizontal component of earth's field at the place ?

1) 2.86×10^{-5} T 2) 8×10^{-5} T 3) 4×10^{-5} T 4) 5.72 10^{-5} T

42. A very long magnet is placed vertically with one pole on the table. A neutral point is found at 20cm from the pole. If B_H = 0.4 gauss the pole strength of the pole is

1) 4 A-m 2) 16 A-m 3) 8 A-m 4) 2 A-m

TANGENT LAW AND DEFLECTION MAGNETOMETER(DMM)

43. A small magnet produces a deflection of 45^0 in Tan A position when used at 20 cm from the magnetic needle. If magnetic moment is 1.44 Am^2, then the value of 'B_H' is

1) $0.36 \times 10^{-4}T$ 2) $0.36 \times 10^{-5}T$ 3) $0.36 \times 10^{-6}T$ 4) $0.36 \times 10^{-7}T$

44. A short bar magnet produces a deflection of 37^0 in a deflection magnetometer in Tan A position, when placed at a distance of 10 cm from the needle. Where should the same short bar magnet be placed in Tan B position of the magnetometer to produce the same deflection 37^0 (Take $2^{1/3} = 1.26$)

1) 9.7cm 2) 6.5cm 3) 7.9 cm 4) 8.5cm

45. **A short magnet produced a deflection of 45⁰ when placed at some distance in Tan A position of the deflection magnetometer. If another magnet of same length and pole strength is kept at the same distance in Tan B position, the deflection produced is**

1) $26^0\ 34^1$ 2) 30^0 3) 45^0 4) 60^0

46. **A magnetometer is in Tan A position. A magnet produces a deflection of 30^0 in the needle. How many such similar magnets must be tied together such that like poles are touching each other to increase the deflection to 60^0 ?**

1) 3 2) 1 3) 4 4) 2

47. **Two short magnets are arranged in Tan A position at equal distances. If they produce deflections of 30^0 and 60^0, the ratio of their magnetic moment is**

1)3:1 2)1:2 3)1:3 4)1:1

48. **A magnet shows a deflection of 45^0 when it is arranged properly in Tan A position. Magnet lies on eastern arm with its north pole towards magnetic needle.If magnet is rotated by 90^0, in clock wise direction,the new deflection will be**

1) 0^0 2) 180^0 3) 45^0 4) Needle can stay in any position

49. **The needle of a deflection magnetometer shows a deflection of 60^0 , due to a short bar magnet at a certain distance in tan A position.If the disatnce is doubled, the deflection is**

1) $\tan^{-1}\left(\frac{3}{8}\right)$ 2) $\tan^{-1}\left(\frac{\sqrt{3}}{8}\right)$ 3) $\tan^{-1}\left(\frac{8}{\sqrt{3}}\right)$ 4) $\tan^{-1}\left(\frac{8}{3}\right)$

50. **A small magnet produces a deflection of 25^0 in Tan B position.The deflection, if a small magnet of magnetic moment 4M be placed at double distance in TanA position is**

1) 50^0 2)25^0 3)35^0 4)53^0

51. **A small magnet of moment M_1 produces a deflection of 30^0 when placed at some distance from the needle of a deflection magnetometer in Tan A position.Another small magnet of double the length and thrice the pole strength is kept at same distance in Tan B position. The deflecton produced is**

1) 30^0 2)40^0 3)60^0 4)45^0

52. **Two short bar magnets of magnetic moments M_1 & M_2 are arranged one over the other with similar poles at same end and produces a deflection 60^0 at certain distance in tan A position. If same magnets are placed in tan B position at same distance but dissimilar poles are at same end produces a deflection 30^0.The value of M_1/M_2 is**

1)1:1 2)1:5 3)3:5 4)5:1

53. **A short bar magnet when placed at a distance of 20cm from the magnetic needle of a deflection magnetometer in Tan A position produces a deflection of 60^0 .If the magnet is cut into three equal parts and one of them is placed at the same distance the deflection will be**

1) 20^0 2) 60^0 3) 30^0 4) 45^0

VIBRATION MAGNETO METER (VMM)

54. **The period of oscillation of a suspended thin cylindrical magnet is 2 seconds. It is broken into exactly three equal parts perpendicular to its length and three parts are then placed on each other with their like poles together, the time period of combination is**

1) 2S 2) $\frac{2}{3}S$ 3) $2\sqrt{3}S$ 4) $\frac{2}{\sqrt{3}}S$

55. **Two bar magnets are bound together side by side and suspended. They swing in 12s when their like poles are together and in 16s when their unlike poles are together, the magnetic moments of these magnets are in the ratio**

1)27:5 2)25:7 3)7:25 4)24:7

56. **A bar magnet has moment of inertia $49x10^{-2}kgm^2$ vibrates in a magnetic field of induction $0.5x10^{-4}$ Tesla. The time period of vibration is 8.8sec . The magnetic moment of the bar magnet is**

1) $350Am^2$ 2) $490Am^2$ 3) $490Am^2$ 4)5000Am^2

57. **A thin rod 30 cm long is uniformly magnetised and its period of oscillation is 4 s. It is broken into three equal parts normal to its length. The period of oscillation of each part is**

1)12 s 2)6 s 3)1.33 s 4)2.66 s

58. **A short bar magnet is oscillating in a magnetic field and its time period is 2 seconds . If another piece of brass of double moment of inertia be placed over that magnet the time period of that combination in that field is**

1) $2\sqrt{3}$ S 2) $2\sqrt{2}$ S 3) 2 S 4) $1/\sqrt{2}$ S

59. **The moment of inertia as well as the frequencies of two magnets are in the ratio 3:4 the ratio of their magnetic moments is**

1) 27 : 64 2) 64 : 27 3) 9 : 16 4) 16 : 9

60. **A magnet freely suspended in a vibration magnetometer makes 40 oscillations per minute at a place A and 20 oscillations per min at a place B. If the horizontal component of earth's magnetic field at A is $36\times10^{-6}T$, then its value at 'B' is (2001E)**

1) $36\times10^{-6}T$ 2) $9\times10^{-6}T$ 3) $144\times10^{-6}T$ 4) $288\times10^{-6}T$

61. **A vibration magnetometer consists of two identical bar magnets placed one above the other, such that they are mutually perpendicular and bisect each other. The time period of oscillation in a horizontal magnetic field is 4 seconds. If one of the magnets is removed the time period of the other in the same field ($2^{1/4}$=1.189)**

1)1.34sec 2) 2.34sec 3) 3.36sec 4) 4.34sec

62. **A bar magnet suspended freely in uniform magnetic field is vibrating with a time period of 3 seconds.If the initial field strength is 2T, then the final field strength, for which time period becomes 4 seconds is**

1) 1.125Tesla 2) 0.625Tesla 3) 3.55 Tesla 4) 0.75 Tesla

63. **A short bar magnet of magnetic moment 2Am² and moment of inertia 6x10²kgm² is freely suspended such that the magnetic axial line is in the direction of magnetic meridian. If the magnet is displaced by a very small angle (3^0), the angular acceleration is — x10^{-6}rad/sec² (Magnetic induction of earth's horizontal field = 4x 10^{-4}T).**

1) $\pi/20$ 2) $\pi/45$ 3) $\pi/60$ 4) $\pi/75$

64. **The period of oscillation of a magnet at a place is 4 seconds. When it is remagnetised, so that the pole strength becomes 1/9th of initial value, the period of oscillation in seconds is**

1) 3 2) 12 3) 5 4) 4

65. **A magnetic needle pivoted through its centre of mass and is free to rotate in a plane containing uniform magnetic field 200 x 10^{-4}T. When it is displaced slightly from the equilibrium it makes 2 oscillations per second. If the moment of inertia of the needle about the axis of oscillation is 0.75 x 10^{-5} kg m², the magnetic moment of the needle is**

1) 0.06 J/T 2) 0.03 J/T 3) 0.12 J/T 4) 0.6 J/T

PROPERTIES OF MAGNETIC MATERIALS

66. The magnetic susceptibility of a medium is 0.825. Its relative permeability is

1) 1.825 2) 825 3) 285 4) 1825

67. A magnetic field strength (H) $3x10^3\ A/m$ produces a magnetic field of Induction (B) of 12π telsa in an iron rod. The relative permeability of iron is

1) 10^5 2) 10^4 3) 10^3 4) 10^2

68. The magnetic moment of magnet of mass 75 gm is $9\times10^{-7}\ Am^2$. If the density of the material of magnet is $7.5\times10^3\ \frac{kg}{m^3}$ then intensity of magnetisation will be

1) $0.9\ A/m$ 2) $0.09\ A/m$ 3) $9\ A/m$ 4) $90\ A/m$

69. A magnetising field of 5000 A/m produces a magnetic flux of 5×10^{-5} weber in an iron rod. If the area of cross section of the rod is 0.5 cm^2, then the permeability of the rod will be

1) 1×10^{-3} 2) 2×10^{-4} 3) 3×10^{-5} 4) 4×10^{-6}

70. A short bar magnet of magnetic moment $20Am^2$ has a cross sectional area of $1.5\times10^{-4}m^2$. If the intensity of magnetisation of the magnet is 10^5A/m. The length of magnet is

1) 0.33m 2) 0.13cm 3) 1.33m 4) 1.33cm

ANSWERS

1) 3	2) 4	3) 2	4) 4	5) 4	6) 3	7) 1	8) 3	9) 3	10) 3
11) 3	12) 4	13) 2	14) 2	15) 2	16) 2	17) 1	18) 2	19) 2	20) 4
21) 1	22) 3	23) 1	24) 1	25) 1	26) 2	27) 1	28) 1	29) 4	30) 3
31) 3	32) 1	33) 3	34) 3	35) 4	36) 3	37) 2	38) 1	39) 3	40) 4
41) 1	42) 2	43) 1	44) 3	45) 1	46) 1	47) 3	48) 1	49) 2	50) 2
51) 3	52) 4	53) 3	54) 2	55) 2	56) 4	57) 3	58) 1	59) 1	60) 2
61) 3	62) 1	63) 2	64) 2	65) 1	66) 1	67) 2	68) 2	69) 2	70) 3

LEVEL -III

MAGNETIC MOMENT

1. Find the resultant magnetic moment for the following arrangement

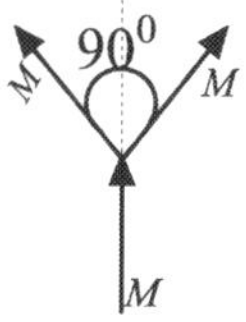

1) $\sqrt{2}M$

2) $\left(\sqrt{2}+1\right)M$

3) $\left(\sqrt{2}-1\right)M$

4) M

2. A bar magnet of magnetic moment 'M_1' is axially cut into two equal parts. If these two parts are arranged perpendicular to each other, the resultant magnetic moment is M_2. Then the value of M_1/M_2 is

1) $\frac{1}{2\sqrt{2}}$ 2) 1 3) $\frac{1}{\sqrt{2}}$ 4) $\sqrt{2}$

3. The magnetised wire of moment 'M' and length 'ℓ' is bent in the form of semicircle of radius 'r'. Then its magnetic moment is

1) $\frac{2M}{\pi}$ 2) 2M 3) $\frac{M}{\pi}$ 4) zero

4. A magnet of length L and moment M is cut into two halves (A and B) perpendicular to its axis. One piece A is bent into a semicircle of radius R and is joined to the other piece at the poles as shown in the figure below.

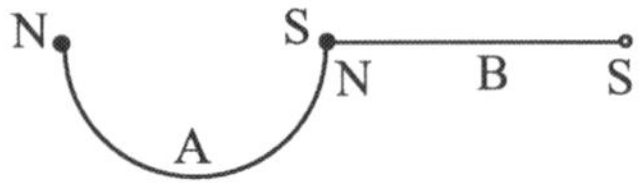

Assuming that the magnet is in the form of a thin wire initially, the moment of the resulting magnet is given by

1) $\frac{M}{2\pi}$ 2) $\frac{M}{\pi}$ 3) $\frac{M(2+\pi)}{2\pi}$ 4) $\frac{M\pi}{2+\pi}$

COULOMB FORCE

5. The force between two magnetic poles is F in air. If the pole strength of first pole is doubled, that of second pole is tripled and the distance between the poles is halved the force between the poles becomes and increases by

1) 24F and 2300% 2) 24F and 2400% 3) $\frac{3}{2}F$ and 150% 4) 2F and 100%

6. The force between two poles is reduced to P newton, when their original separation is increased to x times, it is increased to Q newton, when their separation is made 1/xth of their original value. Then the relation between P and Q is

1) Q = XP 2) $Q = X^2P$ 3) $Q = X^3P$ 4) $Q = X^4P$

7. Two magnetic poles when placed in air separated by a distance 2m experience a force of repulsion of 10^{-6}N. If the difference of their pole strengths is 3Am their pole strengths in Am are

1) 5, 2 2) 7, 4 3) 6, 3 4) 8, 5

8. An isolated north pole of mass 4gm and pole strength 20 A-m is kept in a uniform magnetic field of induction $2\times10^{-3}T$. The acceleration acquired by the pole is

1) $2ms^{-2}$ 2) $4ms^{-2}$ 3) $5ms^{-2}$ 4) $10ms^{-2}$

9. Four like poles each of strength 'm' are kept at the four corners of a square of side 'd'. The net magnetic force on the pole at any one corner is

1) zero 2) $\frac{\mu_0}{4\pi}\frac{m^2}{d^2}$ 3) $\left[\frac{2\sqrt{2}+1}{2}\right]\frac{\mu_0}{4\pi}\frac{m^2}{d^2}$ 4) $\left[\frac{2\sqrt{2}-1}{2}\right]\frac{\mu_0}{4\pi}\frac{m^2}{d^2}$

10. The magnetic induction at a distance 'd' from the magnetic pole of unknown strength 'm' is B. If an identical pole is now placed at a distance of 2d from the first pole, the force between the two poles is (2006M)

1) mB 2) **mB/2** 3) mB/4 4) 2mB

11. The force between two short magnets is 'F'. When the pole strengths of the magnets are tripled and the distance between them is also doubled, the force between them is (Take magnets to be parallel to each other)

1) 0.5625F 2) 7.56F 3) 0.375F 4) 0.75F

COUPLE

12. Two magnets of moments M_1 and M_2 are rigidly fixed together at their centres so that their axes are inclined to each other. This system is suspended in a magnetic field of induction 'B' so that M_1 makes an angles θ_1 and M_2 makes an angles θ_2 with the field direction and unlike poles on either side of the field direction. The resultant torque on the rigid system is

    1) $B(M_1 \sin\theta_1 + M_2 \sin\theta_2)$ 2) $B(M_1 \cos\theta_1 + M_2 \cos\theta_2)$

    3) $B(M_1 \sin\theta_2 + M_2 \sin\theta_1)$ 4) $B(M_1 \cos\theta_2 + M_2 \cos\theta_1)$

13. A short magnet placed with its axis making an angle with a uniform external magnetic field of induction B experiences a torque (τ). If the magnet is free to rotate, which orientation would correspond to its stable and unstable equilibrium.

    1) $\theta = 0^0$, $\theta = 90^0$ 2) $\theta = 0^0$, $\theta = 180^0$

    3) $\theta = 45^0$, $\theta = 135^0$ 4) $\theta = 0^0$, $\theta = 270^0$

14. A magnet is suspended in the magnetic meridian with an untwisted wire. The upper end of the wire is rotated through 180^0 to deflect the magnet by 30^0 from magnetic meridian. Now this magnet is replaced by another magnet. Now the upper end of the wire is rotated through 270^0 to deflect the magnet 30^0 from the magnetic meridian. The ratio of the magnetic moments of the two magnets is

    1) 3 : 4 2) 1 : 2 3) 4 : 7 4) 5 : 8

15. A magnet is suspended in a uniform magnetic field by a thin wire. On twisting the wire through half revolution, the magnet twists through 30^0 from the original position. How much should we rotate the wire in order to twist the magnet through 45^0 from its original position

    1) 257^0 2) 252^0 3) 275^0 4) 127^0

16. A magnetic dipole is under the influence of two magnetic fields. The angle between the field directions is 60^0 and one of the fields has a magnitude of 1.2×10^{-2}T. If the dipole comes to stable equilibrium at angle of 15^0 with this field, then the magnitude of the other field ($\sin 15^0 = 0.2588$)

    1) 1.39×10^{-3}T 2) 2.39×10^{-3}T 3) 3.39×10^{-3}T 4) 4.39×10^{-3}T

17. Two small magnets X and Y of dipole moments M_1 and M_2 are fixed perpendicular to each other with their north poles in contact. This arrangement is placed on a floating body so as to move freeely in earth's magnetic field as shown in figure then the ratio of magnetic moment is

    1) $1 : \sqrt{3}$

    2) $2 : \sqrt{3}$

    3) $\sqrt{3} : 2$

    4) $\sqrt{3} : 1$

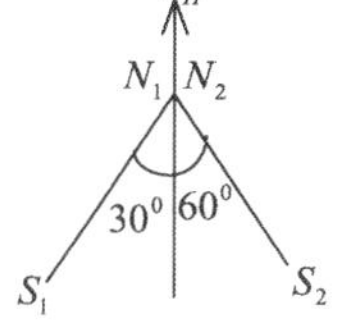

MAGNETIC INDUCTION

18. The poles of a horse-shoe magnet each of pole strength 2A-m are at 4cm apart. The magnetic induction at the mid point between the poles is

    1) $0.25 \times 10^{-3}T$ 2) $0.5 \times 10^{-3}T$ 3) $0.125 \times 10^{-3}T$ 4) $10^{-3}T$

19. Two unlike magnetic poles each of pole strength 10 A-m are at 20 cm apart. The resultant magnetic induction at the mid point on the line joining them is

    1) 1.54×10^{-5} T 2) 2×10^{-4} T 3) 15×10^{-4} T 4) 15×10^{-6} T

20. The length of a magnet of moment $5Am^2$ is 14cm. The magnetic induction at a point, equidistant from both the poles is 3.2×10^{-5} wb/m². The distance of the point from each pole is

1) 5cm 2) 10cm 3) 25cm 4) 15cm

21. A magnet of length 10cm and magnetic moment $1Am^2$ is placed along the side AB of an equilateral triangle ABC. If the length of the side AB is 10cm, the magnetic induction at the point C is

1) 10^{-9}T 2) 10^{-7}T 3) 10^{-5}T 4) 10^{-4}T

22. Two identical magnetic poles each of strength 'm' are placed at the vertices of an equilateral triangle of side 'd' then resultant magnetic induction at the third vertex is

1) $\frac{\mu_0}{4\pi}\frac{m}{d^2}$ 2) $\sqrt{2}\frac{\mu_0}{4\pi}\frac{m}{d^2}$ 3) $\sqrt{3}\frac{\mu_0}{4\pi}\frac{m}{d^2}$ 4) zero

23. The pole strength of a 12cm long bar magnet is 20Am. The magnetic induction at a point 10cm away from the centre of the magnet on its axial line is

1) 1.1×10^{-3}T 2) 2.2×10^{-3}T 3) 1.1×10^{-2}T 4) 2.2×10^{-2}T

24. Two short bar magnet of magnetic moment M each are arranged at the opposite corners of a square of side 'd' such that their centres coincide with the corners and their axes are parallel. If the like poles are in same direction the magnetic induction at any of the other corner of the square is

1) $\frac{\mu_0}{4\pi}\frac{M}{d^3}$ 2) $\frac{\mu_0}{4\pi}\frac{2M}{d^3}$ 3) zero 4) $\frac{\mu_0}{4\pi}\frac{3M}{d^3}$

25. The magnetic induction at a point on the axial line of a short magnet is 10 Tesla east. The magnetic induction at the same point when the magnet is rotated in the anticlock wise direction by 90° and 180° are

1) 10T south and 5T west 2) 5T south and 10T west

3) 20T south and 10T west 4) 10T south and 20T west

26. Two short magnets each of moment M are placed one over the other at right angles. The resultant magnetic induction at a point 'P', that lies at a distance 'd' from the common centre is

1) $\frac{\mu_0}{4\pi}\frac{3M}{d^3}$ 2) $\sqrt{3}\frac{\mu_0}{4\pi}\frac{3M}{d^3}$

3) $\sqrt{5}\frac{\mu_0}{4\pi}\frac{M}{d^3}$ 4) $\frac{\mu_0}{4\pi}\frac{M}{d^3}$

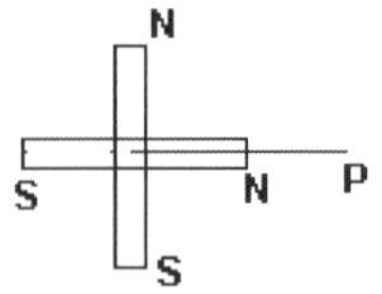

27. Two short magnets of magnetic moment $2Am^2$ and $5Am^2$ are placed along two lines drawn at right angle to each other on the sheet of paper as shown in the figure. What is the magnetic field at the point of intersection of their axes

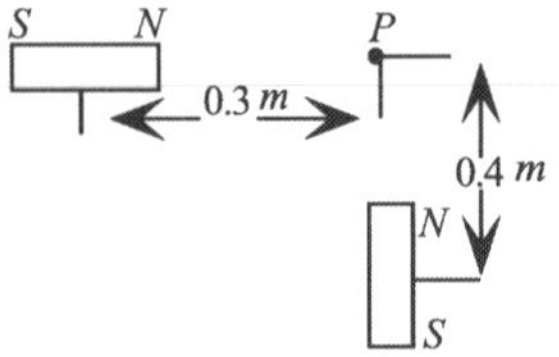

1) 2.15×10^{-5}T 2) 215×10^{-5}T 3) 2.15×10^{-3}T 4) 21.5×10^{-5}T

28. **Two short bar magnets P and Q are arranged such that their centres are on the X-axis and are separated by a large distance. The magnetic axes of P and Q are along X, Y-axis respectively. R lies midway between their centers. If B is the magnitude of induction due to Q at R, then magnitude of total induction at R due to both the magnets is**

1) 3B 2) $\sqrt{5}B$ 3) $\frac{\sqrt{5}}{2}B$ 4) B

29. **Two short magents AB and CD are in the X – Y plane and are parallel to x – axis and the coordinates of their centers respectively are (0, 2) and (2, 0). Line joining the North - South poles of CD is opposite to that of AB and lies along the positive X - axis. The resultant field induction due to AB and CD at a point P(2, 2) is 100×10^{-7}T. When the poles of the magnet CD are reversed the resultant field induction is 50 $\times10^{-7}$T. The values of magnetic moments of AB and CD (in Am^2) are**

1) 300, 200 2) 600, 400 3) 200, 100 4) 300, 150

30. **The ratio of magnetic inductions on the axial line of a bar magnet at distances 10cm and 15cm from the centre of the magnet are in the ratio 5.8 :1. The distance between the two poles of the magnet is**

1) 24cm 2) 12cm 3) 2.4cm 4) 1.2cm

NEUTRAL POINT

31. **Two short magnets of magnetic moments 0.8A-m^2 and 2.7A-m^2 are placed along the same straight line with their like poles towards each other. The distance between their centres is 40 cm. The distance of the zero induction point on their common axial line from the centre of the weaker magnet is**

1) 16cm between the magnets and 80cm out side the magnets
2) 24cm between the magnets and 120cm out side the magnets
3) 20cm between the magnets 60cm outside the magnets
4) zero induction is not possible

32. **A short bar magnet is kept along the magnetic meridian with its north pole pointing north. A neutral point is located at distance 'd' from the centre. The ratio of magnetic inductions at a distance d/2 and 2d from the centre of the magnet on the equitorial line is**

1) 1 : 4 2) 4 :1 3) 8 : 1 4) 1 : 8

33. **A bar magnet of 10 cm long is kept with its N - pole pointing north. A neutral point is formed at a distance of 15 cm from each pole. Given the horizontal component of earth's field is 0.4 Gauss, the pole strength of the magnet is**

1) 9 amps – m 2) 6.75 amps – m 3) 27 amps – m 4) 13.5 amps – m

WORK DONE

34. **If 'W' Joule is the work done in rotating the magnet from the field direction to a position that makes an angle 60^0 with the direction of the field, the torque required to hold the magnet in that direction is**

1) W N-m 2) $\frac{W}{2}N-m$ 3) $\frac{W}{\sqrt{3}}N-m$ 4) $\sqrt{3}WN-m$

35. **The work done in rotating the magnet from the direction of uniform magnetic field to the opposite direction to the field is W. The work done in rotating the magnet from the field direction to half the maximum couple position is**

1) 2W 2) $\frac{\sqrt{3}W}{2}$ 3) $\frac{W}{4}\left(2-\sqrt{3}\right)$ 4) $\frac{W}{2}\left(1-\sqrt{3}\right)$

36. A bar magnet of magnetic moment $2Am^2$ is free to rotate about a vertical axis passing through its centre. The magnet is released from rest from east - west position. Then the K.E of the magnet as it takes north - south position is ($B_H = 25\,\mu T$)

1) $25\mu J$ 2) $50\mu J$ 3) $100\mu J$ 4) $12.5\mu J$

37. A short bar magnet placed with its axis at 30^0 with a uniform external magnetic field of 0.16T experiences a torque of magnitude 0.032Nm. If the bar magnet were free to rotate, its P.E when it is in stable and unstable equilibrium are respectively.

1) –0.064J, +0.064J 2) –0.032J, +0.032J
3) +0.064J, –0.064J 4) +0.032J , -0.032J

TANGENT LAW AND DMM

38. A short magnet produces a deflection of 30° when placed at a certain distance in tan A position of magnetometer. If another short magnet of double the length and thrice the pole strength is placed at the same distance in tan B position,the deflection produced is

1) 45° 2) 90° 3) 60° 4) 30°

39. A short magnet placed broad side to a compass needle with its center at 0.15 m from the needle deflects through 45^0. By how far the magnet should be moved towards the needle so that the deflection increases by 15^0($\left(\frac{1}{\sqrt{3}}\right)^{1/3} = 0.83$)

1)6cm 2)3cm 3)2.5cm 4)4cm

40. A short magnet produces a neutral point at end on position at 15cm.The deflection produced by the magnet in tan B position of deflection magnetometer when used at the same distance is (take 2 tan 27°=1)

1) 45° 2) 30° 3) 27° 4) 54°

41. A deflection magnetometer is placed with its arms in north south direction.How and where should a short magnet having $\frac{M}{B_H} = 40\frac{Am^2}{T}$ be placed so that the needle can stay in any position.

1)2cm from the needle ; north pole pointing north 2)2cm from the needle ; north pole pointing south
3)4cm from the needle ; north pole pointing south 4)8cm from the needle ; north pole pointing north

42. A very long magnet is kept standing on its south pole at a distance of 10 cm from the center of the magnetic needle in the D.M.M, arranged in Tan A position.The deflection is 60^0.The deflection in the needle, if D.M.M is arranged in Tan B position and magnet is kept standing on its N-pole at a distance 10 cm from center of magnetic needle is

1)0^0 2)30^0 3)180^0 4)0 or180^0

43. A short magnet produces a deflection of 37^0 in a DMM in tanB position when placed at a separation of 10cm from the needle. The ratio of the magnetic moment of the magnet to the earth's horizontal magnetic field is ---$x10^3 Am^2/T$($Tan37^0=3/4$)

1) 7.5 2) 3.6 3) 4.75 4) 3.75

44. Using the same bar magnet at the same distance from the magnetic needle at two places A and B deflections in the deflection magnetometer were found to be 30^0 and 60^0.The ratio of the horizontal components of earth magnetic field induction at these two places will be

1)1:3 2)3:1 3)2:1 4)1:2

45. **A small bar magnet produces a deflection of 45^0 in Tan A position . What will be the deflection when the arms are rotated through 90^0 in anti clockwise direction**

1) 45^0 2) zero 3) 60^0 4) zero or 180^0

46. **A small bar magnet produces a deflection of 45^0 in Tan A position. The north pole of magnet is towards magnetic needle. What will be the deflection when the arms are rotated through 30^0 in anti clock wise direction**

1) 45^0 2) zero 3) 0^0 or 180^0 4) 30^0 or 60^0

47. **A deflection magnetiometer is adjusted in the usual way. When a magnet is introduced, the deflection observed is θ, and the period of oscilla-tion of the needle in the magnetometer is T. When the magnet is removed the period of oscillation is T_0. The relation between T and T_0 is**

1) $T^2 = T_0^2 \cos\theta$ 2) $T = T_0 \cos\theta$ 3) $T = \frac{T_0}{\cos\theta}$ 4) $T^2 = \frac{T_0^2}{\cos\theta}$

48. **A short bar magnet placed at a certain distance from a deflection magnetometer in tan A position produces a deflecftion of 60^0. The magnet is now cut into three equal pieces. If one piece is kept at the same distance in 'tan A' position from the deflection magnetometer, then the deflection produced is (Eam(M) 2008)**

1) 10^0 2) 20^0 3) 30^0 4) 60^0

VIBRATION MAGNETOMETER (VMM)

49. **A bar magnet takes $\pi/10$ seconds to complete one oscillation in vibration magnetometer. The moment of inertia of the magnet about the axis of rotation is $1.2x10^{-4} kgm^2$ and the earth's horizontal magnetic field is 30μ T. The magnetic moment of the magnet is**

1) $1200 Am^2$ 2) $800 Am^2$ 3) $1600 Am^2$ 4) $1800 Am^2$

50. **A small magnet of moment $4.8x10^{-2}$ J/T is suspended freely in the plane of a uniform magnetic field of magnitude $3x10^{-2}$ T. If the magnet is slightly displaced from its stable equilibrium and released then the angular frequency of its oscillations in rad/sec. (Moment of inertia of the magnet about the axis of rotation is $2.25x10^{-5} kgm^2$)**

1) 8 2) 4 3) 3 4) 2

51. **A short bar magnet used in a vibration magnetometer is heated so as to reduce its magnetic moment by 36% , then the frequency of magnet is (Neglecting the Changes in the dimensions of the magnet) (2004M)**

1) Increase by 36% 2) Decreases by 36% 3) Increase by 20% 4) Decreases by 20%

52. **A compass needle makes 10 oscillations per minute in the earth's field. It makes 20 oscillations per minute when the field of a bar magnet supports the earth's field. The frequency of oscillations of the needle when the field of the bar magnet oppose the earth's field is**

1) $15\ 0scillations/minute$ 2) $10\sqrt{2}\ 0scillations/minute$

3) $18\ 0scillations/minute$ 4) Zero

53. **The magnetic needle of a vibration magnetometer makes 12 oscillations per minute in the horizontal component of earth's magnetic field. When an external short bar magnet is placed at some distance along the axis of the needle in the same line it makes 15 oscillations per minute. If the poles of the bar magnet are inter changed , the number of oscillations it takes per minute is (2004E)**

1) $\sqrt{61}$ 2) $\sqrt{63}$ 3) $\sqrt{65}$ 4) $\sqrt{67}$

54. A vibration magnetometer consists of two identical bar magnets placed one over the other such that they are perpendicular and bisect each other.The time period of oscillation in a horizontal magnetic field is $2^{5/4}$ seconds.One of the magnets is removed and if the other magnet oscillates in the same field, then the time period in seconds is (2003E)

1) $2^{1/4}$ 2) $2^{1/2}$ 3) 2 4) $2^{5/4}$

55. The magnetic needle of a V.M.M completes 10 oscillations in 92seconds. When a small magnet is placed in the magnetic meridian 10cm due north of needle with north pole towards south completes 15 oscillations in 69seconds. The magnetic moment of magnet $(B_H = 0.3G)$ is

1) $4.5Am^2$ 2) $0.45Am^2$ 3) $0.75\ Am^2$ 4) $0.225Am^2$

56. The frequency of oscillation of a magnet at a place is 8 Hz. When it is remagnetised so that the pole strength becomes 6 times the initial value and placed at a place where horizontal magnetic field is twice that of initial place, the new frequency of oscillation is

1) 16Hz 2) $4\sqrt{3}$ Hz 3) $16\sqrt{3}$ Hz 4) 4Hz

57. A magnetic needle has a frequency of 20 oscillations per minute in the earth's horizontal field. when the field of a magnet supports the earths horizontal field, the frequency increases to 30 oscillations per minute. The ratio of the field of the magnet to that of the earth is

1)4:7 2)7:4 3)5:4 4)4:5

58. Two long bar magnet of magnetic moment $6Am^2$ and $8Am^2$ can be fastened together with their axes horizontal and their centres in a vertical line about which the system is free to oscillate.Compare the time periods in the earth's field when the axes of the magnets are (i) in the same direction.(ii) in opposite direction(iii) at right angle

1)1:7:5 2)$1:\sqrt{7}:\sqrt{5}$ 3)$1:\sqrt{7}:\sqrt{7/5}$ 4)$1:\sqrt{7/5}:\sqrt{14/5}$

59. A compass needle makes 10 oscillations per minute in the earths horizontal field.A bar magnet deflects the needle by 60^0 from the magnetic meridian. The frequency of oscillation in the deflected position in oscillations per minute is (field due to magnet is perpendicular to B_H)

1) $5\sqrt{2}$ 2) $20\sqrt{2}$ 3) $10\sqrt{2}$ 4) 10

60. A magnet makes 12 oscillation per minute at a place where horizontal component of earth's field is $6.4x10^{-3}T$. It is found to require 8 seconds per oscillation at another place X. The vertical component of earths field at X where resultant field makes angle 60^0 with horizontal is ---$x10^{-4}T$

1) $\frac{25}{\sqrt{3}}$ 2) $\sqrt{3}$ 3) $25\sqrt{3}$ 4) 25

61. A short magnet oscillates in vibration magnetometer with a frequency 10Hz where horizontal component of earth's magnetic field is $12\mu T$. A downward current of 15A is established in the vertical wire placed 20cm west of the magnet.New frequency is

1)4Hz. 2)2.5Hz. 3)9Hz. 4) 5Hz.

62. In an experiment with vibration magnetometer the value of $\frac{4\pi^2 I}{T^2}$ for a short bar magnet is observed as $36x10^{-4}$. In the experiment with deflection magnetometer with the same magnet, the value of $\left(\frac{4\pi d^3}{2\mu_o}\right)$ tan θ is observed as $\frac{10^8}{36}$. The magnetic moment of the magnet used is (2002M)

1) $50A-m^2$ 2) $100A-m^2$ 3) $200A-m^2$ 4) $1000A-m^2$

63. Two bar magnets are placed in vibration magnetometer and allowed to vibrate. They make 20 oscillations per minute when their similar poles are on the same side, while they make 15 osillations per minute when their opposite poles lie on the same side. The ratio of their magnetic moments is

1) 7 : 25 2) 25 : 7 3) 25 : 16 4) 16 : 25

64. With a standard rectangular bar magnet the time period of a vibration magnetometer is 4 seconds. The bar magnet is cut parallel to its length into four equal pieces. The time period of vibration magnetometer when one piece is used (in seconds) (bar magent breadth is small.) is

1) 16 2) 8 3) 4 4) 2

PROPERTIES OF MAGNETIC MATERIALS AND EARTH MAGNETIC FIELD :-

65. A rod of ferromegnetic material with dimensions 10cm x 0.5cm x 2cm is placed in a magnetising field of intensity $2x10^5$A/m. The magnetic moment produced due it is 6 $amp-m^2$. The value of magnetic induction will be-----10^{-2}T.

1)100.48 2)200.28 3)50.24 4)300.48

66. A magnetic material of volume $30cm^3$ is placed in a magnetic field of intensity 5 oersted.The magnetic moment produced due it is 6 $amp-m^2$.The value of magnetic induction will be.

1)0.2517 Tesla 2)0.025 Tesla 3)0.0025 Tesla 4)25 Tesla

67. The total magnetic flux in a material, which produces a pole of strength m_p when a magnetic material of cross- sectional area A is placed in a magnetic field of strength H, will be

1) $\mu_0\left(AH+m_p\right)$ 2) $\mu_0 AH$ 3) $\mu_0 m_p$ 4) $\mu_0\left(m_p AH+A\right)$

68. Relative permittivity and permeability of a material are ε_r and μ_r respectively. Which of the following values of these quantities are allowed for a diamagnetic material ? (AIEEE 2008)

1) $\varepsilon_r=0.5, \mu_r=1.5$ 2) $\varepsilon_r=1.5, \mu_r=0.5$

3) $\varepsilon_r=0.5, \mu_r=0.5$ 4) $\varepsilon_r=1.5, \mu_r=1.5$

ANSWERS

1) 2	2) 4	3) 1	4) 3	5) 1	6) 4	7) 4	8) 4	9) 3	10) 3
11) 1	12) 1	13) 2	14) 4	15) 1	16) 4	17) 4	18) 4	19) 2	20) 3
21) 4	22) 3	23) 1	24) 1	25) 2	26) 3	27) 1	28) 2	29) 1	30) 2
31) 1	32) 3	33) 4	34) 4	35) 3	36) 2	37) 1	38) 3	39) 3	40) 3
41) 2	42) 4	43) 1	44) 2	45) 2	46) 4	47) 1	48) 3	49) 3	50) 1
51) 4	52) 2	53) 2	54) 3	55) 3	56) 3	57) 3	58) 3	59) 3	60) 3
61) 4	62) 2	63) 2	64) 3	65) 1	66) 1	67) 1	68) 2		

COMPREHENSION QUESTIONS

Comprehension-1 :

A short bar magnet of moment 4 Am^2 is placed in the magnetic meredian with its south-pole pointing geographic North. The distance between two null points is found to be 20 cm.

1. **The value of horizontal component of the earths' magnetic field at the place is**

   1) 4G　　2) 2G　　3) 6T　　4) 8G

2. **In the above problem, if the magnet is replaced by another one of double the length and double the pole strength, the value of the horizontal component of the earths' magnetic field at the place is :**

   1) 6G　　2) 8G　　3) 32G　　4) 4G

3. **If the magnet is turned through 90^0 in the above problem, the resultant field at the location of the null point is :**

   1) 8G　　2) $8\sqrt{5}$ G　　3) $4\sqrt{5}$ G　　4) $8\sqrt{2}$ G

Comprehension - 2 :

Deflection magnetometer (DMM) is an instrument that can be used to compare magnetic moments & to find the magnetic moment of the unknown magnet and also to verify inverse square law. The experiments are conducted keeping the instrument either in Tan A (or) Tan B positions.

4. **While comparing the magnetic moments, the magnet is arranged along East-West on the arms in Tan A position. In TanB position the magnet held along**

   1) North-South　　2) East-West

   3) Along North-East　　4) Along the Sourth - West

5. **The accuracy of measurement of magnetic moment with DMM maximum if the deflection is**

   1) 30^0　　2) 60^0　　3) 90^0　　4) 45^0

6. **The best combination preferred for comparing the magnetic moments with DMM**

   1) Equal distance method, tan A position　　2) Null deflection method, tan B position

   3) Null deflection method tan A position　　4) Equal distance method, tan B position

Comprehension - 3 :

When a magnet is suspended in uniform magnetic field it experiences only torque. Hence the magnet makes pure rotatory motion. However, if the magnet is suspended in a non-uniform magnetic field, it experiences both a torque and force. Hence it makes both rotatory and translatory motion.

7. **The torque acting on a bar magnet suspended in a uniform magnetic field is maximum if the angle made by the magnet with the field is**

   1) 30^0　　2) 45^0　　3) 60^0　　4) 90^0

8. **A magnet is in equillibrium position in an uniform magnetic field, the angle through which it should be turned such that the couple acting on it is half of the maximum couple is**

   1) 30^0　　2) 60^0　　3) 45^0　　4) 22^0

9. **A magnet is in external uniform magnetic field such that the couple acting on it is maximum. The angle through which it should be turned such that the couple acting on it is half of the maximum.**

   1) 30^0　　2) 45^0　　3) 22^0　　4) 60^0

Comprehension - 4 :

The magnetic length is the distance from pole to pole of a magnet. Pole strength represents the degree of concentration of magnetism at the poles. The product of magnetic length and pole strength is termed as magnetic moment.

10. **A bar magnet moment of 4 Am^2 is sliced into two equal parts parallel to its length. The new pole strength would be**

1) Half of that of the given magnet　　2) Same as that of the given magnet
3) Twice of that of the given magnet　　4) 1/4 th of the given magnet

11. **A magnet of moment 6Am^2 is sliced into two equal parts parallel to its breadth. Then its new magnetic moment would be**

1) 6Am^2　　2) 1Am^2　　3) 4Am^2　　4) 3Am^2

12. **A thin bar magnet of moment 3.14Am^2 is bent into a semi circle. Its new magnetic moment is**

1) 6.28Am^2　　2) 2Am^2　　3) 3.14Am^2　　4) 1.57Am^2

Comprehension - 5 :

Some of the substances are magnetic materials while some other kind of substances are non-magnetic. For example paper, wood.... are non-magnetic while Nickel, Steel are magnetic. Magnetic materials are again classified as Dia, para & Ferro magnetic materials. Some of the important properties of these materials such as susceptibility, permiability, ... etc vary widely among these three types.

13. **For a paramagnetic susbstance, permiability is**

1) Negative　　2) positive & very high
3) positive but slightly greater than one　　4) zero

14. **Which of the following substances is paramagnetic in nature**

1) Water　　2) Nickel　　3) Aluminium　　4) Copper

15. **Suceptibility and permiability values are very very high for one of the following substances.**

1) Oxygen　　2) Cromium　　3) Mercury　　4) Nickel

ANSWERS

1) 4　2) 2　3) 3　4) 2　5) 4　6) 3　7) 4　8) 1　9) 4　10) 1
11) 4　12) 2　13) 3　14) 3　15) 4

LEVEL - IV

1. **Soft iron is used to manufacture electromagnets because their**

1) Magnetic saturation limit is high while retentivity and coercive force are small
2) Retentivity is high but coercive force is small
3) Retentivity is small but coercive force is high　　4) Area of hysteresis loop is large

2. **Which of the following statements is incorrect about hysteresis**

1) This effect is common to all ferromagnetic substances
2) The hysteresis loop area is proportional to the thermal energy developed per unit volume of the material
3) The hysteresis loop area is independent of the thermal energy developed per unit volume of the material
4) The shape of the hysteresis loop is characteristic of the material

3. **The area enclosed by hysteresis loop is a measure of**

1) Retentivity 2) Susceptibility 3) Permeability 4) Energy loss per cycle

4. **In the hysteresis cycle, the value of H needed to make the intensity of magnetisation zero is called**

1) Retentivity 2) Coercive force 3) Lorentz force 4) None of the above

5. **The hysteresis cycle for the material of temporary magnet is**

1) Short and wide 2) Tall and narrow

3) Tall and wide 4) Short and narrow

6. **The hysteresis cycle for the material of permanent magnet is**

1) Short and wide 2) Tall and narrow

3) Tall and wide 4) Short and narrow

7. **The hysteresis cycle for the material of a transformer core is**

1) Short and wide 2) Tall and narrow 3) Tall and wide 4) Short and narrow

8. **If I is the intensity of earth's magnetic field, H its horizontal component and V the vertical component, then these are related as**

1) $I = V + H$ 2) $I = \sqrt{(H^2 + V^2)}$ 3) $I = \sqrt{(H^2 - V^2)}$ 4) $I^2 = V^2 - H^2$

9. **The angle of dip at a place on the earth gives.**

1) The horizontal compnent of the earth's magnetic field

2) The location of the geographic meridian

3) The vertical component of the earth's field 4) The direction of the earth's magnetic field

10. **A dip needle arraged to move freely in the magnetic meridan dips by an angle θ. If the vertical plane in which the needle moves is rotated through an angle α to the magnetic meridian than the needle will dip by an angle.**

1) θ 2) α 3) More than θ 4) Less than θ

11. **Tha angle of dip is the angle**

1) Between the vertical component of earth's magnetic field and magnetic meridian

2) Between the vertical component of earth's magnetic field and geographical meridian

3) Between the earth's magnetic field direction and horizontal direction

4) Between the magnetic meridian and the geographical meridian

12. **A uniform magnetic field is obtained in**

1) a horse - shoe magnet 2) a bar magnet

3) a solenoid carring current 4) a circular coil carrying current

13. **A dip circle is at right angles to the magnetic meridian. The apparent dip angle is**

1) 0^0 2) 30^0 3) 60^0 4) 90^0

14. **A Dip circle lying initially in the magnetic meridian, the angle of dip is 45^0. Now dip circle is rotated through 30^0 in the horizontal plane. The apparent angle of dip is**

1) $Tan^{-1}\frac{2}{\sqrt{3}}$ 2) $Tan^{-1}\frac{\sqrt{3}}{2}$ 3) $Tan^{-1}2$ 4) $Tan^{-1}\frac{1}{2}$

15. The plane of the dip circle is set in the geographical meridian and the apparent dip is θ_1. It is then set in a vertical plane perpendicular to the geographical meridian , the apparent dip becomes θ_2. The angle of declanation α at that place is given by

1) $\tan\alpha = \sqrt{\tan\theta_1 \tan\theta_2}$ 2) $\alpha = \sqrt{\tan^2\theta_1 \tan^2\theta_2}$

3) $\tan\alpha = \dfrac{\tan\theta_1}{\tan\theta_2}$ 4) $\tan\alpha = \dfrac{\tan\theta_2}{\tan\theta_1}$

16. A bar magnet has a coercivity $4x10^3\ Am^{-1}$. It is desired to demagnetise it by inserting it inside a solenoid 12cm long and having 60 turns. The current that should be sent through the solenoid is

1) 2A 2) 4A 3) 6A 4) 8A

17. A dip circle lying initially in the magnetic meridian is rotated through angle θ in the horizontal plane. The ratio of tangent of apparent angle of dip to true angle of dip is

1) $\cos\theta : 1$ 2) $\sin\theta : 1$ 3) $1 : \cos\theta$ 4) $1 : \sin\theta$

18. If $\phi_1 and \phi_2$ be the angles of dip observed in two vertical planes at right angles to each other and ϕ be the true angle of dip, then

1) $\cos^2\phi = \cos^2\phi_1 + \cos^2\phi_2$ 2) $\sec^2\phi = \sec^2\phi_1 + \sec^2\phi_2$

3) $\tan^2\phi = \tan^2\phi_1 + \tan^2\phi_2$ 4) $\cot^2\phi = \cot^2\phi_1 + \cot^2\phi_2$

19. Consider the earth as a short magnet with its center coinciding with the center of earth and its dipole moment is M. The angle of dip ϕ at a place where the magnetic latitude is 'λ' will be

1) $Tan^{-1}(2Tan\lambda)$ 2) $Tan^{-1}(Tan\lambda)$ 3) $Tan^{-1}(Tan\frac{\lambda}{2})$ 4) $Tan^{-1}(4Tan\lambda)$

20. The horizontal component of earth's magnetic field at place is $0.36x10^{-4}\ weber/m^2$. If the angle of dip at that place is 60° then the value of vertical component of earth'smagnetic field will be (in wb/m^2)

1) $6x10^{-5}T$ 2) $6\sqrt{2}x10^{-5}T$ 3) $3.6\sqrt{3}\times10^{-5}T$ 4) $\sqrt{2}x10^{-5}T$

21. At a place the value of H and V are 0.4 $\times 10^{-4}$T and 0.3 $\times 10^{-4}$T respectively. The resultant earth's magnetic field is

1) 0.5 $\times$ 10^{-4}T 2) 10^{-4}T 3) 2 $\times$ 10^{-4} T 4) None of these

22. An iron rod is subjected to cycles of magnetisation at the rate of 50Hz. Given the density of the rod is 8 $\times$ 10^3kg/m^3 and specific heat is 0.11 $\times$ 10^{-3} cal/kg^{0}C. The rise in temperature per minute, if the area enclosed by the B – H loop corresponds to energy of 10^{-2} J is (Assume there is no radiation losses)

1) 78^{0}C 2) 88^{0}C 3) 8.1^{0}C 4) none of these

ANSWERS

1) 1	2) 3	3) 4	4) 2	5) 2	6) 3	7) 2	8) 2	9) 1	10) 3
11) 3	12) 3	13) 4	14) 1	15) 3	16) 4	17) 3	18) 4	19) 1	20) 3
21) 1	22) 3								

Made in the USA
Monee, IL
05 November 2026

67811026R10227